LECTURES

NaNON THE

ELEMENTS OF CHEMISTRY

DELIVERED

IN THE UNIVERSITY OF EDINBURGH;

BY THE LATE

JOSEPH BLACK, M.D.

PROFESSOR OF CHEMISTRY IN THAT UNIVERSITY,

PHYSICIAN TO HIS MAJESTY FOR SCOTLAND; MEMBER OF THE ROYAL SOCIETY OF EDIN-
BURGH, OF THE ROYAL ACADEMY OF SCIENCES AT PARIS, AND THE IMPERIAL
ACADEMY OF SCIENCES AT ST. PETERSBURGH.

NOW PUBLISHED FROM HIS MANUSCRIPTS,

BY

JOHN ROBISON, LLD.

PROFESSOR OF NATURAL PHILOSOPHY IN THE UNIVERSITY OF EDINBURGH.

VOL. II.

EDINBURGH:

PRINTED BY MUNDELL AND SON,
FOR LONGMAN AND REES LONDON, AND WILLIAM CREECH EDINBURGH.

1803.

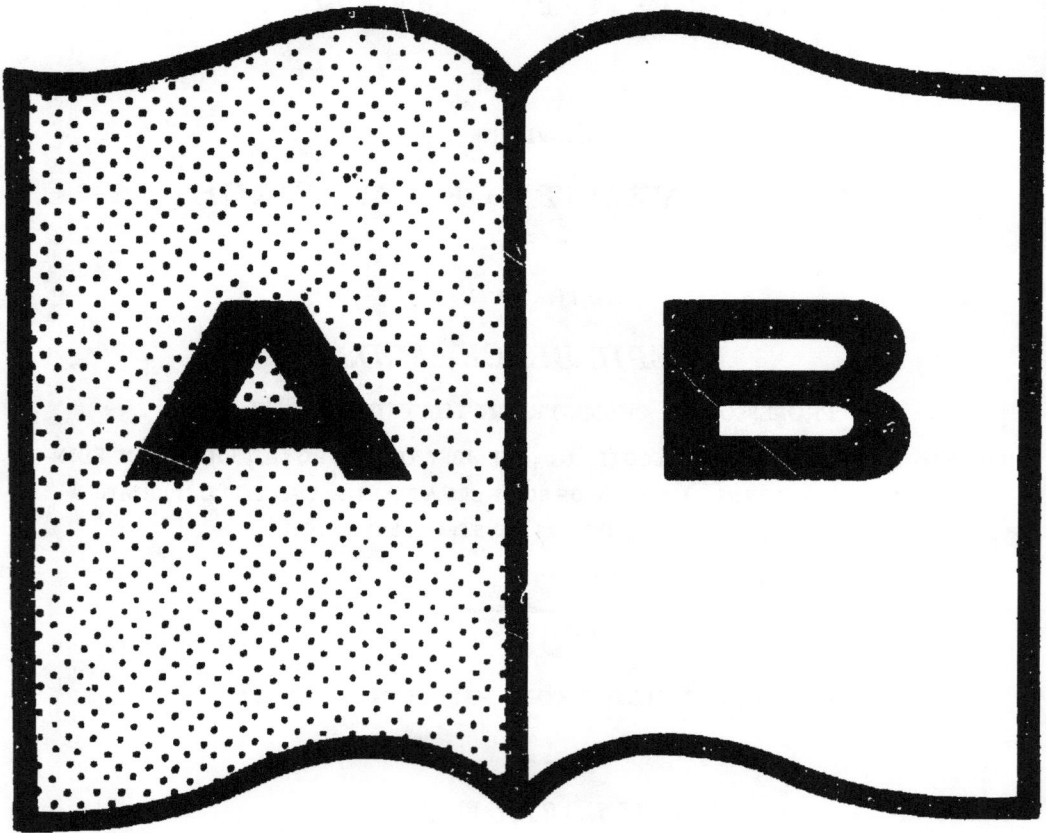

Contraste insuffisant

NF Z 43-120-14

LECTURES

ON

CHEMISTRY.

CHEMICAL HISTORY.

CLASS II.

EARTHS.

THE nature of the falts in general, and that of the principal varieties of them, have been already explained to you. It was proper to prefent that clafs of natural fubftances to your confideration the firft of all, becaufe the falts are the moft active of the objects of chemiftry. They fhew an attraction, or a diffolving and active power, in mixture with a great multitude of other fubftances. We cannot therefore defcribe conveniently the nature of any of the other objects of chemiftry, without taking in the relation of fuch objects to the different falts, or to fome of the principal fpecies of them.

Next after the falts, I have found it moft convenient to defcribe the earthy bodies; and I have two reafons for choofing this arrangement. The firft is, that fome of the earths refemble the falts fo much by their chemical relations, that the tranfition from the one to the other is perfectly eafy; and the knowledge of the falts muft therefore facilitate the apprehenſion of what is to be ſaid of the earths. The ſecond reaſon is, that we muſt have ſome knowledge of the earths, before we can proceed conveniently to the ſtudy of the other objects of chemiſtry which have not yet been conſidered.

That I may prepare you for underſtanding ſome terms and expreſſions, which I ſhall have ocaſion to uſe, in ſpeaking of the different ſtates and conditions in which we find the different earthy and ſtony ſubſtances, I am under the neceſſity of ſtating or deſcribing a number of facts or appearances, which, to be clearly apprehended, require a more than uſual exertion of the attention. I truſt, therefore, that you will favour me with your attention, which, at any rate, will be well employed, as, in thus preparing you, I muſt make known to you many of the moſt important and inſtructive phenomena that are connected with the natural hiſtory of this globe.

The common definition of earthy and ſtony ſubſtances is, that they are bodies not ſoluble in water, not inflammable, and whoſe denſity or ſpecific gravity, compared with that of water, does not exceed the proportion of four to one. Theſe characters are choſen to diſtinguiſh them from ſalts, and from ice, by the want of ſolubility in water ; from inflammables by the want of inflammability ; and from the metallic ſubſtances by inferior weight.

This definition, however, as well as that commonly given of the ſalts, is not unexceptionable. The diſtinction taken from the want of ſolubility may be objected to, as not applicable to ſome of the earths, which are perfectly and totally ſoluble in water. Such is calcareous earth, and barytes, and ſtrontian ſpar. And it muſt be confeſſed, that ſome of theſe earths, or natural ſubſtances, which appear to belong more properly to the claſs of earths than to any other, are truly of an intermediate nature between ſalts and earths, partaking of the properties of each, in ſuch a manner, that it is difficult to ſay which of the two claſſes they reſemble the moſt. Mr. Fourcroy, thinking that they reſemble the ſalts more than the earths, has removed them from the clafs of earths, and defcribed them as falts. But it is of little importance under what name a thing is defcribed, provided it be well defcribed, and the nature of it thoroughly made known.

There is, in every part of nature, a tranfition from one fet of her productions to another; fo gradual, that we cannot find a precife limit between the claffes or affemblages of them which we choofe to make.

Befide thofe qualities of the earths which are commonly given, to diftinguifh them as a clafs from other bodies, they have one which, though not taken into the definition, does, I believe, always make a part of the general idea of an earthy fubftance in the mind of a chemift. This quality is the great degree of fixednefs in the fire, by which the earthy and ftony bodies, in general, excel all the other claffes of the objects of chemiftry.

The fubftances which belong to this clafs are well known to exceed greatly in quantity all the other objects of our fcience. They compofe almoft the whole of the folid materials of this globe which we inhabit. It is, therefore, proper to take fome notice here of the fituation and arrangement of the earthy bodies in the interior part of the globe, fo far at leaft as we have yet been able to penetrate into it, which we muft confefs is only to a very fmall depth. This ftep is neceffary, that we may have an opportunity to explain fome terms which are in general ufe, in defcribing the different ftates and conditions in which the principal fpecies of earths are found in nature. But my obfervations on this fubject have a more important aim. The defcription to be given of the natural fituation and condition of the earthy fubftances, and of thofe which they inclofe and contain, give the moft evident marks of a gradual change, both in the fituations, forms, and even properties, of thofe fubftances. As all thefe are the effects of heat and mixture, operating by means of chemical forces, it is plain that the changes exhibit a feries of inftructive obfervations, and muft furnifh us with informations which nothing but the length of time, and the enormous preffures competent to thefe fituations, could procure. This is a valuable addition to all the momentary labours of our little laboratories and experiments.   And, on the other hand, our chemical knowledge enables us to underſtand many of thoſe changes which muſt otherwiſe have been myſterious and unaccountable.   Thus will chemiſtry and natural hiſtory derive mutual aid and improvement.

We have acquired ſome knowledge of the ſtate of the interior parts of the earth, by obſerving the excavations formed accidentally by torrents of water, and by digging deep pits in many different places for the purpoſes of mining, or, on ſome occaſions, for procuring water.

Whenever we have had opportunities, in theſe different ways, for obſerving the conſtruction of the interior parts of the globe, it has been perceived that the matter of which it is compoſed is arranged, for the moſt part, in extenſive beds, layers, or ſtrata, compoſed of different materials, and very various in their thickneſs, but generally parallel to one another, and in which we ſee many parallel diviſions, like the leaves of a book.   Where the matter of the earth is arranged in this manner, it is called STRATIFIED MATTER, or STRATA.

This arrangement, however, is perceived in few places only, at the ſurface of the ground.   The ſurface is covered almoſt every where, to different depths, with a confuſed rubbiſh and looſe earth, compoſed moſtly of the materials which lie more ſolid and regular below; a part of theſe having mouldered down into this rubbiſh and earth, and, in moſt caſes, having been removed by the action of water to a conſiderable diſtance from the place to which they originally belonged.

Before we can ſee the appearances of the ſtratified matter, we muſt therefore penetrate through this rubbiſh, or find places where it has been broken through, or ſwept away, for a depth of ſome feet or fathoms, or, in many caſes, for a much greater depth; and then we ſeldom fail to perceive them.

The parallel poſition with reſpect to one another, and the great extent and equal thickneſs of theſe ſtrata, have been moſt diſtinctly ſeen in mining countries, where frequent pits have been dug, to reach the objects of the miner; also in digging of wells in populous countries.

Derbyshire is an example of a mining country, in which a very great number of pits have been dug, on account of numerous metallic veins of lead ore which it contains: And Mr. Whitehurst has given an account of the general experience and knowledge that has been acquired by the frequent perforations of the earth in that country.

The country around Newcastle is another example, where, on account of the abundance and numerous strata of coal which are found there, the interior parts of the earth have been diligently examined by numerous pits, and by other means, as boring, &c.

From the insight that has been got on all such occasions, it has appeared that the strata lie in very different positions with respect to the horizon in different places; that in some districts they are level or horizontal; that in others, and these the greater number, they decline from the horizontal position, or have what is called a DIP; that in others, they are somewhat waved, or alternately rise and sink in an irregular manner; and in others, are thrown into *disorder* by interruptions, sudden bendings, and fractures, and by very steep declinations from the plane of the horizon. There are even examples of two sets of strata in different positions, the one over the other; the undermost set commonly erected, more or less, and the hardest; the uppermost less erected, or horizontal, and softer. This is seen in the banks of the river Jedd, near Jedburgh.

Such are the different states of the stratified matter which is found in almost every place where we have opportunities to examine the interior parts of the earth.

But I must farther remark, that in many places, and especially in those countries in which we find the strata the most disordered, we generally also find large quantities of stony matter, in which we cannot perceive that stratified arrangement of parts, but which constitutes enormous masses, irregular in their form and size, sometimes spread

out or interpofed between the ftrata, but never of fuch an equal thicknefs, or difpofed in fuch a regular manner ; and often fplit into numerous pieces.

The ftony matter which has this appearance, may be diftinguifhed
from ftratified matter by the name of rock, or rocky matter.

An example of the difference between the rocky and the ftratified
matter, is to be feen in Salifbury Craig, which is an extended mafs of
rocky matter, with ftratified matter below it, and above fome parts of
it : And Arthur's Seat is a mafs of rocky matter, but has ftratified
matter under it.

Thefe maffes of rocky matter, in which we cannot perceive any
ftratified arrangement, compofe in fome countries very high mountains.   In other places, the ridges, or chains of mountains, are formed
more or lefs by ftratified matter that is uncommonly hard, and the
ftrata of which are much declined from the horizontal pofition ; in
confequence of which, their edges projeƈt above the general furface of
the globe, to form thofe mountainous ridges.   Frequently, however,
the very high mountains are formed by an intermixture of *rock* with
thefe hard and erected ftrata, which, along with it, have fuftained the
injuries of time and of the weather, better than the fofter matter which
furrounded them ; and therefore part of them now remains, and forms
thofe projeƈtions, after the fofter matter has been demolifhed around,
and wafhed away to a great depth.

That the furface of the globe has every where undergone great
changes in this way, by demolition, and removal of the materials to
very great diftances, will be clearly feen by any perfon who furveys
it with a difcerning eye.   When we afcend very high mountains,
the rocky tops of them are all fhattered and ruinous, and their fides
are covered in many places to a great depth with the rubbifh which
has fallen from the upper parts of them.   In this rubbifh, we find
deep fcars and ravins, formed in heavy rains and ftorms, by torrents
of water, which, by their repeated effeƈts, gradually wafh down that

rubbifh into the lower grounds. During the great length of time required for its complete defcent, it undergoes further demolition and decay, into gravel, fand, and earth ; and, under thefe forms, is at laft fpread out by the brooks and rivers, to form the plains and the fands found on the fhores of the fea. For, as the rivers, in confequence of the unequal refiftance of the different foils through which they run, have in general a ferpentine courfe, and are gradually undermining their banks, and fhifting their channel in different places, they have the natural effect to undermine, and bring to a level, the eminences and accumulations of foil and rubbifh which have been formed in particular places ; and, for this reafon, all great rivers are in general furrounded in their courfe toward the fea with immenfely extenfive plains, which have been formed by themfelves.

Any perfon who wifhes to be fatisfied of what I have now faid of the demolition that is going on in the elevated parts of the earth's furface, muft vifit very high, mountainous, and Alpine countries, or read the accounts of thofe who have vifited them. Common mountains and hills do not fhew it fo evidently, for although thefe too are fuffering a gradual demolition and wafte, it is fo flow, on account of their being lefs expofed to the violence of the elements, that it is not obvious. They become covered with rubbifh, and foil, and vegetables,—all which defend them in fome meafure from the lefs violent impreffion of the ftorms ; while the tops of very high mountains are fo much expofed to their fury, that foil or rubbifh cannot remain on them, but is fwept away as faft as it is produced. There is a fine example of this in the mountains of Arran, in the weft, which are among the higheft in thefe iflands. The fummits of thofe mountains are all ruinous. They prefent, in fome places, lofty precipices, which terrify the beholder by their awful heights, and by the mountainous blocks of ftone that are prominent from them, and feem ready to fall and to crufh every thing below them. They are acceffible only to eagles and other birds of high flight, who choofe them for their habitation. The foot of thefe precipices terminates in a fteep flope of great extent, forming a great part of the fide of the mountain. This flope is compofed entirely of blocks of ftone and of rubbifh, which, in the lapfe of time, have fallen from the top of the precipice.

For more examples of fimilar appearances, I refer you to the defcription of the higher parts of the Alps of Savoy, as given by Mr. Sauffure of Geneva, in his account of different excurfions and journeys he made among them, and in fome of which he was expofed to great danger. His accuracy in obferving, and fidelity in relating what he faw, may be entirely relied on : And he prefents his reader with a number of the moft fublime objects and inftructive facts in natural hiftory.

You will alfo find proofs of the decay and demolition of mountains of granite, defcribed by Mr. Haffenfratz, in the eleventh volume of the Annales de Chimie. He faw them on the road from St. Flour to Montpellier ; and it is a very ftriking phenomenon.

But I have wandered I may fay into a digreffion. I was led into it by endeavouring to illuftrate the remark I made on the nature of many mountains and chains of mountains ; that they are compofed of hard ftratified or rocky matter, or of a mixture of both, which now is eminent above the furface of the globe around them, in confequence of the gradual demolition, and removal to a great diftance, of the fofter matter which formerly furrounded thefe harder materials, which are even themfelves liable to a more flow demolition and wafte.

To proceed,—It is neceffary further to be underftood, that both the ftratified and rocky matter are found fplit or divided, in innumerable places, by fiffures nearly perpendicular, or acrofs the ftrata ; and that the parts thus broken afunder are often confiderably difplaced with refpect to one another : The materials on the one fide of the fracture being frequently funk below the level of thofe on the other, or removed to fome diftance from them, fo as to produce in the ftratified matter *discontinuities* of various widenefs. Thefe rents and intervals of the fractured ftrata, or rock, are in general filled with a matter different from that of the rock or ftratum in which the rents are found. And rents of a certain widenefs, thus filled up with extraneous matter, are called *veins*, or in Cornwall *lodes*. Thefe veins are fubject to all the variations of widenefs, and irregularity of direction, which you can eafily imagine in fractures and rents of hard ftrata and rock; and they go down to a depth far below any to which we can reach. The matter which fills them is generally a whitifh ftony matter named *fpar*. It is more pure and more fimple in its compofition than moft other ftones, but there are feveral kinds of it; and intermixed with it, we often find alfo different metallic and mineral fubftances. It is indeed in thofe veins chiefly that the ores of metals are found.

And wherever there are vacuities left in the vein, by its being imperfectly filled, there we are fure to meet with cryftalfized matter of the feveral ftony and mineral fubftances which the vein contains.

In a great many examples, however, the rents and difcontinuities I have defcribed, are of greater widenefs than what are commonly called veins. The widenefs of veins is from an inch, or a fraction of an inch, to feveral feet, or even yards in fome cafes. But we have examples of rents, the widenefs of which is equal to many fathoms; and in this cafe, rock or rubbifh, or other extraneous matter, fills up the interval. When this occurs, the interval and materials which it contains is not called a vein; our miners give it the name of a *dyke*, or *partition*. The extent to which thefe dykes, as well as the veins, are continued downwards into the bowels of the earth, is hitherto unknown. Their depth is plainly much greater than what human art and labour can reach.

Thefe rents and diflocations of the ftrata, with the intervals thus formed, and great quantities of rock or rubbifh which fometimes fill up thefe intervals, are well known to our colliers, who meet with them often in cutting out the coal; and they call them in general *troubles*; becaufe there they meet with an interruption in the coal, in confequence of a divifion and fhift of the ftrata, which they call a *hitch*. The coal is continued beyond this break or hitch, but upon a different level, which the miners muft find out, fometimes at great expence. And often they find interpofed, between the edges of the interrupted ftrata, a thick partition of rocky or other matter, called a *dyke*.

You will fee numerous examples of fmall veins in Salifbury rock; and examples of fmall dykes in the bed of Leith water or river; between Canon-mills and the mills above them.

I have only one general fact more to relate, which belongs to this fubject: It is the difcovery of fea-fhells, corals, and coraline bodies; and of the bones and fkeletons of fifh, and other relics of marine productions, which are found in moft places intermixed with the ftratified matter. Thefe have appeared on innumerable occafions, and in all parts of the world, in which ftratified matter, efpecially fome kinds of it, have been examined;—on the fides of the higheft mountains, as well as at the greateft depths to which we have gone in digging the earth. Sea-fhells have been found at Amfterdam 100 feet deep, or under the furface; at Marly-la-Ville, fix leagues from Paris, at 75 feet deep: Sometimes in mines under a cover of 50, 100, 200, 1000 feet; as in the Alps and Pyrenees. They have alfo been found in mountainous countries at a very great height above the level of the fea; as in the mountains of Spain, the Pyrenees, the mountains of France and of England, the marble quarries of Flanders, the mountains of Gueldres, and in the heights round Paris; alfo in the Alps, particularly Mount Cenis: And M. Sauffure, in his inftructive journeys among the Alps, defcribes oyfter-fhells which he found in a part of the Alps to the fouth-eaft of Geneva, at the height of 1172 fathoms above the fea; that is, 7032 feet. (*Sauffure Voyages dans les Alpes*, p. 393.) They have been feen alfo in the Appenines, and in moft of the ftone nd marble quarries of Italy. (*Vide Buffon, tom.* 1. p. 408.)

We have evidence also of their having been found in the highest mountains of the world, the Andes of South America. Don Antonio de Ulloa found petrified shells upon the *Cordeliers des Andes dans le gouvernment de Wanca-Vellica*, in 13 or 14 degrees south latitude. These shells were found at such a height, that the barometer stood there at 17 inches 1¼ line French measure; that is, at 2200⅓ fathoms above the level of the sea, or 13,202 feet. (*Mem. de l' Acad. 1771.*)

And, in all these different examples, it is in the stratified matter only that they are found.

We also find frequently, in some kinds of stratified matter, impressions of the leaves and other parts of vegetables, and pieces of wood, penetrated with stony matter of different kinds.

From the whole of these phenomena, it is evident that the materials of this globe have been affected, in many places, with violent concussions and derangement, and have also undergone very great changes in their position with respect to one another: That what was once covered by the sea is now land or mountains, and what was once at the surface, or covered only with water, is now under cover of solid matter to very great depths; and that these changes of situation have even been often repeated; although the successions of the one to the other must certainly have been extremely slow, and must have required a length of time, the limits of which cannot be assigned from the phenomena themselves.

Such are therefore, in general, the appearances which present themselves, when we examine the state of the solid materials of which this globe is composed, and which mostly belong to the class of earths, which we are now to consider. Various conjectures and hypotheses have occurred to the human imagination, to account for the manner in which all this has happened.

To give a short account of the attempts to explain these phenomena, I shall first remark; that all these attempts agree in supposing the materials of the strata to have been arranged by water, depositing or arranging them one over the other, in fucceffion. This is inferred not only from their general appearance, fimilar to that we perceive in lakes, pools, and other fmall collections of water; at the bottom of which, when they dry up, we find the fand, mud, and other matters depofited, or arranged in parallel layers; but alfo from the appearance of fhells and other marine productions, which abound in the compofition of a great number of ftrata. Thefe fhells are in many places found fo entire, that they are perfectly well known. The different fpecies of them are very numerous, and in fome places they are collected together in immenfe quantities; in quantities fo great, that we cannot fuppofe any other origin for them, but that the place where they lie was once the bottom of the fea. The ftrata are fuppofed to have been originally in an horizontal, or nearly horizontal pofition. The flope or declination from the horizon, which many of them now have, and the apparent diforder in other refpects, by rents, changes of level, and interruptions, have been imputed to different caufes by different theorifts. Some have imagined that a general convulfion muft have fhaken the whole of the globe: Others have fuppofed that partial ones have gradually affected the different parts of its furface in fucceffion.

The general convulfions which have been fuppofed fufficient to produce thefe great effects are, firft, the cataftrophe of the deluge, defcribed in the fcripture, and by the poets: And Dr. Burnet, in his Theory of the Earth, by giving full liberty to his imagination, has fhewn fome ingenuity in the contrivance, I may fay, of a deluge equal to thefe effects.

The great Dr. Halley, whofe mind was more occupied with celeftial objects than with any other, propofed a fuppofition, that this globe had perhaps more than once received a fhock from a comet, which, impinging on it obliquely, gave it its diurnal rotation; and, by the fudden communication of fuch a violent motion, fhattered and deranged all the materials of it that were near its furface.

But the greater number of philofophers appear now inclined to believe that the convulfions which have deranged the ftrata have been
partial or local only, but with fuch a fucceffion as gradually to affect
every part, or almoft every part of the earth's furface; and that they
have been produced by fubterranean fires.

» That volcanic and fubterranean fires muft have been a very common caufe of the inequalities of the earth's furface, and diforder of its
materials, is plain from the natural hiftory of this globe.   We know
that in fome cafes they have produced fudden and violent elevations
of fome parts of it, and as fudden finkings and depreffions of others.
We know that fome of the higheft mountains are compofed chiefly
of materials thrown out of the craters of volcanoes, and that the moft
convincing proofs are vifible, in innumerable places, of formerly exifting volcanoes, and craters, the eruptions of which have ceafed long
fince, while others have fucceeded to them, and are at prefent in different ftates of activity.

We know farther, that the action of thofe fubterranean fires is not
confined to the derangement of the dry land.   The agitations which
they produce are in many cafes felt with great violence even at fea;
and they are fo very extenfive, that it is plain they muft act at an
immenfe depth below the furface; and therefore far below the bottom
of the ocean: A confequence of which is, that they elevate that bottom in many places fo as to make it rile above the furface of the
waters, where it remains in that elevated ftate.   A great number of
iflands in different parts of the globe fhew the moft undeniable proofs
that they have had this origin.   Such are the iflands of the Archipelago in the Mediterranean; the Azores, Teneriffe; the iflands of the
Weft Indies.   In all thefe we find lavas, and other figns of the action
of fubterranean fire; and in fome of them fprings of hot water, and
fulphureous vapours, breaking out at the furface, fhew that the internal
fires, although fmothered, are not completely extinguifhed.

We find the fame fymptoms of internal fire in almoft all the great ridges of mountains in the world; and many of the greateſt and moſt active volcanoes are found in thoſe ridges. Such are the Appenines in Italy; the mountains of Sicily; thoſe of France in Auvergne and Dauphigny; the Pyrenean ridge between France and Spain; the immenſe ridge of South America named the Andes, or Cordilleras; the Ruſſian peninſula of Kamtchatka; the mountains of Iceland. In all theſe we find either exiſting volcanoes, or indubitable proofs of their former exiſtence in places where their fires and vapours are not now ſeen. Even in this country, there are ſome ſigns of the former action of ſubterranean fire. Many gentlemen who have viſited volcanic regions, and who have given particular attention to the effects and productions of ſubterranean fire, find ſo much reſemblance between our whinſtone and ſome of the lavas of exiſting volcanoes, that they are perſuaded our whinſtone has been a ſort of lava. And, even ſetting aſide that reſemblance, there are many other reaſons for believing that our whinſtone has been a matter melted by ſubterranean heat *. This, however, does not infer actual volcanoes.

Subterranean fires, therefore, and the internal convulſions, the eruptions and exploſions which they have produced, have evidently ſhaken and deranged the materials of the globe in ſuch a very great number of the parts of its ſurface, that there is very great reaſon to think all the derangement we find in other places has been produced in this way; and this is now the moſt general opinion.

The author, however, who has been the moſt celebrated for his attempt to give a natural hiſtory of the globe, I mean the late M.

---

* In digging the cellars and common ſewers of the New Town of Edinburgh, freeſtone ſtrata are found to commence a few feet under the ſurface, and occupy the whole ridge on which the buildings are erected. A fiſſure, or rent was found croſſing all theſe ſtrata, to an unknown depth. This fiſſure is completely filled with whinſtone; and in ſome places, the whinſtone ſpreads out a little above, on each ſide of the fiſſure, in the ſame manner as mortar will ſpread out over the joint between two bricks when they are preſſed together. It appears to have been in a fluid, or at leaſt muddy ſtate, and to have been forced through from below.

EDITOR.

Buffon, formed for himself a different fyftem, which is the moft fully difplayed in his *Epoques de la Nature*, and in his *Hiftoire Naturelle*. He admits that confiderable changes have been produced by fubterranean fires; but is of opinion that the principal phenomena are to be accounted for in another way.

He undertook by his fyftem to point out the very origin of this globe, as well as that of the other planets which revolve along with it around the fun. In his opinion, there was a time when the fun ftood alone, without any planet, and was vifited only now and then by the comets. One of thefe, coming too near, was drawn, by the attraction of gravitation, into the body of the fun, and falling obliquely, and with immenfe velocity, againft his furface, dafhed off great maffes of the melted and burning matter, of which Mr. Buffon fuppofed the external parts of the fun are compofed. Thefe maffes were thrown off with different velocities, and flew to different diftances, before the progreffive motion was overcome by the attraction of the fun. At laft, however, the velocities which thefe maffes had received were fo much diminifhed by this attraction, that it became an equal antagonift to them; and thefe maffes thus became planets, and revolved in the orbits in which they ftill continue to move.

The fame powerful impulfe by which they were driven off from the furface of the fun, gave each of them alfo a rotatory motion round its own centre or axis; and being at firft liquid or melted matter, the attraction of their parts for one another gave them a fpheroidal form, which, at the fame time, became a little oblate, by the greater centrifugal tendency from their own centre of fome parts of them than of others, in confequence of the rotatory motion they had acquired.

Mr. Buffon further fuppofed that, in flying off from the fun, they carried with them a part of his atmofphere, which he imagined to be loaded with a vaft quantity of vapours, produced from every fort of matter which can be converted into vapour by intense heat, but especially a very great quantity of watery vapours.

The planets continued for some time to revolve round the sun, each of them surrounded with its vaporous atmosphere; for their heat at first was too great to admit of any condensation of these vapours on their surface. But they cooled by degrees, and the first effect of this cooling was, that the external parts of them became solid or congealed. The solid surface, which was formed in this manner, he supposed to have had all the roughness and inequality which is commonly seen on the surface of earthy, stony, and mineral substances, which have been all melted together and congealed. There were eminences in some places, and depressions in others; the matter in some parts of it was cavernous, and spongy and more solid in others; and, by the contraction of the cooling matter, a multitude of rents, and chasms, and vast caverns were formed.

Such, he supposes, was the surface of our globe before the water descended from the atmosphere. This did not happen until the heat was so much abated, that the vapours, which were contained in such great quantity in the atmosphere, began to condense there, and to fall in rain. His lively imagination paints here a dreadful scene, by presenting the immense quantity and violence of those original rains which were destined to fill the whole bed of the ocean. They were attended also with violent explosions, and other effects which the water produced when it penetrated into the crevices and caverns of the new earth, the heat of which was still so great as to convert very quickly into vapours again the water which first fell on its surface.

This period of violence and convulsion at the surface of the earth, and in the atmosphere, had a certain duration; but the heat continuing constantly to decrease, the watery vapours were at last more completely condensed, and the fluid element gained possession of the surface, and lodged itself in the depressions and cavities which had been formed either originally, during the change of the globe from a fluid to a folid ftate, or afterwards, in confequence of the convulfions I have juft now mentioned. Still, however, the waters on the furface of the globe were for a long time exceffively hot, and in fome places even boiling. This was neceffarily attended with great evaporation, exceffive rains, and violent commotions of the atmofphere. And during this period, and the one which immediately preceded it, a great part of the folid and hard materials, at the furface of the earth, were penetrated by the hot water and vapours, and by more active fub-ftances, which had condenfed along with the water; and thus thefe hard materials were mouldered down into rubbifh and mud, which formed the firft foil on which vegetables and animals were to be nou-rifhed.

Thus Mr. Buffon thought he had accounted for many of the ap-pearances which are obfervable in examining this globe.

In order to explain the reft, and, particularly, the formation of the ftrata, and the abundance of fhells and other relics of marine produc-tions which abound in the compofitions of many of them, he fuppofed that it afterwards underwent more flow and gradual changes.

His opinion was, that the fea is making a conftant, but very flow pro-grefs, over the furface of the globe from the caft to the weft: That this is the confequence of the action of the trade-winds, and of the general direction of the tides in the ocean, which is from eaft to weft. While the trade-winds brufh the furface of the fea, they affect the water, and in fact produce a current weftward in the Atlantic, which, by its in-ceffant action on the coaft of America, has formed, as Mr. Buffon fuppofed, the great Gulf of Mexico. The water, which in confe-quence of this, would be accumulated there, had it not found an out-let, is well known to turn back again in a north-eafterly direction, and to run in that direction along the coaft of North America, where it forms what is called the gulph ftream, which is diftinguifhed by its being warmer than the reft of the ocean in that latitude. A current westward, similar to that I just now described, is known to exist also in the Pacific Ocean, and in a part of the Indian Ocean, except where it meets with the coast and mountains of Africa, which reverberate it, and give it a turn the other way. By this continual impression of the sea on the eastern coasts of the continents and islands, Mr. Buffon imagined that the shores are worn away; the sea encroaches on them, and by degrees takes possession of their place; the materials of them being deposited at the bottom, either in a horizontal position, or formed by the currents and tides into risings and depressions, or long submarine ridges and extensive valleys, (if they may be so called). And thus he imagined the whole face of the globe has been modelled some time or other by the action of the water, and retains every where the relics of the productions of the sea, from which it received its form.

This splendid system has in some parts of it an air of sublimity and grandeur, especially as it is embellished by the eloquence of Mr. Buffon. But it certainly shews a degree of presumption and temerity in the author of it, which excite in the mind very different emotions from those that arise when the phenomena of nature are explained in a satisfactory manner, and with strict attention to principles of reasoning that are well founded and just.

The very first supposition in this grand system is totally inadmissible,—I mean the supposition that the planets received the projectile motion by means of which they revolve in their orbits, in consequence of their being dashed off from the surface of the sun. According to this supposition, they could have made no more than one very eccentric revolution. As soon as by the constant action of gravitation, they were made to bend their course again towards the sun, they could not do otherwise than to complete the ellipse, the one half of which they had described in flying off to their greatest distance; and in completing this elliptic orbit, they must necessarily have fallen again into the surface of the sun.

Nor is the other part of his system, by which he endeavours to account for the relics of fea-productions, which occur in every part of the globe, at all fatisfactory. We cannot underſtand by it how the fea ſhould have carried thofe relics to fuch a very great height above its prefent level, as 10,000 feet or more. And had the diftribution of land and fea depended on this principle, there would have been no land under the line : All the land there would have been worn away and depofited.

Other attempts have been made more lately to explain this fubject, fuch as Mr. Whitehurſt's, in his *Inquiry into the Original State and Formation of the Earth.* Another by Mr. De Luc, of Geneva, in his *Letters to the Queen.* But I cannot enter into all thefe fyſtems : it would take up too much of our time. They are all conjectures, and liable to great objections.

A fyſtem has lately been communicated to the public in this country, by Dr. James Hutton, which appears to me much more comprehenfive and fatisfactory than thofe. But I think I have done enough when I have prepared you for underftanding and judging of fuch attempts, by defcribing the general facts or appearances to which they refer : While, at the fame time, I had an opportunity of explaining fome terms which I ſhall have occafion to ufe, in mentioning the different ſtates and conditions under which the principal fpecies of earths are found in nature.

And being thus prepared for a furvey of the variety that occurs among the earthy and ſtony fubſtances, we ſhall next proceed to the examination of them.

When we begin to do this, however, we find the variety of them to be fo immenfely great, that to give a full account of every particular would take up a great deal too much of your time; and would be otherwife improper here, as being the province of a feparate inſtitution,—NATURAL HISTORY. I ſhall not therefore attempt a particular defcription of all the varieties of earthy bodies, but refer you to the authors who have publifhed fyſtems on this branch of knowledge.

The moſt eminent are, Wallerius, Cronſtedt, and the late celebrated Profeſſor Bergmann, who made great improvements in our knowledge of many earthy and ſtony ſubſtances, by his accurate and ſkilful analyſis of them. His work on foſſils is tranſlated into our language by Dr. Withering. Mr. Kirwan of Dublin has alſo publiſhed a mineralogy, or ſyſtematic arrangement and deſcription of foſſils, in which he takes advantage of the lights which have been thrown on this ſubjeſt by Bergmann and others; and has added much from his own experiments, ſo as to give in his book the moſt complete enumeration and deſcription of foſſils that has yet appeared.

Theſe authors, and my colleague, the Profeſſor of Natural Hiſtory, will give you information of all the particulars. In this courſe we have not time to follow the ſubjeſt ſo far, but muſt confine our attention to the moſt remarkable and diſtinguiſhed chemical varieties which occur in nature.

Theſe, I think, may be referred to five orders; the denominations and charaſters of which I ſhall now give you:

The *firſt* of theſe I ſhall call the order of the ALKALINE EARTHS: They have been commonly called ABSORBENT EARTHS. The ſtate in which theſe are found in nature, and that under which they are commonly called abſorbent earths, is not a ſtate of purity, but of combination with a particular ſubſtance with which I ſhall ſoon make you acquainted. But as there is hardly any example of their being found in a pure ſtate, and as the ſtate juſt now mentioned is the ſtate of them to which we are moſt familiariſed, I ſhall conſider and deſcribe them, firſt, as appearing in this ſtate, and afterwards ſhew how we render them pure, and what properties they have in their pure ſtate.

The charaſteriſtic of them in their natural ſtate is to efferveſce with acids. They appear under many different modes of concretion; from that of looſe earth to the hardneſs of ſtone, but are never ſo hard as to ſcratch glaſs, or give fire with ſteel. On the contrary, a knife can be made to ſcratch them, or make an impreſſion on them.

The *second* order I call the PLASTIC EARTHS.   They are commonly called the ARGILLACEOUS, or CLAYEY.

These, when mixed and kneaded with a moderate quantity of water, become a tough and ductile paste, which is easily modelled into any form, and which, if it be afterwards well dried, and then burned with a strong fire, contracts in its dimensions, and becomes compact and hard.

The *third* order I call the HARD earths or stones.   They are commonly called the SILICEOUS, or FLINTY.

They are, in general, stony masses of a larger or smaller size, the most eminent quality of which is hardness.   Their hardness is such that they scratch or cut glass, and give fire with steel; and when a knife or file is applied to them, it does not make the least impression. It is also a general quality of them to resist the most violent heat of common fires and furnaces without being melted.

The *fourth* order I name the FUSIBLE.

These are commonly stony concretions which have considerable hardness.   Some are so hard as to scratch glass, and give fire with steel; but they are not quite so hard as the hard stones.   Their most remarkable quality is *fusibility*, or a disposition to be melted in a strong fire.

The *fifth* order I name the FLEXIBLE.

They are concretions of a scaly, a plated, or a fibrous structure, easily divisible into parts which are *flexible*, and often *elastic*.   They are so soft as to be easily scraped with a knife; and do not suffer much change of their appearance and qualities when exposed to a moderately red heat.

We shall now, therefore, proceed to describe the earthy and stony bodies which come under these several divisions, and begin with the first, the alkaline earths, or, as they are commonly called, the absorbent.

GENUS I.

———————

ALKALINE EARTHS.

OF the alkaline earths and ftones there are feveral kinds, which, in their natural ftate, are mild infipid fubftances : And it is difficult to perceive in them, by experiments, any fenfible degree of folubility in water.

Their characteriftic in that ftate is, that they effervefce with acids, and are diffolved by thofe falts; and fuch is their nature with regard to acids, that they produce fimilar effects in mixture with them to what the alkalis produce: Upon which account I call them ALKALINE EARTHS. Thus, we find that in general they unite with acids in one certain proportion only. If there is too much of the earth, it is not all diffolved; if too little, it is united with a part of the acid only, the reft of the acid remaining free. There is, therefore, the fame mutual faturation here as with alkaline falts. The compound produced is likewife analogous to a neutral or compound falt, and in moft cafes diffolves eafily in water. The acid is rendered lefs volatile by its cohefion with the earth; its acrimony is greatly diminifhed, and its acid tafte is not perceivable. And further, thefe earths attract different acids with different degrees of force, and this in much the fame order as that in which they are attracted by alkalis: So that a compound earthy falt, which contains the nitric acid, can be decompoſed by the vitriolic acid ; and one which contains the muriatic acid, by vitriolic acid or nitric acid ; and one that contains the acetous acid, by any foſſil acid, &c.

They, therefore, reſemble the alkalis, both in their attraction for acids, and by many particulars in the nature or manner of this attraction : But, on the whole, its force is leſs than that of the alkalis. Moſt compounds of acids with the alkaline earths may be decompoſed by the alkalis, and eſpecially by the fixed alkalis. Theſe unite themſelves to the acid with more force, and cauſe the earths to ſeparate and fall down in the form of white powder.

Such is the general nature of theſe earths : Let us now attend to the different kinds of them.

SPECIES I.—CALCAREOUS EARTHS.

The principal ſpecies, which occurs the moſt abundantly in nature, and is at the ſame time the moſt uſeful, is called the CALCAREOUS EARTH. It is diſtinguiſhed from other alkaline earths by the effects of a moderately ſtrong heat, which changes it into common QUICKLIME.

It is found in a variety of forms or conditions,

1mo, Very numerous and extenſive ſtrata are compoſed entirely, or principally, of this kind of earth. Such are the MARBLES, LIME-STONE, and CHALK, which differ from one another only by the degree of purity in which they contain this earth, or by the manner in which it is concreted together. Both chalk and marble are often burnt to make lime, and afford excellent lime. Marbles and lime-ſtones are found in all parts of the world; and form numerous and extenſive ſtrata,—and ſo does chalk, in ſome countries.

2do, This earth is often found in veins, or filling up rents and crevices in rocky mountains, and hard ſtrata. In this ſtate, this earth is called SPAR, SPATHUM, or, more particularly, CALCA-REOUS SPAR, to distinguish it from some other stony substances called spars, on account of a resemblance of form, and of their native situation, but which are of a different nature.   Its appearance in this state is that of a whitish stone, more or less transparent, and which, when broken in pieces by the strokes of a hammer, is shivered into fragments of a rhomboidal figure.   In some veins it is perfectly transparent, and is then called ICELAND CRYSTAL.   This has caught the attention of opticians, in consequence of its refracting light in a particular manner, so as to make objects appear double.   In veins that contain calcareous spar, wherever a vacuity remains not filled, the inner surface is beset with crystals of the same matter, which are columnar or pyramidal; and sometimes composed of two pyramids joined together, and forming one crystal.

3tio, It is commonly this earth which forms the stony masses called STALACTITES.   These are stony concretions, formed chiefly in the roofs of subterranean caves, and cavities of some extent. They are produced by waters which contain a small quantity of this earth dissolved in them, and which, penetrating into such caves, drip from their roofs, or ooze out from their sides and floors, and deposit this earth.   While it separates slowly from the water, it concretes, and forms gradually large pendulous cones and columns, and a variety of fantastic figures, which are found hanging from the roofs, or projecting from the sides and floors of such caverns, and somewhat resemble the icicles more quickly formed from dropping water in frosty weather.   The famous cave at the Peak, in Derbyshire, is a well known example of this.   There are similar examples near Slains Castle, north of Aberdeen, and in many other parts of the world.

The waters which produce this effect are called PETRIFYING WATERS, on account of their forming these stony concretions, and sometimes penetrating and encrusting vegetable and animal substances with the fame ftony matter, fo as to convert them, as it were, into ftone. Such is the fpring called the Dropping-well, at Knaref-burgh in Yorkfhire, which has formed a fort of quarry of ftone, by petrifying the mofs and other vegetables which its water paffes over. There is a fimilar fpring in the Duke of Hamilton's deer park, and at Matlock in Derbyfhire.

Waters of this kind are very common in Italy, in confequence of the great quantities of limeftones and marbles with which that country abounds. There is a marfh not far from Rome, the water of which incrufts the reeds which grow in it ; and by means of thefe waters they have made very pretty medallions.

The formation of thefe calcareous petrifactions and ftalactites, fhews plainly that this earth is in fome degree foluble in water ; and we have many other proofs of this fact. Calcareous ftalactites and incruftations, and the petrifying waters which produce them, are in general found in countries which abound with limeftone or marble, or other forms of the calcareous earth. Wherever any parts of the calcareous ftrata are expofed to the rain, they fhew moft evident figns of their being fubject to a flow diffolution and wafte. (A calcareous mafs at Lord Elgin's quarry is diffolved at top, and incrufted at the bottom). And when more water gains admittance into rents and chafms of fuch ftrata, it gradually enlarges the paffages through which it runs, and forms at laft extenfive caverns, in which it is common to find abundance of ftalactites. The quantity of this earth, however, which pure water can diffolve, is exceedingly fmall, as I learned by experiment. But water fometimes contains other things combined with it, which increafe its power as a folvent of this earth.

4to, The ftony fhells of all cruftaceous animals, among fifh and rep-tiles, from the coarfeft to the pearl, are made up of this earth, and a fmall quantity of animal glue. And of the fame materials are egg-fhells formed, as alfo thofe marine bodies, which, on account of their refem-blance to the external form of plants, with the hardnefs of ftone, were formerly called LITHOPHYTA, or ſtony plants,—as coral, and a variety of other bodies of the coral kind, or allied to it, which later naturaliſts ſuppoſe to be the work of animals, or to be produced in the ſame manner as ſhells. And a late author, Pallas, who has written particularly upon this ſet of productions, repreſents them in a ſtill more curious light. Hence they are called ZOOPHYTA, or animated plants. Whatever may be their nature, and in whatever manner they may be formed, they are compoſed, like the ſhells, principally of calcareous earth ; and there is a very great variety of them in reſpect of form and appearance. Some are ſpread like a cruſt upon ſtones and other ſolid bodies, and full of microſcopic pores inhabited by animals,—MILLEPORA ; others are protuberant maſſes ; others an aſſemblage of pipes,—MADRIPORA ; others are branched,—CORALS. Theſe are full of animals.

Under. theſe four forms now deſcribed, the calcareous earth has generally a great degree of purity ; and in ſome of them, perhaps, is perfectly pure,—that is, unmixed with any other earth. But there are alſo ſome foſſil earths, which, though they be not compoſed entirely of calcareous earth, but, on the contrary, contain ſometimes but a moderate portion of it, deſerve, however, to be mentioned here, on account of their uſefulneſs and importance. Theſe are the earths called MARLES, which have been long ſucceſsfully employed as manure to improve ſoil. Every ſubſtance to which the name of marle is properly applied, and which proves of general uſefulneſs as a manure, contains more or leſs calcareous earth ; and is uſeful and valuable in proportion to the quantity it contains.

Marles are commonly divided into three kinds,—SHELL MARLE, CLAY MARLE, and STONE MARLE.

1ſt, Shell marle is compoſed of the ſhells of ſhell-fiſh, or other cruſtaceous aquatic animals, lying together in immenſe quantities. Many are entire, but in general they are decayed, and moſtly mouldered down to duſt, and intermixed with more or leſs ſand, or other earthy ſubſtances. When we examine this matter, as occurring in different places, it may be diftinguifhed into two kinds,—frefh water marle, or bog marle, and fea-fhell marle. We have an example of bog marle in the Meadow, compofed of the fhells of the fmall frefh water wilk or fnail, which multiply greatly in lakes or ponds, or brooks of frefh water, and of other fhells, which are gradually depofited in great collections by the water. The other kind of fhell marle, the fea-fhell marle, conftitutes much greater collections, which are found in many places, though at prefent far removed from the fea. One of the moft noted examples has been defcribed by Mr. Reaumur, in the Memoirs of the Academy *.

2*dly*, Clay marles confift of earths of different colours, which more or lefs refemble, and actually contain clay; but with it fome calcareous earth is mixed, in fine powder like chalk.

3*dly*, Stone marles are of very great variety in colour and appearance. They are harder and more ftony than clay marles, and that is the only diftinction between them; but they differ from maffes properly ftony in this, that by being only a few weeks or months expofed to the air, they fplit into pieces, and crumble down into earth, or a matter like clay. Some, however, take a long time to break down in this manner; and there is all the variety poffible in this refpect between them and the clay marles. This difpofition to moulder down by the weather, depends on the admixture of clay, which they contain. A confequence of the gradual, though very flow diffolution of this earth by water, has been obferved in the ftrata of hard marles. It

---

* Of the Fahlun or marle of Touraine. This diftrict, of 80 fquare miles, is more than 36 leagues diftant from the fea, and is every where filled with marle. It is found about eight or nine feet from the furface, and they dig it to the depth of 20 feet. The marle is ftill deeper, but it is too expenfive to follow it. One gentleman, from curiofity, penetrated 18 feet, and did not know how much farther it might extend. Suppofe it is of no greater depth, it would contain above 130 millions of cubic fathoms. It is totally compofed of fhells, fome of which are entire, but they are generally decayed, or broken into fragments, and mixed with other marine productions, fuch as coralines, millepores, madrepores, &c.

has been found by experience, that such strata do not contain any calcareous earth where they crop out, on the surface of the ground, nor even at the depth of a few feet below this. Many gentlemen who were in possession of a stratum, which they knew, from the neighbourhood, to be a rich stone marle, have, without fear, expended great sums of money, and covered much of their land with this crop marle. They not only lost their money, but, in many cases, spoiled their land, by filling it with a hard baking clay. This effect upon beds of marle is without exception; and it is even found that the upper plates of single strata of limestone are deteriorated in the same way. This fact also explains, by the way, the wearing out of lime used as a manure upon land.

When we now examine with attention many of these natural collections of this sort of earth, we are led to a conclusion which may appear surprising at the first hearing of it; but which is founded on a multitude of the most authentic facts.

The inference I mean is, that all the large collections of calcareous earth have derived their origin from shells and lithophyta, or that they were once in that form. The proofs of this are so numerous and striking, that they cannot be resisted.

1. Beds of sea shells, or other calcareous productions of the sea, are found in many parts of the world, of great extent, and in all the different states of decay.

2. We find abundance of shells, or fragments of shells, and of zoophytes, in the greatest number of the calcareous strata of marbles, limestones, and chalk; which are the greatest collections of calcareous earth.

We do not find shells in calcareous spar, nor in stalactites; but it is evident that these forms of the calcareous earth have been completely fluid, so that all the organic structure of the original matter is necessarily lost.

There are also many marbles and limestones, in which we do not find any of the appearances I speak of. But this too can be explained, by examining their structure or aggregation, which shews that the matter of them has been dissolved or liquefied by other operations of nature *. Another cause of the non-appearance of shells and lithophyta in some collections of calcareous earth is, that the shells have been broken and worn down into fragments like sand, or even reduced to powder, by agitation and attrition, upon the coasts, or by other causes. This seems to be the case of chalk. There is another example of this kind in England, in the Bath stone and Portland stone, which are calcareous: And similar stones occur in many parts of the world. There are some few entire shells found in these stones; but the greater part of it does not exhibit any.

The abundance and great extent of the strata of calcareous earth are, therefore, a proof of the great antiquity of this globe, and of the great changes which have happened in its surface.

In some few places the bones of land-animals have also been found, in very considerable quantities, mixed with calcareous matter; as in Dalmatia, and some islands of the Adriatic; at Gibraltar, in Spain, and in France †.

In whatever condition the calcareous earth is found, it is easily distinguished from other earthy substances, by effervescing with acids; and, when it is not very impure, by becoming common quicklime, when exposed for some time to the action of a strong fire.

In making the trial with acids, however, it is necessary that the person who makes it should understand it well, as I have known many who never had seen or attended to real effervescences imposed upon by the appearance of it. I have often had earths sent me by farmers,

---

* The rapidity with which the coral rocks in the Pacific Ocean are formed, by the little animals, is altogether astonishing; and it is evident to the sight.

† *Memoires sur les ossemens fossils qui ont appartenu à des grands animaux : Observations de Physique, Mai* 1781. The author is of opinion that they are not the bones of elephants, but of some large aquatic animals which have inhabited the lakes and great rivers, or rather the sea. His observations render this very probable. The memoir is curious, and deserves a careful perusal.

or country gentlemen, as calcareous marles, which they affured me effervefced, or fermented with vinegar, and diffolved in it too ; but which, when I tried them, fhewed not the fmalleft effervefcence. Upon defiring them, therefore, to fhew me the fermentation and dif-folution they fpoke of, I found that they had throwu the earth dry into the vinegar, and that what was taken for effervefcence, was the expulfion of a fmall quantity of common air which was contained in the pores and vacuities of the dry earth, and which arofe flowly in little bubbles, when the fluid was fucked into thefe pores ; and the fuppofed diffolution was the diffufion of the earth. Remember, therefore, always to foak and moiften with water before the acid be applied.

In defcribing the properties of the objects of chemiftry, I begin commonly with the effects of heat ; but, as this earth, in confe-quence of the ftate in which it is found in nature, undergoes a very great change in all its chemical relations by the action of a ftrong fire, I fhall firft notice its properties with regard to mixture, in its ordinary or native ftate.

I have already remarked that the action of acids on it is very fimi-lar to their action on alkaline falts. Let us, therefore, attend to the manner in which it unites with each of the different acids.

The firft of thefe, therefore, the fulphuric acid, attaches itfelf to this earth with great rapidity, and violent effervefcence. Thus, when we throw powdered chalk into diluted fulphuric acid, violent ef-fervefcence is produced ; but the chalk, however, is not diffolved. It renders the liquor thick and muddy ; and, if allowed to reft, falls to the bottom in the form of a fediment much more bulky than the chalk alone. The reafon of this is, that the compound produ-ced by the union of the fulphuric acid with calcareous earth has very little folubility in water. It requires a very large quantity of water to diffolve it, 500 times its weight at leaft. Hence, not find-ing water enough in the mixture, the greateft part remains undiffolved, and is more bulky than the chalk, becaufe it confifts of that chalk and the falt which it has formed with the fulphuric acid. This compound was formerly called Gypsum or Selenites, or the Selenitic Salt. The name now given to it, by the French chemifts, is Sulphat of Calcareous Earth, or Sulphat of Lime. It is formed naturally in confiderable quantities in the bowels of the earth, and is valued on account of feveral ufeful properties which belong to it. Thefe we fhall mention more fully afterwards.

Although this compound is difficult to diffolve, and requires much water to its diffolution, it can, however, be completely diffolved, when enough of water is applied to it, viz. one ounce for each grain; and when the water is evaporated, the compound concretes into numerous flender and fmall cryftals, as fine as hairs, which fubfide and form a white fediment. A fmall portion of this falt is the general impurity of thofe fpring waters which are called hard waters.

Such is the action of the fulphuric acid upon this earth, in the form of powder. But it does not operate upon it fo readily in its folid and compact form, becaufe the furface is neutralized, and protects the interior parts from the corrofion of the acid.

None of the other acids, hitherto defcribed, form infoluble or difficultly foluble compounds with this earth, except the tartarous acid, which forms with it the compound called by Mr. Scheele *felenites tartareus*, now named the tartrite of lime *.

The other acids, viz. the nitric, muriatic, and acetous acids, form with it very foluble compounds; and all of thefe acids too, as I before obferved, are attracted by the earth in the fame order as they are attracted by the alkali; and therefore, if the fulphuric acid be added to the folutions or compounds formed by any other acid, it immediately

* It forms infoluble compounds alfo with the phofphoric, malic, and fome other acids.

disengages the other acid, and unites itself to the earth, in consequence of its stronger attraction, forming with it the sulphat of lime already described, which therefore precipitates or falls to the bottom.

The calcareous earth not only has a stronger attraction for the sulphuric than for other acids, but, further, some experiments shew that it has a greater partiality (so to call it) for this acid above others, than the alkalis have. This I think appears from what happens when we mix any compound salt containing the sulphuric acid, with a compound or solution of the calcareous earth in any other acid than the sulphuric. Thus, if we mix a solution of Glauber's salt, or sulphat of soda, in hot water, with some of the solution of calcareous earth in nitric acid greatly diluted, the mixture will become muddy, and deposit a sediment. There is therefore a double elective attraction. *(See Bergmann)*.

There are only two of the acids already described, which, as I learned by some experiments made on purpose, have not power to act on the calcareous earth in its natural state. These two acids are the sedative salt or boracic acid, and the sulphurous acid *(acidum sulphurosum)*, or acid of sulphur in its volatile suffocating state. And to these we may add sulphur itself, which was spoken of in treating of the sulphuric acid. These three chemical bodies, the boracic acid, the sulphurous acid, and sulphur itself, are but weakly attracted by alkaline substances in general.

With regard to the other class of simpler salts, the alkalis shew as little disposition to act on calcareous earth as on one another. The only observation that belongs to this head is, that they have a stronger attraction for acids; and when mixed with solutions of the calcareous earth, immediately precipitate it,—uniting themselves to the acid in its place. And the effect is the same whether we use a fixed or a volatile alkali in their ordinary state. It should also be remarked, that when an alkali is employed to dislodge calcareous earth from an acid, *we have no effervescence.*

Such is the relation of the calcareous earth to the acids and alkalis.

Of the compound falts, none are affected by it except tartar, and the ammoniacal falts. Tartar, when boiled with it, is neutralized, or, to exprefs it more properly, it is deprived of its fuperfluous acid. And the ammoniacal falts can be decompounded by it, but only when it is affifted by heat.

If, for example, we mix chalk with common fal ammoniac,—the muriat of ammoniac; either taking this falt as diffolved in water, or in the form of very fine powder, we cannot perceive any change or effect produced, however long they remain fimply mixed together. But if they be mixed in the form of fine powders, and we then apply heat to this mixture, as foon as the heat increafes to a certain degree, the ammoniacal falt begins to be decompounded. The acid of it is joined to the calcareous earth; and the volatile alkali arifes into the receiver in large quantity, and forms a very folid faline cruft on its internal furface. The procefs is as eafy as the diftilling of water. The veffels are to be luted with chalk made into a thin pafte, with mucilage or gum-water. The heat may be flowly raifed till the fublimation begins to dim the veffels, and then kept in that ftate; and the operation ftopped whenever a dimnefs appears in the top or neck of the retort. This indicates a commencing fublimation of the fal ammoniac, if too little chalk has been employed.

This procefs for decompounding fal ammoniac by the action of chalk and a ftrong heat, is often performed in England, to obtain a volatile alkali for the purpofes of medicine. The alkali thus obtained is remarkably folid and dry, and in larger quantity than that which is got from the fame quantity of fal ammoniac, when it is decompounded by a fixed alkali.

Thefe are the properties of the calcareous earth with regard to mixture in its natural ftate.

But it fuffers a remarkable change by the action of fire on it, which is next to be defcribed.

VOL. II.   E

If it be expofed to the action of a ftrong red heat for fome time, it becomes what is well known by the name of QUICKLIME. For this purpofe, the marble, limeftone, or chalk, is broken into pieces of fmall fize, and piled up in a kiln built in the form of an inverted fruftum of a cone, mixed with a proper quantity of fuel, taking care to have a fufficient quantity of fuel at the bottom, to raife a heat in the beginning ; and alfo taking care that the whole may be fo piled up, with fpiracles properly conducted through it, by laying the materials open in thofe places, that the heated air may have a current pretty equably diftributed through the whole mafs. This foon kindles the interfperfed fuel all over the kiln ; and it would foon become fo hot as to rifk the melting of impure limeftone, were it not checked by contracting the apertures by which the air gets in below, or (if it be a fimple clamp kiln) by covering the whole at top with earth. The proper degree of heat muft be learned by trials ; it being different according to the nature of the limeftone and alfo of the fuel. Some pit-coals have a great tendency to vitrify even the pureft limeftone ; and limeftone containing iron in any ftate is very much difpofed to vitrification. It is only experience which can teach how much fuel is required for raifing the due heat, and for continuing it till all fhall be calcined. If it be a clamp kiln, the fire is allowed to burn out, and the quicklime is taken out when the kiln is fufficiently cooled. But limekilns are more artificially conftructed, fo that the work is never ftopped. The calcined lime is taken out at an opening properly contrived below, by which the whole fubfides ; and the limeftone is continually added above, mixed with bafkets of fuel. The fuel is entirely confumed in the lower and narrower part of the cone ; and the great heat of the glowing blocks of quicklime maintains the current of frefh air through the interftices, and caufes it to arrive at the unconfumed fuel above, in a ftate fit for continuing the combuftion. The confumption of the fuel is vaftly flower than would happen were it interfperfed in the fame manner among lumps of free-ſtone or pebbles, for reaſons which you will ſoon underſtand.

When we want to prepare a ſmall quantity for experiments, we juſt mix it with abundance of fuel, in any ordinary furnace, and urge it with the greateſt heat, which will not vitrify it when impure. Pure limeſtone, marble, or chalk, may even be calcined in a crucible, without touching the fuel; but this requires a very great heat, long continued.

Quicklime has a number of qualities different from thoſe of the natural calcareous earth. If it have been hard before, it becomes more friable and porous, eſpecially the pureſt kinds. It alſo loſes a very great part of its weight, 40 per cent. in the pureſt kind. This great loſs is well known to thoſe who have occaſion to fetch lime from a diſtance; and it is the beſt mark of complete calcination. Further, from being a mild, inſipid, and inert earth, we find it now an active and acrid ſubſtance, like the alkaline ſalts. It ſhews a very conſiderable degree of activity in mixture with a variety of other bodies. Applied to the tongue, it is extremely acrid, giving a taſte very much like an alkaline ſalt. If left any time upon a moiſt ſucculent part of animal or vegetable ſubſtance, it either corrodes and diſſolves it to a pulp, or weakens the coheſion of it to a great degree; and it has the ſame effect on the moſt ſolid and firm parts, if it be made into a ſoft paſte with water before it is applied. This ſhews that it has a ſtrong chemical attraction for the matter of thoſe ſubſtances. We diſcover in it alſo a ſtrong attraction for water. This attraction is one of the moſt remarkable qualities by which it differs from calcareous earth in its native ſtate. Thus, when we pour water upon the lime, a quantity of it is quickly ſucked up into the pores of the ſtone; and, after a ſhort time, the maſſes of quicklime which we have moiſtened begin to grow warm, and to ſmoke. They ſwell, ſplit, and crumble down into pieces; and theſe are affected in the ſame manner, until the whole, in a few minutes, is converted into a ſubtile white powder,

E 2

greatly more bulky, and which, if too much water have not been ufed, is perfectly dry and dufty. While this is going on, it becomes fo hot, that a part of the water is evaporated in boiling hot fteams; and if the quantity of lime fo wetted at once is much greater than this, the heat increafes to a degree capable of inflaming combuftible bodies. If a ftick, for example, be thruft into a large heap of it, the extremity will be burnt to a black coal. Leaky fhips have fometimes been unfortunately fet on fire at fea by a cargo of quicklime. This is fo well known, that when they venture to load with it, they take every precaution to prevent fuch a misfortune.

As foon as the lime has been thus reduced to a fubtile powder, by a fufficient quantity of water, no more heat is produced. It cools and does not produce heat again if mixed with water. It is called SLAK- ED LIME,—CALX EXTINCTA. If we weigh it, we find it confidera- bly heavier than before the water was added to it. This addition of weight is from a part of the water now clofely united with it; every particle of the lime having its due proportion. The water thus united with the lime is retained by it with a ftrong attraction; and cannot be feparated again without a red heat. Being in a folid form, the emerfion of the latent heat, which occafioned its fluidity, is the fource of that great heat which is produced during the flaking of the lime, and affords one of the moft remarkable inftances of this emerfion. But if fuch a heat is applied to the flaked lime in a retort and re- ceiver, we recover the water in its former ftate. The lime from which it is feparated becomes quicklime again, and is difpofed to pro- duce the fame effects with water as before. Thefe effects, therefore, evidently depend on a ftrong attraction which the lime and water have for one another. And flaked lime is lime faturated with wa- ter, or combined with as much water as can unite with it, fo as ftill to form a dry compound. The confequences of this union with wa- ter, however, are not fuch as to take away the attraction of the lime for other fubftances. It has the fame acrid tafte as before, and the fame, or even a greater corrofive and diffolving power, with refpect to a variety of different fubftances. Nor is its attraction for water faturated by this folid combination. It ftill manifefts a folubility in that fluid, perfectly refembling, in this refpect, the relation of cryftallizable falts to water. They, after having combined wit ha certain quantity, are diffolved in a larger quantity. This folution is perceived when we mix flaked lime with a much larger quantity of water, or dilute it largely in water. If we examine this water, we fhall find that it has either diffolved a part of the lime, or extracted fomething from it, fo as to form a clear folution like that of the alkaline falts, and refembling them much in tafte.

This is the fluid called LIME-WATER, the nature and beft preparation of which was, fome time fince, a fubject of much difputation in this place. The general opinion was, that the production of the lime-water depended on an active and fubtile principle, extracted by the water from the lime, which was imagined to contain only a fmall portion of it; and that the lime could therefore make only a moderate quantity of good lime-water, fuch as 10 or 12, or at moft 20 times its weight.

It is known now that a much greater quantity of good lime water can be made of lime,—that is of a good quality; perhaps 300 or 400 times its weight. And the procefs for making good lime-water is fo fimple, that it was thought unneceffary to infert it in the former editions of the Edinburgh Pharmacopœia. But a procefs was given for it in the laft edition, viz. 1mo, Slake the lime with a fmall quantity of water, about one-fourth or one-third of its weight, and let this be done in a veffel of earthen ware or glafs, keeping it covered while the lime is flaking: 2do, Pour on it 30 or 40 times its weight of more water, and after agitating to diffufe the lime through the water, cover up the veffel again. The agitation being repeated afterwards 10 or 12 times to promote the diffolution, we may then allow the remaining lime to fubfide, and filtrate the lime-water, to have it clear. Thus lime-water

is made as ftrong as poffible.   But, in its ftrongeft ftate it contains
but a fmall proportion of the lime, or limy matter.   I made experiments to afcertain this point, and found, that the ftrongeft lime-water
we can make does not contain a larger proportion of lime than one
part in 500 parts of the water.   It has, however, a pungent, acrid,
difagreeable tafte, refembling that of the alkaline falts ; and has alfo
the power to change the vegetable colours like thofe falts.   A moderate
quantity produces firft a fea-green ; a little more, a grafs green ; and
more ftill, a yellow, or fillemot, attended with muddinefs ; the colouring matter being corroded and deftroyed.   It is the moft tranfparent
fluid known.

Such are the effects of water mixed with lime in different ways.

When we next try the effects of acids, we find that lime is ftill more
difpofed to unite with thefe, than the calcareous earth was when in
its natural ftate.   It unites much more quickly with the vegetable
acids.   Diftilled vinegar, which diffolves the calcareous earth but
flowly, diffolves lime readily and quickly.   And when we boil lime
with a folution of tartar in hot water, the lime takes to itfelf the whole
acid of the tartar, not the fuperfluous acid alone, which only is feparated from the tartar by the calcareous earth in its natural ftate.

Further, lime can be joined with the boracic and fulphurous acids,
fo as to form a borat and a fulphite of lime ; neither of which can be
produced from the natural calcareous earth.   And when we mix lime
with the muriat of ammonia, it fhews fuch a ftrong propenfity to
unite with the muriatic acid, that it decompounds the muriat the moment they are mixed together, not requiring the affiftance of a ftrong
heat, which is required by the natural calcareous earth.

Befide the quicknefs and facility with which lime decompounds
the ammoniacal falts, the volatile alkali, which we obtain from the
mixture by diftillation, is in an uncommon and very remarkable ftate.
When heat is applied, this alkali arifes in vapours which are in the
higheft degree elaftic, pungent, and penetrating.   They cannot be

condenfed without the aid of fome water to reprefs their volatility by its attraction. Were we to neglect the proper application of water to this mixture, for the purpofe of condenfation, this alkali would continue in the ftate of an incondenfible elaftic fluid or gas. This is what Dr. Prieftley called alkaline air, and made the fubject of many experiments. One of them is very pretty and amufing. If fome of this vapour be mixed with that of the muriatic acid, we have immediately a precipitation of fal ammoniac. If this experiment be made in the modern pneumatic apparatus, a difagreeable accident fometimes happens. The vapour of either being fuddenly mixed with the other, in the upper part of a jar, ftanding in water or mercury, the whole collapfes in an inftant, and the water or mercury dafhes againft the top of the jar with fuch force, as to beat it up to a confiderable height, and with rifk to the operator.

When we do employ water, though in no greater quantity than what is abfolutely neceffary, no part of the alkali ever is condenfed into a folid form. It always conftitutes, with the water, an highly pungent, acrid, volatile, alkaline fluid.

A volatile alkali is ordered to be prepared in this ftate, in both the London and Edinburgh Pharmacopœias. It was formerly named *Spiritus Salis Ammoniaci cum calce vivâ*. Now it is named *Aqua Ammoniæ puræ*, for a reafon which you will foon underftand.

Various proceffes have been recommended, by different authors, for preparing it in the beft and moft convenient manner. Some have advifed to put the quicklime into the retort unflaked, and then pour on it the fal ammoniac diffolved in water. But this mode is inconvenient, caufing fudden and violent heat, which makes the volatile alkali rife fo impetuoufly, and in fuch incoercible vapours, that the greater part of it is loft. In diftilling it, a very gentle heat is abfolutely neceffary, that time may be allowed for its condenfation, by means of its union with the water that is to affift in condenfing it. It requires much caution and patience to accomplifh this, becaufe the alkali rifes in vapour more readily than the water, and does not carry off with it a very confiderable portion. And, fhould we attempt to promote the condenfation by putting water into the receiver, we fhall ftill be difappointed, becaufe the alkali combined with the water merely fufficient for its condenfation, forms a fluid confiderably lighter than water, which will, therefore, float on the furface of that in the receiver, and this, being already faturated with alkaline vapour, will condenfe no more. I mention this particularly, becaufe fimilar things frequently occur in diftillations; and muft be carefully attended to, to avoid accidents. Mr. Woulfe's method obviates this difficulty; but is beft adapted to the diftillation of large quantities.

After having tried different methods, I now practife one which does not require this complicated apparatus, and ferves very well for preparing a fmall quantity of this alkali. Having put the fal ammoniac, in lumps, into the retort, pour on it a mixture of an equal quantity of lime, and thrice its weight of water, ftirred till uniformly mixed, and kept till cold. The volatile alkali begins immediately to be feparated in pungent vapour: But, as there is enough of water prefent to reprefs and moderate its volatility, the vapours are not troublefome. Now apply the receiver, which has firft been well warmed, to rarefy the air, and thus expel a part of it; and immediately clofe up the joining of the two veffels, fo as to make that joining perfectly air-tight. This is done by firft applying common putty, made of chalk and lintfeed oil. But as this pafte, though impenetrable to air, is in fome degree foft, and would be forced open by the preffure of the elaftic vapour from within, we may fecure it from this accident by putting over it another luting, which will foon grow quite firm and hard, and will bind the putty ftrongly in its place. This other luting is a mixture of chalk and gum-arabic,— nine parts of the chalk to one of the gum. It muft be made into a fluid pafte with water, and fpread on flips of paper, and thefe put on the joint. Being then brought into clofe contact with every part of it by a ligature or bandage, it makes the joint fo clofe and ftrong, that the veffels would be burft open in any other part of them fooner than in this. The joining being thus made perfectly air-tight, it is evident that we muft be very cautious in applying the heat to perform the diftillation. Too much heat would certainly burft the veffels, by giving too much elafticity to the air and vapours; and it would alfo bring over too much of the water, fuppofing the veffels were ftrong enough to withftand it. We, therefore, apply a very gentle heat, and apply it fo long only as may be neceffary for obtaining the product in proper quantity. Thus the alkali, and a portion of the water, will diftil over together, flowly and imperceptibly, until we have enough of it, which will be the cafe when the heat fhall have continued 24 hours. *(See Note 1. at the end of the Volume.)*

Befide this action of quicklime on muriat of ammoniac, and other qualities of it which we mentioned, a further particular, which fhews that its powers as an alkaline fubftance are increafed, is its effect on fulphur. It is capable of uniting with it in the dry way, by mixing it in powder with the fulphur, and applying fuch a heat as converts the fulphur into vapour, or melts it.

But a more convenient way is to join them in the way of folution, or *vià humidà*. Equal weights of fulphur and quicklime muft be mixed in a flafk, and a quantity of water added boiling hot. The mixture muft be boiled for fome time, and then more boiling water added, and the boiling repeated. After ftanding a proper time for fubfidence, the clear liquor muft be decanted into a phial, which muft be carefully ftopped. This gives a perfect compound of fulphur and lime.

The compound we thus obtain is called the fulphuret of lime,— *fulphuratum calcis.* It is incomparably more foluble in water than the lime alone, and it has the fame yellow colour and difagreeable odour, that is fo ftrong in the compounds of fulphur with alkaline falts. The union of the two ingredients of this compound is fimilar to that of the alkaline fulphuret.   It is decompounded by expofition to the air, and in its place we obtain gypfum or felenite, compofed of the fulphuric acid and calcareous earth.   Moreover, this decompofition is accompanied by the abforption of oxygenous gas, leaving only the mephitic air of Scheele, or the phlogifticated air of Prieftley.   Indeed, it was with this fulphuret that Scheele made his firft experiment. Like the alkaline liver of fulphur, too, it is decompounded by any acid.   The nitrous acid does not difengage from it the hepatic or ftinking gas, without fome particular attentions, which will be mentioned afterwards.

All thefe facts and experiments fhew that lime has a ftronger tendency to unite with other bodies than the calcareous earth has when in its natural ftate.

The moft remarkable property of lime appears in the change which it fuftains by mixture with the alkaline falts, and the change which it induces on them.   This conftitutes the chief difference between quicklime and the natural or crude calcareous earth.   If we mix it properly with an alkaline falt, the lime immediately lofes its activity and corrofive nature, and becomes quite mild, infipid, and no longer foluble in water.   To illuftrate this, we may notice the effects of an alkali upon lime-water.   This fluid, as I obferved before, contains but a fmall proportion of limy matter diffolved in the water, about one grain to the ounce.   Yet, upon dropping into it a fmall quantity of diffolved alkaline falt, it inftantly becomes muddy, and depofits the lime in the form of mild infipid calcareous earth.   And the effect is fimilar when we ufe, inftead of lime-water, a muddy mixture of lime and water, or a quantity of water with more lime in it than it can diffolve into lime-water.   Provided we add to fuch a mixture a fufficient quantity of the alkaline falt, both the diffolved part of the lime, and that which, for want of a fufficient quantity of water, is not diffolved, will be immediately rendered mild, inactive, and incapable of being diffolved by pure water.

Here is, therefore, a fudden reftitution of the lime to its original ftate of a mild, inactive, infoluble earth; for when we collect and dry the mild earth thus obtained, we find it has every property of the calcareous earth, in its natural or ordinary ftate, and may again become lime by calcination.

But, further, when this change is produced in the lime, the alkali fuffers one which is as remarkable. This is perceived when we feparate the alkaline liquor again from the lime, and examine this liquor, or by evaporation obtain the alkali again in a dry ftate.

We find it incomparably more fufible than before. It is melted long before the heat of it is raifed to the degree of ignition.

Moreover, it becomes more fufceptible of vaporifation. A very fenfible quantity is carried off by hafty boiling the folution of it in water; and if long continued in a red heat, with fpongy matters to prevent its melting, it evaporates very faft, and is loft.

We find alfo that its acrimony, or activity to diffolve different fubftances, is very greatly increafed. In the treatment juft now mentioned,—the melting of it by heat,—it often happens, that by raifing it to a red heat, it diffolves the earthen veffel in which it is melted.

Alkali, rendered more active in this manner by lime, is much employed in fome of the arts, and, being alfo ufed for fome purpofes in medicine, is ordered to be prepared in both our pharmacopœias.

Three parts of alkaline falt, and four parts of perfect quicklime, are to be fuddenly mixed in twice their weight of water. This will produce great heat and ebullition. The veffel muft be covered, and the mixture agitated repeatedly. It muft then be poured into a filtrating funnel, which is fet into a phial, or other veffel, proper for holding the lixivium. After it has ceafed running through the filtre, water muft be poured on a-top, and allowed alfo to run through the filtre, till the whole lixivium is about thrice the weight of the falt and lime. It will be without colour or smell, if these have been pure, and must be kept in bottles closely stopped.   The water may be expelled by heat.   When the boiling has ceased a little while, the purely saline part remains fluid like sluggish oil, and the vessel is, at this time, almost red hot.   When poured out on an iron plate, it quickly congeals, and may be divided, while very hot, into pieces fit to go into the mouth of a phial, where it may be kept shut up from the air. This is the *alkali caufticum acerrimum* of the Dispensatory.   The *caufticum mitius* is made by pouring the melting salt on quicklime, and mixing before it cools.

The advantages of this process are,   1*st*, The lime, being slacked with unusual violence, is the more subtilely divided.   2*d*, The heat promotes the action of the lime and alkali.   3*d*, The small quantity of water used also promotes their action; the lime not having room to subside.

Here I must mention a circumstance which I am not yet able to explain; namely, that this caustic ley has a singular effect on the earthen-ware vessels in which it is kept.   I always used vessels of that kind of pottery called in this country stone-ware, which is a species of porcelain, far superior to the soft bibulous pottery.   I have very often observed that, after two or three months keeping in such vessels, they split all over, with a loud crack.   The same observation has been made by other persons of my acquaintance.   Nor is this confined to this particular material; it happens also to glass bottles, though not so frequently.   The only conjecture that I can form on the subject is, that the alkalis, acting on the inner surface of a vessel not well annealed (and it is scarcely possible that any can be *perfectly* annealed) corrodes the stratum which serves to support the whole in its state of unequable contraction.   Yet I could not see any such corrosion.

I have also observed, and my observation was confirmed by Dr. Irvin, that caustic ley, very long kept, loses its acrimony, and that its alkaline properties are really impaired, as if the salt had evaporated.

from it, or that it was neutralifed by fome other fubftance. You will notice that I fuppofe it all the while to have no communication with the air.

Dr. Alfton made an obfervation on cauftic leys, which is curious, and of which I do not fee any explanation. The fpecific gravity of an alkaline ley is confiderably diminifhed by adding lime to them. This effect is very remarkable in the folution of volatile alkali.

The activity of the alkali to diffolve animal and vegetable fubftances is now increafed to a furprifing degree; fome of the firmeft parts of animals being foon diffolved by it into a pulp. Hence it is often ufed externally in furgery, for opening abfceffes, deftroying fungous flefh, or making iffues in the fkin, when the patient is afraid of the knife. It acts with burning pain : Hence it is called the POTENTIAL CAUTERY, or the CAUSTIC ALKALI. Notwithftanding this high degree of acrimony, it has been ufed of late internally; but it muft be greatly diluted and obtunded, by which means the foffil acids themfelves may be taken internally with fafety.

Such is the activity of the cauftic alkali with refpect to animal fubftances. Of the vegetable there are many which it will penetrate and diffolve as readily. But the dry or pure woody fibres of vegetables fuftain its action for fome time. Of this woody or fibrous matter, cotton, bleached linen, paper, and the like fubftances, are compofed. Even thefe, however, though they are not diffolved, are corroded, or made tender by it, and their cohefion much diminifhed, as has often happened to linen cloth in bleaching. The two fixed alkalis acquire juft the fame activity from lime, and have the fame effect upon lime. The volatile alkali likewife produces the fame effect upon the lime, in reftoring it to its mild ftate; and is affected itfelf much in the fame way as the fixed alkalis are, if allowance be made for its different nature. Its volatility is prodigioufly increafed, as I have already obferved; and it is now incomparably more acid and pungent, fo that it muft be fmelt to with great caution, otherwife it will excoriate and deftroy the noftrils. Like the cauftic fixed alkalis, it diffolves animal and vegetable fubftances into a pulp ; and acts on oils and fats in the fame manner, making them mifcible with water ; in fhort, rendering them foaps. It becomes precifely the fame with the *fpiritus falis ammoniaci cum calce vivâ.* As in that preparation the vapours are incondenfible without water, fo, when as little as poffible is ufed, they are infeparable by diftillation. Mr. Du Hamel imagined that the volatile alkali was deftroyed by repeated additions of lime ; but this is a miftake, owing to the efcape of part of it at each addition, by the fudden eruption of incoercible vapours, in confequence of the heat produced by the addition of the lime.

I have now defcribed the more remarkable properties of quicklime in the way of mixture. I muft now obferve that when it is expofed to the air, it begins, after fome time, to fwell, fplit, and crumble down, until it is flowly reduced to as fine a powder as when it is flaked with water. This happens to it plainly in confequence of its gradually attracting humidity from the air, until it has received enough to faturate it, or to convert it into flaked lime. But the lime flaked in this manner is not fo acrid and active as that which is flaked immediately with water, and without being much expofed to the air. The air, at the fame time that it fupplies humidity to it, is alfo found to weaken its force by degrees, and continues to weaken it more after it has been completely flaked, until, in a certain length of time, it deprives it of the qualities of quicklime altogether, and reduces it to the ftate of infipid calcareous earth.

The lime does not undergo this change when it is preferved in clofe veffels. It fuffers it only when we expofe it to the free air, or when we keep it in veffels that are left open. And we further find, that the more wide and loofely it is fpread out to the air, it becomes inert fo much the fooner. It lofes its activity fafter, in proportion to the more extenfive contact or communication of the free air with it.

But, if the lime be made into a pafte with water, and this pafte be collected and compacted together in the form of a round mass, it then retains its activity, and the other qualities of lime, much longer than when it is expofed to the ftate of a dry powder, and fpread out to the air: And the reafon is plain. When it is in the ftate of a pafte, the water, filling up the interftices between the particles, excludes the air from among them; and only thofe that are at the outfide of the wet mafs can be affected by it.

Lime-water is alfo liable to a fimilar change by expofition. It may be preferved in the fame manner for a long time in clofe veffels, without alteration of its qualities. But if it be left in an open veffel, we may fee, in a few minutes, a thin film produced upon its upper furface, where it is in contact with the air. This film continues to increafe in thicknefs, until, after a number of hours, or perhaps a few days, it will form a thin ftony cruft. In proportion as this cruft is formed on the top, the water below lofes its tafte and the other qualities of lime-water, and at laft becomes a mere infipid water. The crufty matter itfelf, on examination, proves a mild calcareous earth, like the calcareous earth in its natural ftate. Thefe feveral particulars occafioned lime-water to be confidered as a furprifing unaccountable fluid. When frefh drawn from the lime, it has a pungent, acrid, alkaline tafte, and feveral other qualities, as if it contained an alkaline falt. And it retains thefe qualities if it be fhut up in clofe veffels; but if expofed to the air, it becomes quite infipid: And yet we can obferve no caufe of this change, but the feparation or rejection of an earthy matter from it, which is alfo perfectly infipid, and which cannot again be diffolved.

There remains to be mentioned one other property only of lime, and it is one of its moft ufeful properties; I mean that of making good MORTAR, when it is fkilfully mixed with fand and water. The qualities of *good mortar* are, to adhere readily to ftones or other folid bodies; to cement them together; and to acquire, by time, a ftony hardnefs. Thefe qualities, in fuch a mixture, have been thought the more furprifing, for thefe reafons, that lime and water without fand has not all the qualities of good mortar, and that nothing is fo little difpofed to cohere as fand by itfelf.

After all, however, the effect of the fand in giving cohefion and hardnefs to the lime and water has been exaggerated. And Mr. Macquer errs a little from the truth, when he fays that lime and water without fand do not concrete in the leaft.

If we give to a pafte of lime and water the form of a fmall ball, and allow it time enough, that it may dry flowly, it will contract its fize, and become very hard,—harder, I believe, than if fand had been added. But that is not the way in which lime mortar is ufed. It is not made into balls or other maffes that are to dry and harden without being connected with other matter; it is employed for ce- menting ftones or bricks together, and for plaftering walls and other fuch parts of buildings. And, if it were made up fo as to be liable to contract much in drying, the confequence would be, that it muft neceffarily feparate from thofe bodies to which it is applied, and which cannot contract along with it; or, if continued adhering to them, it muft fplit and divide itfelf into innumerable fmall parts, and fo become quite fhattered and ufelefs. The reafon why lime alone, when made into a pafte with water, contracts fo much, is the great quantity of water which fuch a pafte contains; for a good quantity of water is required to make flaked lime into a foft pafte. But this contraction and its confequences are prevented almoft entirely by the addition of fand to the lime.

Some fkill, however, is required to add the fand in the moft pro- per manner. If too much be added, the mortar will not be very hard and firm when it dries. The fand will give it an opennefs of texture, which both diminifhes the cohefion of its parts, and admits water too readily into its pores, and expofes it to the effects of froft.

Some time ago, a French engineer, Loriot, propofed a contrivance, which he afferted to be a great improvement in the compofition of lime-mortar for many purpofes, and by the ufe of which lefs fand was neceffary to prevent its contraction. He made common mortar with the ufual proportion of fand, but with a little more water than ordinary, fo as to be almoft fluid; and juft before it is ufed, he added to it one-fourth or one-third of quicklime, not flaked, but reduced to powder mechanically. When quicklime is thus mixed with a good deal of cold water, it does not flake fo fuddenly as when no more water is ufed than is neceffary to flake it. The quicklime put in does not fwell and flake until the mortar is ufed, and put in its place. The powder of the quicklime, then flaking throughout the whole of the mortar, abforbs and confolidates a part of the water which it contains; and the mortar becomes hard and dry remarkably foon, and without contracting. And when the reft of the humidity is evaporated, it acquires a degree of hardnefs and compactnefs far fuperior to that of ordinary mortar. It is impenetrable by water. Vafes of it filled with water do not imbibe a fenfible quantity. The hardnefs and ftrength of this mortar, is undoubtedly owing to its denfity and more perfect fetting. Mr. Loriot was of opinion that the Romans made their mortar in fome fuch manner. Lime mortar fets in time, if left at reft. This fetting is fomewhat analogous to the concretion of falt and water in the act of cryftallization. Mortar will fet again and again, after kneading or beating up, but more weakly than before. Loriot's mortar, therefore, muft fet the moft ftrongly of all, becaufe it is not afterwards difturbed.

Since Loriot, Dr. Higgins of London publifhed a little tract on this fubject, containing a defeription of a great number of experiments, made to difcover the beft compofitions for making mortar or lime cements of different kinds. His experiments agree in fhewing that the hardnefs and ftrength which mortar or lime cements acquire depend very much on the frefhnefs and activity of the lime employed. And, in his receipt for the beft compofition, he infifts particularly on a method of ufing the whole lime in its moft active poffible ftate, fo as it may be expofed to the air as little as poffible, after being burnt and flaked. And he alfo found that much advantage was gained by ufing fands of different fizes mixed together, and even a fmall addition of fome other ingredients ftill finer than the fineft fand. The particulars of all this are in his pamphlet, which I think judicious.

There is a way of making a lime-mortar or cement, the concretion and induration of which does not depend on its being dried. It is found to concrete and become hard and very durable, even under water, and is therefore ufed in buildings or works that ftand in water, or are very much expofed to be wafhed by it ; and fuch mortar is named water-mortar. Mr. Smeaton having had occafion to ufe this mortar in building a moft ufeful and important light-houfe, which ftands on a low narrow rock in the mouth of the Channel, where it is often ex-pofed to the utmoft fury of the waves, exerted all his ingenuity to make the building durable ; and among many other points, made a number of experiments to afcertain the beft compofition for water-mortar. And he was fo accurate and induftrious in his inveftigation of it, that he is the beft author to be confulted on this fubject. The detail of his experiments is given in the account he publifhed of the building of that light-houfe, which was a moft arduous work.

I have now proceeded fo far in the hiftory of the calcareous earth, as to have related the principal facts upon which were founded the firft attempts to explain this fubject, and to account for the difference between the earth in its natural ftate, and in the ftate of quicklime.

In general, thofe who attempted to give an explication applicable to the whole, or to the greater part of this fubject, entertained a notion, or rather formed a fuppofition, that the calcareous earth, when burnt to quicklime, receives fome fubtile and active principle, communicated to it by the fire or heat ; that from this principle it derives its acrimony, activity, and folubility in water. They imagined that lime-water is produced by diffolving fuch a part of the lime as has the largeft proportion of this principle combined with it, and that this principle evaporates from the lime or lime-water expofed to the air. Further, fome fuppofed that this principle is attracted from the lime by alkaline falts, and that thus are produced the cauftic alkalis. The greater number, however, feem rather to have confidered thefe as compounds of the alkali and the quicklime in fubftance, or the more acrid part of the quicklime. But neither has the exiftence of this active principle ever been demonftrated, nor is the fuppofition of its combining with lime confiftent with many capital facts which are known to all. Indeed it is inconfiftent with our general experience in chemiftry. When limeftone is burnt to quicklime in the fire, inftead of receiving any addition, it fuffers, on the contrary, a very confiderable lofs: And the alkali is alfo diminifhed in weight when rendered active by the quicklime. Moreover, we generally find, that the combination purely chemical, of the moft active fubftances, generally diminifhes their activity. It is thus that neutral falts are lefs active than their ingredients.

Other theories have been attempted, as I faid, chiefly to explain fome parts of the fubject. Such was Dr. Stahl's opinion. He was one of thofe who believed falts to be compounds of earth and water; and he fuppofed flaked lime to be a fort of falt haftily and imperfectly formed; and therefore liable to be eafily decompounded again into its conftituent parts,—calcareous earth and water.

Mr. Macquer again, in order to explain fome particular phenomena, fuppofed that the calcareous earth acquires a fmall quantity of fome acid principle in the fire, juft enough to give it folubility in water, but in too fmall quantity to faturate it; and therefore lime is ftill alkaline. By this fuppofition he endeavours to explain the action of lime and tartar on one another; but at the time when he wrote, the nature of tartar was but imperfectly known, and ill underftood. And neither his theory nor Dr. Stahl's are at all fatisfactory when clofely examined. You may fee an account and examination of thofe theories, in the fecond edition of the Englifh tranflation of Macquer's Dictionary, under the article *Caufticity*.

Upon the whole, the chemifts had been, at that time, but little attentive to the variety of earthy bodies, and had but an indiftinct knowledge of any alkaline earth, except the calcareous.

They only knew that there were two or three earths befide the calcareous ; and they knew fome faline compounds which held earths combined with acids ; but they had not turned their attention to them, fo as to examine by experiments how they were diftinguifhed from one another. They fuppofed them to be all more or lefs fimilar to the calcareous, or to have a great deal of affinity with it, if they be not the fame thing.

Dr. Hoffmann was the firft who gave diftinct notice of an alkaline earth, which he fhewed to be different from the calcareous. It was called *Magnefia*. When I firft began to make chemical experiments, I had the curiofity to examine more carefully this earth indicated by Hoffmann. I made a number of experiments to learn its properties, and how it was diftinguifhed from the calcareous and fome other earths. The refult of thefe experiments fuggefted to me, fome time after, an explication of the nature of lime, and of its effects on the alkaline falts, and engaged me in an inquiry, by which a clear light was thrown on this fubject, and on many other important parts of the fcience of chemiftry.

I propofe, therefore, to give an account of that inquiry. It is a long ftory. But when I began to give thefe lectures, and for a long time after, I was in fome meafure under the neceffity of following this method, giving, at full length, the invefligation of the fubject of quicklime, and the foundation of the opinion and fyftem which I formed upon that fubject ; for this reafon, that it was not generally underftood and approved of by the chemifts abroad. Some chofe to doubt, or explain in a different way, the proofs which I had adduced ; and they adhered to their old opinions. Others fet up, or fupported and extolled another new fyftem, which was directly contrary to mine. So long as the opinion of a number of reputable chemifts was unfavourable to my views of the fubject, I thought it became me to fhew the grounds of my opinions; and to leave thofe to whom I addreffed myfelf at liberty to judge for themfelves. Of late, however, moft foreign chemifts and philofophers agree on this fubject with me.

But I think it is ftill worth your while to hear a hiftory of the inveftigation. The fame facts and doctrines, when delivered in a dogmatic or didactic manner, do not make fuch an impreffion as when made the fubjects of hiftorical narration. The mind is then led through the moft natural chain of ideas; it being that which took place in the mind of the perfon who followed out the inveftigation.

I faid juft now, that a differtation or effay of Dr. Hoffmann's, on an earth called magnefia, occafioned the beginning of the inquiry, the hiftory of which I am going to give you. But I was peculiarly excited to it by the then recent difcoveries of the power of lime-water to give relief in cafes of the ftone and gravel, in which it was fuppofed to act by diffolving thofe concretions, and expelling them out of the body. Dr. Whytt and Dr. Alfton, profeffors in this univerfity, were then engaged in a difpute on this fubject. They both believed that it had efficacy; but Dr. Whytt imagined that he had difcovered that the lime-water of oyfter-fhell lime had more power as a folvent, than the lime-water of common ftone lime. I therefore conceived hopes that, by trying a greater variety of the alkaline earths, fome kinds might be found ftill more different by their qualities from the common kind; and perhaps yielding a lime-water ftill more powerful than that of oyfter-fhell lime.

I therefore began by examining the alkaline earth which Hoffmann had mentioned or defcribed; and for the preparation of which there was a procefs given in our pharmacopœia.

SPECIES II.—MAGNESIA.

We learn from Dr. Hoffmann, that it was a white earthy powder, which had gained reputation as a secret remedy, under the name of the POWDER OF COUNT PALMA, or MAGNESIA ALBA. The Doctor having learned the manner of preparing it, which he describes, made several trials to assure himself of its medical virtues, and to acquire some knowledge of its nature in other respects. And he reports that it is an alkaline or absorbent earth, which effervesces with all the acids, and neutralises them; that it is, therefore, proper for neutralising acids in the stomach, with this additional good quality, by which it is distinguished from the calcareous earth commonly employed at that time by physicians, that the magnesia has purgative or laxative powers, and therefore clears the bowels from the undigested matter which had produced the acid. And he supposed that the purgative power was a quality of the compound of the magnesia and acid after they were united together in the stomach.

The subjects from which Dr. Hoffmann prepared this earth, were some saline compounds, in which it is combined with acids; and these saline compounds are obtained in preparing nitre and common salt. He first made use of that which is got in preparing nitre, or in extracting the nitre from nitrous earths and composts. It is called the MOTHER LEY OF NITRE *.

But, afterwards, having learned, that, in preparing common salt from the waters that contain it, a similar liquor is obtained, he made use of this also in some of his processes for preparing magnesia. This

---

* Meder is the German word for scum, mud, or dregs, which separates from vinegar, ale, or other liquids, by careless keeping. This term is still in use in the northern parts of these islands, and the liquor is said to mother. The French translate it ignorantly eau mere.   EDITOR.

bitter faline liquor, produced in manufacturing common falt, is well known to our falt boilers who prepare falt from fea-water : And they call it *Bittern* in England,—in this country commonly *Oil of Salt.*

Dr. Hoffmann defcribed two ways of obtaining this earth from thefe faline liquors. The firft was by evaporating the faline liquor to drynefs, and then expofing the dry matter to a red heat in a crucible ; by which heat fome acid vapours were expelled, and the earth remained in the crucible. The fecond was by adding to the faline liquor a folution of an alkaline falt, which immediately precipitated the earth, by uniting with the acid with which this earth was combined. And he prefers this procefs to the other.

In this country, as we cannot procure any of the nitrous liquor, we have always prepared our magnefia from the *bittern* of common falt, or of fea water. Or, which is the fame, we prepare it from the faline compound which is got from the bittern by cryftallization.

The faline compound has been long ufed in the practice of medicine, under the name of *Sal catharticus amarus,* the bitter purging falt. It was firft obtained from the waters of fome purging mineral fprings at Epfom, in Surrey, and fold at an high price, on account of the fmall quantity of it which thofe waters contained. But afterwards, the manufacturers of common falt, finding they could get it in plenty from the bittern of fea water, and very cheap, prepared large quantities of it, at Lymington efpecially, and exported it to the Continent, where great quantities were fold.

The appearance of it, as commonly prepared from the bittern, is that of fmall prifmatic or columnar cryftals. It is rather more purgative than Glauber's falt, or fulphat of foda, and of a bitterifh, and vaftly more difagreeable tafte. Thefe cryftals contain about one half of their weight of water, and therefore readily undergo the watery fufion when fuddenly heated. The reft of their fubftance is a compound of the fulphuric acid and an alkaline earth, which becomes evident by the action of an alkali on this falt. The alkaline earth, which is the magnesia, is precipitated : And, by the same experiment, we learn that the acid with which it was combined is the sulphuric. It forms, with the vegetable alkali added, a sulphat of potash. The Epsom salt is, therefore, a sulphat of magnesia.

This saline compound, being crystallized with a little more care than is commonly employed in manufacturing it, concretes into much larger crystals, which, as they resemble and even excel Glauber's salt by their purgative quality, have also a good deal of resemblance to it in their appearance. This became known to the manufacturers, who soon crystallized large quantities of it in this manner, and sold it for Glauber's salt for some time. But the fraud was detected, and Glauber's salt has been sold so cheap since that time, that it is not worth while to attempt such a fraud, especially since the method for distinguishing the one of these salts from the other is generally known. We need only to dissolve a little of the salt in water, and add an alkali. The solution of the sulphat of soda remains transparent. The other becomes turbid and thick by the precipitation of the magnesia.

By not having knowledge of this distinction, Mr. Pott, of the Berlin Academy, maintained a dispute with Monf. Du Hamel, of the Royal Academy of Sciences, about the alkaline basis of common salt. Mr. Pott asserted that common salt contained an earth for its basis. Mr. Du Hamel demonstrated by experiments, that it contains an alkaline salt. Pott had drawn his conclusions from experiments made with the sulphat of magnesia, which was sold at that time in Germany as Glauber's salt ; and he did not suspect that it was any thing else. But, as genuine Glauber's salt is prepared from common salt and sulphuric acid, it must necessarily have the same alkaline basis as common salt. And finding that the salt which passed for Glauber's salt in Germany had an earth for its basis, he concluded that common salt must contain the same earth.

The process I used for preparing magnesia for my experiments, was, that we have now in the Edinburgh Pharmacopœia, or much the same. I diſſolved in water ſeparately equal quantities of Epſom ſalt, (ſulphat of magneſia) and of potaſh, ſuch as is commonly ſold under the name of pearl aſh. The Epſom ſalt muſt be diſſolved in twice its weight of the pureſt water, and an equal quantity of dry pearl-aſhes muſt be diſſolved in four or five times its weight of pure water. Each of theſe ſolutions muſt be carefully purified from all admixture or foulneſs. The alkali may be purified by ſubſidence and decantation. It is adviſeable to clarify the ſolution of Epſom ſalt with the white of egg, violently agitated with it when juſt hot enough to coagulate the egg. This produces a fine net–work, which will entangle every thing that is not chemically diſſolved, and the clear ſaline ſolution will then drain ſlowly through a linen cloth.

Equal quantities of the two ſalts having been thus diſſolved, the clarified ſolutions muſt be mixed together by violent agitation, in order that the ſalts may act quickly on each other, and the decompoſition be as perfect as poſſible. This muſt be further promoted by making the mixture juſt boil over a briſk fire.

Now, add to the mixture four times its bulk of boiling hot water, and again agitate briſkly. Let it now ſtand to ſettle. The magneſia is ſcattered over the whole liquor in a moſt impalpable powder, and will ſubſide with extreme ſlowneſs. When ſettled, decant off the clear liquor, and again pour on it the ſame quantity of cold water, and, when it has completely ſettled, again decant the clear liquor. This edulcoration muſt be repeated ten or twelve times before the magneſia can be cleared of the vitriolated tartar, formed by the alkali and the ſulphuric acid of the Epſom ſalt, which, you know, requires a prodigious quantity of water for its ſolution, eſpecially when cold ;— and in theſe ablutions it is found that hot water occaſions a much longer ſuſpenſion of the magneſia. But theſe repeated effuſions will effectually clear it of all ſaline admixture, leaving the magneſia pure, and in a fit ſtate for philoſophical examination. Having decanted off the tafteless water for the laft time, ftrain the fediment through a linen cloth, ufing gentle compreffion at laft,—for very little of the powder will come through. N. B. We fhall never fucceed by ufing the finer filtering papers. For, although magnefia is a fine hard powder, yet, while wet, it has more the appearance, the feel, and even the tranfparency, of a jelly *.

There are feveral niceties in this procefs, which feem fomewhat capricious, but are abundantly plain when we attend to the properties of the fubftances employed. The vitriolated tartar, being of difficult folution, impedes by its very formation the progrefs of the decompofition; therefore boiling brifkly is abfolutely neceffary. Alfo, the magnefia has a tendency to a peculiar mode of cryftallization in little round grains formed, like zeolyte, of fpiculæ diverging from a centre. They are hard, and would be very inconvenient for medical preparations. When a very minute quantity of the two falts is diffolved in a great deal of water, and allowed to reft without difturbance, the particles have room and time to affume this arrangement very regularly. Therefore, we muft prevent this by making our firft liquors pretty rich in the falts, and this, with the agitation and boiling, immediately deftroys this regular formation of cryftals, and caufes the whole to become a fine impalpable powder. I find that a quart of water to a pound of Epfom fait, and three quarts to a pound of dry pearl-afhes, are a very good degree of ftrength for the firft folutions. The gelatinous appearance, when wet, arifes from a fpongy adhefion of a great quantity of the water, and when this is evaporated to drynefs, the mafs is an exceedingly light and fpongy fubftance,—and this extreme lightnefs is one of the beft marks of its goodnefs. Mr.

* This tranfparency is unfortunate for the painters; for magnefia would otherwife give a moft dazzling white, fuperior to any they poffefs, and which would not change its colour, as the preparations of white lead do. But when ground with oil, it has hardly any colour, forming an almoft tranfparent varnifh. It will do perfectly well, however, in water colours, becaufe it regains its brilliancy when perfectly dry.

EDITOR.

Henry of Manchester thinks that in this procefs too little water is ufed, faying that thefe gritty particles require a vaft deal of water to diffolve them. But this is a miftake,—no quantity whatever of pure water will diffolve them ; and if the firft liquors are not too much diluted, and the mixture be well agitated at the firft, thefe gritty particles will not be formed.

My prefcription of equal quantities of the two falts, as proper for decompounding the Epfom falt, is founded on an experiment by which I difcovered that it requires 100 grains of dry alkali to faturate the acid faturated by 30 grains of calcined magnefia.

Having thus procured a magnefia, of the purity of which I am certain, I was anxious to examine it as a medicine. I made a falt by diffolving it in vinegar, thinking this acid the moft nearly analogous to that which is formed by vegetable matter, imperfectly digefted in weak ftomachs. Two drachms of this powder proved a good purge for an adult, and half an ounce a very brifk purge, but both operated gently and without any fpafm or fenfible inconvenience. The tafte was in a moderate degree naufeous.

Let us now attend to the chemical relations of this new fubftance ; and, firft, with regard to the acids.

1/t, It effervefces with all the acids, and neutralifes them ; and forms with them compounds remarkably different from thofe formed by the calcareous earth. With the fulphuric acid, for example, it forms a compound eafily foluble in water ;—with the nitric acid a compound foluble and cryftallizable ;—with the muriatic acid a deliquefcent compound, but which differs from the calcareous compound with this acid, by being fubject to decompofition, or feparation of its acid, or of the greater part, by the action of heat alone. Its compound with the acetous acid has been already defcribed.

2dly, Magnefia, prepared in the way now defcribed, feparates the calcareous earth from acids. When a little uncalcined magnefia is put into a folution of the muriat of lime, and the mixture promoted by agitation, and allowed to fettle, we have a powder lying at the bottom, which is not diftinguifhable at firft from the magnefia that we put in. But when it is dried and edulcorated, we find it to be a calcareous earth, for it burns to a true quicklime, wholly foluble in water. If we have added enough of magnefia, and not too much, we fhall find that, after having feparated the powder now fpoken of, if we then pour a little fulphuric acid into this folution, we fhall obferve no precipitation. Now we know that the fulphuric acid, when added to the muriat of lime, forms a felenite hardly foluble in water. Again, if, inftead of the fulphuric acid we put a folution of Glauber's falt into the faline liquor we are now confidering, we fhall obtain an Epfom falt and common falt.

3dly, It produces a very remarkable effect on lime or lime-water. In the firft experiments by which this was difcovered, I added fome magnefia to a muddy mixture of lime and water. The lime was rendered infoluble. In a fecond experiment I added the magnefia to lime already diffolved; that is, to lime-water. After long digeftion and frequent agitation, the water was rendered perfectly taftelefs, and the powder at the bottom was a mixture of magnefia and crude calcareous earth.

I was induced to try thefe experiments with lime and lime-water by the event of the preceding experiment, which made me imagine it was allied to the alkalis.

4thly, When expofed to the action of fire, it does not become a quicklime. Some magnefia, which had been expofed to a ftrong heat, was tried with water, to learn whether it would flake like lime, or form a lime-water. I alfo laid a little of it on my tongue, to learn whether it had acquired any degree of acrimony. But in none of thefe trials did it fhew the qualities of lime. It neither flaked or fhewed any attraction for water like quicklime, nor formed a lime-water; nor was it fenfibly acrid when laid on the tongue. I learned, however, that in fome of its qualities it is confiderably affected and changed by the fire; and I called it calcined or burnt magnefia. The changes induced on it by the action of the fire, when a good red heat has been applied to it for fome hours in a crucible, are thefe:

1ſt, We find that it is much diminifhed in bulk.

2dly, We alfo find that its weight is very confiderably diminifhed, viz. by more than ½;—or, by $\frac{7}{12}$ths, 12 parts are reduced to 5.

3dly, When we try how the magnefia, in this diminifhed ſtate, diffolves in acids, we find it can be diffolved without the leaſt effervefcence. When lime is thrown into ſtrong and brown fulphuric acid, there is fometimes a violent heat produced (even flafhes of flame) which makes the water boil; but this happens only when the acid is applied too ſtrong; and it is quite different from the effervefcence which appears in diffolving magnefia in its ordinary ſtate. And if we dilute the acid properly before we apply it to the calcined magnefia, there is no ebullition whatever, but only a moderate degree of heat produced.

This fuggeſted an improvement of magnefia as a medicine for fome cafes and conſtitutions. Hoffmann remarks that fome of his patients complained of the magnefia, faying that it raifed wind in their bowels; and this he imputes to its effervefcing quality. But by thus calcining it, we deprive it of this quality, although it unites with acids as effectually as before; for by diffolving the calcined magnefia in the different acids, I learned that it neutralifes the acids, and produces compounds entirely fimilar to thofe produced by the magnefia before it has been calcined. The ufe of calcined magnefia in medicine requires, however, fome caution and attention: For magnefia, in the fhops, is not always perfectly pure. Some of it contains a portion of calcareous earth; and if fuch magnefia is calcined, we fhall have a portion of quicklime in it; which is not fafe for internal ufe. The calcareous earth which is often found in common magnefia, proceeds either from a fulphat of lime, which is often contained, in fmall quantity, in the Epfom falt, or from a fulphat and a muriat of lime contained in the water with which the magnefia was wafhed, if this was a hard or a bad water. It very rarely happens that any troublefome degree of flatulence is produced by common magnefia ; and if there fhould, the addition of fome aromatic will be a fufficient antidote.

4*thly*, A fourth peculiarity of the calcined magnefia was, that it did not feparate the calcareous earth from acids. And a

5*th*, That it did not precipitate the lime from the lime-water.

The refult of thefe trials prompted me to make further experiments, to learn in what manner the fire produced thefe changes in the magnefia, and what was the nature of that matter which was feparated from it by heat in fuch great quantity, and by the lofs of which it was fo much reduced in its bulk and weight.

I therefore put a quantity of magnefia into a glafs retort, and joined a receiver to the retort with common lute made of fand and clay. I then placed the retort in a fand-pot, which was very gradually heated until the bottom of the pot began to be ignited. The receiver was kept cool, to promote the condenfation of any volatile matter that might arife from the magnefia ; but I obtained only a very fmall quantity of watery fluid, which contained a fmall portion, fcarcely perceptible, of volatile alkaline matter. The magnefia which remained in the retort had loft a great deal of its weight, (though not fo much as when it is heated to a good red heat in a crucible) ; yet the lofs it had fuffered was equal to more than fix or feven times the weight of the fmall portion of water condenfed in the receiver. This appeared at firft an unaccountable fact ; but it made me recollect fome of Dr. Hales's experiments, defcribed in the effay to which he gave the title of Analyfis of the Air, and in other parts of his works.

Mr. Boyle had before made experiments, in which he extracted air from different fubftances by the ufe of the air-pump, and by other means. But Dr. Hales pufhed fuch experiments a great deal farther. He expelled, by the action of heat, from a variety of different fub-

ftances, large quantities of elaftic incondenfible fluids, which he confidered as different kinds of air, or as air in different ftates, which had
been concealed in thefe bodies, united with their other ingredients in
a denfe and folid form. Chemifts have diftinguifhed this clafs of
elaftic fluids from atmofpherical air, by the term GAS, a German
word ufed to exprefs the eruption from fermentations (*garung*) of
every kind. (*See Note* 2. *at the end of the Volume.*)

The apparatus he ufed in many of thefe experiments was very
fimple, confifting of a gun-barrel clofed up at the breech, and bent
into a fwan neck. The fubftance from which the gas is to be expelled by fire, is put into the barrel, and this end is put into the
fire. The other end is dipped in a tub of water, and a jar filled
with water is inverted over the mouth of the barrel. The preffure
of the atmofphere fupports the water in the jar. The gas expelled
from the gun-barrel rifes through this water, difplaces fome of it
from the jar, and occupies the upper part of it *.

By fuch experiments, and by others, Dr. Hales got great quantities of air, or permanently elaftic fluids, from a great number of
vegetable, animal, and foffil fubftances. And this explained to the
chemifts fome facts which often occurred in their diftillations, and
which they had not been able before to underftand, viz. lofs of
weight, and the burfting of their diftilling apparatus, in fpite of all
attempts to promote condenfation. Obtaining a vaft deal of air in
the fame apparatus without red heat, from peafe, and other fermentable fubftances, gave fome knowledge of what happens in fermentation.

As what had happened in the laft experiment with magnefia appeared to be very fimilar, I began to fufpect that the lofs of weight
which it fuffered in the fire was occafioned, in the fame manner, by
the lofs of a quantity of elaftic aërial matter, or air, which had
efcaped through the lute. And this fuppofition appeared the more

* The fame apparatus was employed by Mayhow in 1674.

probable, as it was very confiftent with one of the qualities of burnt magnefia; I mean its uniting with acids without effervefcence. For I began to fufpect that the effervefcence of the common magnefia proceeded from air which it contains, and which is expelled by the fuperior attraction of the acid; and that the reafon why burnt magnefia did not effervefce was, that it did not contain this air, the air having been expelled from it by the action of heat.

Another fact which fupported the fame opinion, was the fuccefs of an experiment, in which I contrived to reftore this air again, if poffible, to the calcined magnefia.

The contrivance or plan of this experiment was fuggefted by confidering in what manner the magnefia had got this air at the firft. It could not have it while it was joined with the fulphuric acid in Epfom falt. The effervefcence of magnefia with acids, fhews that it cannot be joined with an acid and this air at the fame time. It muft therefore have received it from the alkali which was employed to precipitate it from the acid. This appeared the more probable, from recollecting that Dr. Hales had long before extracted much air from alkaline falts. By heating fome alkali of tartar in his iron retort or gun-barrel, he extracted from it 224 times its bulk of air. In other experiments, he got great quantities of air, by adding different acids to folutions of alkalis.

I therefore fuppofed that magnefia receives air from the alkali employed to precipitate it at firft; that we can expel this air from it by the action of fire; but that it might perhaps be reftored by diffolving the burnt magnefia in an acid, and precipitating again by the addition of an alkaline falt.

To make trial of this, I took 120 grains of common magnefia. I burnt it in a crucible with a fufficient heat, by which it loft 70 grains of its weight. This magnefia, thus calcined, was diffolved in a fufficient quantity of diluted fulphuric acid, in which it diffolved without effervefcence, and it was again precipitated by adding a clear and warm folution of a common fixed alkali, which was boiled with it a little in the ufual way ; and the precipitate was wafhed in the ufual manner. After being carefully dried with a gentle heat, it was found to have regained the whole weight which it had loft in the fire, except a mere trifle. It alfo effervefced with acids as violently as common magnefia. It feparated the calcareous earth from acids, and it made the lime feparate from lime-water. Thus it was reftored to the ftate of common magnefia in every refpect. It was juft the fame as if it never had been calcined.

The event of this experiment, therefore, confirmed me in the perfuafion, that magnefia receives a quantity of air from the alkali employed in preparing it ; and that the precipitation of magnefia from Epfom falt is not a cafe of fingle, but of double elective attraction.

This became ftill more evident, by confidering attentively what happens in precipitating magnefia ; for the alkali, in precipitating it, forms with the acid a compound falt, which is precifely the fame in quantity and quality, as if the fame alkali had been faturated with a pure acid. When an alkali is faturated with a pure acid, we fee plainly that the air of the alkali is expelled. The effervefcence fhews this, and Dr. Hales's experiments. And the compound falt produced is a compound of the acid with the alkali deprived of its air. But as the compound falt produced in precipitating magnefia is precifely fimilar, in quantity and quality, to a compound falt produced from the fame quantity of alkali and a pure acid, the alkali of it cannot have retained its air. This air muft have been expelled by the acid, and yet this expulfion of the air is not apparent by any effervefcence. But as the air is actually found in the magnefia, this accounts for the non-appearance of effervefcence, and fhews that in the precipitation of magnefia there is a double elective attraction or double exchange. The alkali unites with the acid, feparating the magnefia, which in the fame inftant unites with the air as it quits the alkali. We muft therefore conclude that the fum of the forces which tend to unite the alkali with the acid, and the magnefia with the air, is greater than the fum of the forces that tend to unite the magnefia with the acid and the alkali with the air.

Thus, I found reafon to fet afide the common opinion then entertained of the nature of the effervefcence of acids with alkalis and alkaline fubftances. It was generally fuppofed to be a confequence or effect of the violent fhock of the acid and alkaline particles,—but it now appeared to be a feparation of a quantity of air which is prefent in alkalis, but which is expelled by the fuperior attraction which the acid has for the alkali.

When I now reflected on the great lofs of weight which magnefia fuffered in being calcined, it occurred that the quantity of air which it contains muft be very confiderable, and that it and the other alkaline fubftances, muft lofe a proportional part of their weight during their effervefcence with acids,—fince we have reafon to conclude that this effervefcence is nothing elfe but a feparation and difcharge of this air.

This idea led me to examine what lofs of weight magnefia, or the common fixed alkali, would fuffer by their effervefcence with a pure acid, for which purpofe I made the following experiments :

Into a diluted acid contained in a flafk or phial, put a little alkaline falt, or chalk, or magnefia, and immediately ftop the mouth with a cork, through which there paffes a tube bent into a fwan-neck. The other end of the tube is introduced (in Dr. Hales's manner) into an inverted glafs jar filled with water, and ftanding in a tub of water. You will obferve the effervefcence, and the elaftic bubbles rifing copioufly through the water to the top, where they collect, driving the water out of the jar.

Thus it is manifeft that it is not a temporary vapour, but a permanently elaftic fluid, which efcapes in the effervefcence, not condenfible by cold.

Mr. Homberg, therefore, was led into an error in the experiments which he made and published in the memoirs of the Academy, to determine the strength of the different acids, and the proportion of the acid to the alkali in the compound salts. He saturated equal quantities of an alkali with each, and concluded that the weight gained by the alkali, when perfectly dried, was the quantity of solid salt contained in that part of the acid which had completed the saturation. But we now see that he ought to have added the weight of all the air that is lost by the alkali.

We can also now perceive the reason why common and burnt magnesia, which differ by several properties, agree in forming the same salts with the different acids; for, since the different qualities of the magnesia in these two states proceed from the presence or absence of the air, it must happen that when magnesia, in its ordinary state, is dissolved in an acid, the air is expelled,—the compound produced must therefore be the same as if we had dissolved burnt magnesia, or magnesia previously deprived of its air by fire.

The only other experiments which I first published in my inaugural dissertation on magnesia, were a few more on the earth of alum, and on the ashes of animal bones, which were at that time employed in medicine as an alkaline earth. I satisfied myself that neither of them acquired the qualities of quicklime in a strong fire; and that was one principal object of my inquiry. More light has been thrown upon both these earthy substances since that time by eminent chemists in Germany and Sweden; but of this we shall give further notice soon.

Such were the experiments which I then made to investigate the nature of magnesia. And from these you will perceive that this alkaline earth is very distinct from the calcareous. I need not add any thing further with regard to it at present; but shall take some notice of the different states in which we find it in nature.

It is produced by nature in great quantity. The sulphat of magnesia is contained in considerable quantity in the waters of the sea; and occurs also in many spring waters, in which it has been often miflaken for Glauber's falt. And fometimes waters that contain it, when they happen to be expofed in circumftances proper for evaporation, leave this falt behind in a concreted or cryftallized form. Such is this fpecimen which I was told came from a coal-pit in England. And I am perfuaded, that the falt called Glauber's falt, which is faid to be found folid in Siberia in fuch great quantities as to fupply the confumption of the whole empire of Ruffia, is not in truth a Glauber's falt, or fulphat of foda, but a fulphat of magnefia. My reafon for this opinion is, that in Ruffia they do not know the diftinction between this faline compound and fulphat of foda. Mr. Model, the Emprefs's apothecary, in one of his effays, mentions, as a well known property of Glauber's falt, that when we diffolve it in water, and add a folution of the fixed alkali, there is a plentiful precipitation of earth, which certainly happens with Epfom falt only, but not with Glauber's falt. And there are other paffages in his effays, which fhew plainly, that what they called Glauber's falt in Ruffia, was no other than this falt.

We may further remark with refpect to the natural waters which contain fome of this falt, that they alfo often contain a fmall quantity of a muriat of magnefia. There is always a quantity of it contained in fea water.

But thefe are not the only ftates in which magnefia is found. Mr. Margraaf of Berlin made great additions to the natural hiftory of this earth, by difcovering that it exifts alfo feparate from any acid, and fometimes pretty pure. There is a fet of earthy or ftony fubftances, concerning the claffing of which, foffilifts were a long time undecided and difagreed. Moft ranked them among the clays, and Cronftedt among the reft. They have been known by the names fteatites, lapis ferpentinus, lapis nephriticus, and lapis ollaris. In the pureft, the texture is clofe and femitranfparent. In fome fpecies, it is fomewhat plated or fcaly. In general they are foft like foap or fuet; fo foft as to be cut or turned. It hardens in the fire without melting. Hence fome fpecies are turned into veffels. This is the lapis ollaris. Inve-

rary Houfe is built of an impure fpecies of it. Mr. Margraaf firft
fhewed that all thefe contain more or lefs of magnefia, clofely combined with fome other earthy fubftances, and often with much iron,
by which they are tinged with the green colour, more or lefs deep,
that appears in many of them.

Mr. Margraaf alfo found that magnefia is an ingredient in the
compofition of the flexible ftony fubftances, which we fhall hereafter defcribe. And Mr. Lewis got fome of it from the afhes of vegetables, and of the fofter animal fubftances.

It is a confiderable time fince Mr. Margraaf publifhed thefe difcoveries. I can add further, 'that more lately, Monf. Monnet, in
France, has repeated and confirmed the experiments of Margraaf,
and has added many new ones, which are entirely his own, and
which fhew that this earth is produced by nature in much greater
quantity ftill than was imagined. In making experiments with clays,
and marles, and flate, he found that many of thefe contain a quantity of magnefia, mixed with the other earths, which compofe the
principal part of them. There is an extract of his difcoveries on
this fubject in the *Obfervations de Phyfique* for June 1774, which
well deferves your perufal, as a fpecimen of well conducted inveftigation, and as it contains much new information.

There is alfo a limeftone, very abundant in the neighbourhood
of Doncafter, and, I believe, in other parts of England, which, when
burnt, and laid on land as manure, is found greatly to injure its
fertility. This has been examined lately by an intelligent chemift,
and found to contain a very confiderable proportion of magnefia.

———

When I reflected on the experiments already defcribed, they appeared to me to lead to an explication of the nature of lime, which eafily
accounted for the moft remarkable properties which we find in it, and
for many phenomena relating to it and to other alkaline fubftances.

By thefe experiments it was made evident that magnefia and the vegetable alkali, in their ordinary ftate, contain a large quantity of air, in an elaftic, folid, or fixed ftate, which makes up a confiderable part of their bulk and weight; and that their effervefcence with acids is a difcharge or feparation of this air from the alkaline part of thefe fubftances,—the acid acting here in the fame manner as the fulphuric acid does when it expels the lefs powerful acids from the compound falts which contain them.

I therefore concluded, with refpect to the other alkalis and alkaline earths, that their effervefcence with acids depended on the fame caufe; that they all contain a large quantity of fixed air, which is expelled when they unite with acids; and that this air adheres to them with confiderable force, fince, notwithftanding that it is fuch a volatile fubftance, a full red heat is neceffary to feparate it from magnefia, and the fame red heat is not fufficient to expel it entirely from the alkalis, or to deprive them entirely of their power of effervefcing with acid falts.

I further was induced to think that the relation of alkaline fubftances to fixed air refembled, in fome particulars, their relation to acids; that, as the alkaline falts and earths attract acids ftrongly, and, when faturated with them, become mild neutral falts, they, in the fame manner, have an attraction for their fixed air, and, in their ordinary ftate, are in fome meafure *neutralized* by it, appearing on this account milder, or lefs active bodies, than when we have an opportunity to examine them in a pure ftate.

With refpect to the calcareous earth in particular, I imagined that, when it is expofed to the action of a ftrong fire, and thereby converted into quicklime, the change it fuffers depends on the lofs of the large quantity of fixed air which is combined with this earth in its natural ftate; that this earth is expelled by the heat; and that the folubility in water, and the remarkable acrimony which we perceive in quicklime, do not proceed from any fubtile or other matter received in the fire, but are effential properties of this earth, depending on an attraction for water, and for thofe feveral fubftances with which the lime is difpofed to unite ; but that this attractive power or activity remains imperceptible, fo long as the lime or calcareous earth is in its natural ftate, in which it is faturated and neutralized by the air combined with it.

This fuppofition agrees much better with our general experience of the confequence of combining and feparating different bodies in chemiftry, than the opinion which then prevailed concerning the nature of lime. The eftablifhed opinion was, that quicklime was a compound, formed by the union of the calcareous earth with a fubtile and active principle, fuppofed to be communicated to it by the fire or heat. But fubtile and active fubftances, when combined with others, do not in general communicate their activity to thefe. Our general experience fhews us that activity is diminifhed by combination. We have a well known example of this in the combination of acids with alkalir. Both of thefe falts, in their feparate ftate, have great activity and corrofive powers. When united, they form the neutral falts, which, inftead of having the joint activity of their conftituent parts, are mild and inert, if compared with either the acid or the alkali of which they are compofed.

I therefore confidered the calcareous earth as a peculiar, acrid, foluble earth, appearing commonly under a mild and infoluble form, on account of its union with fixed air : and I confidered quicklime as the fame earth deprived of its air; and therefore fhewing its proper folubility and acrimony, or its natural attraction for water, and for various other fubftances with which it is then capable of being combined.

This idea of the manner in which the calcareous earth becomes quicklime, not only agreed with our general experience of the confequences of combination and feparation in chemiftry, but was immediately fupported by fome capital facts belonging to this fubject; and which, at the fame time, were quite inconfiftent with the common opinion. It is well known, that when a calcareous ftone or mafs is burnt to quicklime, it does not acquire any additional weight in the fire; but, on the contrary, fuffers a very great lofs. The lime, when frefh drawn from the furnace, weighs generally no more than 60 per cent. of the weight of the limeftone; and when I confidered fome experiments made fome time before by Mr. Margraaf of Berlin, it appeared plain that the matter feparated from the limeftone by heat, is an elaftic aërial matter, incapable of being condenfed by cold into a palpable form. Thefe experiments are defcribed in the Tranfactions of the Berlin Academy for 1748, and were made upon a particular calcareous fubftance of which he had undertaken the examination: It is called ofteocolla. He put eight ounces of ofteocolla, which is a calcareous earth, into an earthen retort, to which he joined a receiver, and fet the retort in a proper furnace, where it was gradually heated to a violent degree. Nothing, however, was condenfed in the receiver, except two drachms of water, which, by its fmell and properties, fhewed itfelf to be flightly alkaline. He does not tell us the weight of the ofteocolla remaining in the retort: He only fays that it was converted into quicklime; but this alone, and the heat he applied, are fufficient proofs that it had loft about three ounces of its weight; and as no more than the quarter of an ounce of water was found in the receiver, it is plain that this lofs was occafioned chiefly by the feparation of an elaftic aërial matter which could not be condenfed.

The fame thing has fince been more fully afcertained by Profeffor Jaquin of Vienna. In order to fatisfy himfelf of the truth of my theory, he put a quantity of limeftone, broken down to fmall pieces, into an earthen retort, and fet it in a furnace, in which it might be heated to a violent degree. A receiver, with a very fmall hole drilled in the bottom of it, was luted to the retort, perfectly air-tight: Then fire was applied, and heat raifed in a flow and regular manner.

1ft, A very fmall quantity of water was condenfed in the receiver.

This all came over in the beginning of the procefs, and with a very low or moderate degree of red heat.

2*dly*, After this period of the diftillation no more water came, though the heat continued increafing; but an elaftic fluid began to iffue through the little hole of the receiver with a hiffing noife, which continued a long time, but at laft ceafed. The limeftone being then taken out, it was found to be excellent quicklime.

N. B. He repeated this procefs and varied it. When he withdrew the fire, as foon as all the water was expelled, there was no quicklime; and when he continued it until a part only of the air was expelled, a part only of the ftones were changed into lime, viz. from the furface inwards, and to a greater depth in proportion as the heat was continued nearer to that period at which the eruption of the air commonly ceafed. This fupported the idea of quicklime which I had propofed.

But what further tends very much to fupport it, is the facility with which it enables us to explain the greater part, if not the whole of this fubject. That you may be fatisfied of this alfo, I fhall now ftate fome of thefe facts or phenomena which our theory explains.

Thus, to confider in the firft place, the flaking of lime, the formation of lime-water, and fome of the qualities of lime-water: The calcareous earth, in its quicklime ftate, or deprived of its air, as it has an attraction for water, will be found to refemble the falts in feveral particulars in the mode of this attraction. The falts, if we take them in their pureft ftate, are difpofed to combine with water in two different ways. With a certain quantity of water they unite clofely, and with confiderable force, to conftitute the cryftals of falts,—in which the water is joined with the particles of falt in fuch a manner as to become folid along with them. There are fome of the falts which become very hot in uniting with this portion of water: Such are Glauber's falt, Epfom falt, fixed alkali, and feveral others. This heat is fuppofed by moft authors to come out of the falts: I am rather inclined to think it comes from the water. After this, if more water be added, the falt unites with it in a different manner, fo as to become fluid along with it, or form a folution or liquid, in which the falt is diffolved in the water; and, in this part of the procefs, cold is produced. In the fame manner, if water be added to quicklime, a certain quantity of it is attracted by the quicklime, and deprived of its fluidity with violence and heat; and it adheres to the lime with confiderable force, confti-tuting with it a dry powder, which is called flaked lime. But if this flaked lime be mixed with a much larger quantity of water, a part of it is diffolved, and compofes with the water a lime-water.

The heat produced in flaking lime, is juft one of the numberlefs examples of the emerfion of latent heat: And if any perfon fhould think that the heat produced in fome of thefe inftances is too great to be explained in this way, let him confider that the 140 degrees, which efcape from water in congelation, refers only to the difference between the heats neceffary for appearing in the forms of water and ice; but we have no authority to fay that the fame abftraction of heat from the fame quantity of water, will fuit its fubfequent appearance in a cryftal of Glauber's falt, Epfom falt, or nitre. A much greater emer-fion may be neceffary, or a much lefs; therefore, till the experiment be tried, we cannot fay how much heat muft emerge before the water can unite with quicklime in a folid form. And, let it be further re-marked, that the heat extricated in this cryftallization, can be very little diminifhed by the fubfequent folution, becaufe there is very little lime diffolved in the lime-water.

When this fluid is expofed to the open air, the particles of lime which are at the furface gradually attract fixed air, which is mixed with the atmofphere; but while the lime is thus faturated with air, it is thereby reftored to its original ftate of mildnefs and infolubility. And, as the whole of this change muft happen at the furface of the lime-water, the whole of the lime is fucceffively collected there, in its original form of an infipid calcareous earth, called the cream or crufts of lime-water.

In forming this theory, I was neceffarily led to perceive a diftinction between atmofpherical air, or the greater part of it, and that fort of air with which the alkaline fubftances are difpofed to unite. It was plain that the lime of lime-water, for example, is not difpofed to unite with the whole mafs of atmofpherical air that happens to be confined with it, or with every part of it equally. If this were the cafe, it would be impoffible to preferve lime-water in good condition, without extraordinary precautions to keep the air from ever entering the bottles or other veffels in which it is contained. But we find, in fact, that it is not neceffary for the prefervation of the lime-water, that we keep the air out of thefe bottles, or exhauft them of air. They always contain as much air in the part of them that is not filled, as other bottles do in which we keep other fluids; and yet the lime-water may be preferved good in them for a long time. The only circumftances in which lime-water lofes its qualities, and throws up the lime to its upper furface, are, when we expofe it to the open air, or keep it in bottles that are left open. From this it was evident, that the fort of air with which the lime is difpofed to unite, is a particular fpecies, which is mixed in fmall quantity only with the air of the atmofphere. To this particular fpecies I gave the name of FIXED AIR, the only term then ufed to denote any air that is condenfed and fixed in different bodies, and is a part of their conftituent principles.

To return to the explanation of the properties of quicklime.—

When quicklime itfelf is expofed to the open air, it muft gradually attract the humidity and the fixed air which are contained in the atmofphere. And as our atmofphere contains more of humidity than of fixed air, the change which the quicklime, when expofed, undergoes the moft readily, is a change of it into flaked lime. But it attracts alfo fome fixed air, and continues afterwards to attract more, until it is gradually faturated with it, and thus is reftored to its original mild and infoluble ftate.

To explain all the effects which are produced by mixing the alkaline falts with lime or lime-water, we need only to fuppofe that the fixed air is more ftrongly attracted by the lime than by the alkali. The lime in this cafe muft attract the air from the alkali to itfelf, and muft thereby return to a mild and infoluble ftate; while the alkali, on the contrary, becomes more corrofive,—that is, fhews its proper degree of activity, or attraction for water, and its natural action on bodies of the inflammable kind, and thofe of animals and vegetables; which attraction, and confequent action, was neceffarily weaker while the alkali was combined with its air, and in fome meafure neutralized by it. It therefore becomes what we call highly corrofive. And, in like manner, the volatile alkai, when deprived of its air by quicklime, befides fhewing a ftronger attraction for water, fhews alfo its proper degree of volatility, fuch as we fee it in the cauftic volatile alkali prepared with quicklime; which high degree of volatility is diminifhed or repreffed in the common volatile alkali by the air adhering to it,—in the fame manner as it is repreffed in various degrees, by the union of this alkali with the different acids, according to their degrees of fixednefs.

In like manner, the effects of magnefia, applied in its different ftates to lime, or lime-water, are eafily explained, by fuppofing that lime has fuch an attraction for fixed air, that it has the power to take it from the earth of magnefia. When the magnefia is added to lime-water, the air is feparated from the magnefia by the ftronger attraction of the lime; and as the lime, when faturated with air, does not become active, or perceptibly foluble in water, the lime-water becomes infipid,— the lime which it contained being depofited at the bottom of the mixture along with the magnefia. But, if we make this experiment with magnefia which has been deprived of its air by heat before it be added to the lime-water, this fluid fuffers no perceptible change.

This account of the nature of lime recommended itfelf, therefore, by thus affording an eafy explanation of many of the facts relating to quicklime and lime-water; and the effects of mixing thefe with alkaline falts, and with magnefia, in different ftates.

But, while I was employed in considering it with more attention, I found it to be necessarily connected with consequences which were contrary to what were at that time esteemed to be facts, or truths, established upon experience. And the consideration of these began to raise some doubts concerning the solidity of the whole of my system. In every other respect, however, it had so much the appearance of being well founded, that I resolved to consider more particularly these unavoidable consequences of the theory, and not to trust to the common opinion of what was fact, but assure myself of what was really so, by making experiments.

I therefore found, that the consequences I speak of were reducible to these propositions:

1. If we only expel fixed air from the calcareous earth when we burn it to quicklime, the quicklime thus formed must dissolve in acids without effervescence; and, notwithstanding the loss of weight it suffered in the fire, it must saturate the same quantity of acid as the whole of the calcareous earth would have done from which it was made. And the same qualities must also be found in the alkaline salts when rendered caustic by lime.

2. If quicklime be only a calcareous earth deprived of its air, and whose attraction for fixed air is stronger than that of alkalis, it follows, that, by adding to it a sufficient quantity of alkali saturated with air, the lime will recover the whole of its air, and be entirely restored to its original weight and condition. And it also follows, that the earth precipitated from lime-water by an alkali, must be the lime which was dissolved in the water restored again to its original mild and insoluble state, by having attracted the fixed air from the alkali.

3. If it be supposed, that slaked lime is an uniform compound of lime and water, and does not contain any parts which are more fiery, active, subtile, or soluble than the rest, it follows, that as part of it can be dissolved in water, the whole must be capable of dissolution in that fluid.

4. If the exceffive acrimony of the cauftic alkali depends on its being free from air, and not upon a part of the lime adhering to it, a clear cauftic ley will confequently be found free from any admixture of lime, except it fhould happen by accident, when the quantity of lime employed in making it is much greater than what is fufficient to extract the whole air of the alkali ; for then we can imagine, that as much of the fuperfluous lime may be diffolved by the ley as would be diffolved by pure water, or that the ley may contain as much lime as lime-water does.

5. It was proved by the former experiments that alkaline earths lofe their air when they are joined to an acid, but recover it if feparated again from that acid by an ordinary alkali,—the air paffing from the alkali to the earth at the fame time that the acid paffes from the earth to the alkali.

If the cauftic alkali be deftitute of air, it muft, therefore, precipitate magnefia from acids in the form of a magnefia free of air, or which will not effervefce with acids. And the fame cauftic alkali muft precipitate the calcareous earth from acids in the form of a calcareous earth deftitute of air, and faturated with water only, or in the form of flaked lime.

Thefe were the confequences of the theory which required the moft attentive examination, as being either quite new and unheard of, or inconfiftent with the eftablifhed opinions. I was encouraged, however, to proceed in the inquiry by one or two facts which coincided with thefe propofitions. One of thefe was the nature of the cauftic volatile alkali. Mr. Boerhaave having prepared fome of this with the greateft care, was furprifed to find that it did not effervefce with acids. He calls it *liquor omnium acerrimus, neque tamen alcalinus.* Thinking effervefcence with acids an effential character of alkaline fubftances, he thought that he had deftroyed it. And the other was a paffage in Hoffmann, in which he fays, that in making fome experiments with quicklime, he once found that it was diffolved by acids without effervefcence.

I accordingly engaged in a fet of experiments for proving the truth or falfity of thefe propofitions : And the confequence was, that they, in general, proved true, or agreeable to the theory. But it may be proper to give a fhort account of the experiments, that I may have an opportunity not only to explain this fubject more fully, but alfo to mention fome other difcoveries which have been the confequence of this fet of experiments.

Experiments to try the Firft Propofition.

To examine the truth of this propofition, it was neceffary, in the firft place, to learn the quantity of acid required to diffolve and faturate the calcareous earth in its natural ftate, in order to compare this quantity of acid with the quantity required to diffolve and faturate it when in the ftate of lime.

I therefore put 120 grains weight of chalk into a Florentine flafk with a fmall quantity of water, and placing the flafk on the fcale of a balance, I counterpoifed it by putting fand in the other fcale. I then gradually faturated and diffolved the chalk with the muriatic acid diluted, as related in fimilar experiments on the fixed alkali and on magnefia : 421 grains weight of the diluted acid completed the diffolution of the chalk ; and the lofs of weight by the effervefcence was 48 grains.

Being thus inftructed by this previous experiment, I took another bit of chofen chalk, of the fame weight with the former. I expofed it to a proper heat for changing it into perfect quicklime, in a fmall quantity of diftilled water, and then diffolved it in the fame manner as I had diffolved the chalk in the laft experiment, and with fome of the fame diluted muriatic acid : 414 grains of this acid were required to complete the diffolution. This was accomplifhed without the leaft effervefcence or lofs of weight.

This experiment, therefore, established the truth of the first proposition, with respect to the calcareous earth, by shewing that quicklime,
when well prepared, and made as perfect as possible, can be dissolved
in acids, without effervescence, or any loss of weight. It is also a
sufficient proof that quicklime requires as much acid to saturate and
dissolve it as the quantity of calcareous earth of which it was made
would have done. For although the quantity of acid required for dissolving the 68 grains of quicklime in this experiment was not quite so
great as the quantity required for dissolving the 120 grains of chalk, the
difference is so small that it is not worth notice. The difference of
weight between the chalk and the lime was 52 grains in 120. The
difference between the quantity of acid required for dissolving them
was only 7 in 421, and even this difference can be accounted for.
We know by experience, that in separating volatile from fixed substances by the power of heat, a small portion of the fixed is commonly or often carried away by the volatile matter. It is therefore probable that a little of the calcareous earth or lime is carried away when
this last is driven off by the action of a violent fire. This appears, I
think, in Jaquin's experiment. The calcareous earth yielded, in distillation, water which was slightly alkaline. This water would have
saturated some acid. For this reason the lime can be dissolved by a
little less of the acid than if none of it had been lost.

It was, however, the established opinion at the time when I first
made this experiment, that quicklime or slaked lime effervesced with
acids as the calcareous earth does; the experiments before that time
having been mostly made with imperfect lime, in consequence of the
want of knowledge of the true nature of lime, and of what was necessary to make it perfect, and to preserve it in that state.

Having thus ascertained the truth of the first proposition, with regard to lime, I made some experiments with the caustic fixed alkali,
to learn whether these also would agree with what was indicated in
the first proposition.

I prepared a cauftic ley by firft flaking 26 ounces of very good quicklime made of chalk, with nearly feven times its weight, or eleven pounds of boiling water, in a glafs veffel, and then adding 18 ounces of purified pearl afhes, diffolved in two pounds and a half of water. The mouth of the veffel was clofely covered. This warm mixture was fhaken frequently for two hours, at the end of which the action of the lime on the alkali was fuppofed to be over, and nothing remained to be done but to feparate them again from one another : I therefore added 12 pounds of water, ftirred up the lime, and, after allowing it to fubfide again, I poured off as much of the clear ley as poffible, which was immediately corked up in bottles.

The lime and alkali were mixed together in this procefs at firft by the medium of fo much water only as reduced the mixture to the confiftence of thick cream; for this reafon, that they are thus kept in perpetual contact and equal mixture, until they have acted fufficiently on one another. When more water is ufed in the beginning of this operation, the lime fubfides to the bottom; and, though often ftirred up, does not act fo ftrongly on the alkali, which is uniformly diffolved in every part of the liquor. But I added more water afterwards to dilute the mixture, that the lime might fubfide, and the clear liquor containing the alkali might be decanted from it.

The cauftic ley prepared in this manner was found, upon trial, to mix with acids, and to neutralize them, without the leaft effervefcence or lofs of weight: And when fome of it was added to lime-water, it produced only a very fmall diminution of tranfparency, but not a precipitate, like that produced by an alkaline falt in its ordinary ftate. This was a fure fign of its being a perfect, or very nearly perfect cauftic alkali, or of its being deprived of the whole, or very nearly the whole, of its fixed air; by which the common alkalis precipitate the lime from lime-water.

This is the fevereft teft of a cauftic ley. A remainder of air, which will not make any fenfible effervefcence with acids, is fufficient for faturating a minute portion of the lime contained in lime-water, and rendering it perceptible by a flight want of colourlefs tranfparency.

By evaporating a part of the ley, to learn the quantity or weight of the cauftic alkali it contained, I alfo fatisfied myfelf that the alkali was not increafed, but diminifhed in weight, by being made cauftic ; and that it required much more acid to faturate it than an equal quantity of the common fixed alkali does.

I muft obferve that many precautions muft be taken for this eva-poration to drynefs. The cauftic alkali requires a low red heat for expelling all the water ; and is fo acrid in this ftate as to corrode common earthen ware, and even copper and iron. I evaporated it therefore in a thin filver bowl. It is not eafy to know by the look of it when all the water is gone, for the falt is then fluid, and as tranf-parent as the ley. I knew it to be as much cleared of water as I could hope to accomplifh, by obferving the bottom of the bowl beginning to be vifible in the dark : It was then removed from the fire, and quickly congealed into a hard cake.

Thus far the experiments fupported the truth of the firft propofi-tion, in every part, and gave encouragement to proceed to the verifi-cation of the other propofitions, or to try if they would be verified by proper experiments.

Examination of the Second Propofition.

If quicklime be only a calcareous earth deprived of its air, and having an attraction for fixed air ftronger than that of alkalis, then, by adding to it a fufficient quantity of alkali in its ordinary ftate, the lime fhould recover the whole of its air, and be reftored to its original weight and condition. And it is alfo a confequence, that the earth precipitated from lime-water by an alkali muft be the lime which was diffolved in the water, reftored again to its original mild and infoluble ftate, by having attracted the fixed air from the alkali.

With refpect to all thefe points, the following experiments were made:

A piece of perfect quicklime, made from 120 grains of chalk, and which weighed 68 grains, was ground to a fine powder, and thrown into a clear folution of an ounce of alkali of tartar in two ounces of water. This mixture, being digested fome time, was then diluted with more water; and the alkaline liquor was carefully washed away from the lime by repeated affufions of pure water, and fubfequent decantations of it from the fediment. The lime or fediment being then dried, weighed 118 grains, although the piece of quicklime from which it was made weighed only 68 grains. It was quite mild; and fimilar in every trial to a fine powder of common chalk. It effervesced violently with acids; and was therefore faturated with air, which muft have been fupplied by the alkali. The weight of it was not made up completely to the weight it had before it was burnt; but the deficiency is only of two grains, and this can be accounted for upon the principle I mentioned formerly, that a very fmall portion of the lime itfelf is volatilized and carried away by the air in a violent heat.

In order to examine the earth which is precipitated from lime-water by alkaline falts, 60 grains of the alkali of tartar were diffolved in 14 pounds of lime-water, and the earth thereby precipitated was carefully collected on a filtre and dried; it weighed 51 grains. When afterwards expofed to a fufficient heat, it was converted into a true quicklime, and had every other quality of the calcareous earth. And this experiment being repeated with volatile alkali, and alfo with the foffil alkali, the refult was exactly the fame as when the alkali of tartar was ufed; the precipitated earth being always a calcareous earth. This was a fufficient proof that the propofition was true.

Examination of the Third Propofition.

If it be fuppofed that flaked lime is an uniform compound of lime and water, and does not contain any parts which are more fiery, active, fubtile, or foluble than the reft; it follows, that as part of it can be diffolved in water, the whole of it muft be capable of diffolution in that fluid.

This propofition had, at that time, lefs appearance of probability than any of the former. It was univerfally believed that lime was only partially foluble in water; and different opinions were entertained of the proportion of it that could be diffolved. Dr. Alfton contended that a fourth part of it, or perhaps a little more, might be diffolved, provided a very large quantity of water was employed; fuch as 500 times the weight of the lime. But the general opinion was that a much fmaller portion of it only was foluble and active. The queftion had never been decided by an accurate experiment.

I therefore chofe a bit of chalk, which, when heated to a fufficient degree in a crucible, afforded a little mafs of perfect quicklime, weighing eight grains. This little mafs was thrown, while yet warm, into a fmall quantity of warm diftilled water in a phial, in which it was foon flaked, and formed a white mixture like milk. This mixture was immediately poured into a larger glafs veffel, in which I had eighteen ounces of diftilled water.

While the milky mixture was diffufed through the water, the lime was feen to diffolve almoft entirely; nothing remained undiffolved but a very light fæculency, which was almoft tranfparent, or only like thin clouds in the liquor. This fæculency, when it was allowed to fubfide, and was collected with the greateft care on a fmall filtre and dried, weighed only the third part of a grain. In fome repetitions of the experiment it weighed lefs, and in others a little more. Being examined by putting it on a plate of glafs, and adding a drop of diluted nitric acid, it was diffolved in part, with effervefcence, but a part of it remained undiffolved, which was ochre of iron. It appeared, therefore, to be compofed of a minute portion of the lime, which had fomehow recovered fixed air, and of ochre of iron, and perhaps a little clay, which are well known to be often prefent, in fmall quantity, in chalk.

We may therefore reasonably conclude from this experiment, that lime, when it is quite pure and perfect, is totally soluble in pure water. And that the reason why it had appeared hitherto soluble only in part, was, that the experiment had never been made with perfect lime and with pure water. The water tasted strongly of the lime ; it was a true lime-water, and yielded twelve grains of calcareous earth, when some alkali of tartar was added to it.

The event of this experiment was even more favourable than I had expected. I expected a much larger quantity of sediment produced from this lime when dissolved ; chiefly because I suspected that the air, which we all know to be commonly dissolved in water, might be attracted by the alkaline substances, and therefore render the lime mild and insoluble. The result rather surprised me, and, raising new thoughts in my mind, which seemed to lead to very extensive and important consequences, I was anxious to put it to some trial.

To learn, therefore, whether water saturated with lime had given up the air which it usually holds in solution, and whether that air is united with the lime, I made a very strong lime-water, and placed four ounces of it under the air-pump receiver, along with four ounces of common water in another vessel of the same size. The air was taken out of the receiver ; and while this was doing, air bubbles formed and arose in both phials, in equal quantities, and in the same manner, as far as I could judge ; and the lime-water continued perfectly transparent.

Hence it is evident that the air arising from the lime-water had been combined with the water, and not with the lime ; and the air which water commonly holds in solution is of a different nature from that which is attracted by lime and alkalis : For, had it been the same, and combined with the lime, as fixed air is, the removal of the atmospherical pressure would not have been sufficient for occasioning its separation.

Quicklime, therefore, does not attract the air that is ufually contain-ed in common water, nor does it attract the whole of the mafs of at-mofpherical air, as I have already obferved. It attracts only a particu-lar kind of aërial fluid which is mixed with the air of the atmofphere, in the way of common diffufion, in a fmall quantity only. It is mix-ed as fpirits are in water, or one metal with another, without any change of properties; and alkaline fubftances take out this air, as aqua-fortis takes out the filver which alloys a piece of gold.

———

Here a new, and perhaps boundlefs field feemed to open before me. We know not how many different airs may be thus contained in our atmofphere, nor what may be their feparate properties. This particular kind has evidently very curious and important ones. It renders mild and falutary the moft acrid and deftructive fubftances that we know. I refolved to begin the ftudy of them, by a clofer examination of the fpecies which I had fortunately difcovered.

I gave it the name of Fixed Air, for the reafons already mentioned, a term which was then common to denote any elaftic matter, capable of entering into the compofition of bodies, and of being condenfed in them to a folid concrete ftate, by its chemical attraction for fome of their conftituent parts. The name may perhaps be thought to be not very judicioufly chofen, to denote this matter in its elaftic ftate; and accordingly it has now been changed for gas. But I chofe rather to employ a term already familiar, than invent a new name, before I was well informed refpecting the peculiar properties of this fubftance.

It is fomewhat fingular, that when a folution of mild alkali is ren-dered cauftic by lime, the fpecific gravity is confiderably diminifhed. We fhould naturally expect the contrary effect, from the abftraction of fo rare a fluid as air. But this fhews, that in the folution the fixed air is rendered confiderably denfer than water, being reduced to lefs than $\frac{1}{800}$ of its aërial bulk.

I fully intended to make this air, and some other elaftic fluids which frequently occur, the fubject of ferious ftudy. But my attention was then forcibly turned to other objects. A load of new official duties was then laid on me, which divided my attention among a great variety of objects *. In the fame year, however, in which my firft account of thefe experiments was publifhed, namely 1757, I had difcovered that this particular kind of air, attracted by alkaline fubftances, is deadly to all animals that breath it by the mouth and noftrils together; but that if the noftrils were kept fhut, I was led to think that it might be breathed with fafety. I found, for example, that when fparrows died in it in ten or eleven feconds, they would live in it for three or four minutes when the noftrils were fhut by melted fuet. And I convinced myfelf, that the change produced on wholefome air by breathing it, confifted chiefly, if not folely, in the converfion of part of it into fixed air. For I found, that by blowing through a pipe into lime-water, or a folution of cauftic alkali, the lime was precipitated, and the alkali was rendered mild. I was partly led to thefe experiments by fome obfervations of Dr. Hales, in which he fays, that breathing through diaphragms of cloth dipped in alkaline folution, made the air laft longer for the purpofes of life †.

In the fame year I found that fixed air is the chief part of the elaftic matter which is formed in liquids in the vinous fermentation. Van Helmont had indeed faid this, and it was to this that he firft gave the name *gas filveftre*. It could not long be unknown to thofe occupied in brewing or making wines. But it was at random that he faid it was the fame with that of the Grotto del Cane in Italy, (but

---

* Dr. Black was at this time elected Profeffor of Medicine and Chemiftry in the Univerfity of Glafgow. EDITOR.

† In the winter 1764-5, Dr. Black rendered a confiderable quantity of cauftic foffil alkali mild and cryftalline, by caufing it to filtre flowly by rags, in an apparatus which was placed above one of the fpiracles in the cieling of a church, in which a congregation of more than 1500 perfons had continued near ten hours.

EDITOR.

he fuppofed the identity, becaufe both are deadly); for he had examined neither of them chemically, nor did he know that it was the air difengaged in the effervefcence of alkaline fubftances with acids. I convinced myfelf of the fact by going to a brew-houfe with two phials, one filled with diftilled water, and the other with lime-water. I emptied the firft into a vat of wort fermenting brifkly, holding the mouth of the phial clofe to the furface of the wort. I then poured fome of the lime-water into it, fhut it with my finger, and fhook it. The lime-water became turbid immediately.

Van Helmont fays, that the *dunfte*, or deadly vapour of burning charcoal, is the fame gas filveftre; but this was alfo a random conjecture. He does not even fay that it extinguifhes flame; yet this was known to the chemifts of his day. I had now the certain means of deciding the queftion, fince, if the fame, it muft be fixed air. I made feveral indiftinct experiments as foon as the conjecture occurred to my thoughts; but they were with little contrivance or accuracy. In the evening of the fame day that I difcovered that it was fixed air that efcaped from fermenting liquors, I made an experiment which fatiffied me. Unfixing the nozzle of a pair of chamber-bellows, I put a bit of charcoal, juft red hot, into the wide end of it, and then quickly putting it into its place again, I plunged the pipe to the bottom of a phial, and forced the air very flowly through the charcoal, fo as to maintain its combuftion, but not produce a heat too fuddenly for the phial to bear. When I judged that the air of the phial was completely vitiated, I poured lime-water into it, and had the pleafure of feeing it become milky in a moment.

I now admired Van Helmont's fagacity, or his fortunate conjecture; and, for fome years, I took it for granted that all thofe vapours which extinguifh flame, and are deftructive of animal life, without irritating the lungs, or giving warning by their corrofive nature, are the gas filveftre of Van Helmont, or fixed air.

Some time after I had made, and publifhed in my inaugural differtation, the experiments you have feen, the attention of fome other perfons was excited, and keenly engaged with this new and interefting fubject. The late Dr. Macbride of Dublin began to attend to it, in confequence of fome letters which I wrote to my friend Dr. Hutchefon, then lecturer on chemiftry in Trinity College. In thefe letters I defcribed to him fome of my newly contrived experiments and apparatus, of which Dr. Macbride made ufe in his inveftigations.

He made a great number of experiments to fhew that the air emitted from fermenting vegetables is, in all cafes, an air of this kind; that it is attracted by alkaline falts and earths, and precipitates lime from lime-water. He alfo obtained fome of this air from animal fubftances in a putrefying ftate; and he thought that when applied in quantity to putrefying animal fubftances, it ftopped putrefaction, and even reftored putrid fubftances to a found ftate.

Concluding from what he thought was proved by his experiments, he confidered this fort of air as an element neceffary, or of great importance, in the compofition of moft kinds of matter. He imagined that the cohefion of the parts of folid bodies depends on it, and that putrefaction, and the concomitant refolution of bodies into their firft principles, are entirely a confequence of the feparation or lofs of this kind of air, which he fuppofed to be the great cementing principle of folid bodies.

But this fyftem was not well founded, and was not only not fupported, but, in fome meafure, refuted afterwards, by the experiments and difcoveries of other authors. The Doctor having obferved that fome effervefcence or ebullition accompanies many cafes of the diffolution of folid bodies by putrefaction, and fome other natural operations, he fuppofed that in all thefe cafes, air like this was feparated from the materials, and that the feparation of this air was the circumftance moft effectual, or even effential to the diffolution.

But, by a little more knowledge of chemiftry, he would foon have learned that he was wrong with refpect to a great number of fuch cafes; for later experiments on the putrefaction of animal fubftances have fhewn that the claftic fluid matter emitted by thefe is only in part this kind of air,—by much the greater part being an inflammable vapour, and other kinds.

Next after Dr. Macbride, the Honourable Henry Cavendifh publifhed, in the Philofophical Tranfactions 1765, fome neat and ingenious experiments on this fort of air and fome other claftic fluids.

He, in the firft place, fhewed that this air, when feparated from alkalis or earths by acids, is beyond all doubt a permanent elaftic fluid. He kept fome of it twelve months in a veffel inverted into mercury, without any diminution of its elafticity.

Dr. Macbride had before difcovered that water could abforb a quantity of this air, and become thereby capable of precipitating lime from lime-water. Mr. Cavendifh demonftrated this by more decifive experiments, and has determined the full quantity which the water can abforb. When of a middle temperature of heat, or about $55°$, it will abforb rather more than an equal bulk.

From Mr. Cavendifh's experiments, it appears that when the water is warm, it does not abforb the air fo readily, nor fo much of it; and after cold water is faturated with it, if we make it hot to a certain degree, the air is feparated, forming itfelf into bubbles which arife out of the water. It alfo efcapes flowly and imperceptibly, if the water be left in an open veffel, atmofpherical air having rather more attraction for this fort of air than water has.

Mr. Cavendifh alfo difcovered that other fluids befide water can abforb fome of this air; as fpirit of wine, which abforbs more than twice its bulk; and fome of the oils abforb as much as water does. In thefe experiments with water and other fluids, he thought there was reafon to infer that this air is not homogeneous, but that fome parts of it are more abforbable than the reft, and that a certain part of it could not be abforbed. I fufpect, however, that this was a deception, proceeding from the common air which water contains, and which arifes with the fixed air during the extrication of this laft from the alkaline fubftances.

From fome of his experiments, Mr. Cavendifh calculated with great exactnefs the quantities of this air contained in marble, in pearl-afhes or common fixed alkali, in volatile alkali, and in magnefia, when thefe alkaline fubftances are in their ordinary ftate. And he thereby explained fome phenomena which occur in mixing the alkalis with folutions of marble, or of magnefia by acids.

Thus he found that marble contained $\frac{4 0 7}{1 0 0 0 5}$ of fixed air.

Mild volatile alkali	-	$\frac{5 3 1}{1 0 0 0 6}$
Pearl afhes	-	$\frac{2 8 5}{1 0 0 0 0}$
Cryftals of foda	-	$\frac{4 2 1}{1 0 0 0 0}$
Magnefia	-	$\frac{5 9 4}{1 0 0 0 0}$

And that a certain quantity of acid faturated——

Of marble	-	1000 grains.
Mild volatile alkali	-	1661
Pearl afhes	-	1558
Cryftals of foda	-	2035

Cryftals of foda effervefce with a folution of chalk in an acid. They do not precipitate magnefia without heat; and a confiderable effervefcence attends the precipitation. Mild volatile alkali effervefces alfo with a folution of magnefia in an acid; and frequently does not precipitate it, but yet detaches it from the acid, and rediffolves it by the fixed air which is extricated.

He alfo afcertained the precife denfity of this air, which he has fhewn to be greater than that of common air, in the proportion of 157 to 100; and has fhewn that the air of marble and of fermenting vegetables agree in this as well as in other refpects. In confequence of this, it lies at the bottom of a veffel, and may be poured out like water. When thus poured out on a candle, it extinguishes it as water would do. It affords an amusing spectacle by letting a large soap-bubble fall on it in a vessel. The bubble rebounds from it like a football, and seems to rest on nothing. A burning candle may be held in it, having the top of the wick about half an inch under the surface, in which case the flame will continue for a few seconds, but altogether detached from the candle. The wick remains hot enough to cause the tallow still to evaporate; and the vapour kindles at the surface of the fixed air. The floor of the Grotto del Cane, in Italy, is lower than the door; and this hollow is always filled with fixed air, which can rise no higher than the cill, or threshold of the door, but flows out like water. If a dog go in, he is immersed in the fixed air, and dies immediately: But a man goes in with safety, because his mouth is far above the surface of this deleterious air.

He also mixed this air with common air, in different proportions, to discover what effect these mixtures had upon flame or burning bodies. All his experiments are ingeniously contrived, and executed with accuracy.

Immediately after Mr. Cavendish had published his experiments, some of them, particularly those which shew that water is capable of absorbing such a large quantity of this air, recalled the attention of Dr. Brownrigg to an opinion he had long entertained, and had communicated, with his reasons for it, to the Royal Society so early as the year 1741. The Doctor had been at Spa, where the appearance of the waters had drawn his attention. They are among the most remarkable of those waters called *acidulæ*. They have a pleasant light acidity and briskness, and sparkle in the glass like a fermented liquor. They appear as if something very elastic and volatile were contained in them. He also observed, that not far from their fountain, there are caverns which contain the choke-damp, and that something like the choke-damp hovers upon the water, by which ducks are killed.

He therefore supposed that the waters derived these qualities from a quantity of this choke damp combined with the water, the nature of which choke-damp, however, was at that time unknown, excepting its power to kill animals immerfed in it, and to extinguifh flame.

When Dr. Brownrigg faw Mr. Cavendifh's experiments, he became ftill more inclined to this opinion of the nature of the waters at Spa; and very foon after going back to that place, he gave a full demonftration or proof of it, in a number of experiments made with the waters on the fpot, which fhewed that they do in reality contain a confiderable quantity of a fort of air, which, when feparated from the water by heat, kills animals. And it further appeared, by the experiments of others, that common water, when combined with this fort of air, extracted by art from alkaline fubftances, acquires all the remarkable qualities of the acidulous mineral waters. From hence has arifen the art of imitating thofe waters exactly, by an artificial compound of this air with water, by fmall additions of fome of the falts, or alkaline earths, or iron, which, by accurate analyfis, have been found varioufly mixed in the compofition of mineral waters. When this air is combined with pure water, the water acquires the brifknefs and light fourifh pungent tafte of the acidulous mineral waters, and like them, has the power to change the infufion of litmus to a red colour.

Thus, Dr. Brownrigg explained fome of the qualities of the moft remarkable mineral waters, which had never before been well underftood. But our knowledge of fome varieties of natural waters was made ftill more perfect by the further difcoveries of Mr. Cavendifh and Mr. Lane. Many waters are well known to have a petrifying quality. They depofit a calcareous earth in the pores or on the furface of different fubftances which are expofed to them; or at leaft they cover the infides of tea-kettles with a calcareous incruftation. There are other waters which contain a fmall quantity of iron diffolved in them, but which are fure to depofit the whole of it in the

form of ochre, if they are expofed to the air, or are corked up in bottles not fufficiently clofe.   Mr. Cavendifh, while employed in examining a water near London, at Rathbone Place, difcovered that
calcareous earth can be diffolved by aërated water; and that it is depofited when fuch water is deprived of its air; and that the water of
Rathbone Place actually contained calcareous earth diffolved in this
manner.

This difcovery explains the nature of moft petrifying waters. They
contain a fmall quantity of calcareous earth, diffolved in this manner,
and fome perhaps of the earth of magnefia; for this earth alfo can be
diffolved by water charged with fixed air, and even more eafily, and
in greater quantity, than the calcareous earth. When fuch waters are
boiled often in tea-kettles, the air is driven away by the heat; and the
earth feparates from the water, forming the earthy incruftation that is
found on the infide of them. The earth is alfo depofited more flowly
when fuch waters are long expofed to the air, or run along the
furface of the ground, and fuffer evaporation. The fixed air, in this
cafe, evaporates, and the earth often forms incruftations, and ftalactites, and petrifactions.

It may appear to you furprifing that the fame fubftances which, added to lime-water, precipitate the lime, by making it infoluble, fhould
alfo be the caufe of its rediffolution when added in larger quantity. But
the fact is certain, and it is not fingular. There are many other facts
in chemiftry which are fimilar to it: For example, moft of the compound falts can be made more foluble in water than they are in their
perfect or neutral ftate, by adding to them a fuperfluous quantity of
acid.   A certain quantity of this fuperfluous acid joins itfelf to the
compound falt with a weak attraction, and forms an acidulous compound falt, which is more foluble in water than the perfect neutral.
There is no exception to this but in the compounds with acid of tartar, which, in this acidulous ftate, are lefs foluble than in their neutral ftate; and differ alfo in this, that the fuperfluous acid is ftrongly

united. All the other compound falts become more foluble in water by being acidulated with fome fuperfluous acid; which however ad-heres to them with an attraction that is very weak. The phenome-non you have now feen appears to be analogous to this. The firft effect of the aërated water on the lime-water is to precipitate the lime, by fupplying it with that quantity of fixed air which changes the lime into chalk, which, though not perfectly infoluble, is very nearly fo. But after this, if we add more of the aërated water, a fuperfluous quantity of fixed air joins itfelf to this chalk, and forms what may be called *acidulous chalk*, which is more foluble than chalk. But this fuperfluous quantity of fixed air adheres to the chalk with a weak at-traction, and is feparated and driven off from it by the heat of boiling water, and alfo by the attraction of atmofpherical air. We cannot therefore reduce this acidulous chalk to a dry ftate. It is fure to lofe its fuperfluous quantity of fixed air when we attempt to evaporate the water from it; and then it becomes common chalk. The folution of a faline cryftal is alfo analogous to this phenomenon.

From all this, you will eafily underftand the neceffity of one ftep of the procefs, which I recommended, for preparing magnefia from Ep-fom falt. The ftep I mean is the boiling over the fire, for a little while, the mixture which contains the folution of Epfom falt and the folution of potafh, by which the magnefia is precipitated.

The reafon of the neceffity for boiling this mixture is, that a great quantity of fixed alkali is neceffary for the complete precipitation of the magnefia; and this proceeds from the great quantity of acid which is united to the magnefia in Epfom falt. I found that one pound of pearl afhes, or little lefs, is neceffary for precipitating the magnefia completely from a pound of Epfom falt, although that falt contains one half of its weight of water in its cryftals.

When the pound of alkali therefore unites with the acid, although there is no vifible effervefcence, fo great a quantity of air is extricated from it, that this air is fufficient, not only for faturating the magnefia,

but for acidulating both it and the cold water with which the mixture
is made; and confequently, a great part of the acidulated magnefia
is diffolved in the acidulous water. But the mixture being boiled,
the fuperfluous air evaporates from the water and from the magnefia;
and thus the magnefia is completely precipitated. This procedure,
which explains a very extenfive and curious natural phenomenon, was
difcovered by Mr. Cavendifh. The nature of the volatile chalybeate
waters, which, when frefh from the fpring, hold fome iron in folution, and fparkle in the glafs, but lofe their brifknefs and inky tafte
by carelefs keeping, and gradually depofit the iron in form of a red,
or brown, or yellow ochre, were explained in the fame manner by
Mr. Lane. The iron had been diffolved by the acidulous water containing fixed air.

It will perhaps appear to fome of you furprifing that any phyfician
fhould have been fo bold as to think of giving this fubftance internally,
when it is known to extinguifh life fo fuddenly when applied to the
lungs and organs of fmell (for this feems a neceffary condition). But the
truth is, that thefe dangerous and fatal effects of it happen only if applied in that particular manner. It has no fuch effect when applied to
the nerves of the ftomach, or other parts. On the contrary, we have
daily experience that it is grateful to the ftomach, and has a moft agreeable, refrefhing, and cooling effect when applied there. This appears,
both from the agreeable effect of acidulous waters on the ftomach, and
from that of very brifk fermented liquors, fuch as champagne, beer,
&c. which are highly grateful in hot climates. Even to the lungs,
fixed air may be applied, not only with fafety, but even with advantage,
as we are informed by practitioners, who have tried it in confumption
and ulcerated lungs. But it muft be employed with four or five
times its bulk of atmofpherical air, or even in a greater proportion.
Were it to be breathed in confiderable quantity in its pure ftate, I
have no doubt but that it would extinguifh life in a fhort time.

Thofe who ventured to apply fixed air to the lungs, were induced to this by obferving its effects on fome very bad external ulcers, to which it proved an ufeful ftimulant, and a powerful corrector of the putrid and acrid humours which bad ulcers often emit.

The firft phyfician who formed an opinion of the falutary qualities of fixed air, was, I think, the late Dr. Macbride of Dublin. His opinion was formed very much on the theoretical views which I mentioned lately. Dr. Percival of Manchefter efteems fixed air highly medicinal in pulmonic confumptions, and in malignant fevers. The happieft effects have been experienced from the ufe of it, both exter- nal and internal; and he fays that he does not know a more powerful remedy for foul ulcers, as it mitigates pain, promotes a good digeftion of the fore, and corrects the putrid difpofition of the fluids. He thought that he had reafon to infer from feveral experiments that water with fixed air is a folvent of the urinary calculus, and that the urine of a perfon who drinks plentifully of fuch water becomes ftrongly impregnated with the fixed air, and diffolves the calculus. Some calculi are beft diffolved by alkalis, others by acids; but fixed air acts on them all.

The late Dr. Dobfon of Liverpool, afterwards of Bath, has publifhed a number of cafes in which he found it very ufeful and falutary; par- ticularly, in putrid fevers, in the cure of ill-conditioned ulcers, and in certain relaxed and debilitated ftates of the ftomach, occafioning want of appetite and indigeftion. But he did not find it effectual in reliev- ing fymptoms of the ftone and gravel; though it proved often ufeful in ulceration of the organs.

The *aqua acalina aërata* is certainly an excellent medicine in cal- culous cafes,—not as a folvent, but as a moft effectual palliative, afcer- tained by experience. I know no folvent to be relied on. Cauftic alkali is very powerful; but, that it may not act on the bladder itfelf, it muft be fo employed, that its action on the calculus is very flow; and the patient is fatigued and tires. The aërated alkaline water con- tinues agreeable.

I have now confidered, at as great length as was proper, the properties of fixed air, confidered as an object of chemiftry, and have taken notice of its natural hiftory, or the forms in which we meet with it, and alfo of the many fources from which it may be obtained by art. But there remain fome obfervations on it, from which we are led to affign it a more remote origin, fhewing it to be itfelf a compound fubftance. Soon after, its properties and particular nature were fully made known by the gentlemen who occupied themfelves fo ferioufly with this difcovery of mine, various opinions were formed as to its real origin. All thefe opinions were connected with the belief of the exiftence of a phlogifton; and, in one way or another, I believe all of them confidered fixed air as an emanation from inflammable bodies, or as a compound of air with their inflammable principle, or fomething containing it. This was almoft an unavoidable inference from our obferving that all fuel, and inflammable fubftances commonly employed for producing a burning heat, when burned in air, produce a great quantity of fixed air, and diminifh the quantity of air in which they burn; and as all fixed air appeared to be of the fame nature, phlogiftication appeared the only way of producing it from fuch a variety of bodies.

But it was afterwards found that fome bodies, not familiar indeed as fuel, or ever employed as fuch, yet which burn with great vivacity, fpoil the air, making it lethal, and unfit for maintaining flame, yet void of the acid quality of fixed air, without attraction for alkalis or calcareous earth, and caufing no precipitation of lime from lime-water. This is the cafe with fulphur, with phofphorus, with zinc, and fome other metals, which we know burn like a bit of charcoal, or even with flame.

This was, I think, the firft obfervation that made any change in the opinions formed on this fubject, and caufed the chemifts to feek for the circumftance of refemblance among all the fuels whofe combuftion produced fixed air. I cannot fay who was the firft who obferved that all fuch fuels will be changed, in one way or another, by great heats in clofe veffels, into what we call charcoal *. This is the cafe with all animal and vegetable fubftances in their natural ftate. Careful obfervation, and a well conducted chemical analyfis, fhew this to be equally true with refpect to all the fubftances which are can any how extract from them, or form by mixing them. Even fpirit of wine and æther, when properly treated, afford charcoal. One fimple mode of treatment will have this effect on all. This is to mix them with vitriolic acid, fo as to force them to ftand a ftrong red heat along with it. Sulphurous acid is always produced, and fometimes real fulphur; but, at the fame time, there is a black refiduum in the retort, which is found a perfect charcoal. I believe it was among the French chemifts who were affociated with Mr. Lavoifier in his ingenious inveftigations, that the univerfality of this fact was firft obferved.

All fuch fuels produce fixed air by their combuftion. Vegetable and animal fubftances alone are fubject to the vinous and putrefcent fermentations which emit fixed air. Charcoal feems the only common principle among them, diftinguifhing them from other combuftible bodies. It was, therefore, a natural inference that charcoal is the primitive fource of fixed air. Accordingly Mr. Lavoifier affumes charcoal for the radical or characteriftic ingredient of this acid gas. But as common charcoal, from whatever fubftance we obtain it, contains an earthy uninflammable part, Mr. Lavoifier defires it to be underftood, that it is the pure inflammable part only that he confiders as the radical of fixed air; and, to diftinguifh this from any compound, he ufes the word CARBONE. He confiders fixed air, therefore, as a compound of oxygen and carbone, in the fame manner as the vitriolic acid is confidered by him as compounded of oxygen and fulphur. And, as he calls this the fulphuric acid, he calls fixed air the CARBONIC ACID.

* Dr. Hooke fays this in many parts of his Cutlerian Lectures.   EDITOR.

Mr. Lavoisier has made some very ingenious experiments, which seem to demonstrate this composition. He burned small quantities of charcoal in pure oxygen gas, in close vessels, and he found that a part of this gas was converted into fixed air. He separated this from the rest of the oxygen by means of caustic alkali, and weighed the alkali after it had attracted the fixed air. He also expelled the air again by an acid, and examined its bulk. Thus he learned the weight of the air, and what measure of it had been produced. Then, comparing this weight with that lost by the charcoal which had been consumed, he found it to exceed greatly the weight of the charcoal, and was exactly equal to the weight of the charcoal and of that portion of the oxygen gas which had been changed into fixed air. He found that 100 grains of carbonic acid contained 72 grains of oxygen gas and 28 grains of carbone. This composition, and this proportion of the ingredients, have been confirmed by many other direct experiments of the same kind; and they agree surprisingly with the results of more complicated experiments, in which this proportion is taken for granted in the explanation of other phenomena. I therefore readily adopt his denomination of carbonic acid as extremely proper, indicating the nature of the substance.

It appears then, from some experiments which have been mentioned occasionally, that this carbonaceous matter is separated and thrown off from the blood in the lungs in the act of respiration: For air that has been breathed always contains carbonic acid. In proportion to the quantity of this acid which air contains, it is deficient in its due proportion of free oxygen gas, a part of it having been changed into carbonic acid, by meeting with carbone in the lungs. This has been ascertained by the experiments of Scheele, Lavoisier, Dr. Goodwin, and Dr. Menzies; which two last gentlemen have published good experiments on this subject in their inaugural dissertations.

You will, therefore, easily understand what happens when atmospherical air passes through burning fuel. The oxygenous part of that air, and the carbone of fome of the charcoal, unite and form a quantity of carbonic acid; and, when the air arifes from burning fuel, inftead of being a mixture of foul air and oxygen gas, as it was before, it is now a mixture of this foul air, carbonic acid, and the remainder of oxygen gas, which has not yet been faturated with carbone, but would become faturated by a fomewhat longer and more effectual application of the one to the other. The heat, which is produced in great quantity on this occafion, is fuppofed to have come chiefly from the oxygen gas, which, becoming more denfe, and having its capacity for heat diminifhed by this condenfation, muft throw out a confiderable portion of heat which it previoufly contained, and along with the heat a quantity of light, which is perhaps the fame matter, acting or modified in a different manner *.

Other names have been given to this fluid by the many chemifts who were occupied on thefe fubjects. I called it fixed air, becaufe it was found by me fixed in a number of fubftances. Mr. Cavendifh changed this name to fixable air. Many preferred the name gas, and called it acidulous gas. Mr. Henry of Manchefter, and Profeffor Bergmann, called it the aërial acid. Mr. Fourcroy called it *acide crayeux*,—acid of chalk; but all feem now agreed in giving it the fcientific name carbonic acid.

In conformity with the general plan of their reformed chemical language, the French chemifts have named the compounds which contain this acid *carbonats*. Thus, chalk or limeftone is the *carbonat*

---

* This Lavoifierian theory of the combuftion of charcoal is precifely the fame with that publifhed by Dr. Hooke in his Micrographia, in 1664-5, (page 103), as a general theory of combuftion. All combuftion is, according to him, the folution of what we call the combuftible body in the pure nitro-aërial fpirit which makes part of the atmofphere. In the cafe of certain bodies, there is not only no incombuftible or recrementitious matter, but the compound itfelf is volatile, and is difperfed in the air. The great heat proceeds entirely from the nitro-aërial fpirit; and the light is the vibratory pulfe produced in the æther, whofe undulations produce in us the fenfation of light.

EDITOR.

I

of lime ; mild vegetable alkali is the *carbonat of potafh ;* mild foffil alkali is the *carbonat of foda ;* crude magnefia the *carbonat of mag-nefia ;* and mild volatile alkali the *carbonat of ammonia,* &c.

Thus, gentlemen, have I thrown together the chief difcoveries which have been made concerning the nature and chemical proper-ties of the elaftic fluid which I difcovered in 1756. Thefe difcove-ries have been made at very different times, and by many different authors. For the public attention continued for a long while to be very much turned to this fubftance, which comes fo often in our way in chemical proceffes, and alfo in the great operations of nature. The curiofity of philofophers being thus turned to a very novel kind of ob-ject, an elaftic fluid, this gave rife to a new kind of manipulation, and a new apparatus, and a manner of management equally novel,—all which made it give much entertainment. And in this new path a number of other objects of the fame uncommon kind came in their way, and increafed the intereft taken in the ftudy. Curious chemifts even tried to produce new airs, as they were called, by every poffible means, in ex-pectation of fingular refults and difcoveries: And thus has arifen a quite new fpecies of chemiftry, which may be called PNEUMATIC CHEMISTRY, becaufe occupied in the ftudy of fluids permanently elaftic, like air. In the profecution of thefe inveftigations, chemical apparatus was greatly improved ; and we can now manage thofe flippery fubftances as eafily as we formerly managed the folids and liquids in our ordinary veffels. The boundaries of chemiftry have been wonderfully enlarged, and difcoveries have been made of the moft unexpected nature. Com-mon water, which, from the dawn of natural fcience, has been con-fidered as an unchangeable element, is now found to be a compound of two kinds of air. Diamond, feemingly the pureft and moft un-changeable of things, is now found to be coal. And all our former notions of chemical relations are now changed.

Among thefe new chemifts, Dr. Prieftley is certainly one of the moft

eminent. He was one of the firft in refpect of time; and he has fur-
paffed them all in the number and variety of his experiments, and I
may add, in his difcoveries. Dr. Prieftley firft narrated a number of
experiments on fixed air and fome other elaftic fluids, in feveral fucceeding volumes of the Philofophical Tranfactions in 1772, &c. and
then publifhed them in feveral feparate volumes. The elaftic fluids
which engaged his attention were, 1ft, *Fixed Air.* 2d, *Atmofpheric
Air*, in its ordinary ftate, and as changed or vitiated for the purpofes
of life by various means. 3d, What he called *Marine Acid Air*,
which is nothing elfe than the incoercible vapours of pure muriatic
acid without water to condenfe it. 4th, The incoercible vapours
of the fuffocating fulphurous acid, named by him *Vitriolic Acid Air*.
5th, The incoercible vapours of pure volatile alkali, which he calls
Alkaline Air. 6th, The highly inflammable elaftic fluid which we
have long known by the name of *Inflammable Air.* 7th, The incoercible vapours which efcape from a folution of metals in nitrous acid,
which he calls *Nitrous Air.* 8th, That furprifing fort of air in which
inflammable fubftances burn with extraordinary rapidity and brightnefs, and which fupports animal life and flame four or five times
better than common air. This he called *Dephlogifticated Air.* In giving the name air to fome of thefe elaftic fluids, he followed the practice of others. But he was the firft who applied this term to them
all. He has not been followed in this practice by many chemifts.
The moft general practice has been, to denominate all permanently
elaftic fluids, except air, gas,—a name firft given by Van Helmont to
the vapour which is emitted by fluids in the vinous fermentation *.

* This is probably a latinifation of *gefcht*, a vulgar word for leaven, whence we have
our word yeaft: It is likely derived from *gahren*, to ferment. Van Helmont calls the
vapour emitted by fermenting liquors, *gas filveftre*, an epithet borrowed from Paracelfus,
who calls it *fpiritus filveftris.* Van Helmont confiders this as of the fame kind with that
of the Grotto del Cane in Italy, becaufe both kill breathing animals and extinguifh flame.
He mentions, however, feveral fuch vapours which he confiders as of a different nature,
—*gas ventofum,—flammeum,—pingue, &c.*; and he fays, that thefe gafes do not exift in

Dr. Prieftley's writings contain the defcription of various and very curious experiments, by applying water and many other fubftances to thefe airs; by expofing in them living animals and vegetables. or animal and vegetable fubftances under the vinous, acetous, or putrefcent fermentation; by burning bodies, or by calcining or reducing metals in them; or by mixing the different airs together, and treating the mixtures with great heats, or the electrical fpark, &c. He notices many properties of the air which we have been fo minutely confidering; but I fufpect that he has not good authority for fome of his notions on this fubject. He has made many furprifing difcoveries; but his experiments are fo numerous, and the fucceffion in which they were made and publifhed is fuch, that no general account can be given of them. It will be better for you to read his own accounts of them, when you are farther advanced in this courfe: For you muft be acquainted with a great many more chemical fubftances and general facts, before you can perufe Dr. Prieftley's writings fo as to underftand them. I fhall have occafion, in the reft of this courfe, to notice the greateft part of his leading experiments and difcoveries.

Soon after Dr. Prieftley, or much about the fame time, the late Mr. Lavoifier in France, a gentleman diftinguifhed for his love of fcience, and particularly of chemiftry, eminent alfo for a penetrating genius, found judgment, and logical accuracy; and having an ample fortune, which enabled him to execute the moft extenfive and coftly experiments,—this gentleman, I fay, made and publifhed experiments on this very fubject. He began, by repeating with fcrupulous exactnefs, and on a large fcale, and with fine inftruments, the experiments al-

their elaftic form in the bodies from which they proceed,—nor indeed exift at all in them, but arife from new combinations of the ingredients, which cannot exift in a folid or liquid form, but which required particular circumftances to change their former combinations, and enable the ingredients to form thefe new ones. H otions on this fubject are wonderfully fagacious, and even precife, confidering the time in which he wrote, when vague and indiftinct conceptions feemed to be more generally acceptable. (See his *Complexionum atque Mixtionum elementarium Figmentum.*) EDITOR.

ready made by others. But he afterwards pufhed on his inveftigations much farther, and made fome decifive experiments, which led him to great difcoveries and enabled him to explain many general phenomena of nature, and to compofe a new fyftem or theory exceedingly ingenious and ably fupported by numerous and convincing experiments. This theory has changed the whole face of chemiftry. I fhall have an opportunity very foon to take further notice of the outlines at leaft of this new chemical philofophy.

While Dr. Prieftley in England, and others elfewhere, were attending to thefe objects, Dr. Scheele of Sweden, whom I formerly had occafion to mention, engaged in an inquiry into the nature of fire and light and heat; and, without any knowledge of what was done and doing by others, carried on a moft ingenious inveftigation, in which he was led, by the train of his refearches, to make many of the experiments and difcoveries already made by thefe gentlemen. He added many of his own. His work has been tranflated, but very badly indeed, by J. Rheinhold Fofter, and is entitled, " Scheele on Fire."

Among the remarkable difcoveries which thefe experiments on gafes have produced, perhaps the chief is that of VITAL AIR, by Dr. Prieftley. It is this, in combination with Lavoifier's theory of combuftion, that has totally changed our notions of the chemical relations of moft fubftances, and made a revolution in all the leading doctrines of chemiftry. I have already taken notice of its remarkable property, when we were confidering the decompofition of nitre, in order to obtain its acid. We fhall meet with it again, almoft at every ftep; and its various chemical relations will be difcovered as we proceed.

Scarcely inferior to vital air in chemical importance is the *faul air* of Dr. Scheele, which I mentioned on the fame occafion, as that noxious portion of atmofpherical air which remains when the vital air has been abforbed by the *hepar fulphuris*. I muft here obferve, that this portion of our atmofphere was firft obferved in

1772 by my colleague Dr. Rutherford, and published by him in his inaugural differtation. He had then discovered that we were mistaken in supposing that all noxious air was the fixed air which I had discovered. He says, that after this has been removed by caustic alkali or lime, a very large proportion of the air remains, which extinguishes life and flame in an instant.

Soon after this, Dr. Priestley met with this noxious air, which was produced in a variety of experiments, in which bodies were burned, or putrefied, or thickened in certain cases, or metals calcined, or minerals effloresced, &c. &c. In all these cases, he thought that he had reason to believe that phlogiston had quitted the substances under consideration,—had combined with the air,—and had thus vitiated it. Now saturated with phlogiston, the air could take no more, and therefore extinguished flame. He called all these proceffes *phlogisticating proceffes*, and the air thus tainted *phlogisticated air*.

In these proceffes the air was not only spoiled, but also diminished in bulk. This had been observed long before by Boyle and Hales. They ascribed this to a diminution of its elasticity. Dr. Priestley ascribed it to the precipitation of fixed air. Mr. Cavendish ascribed it to the combination of some of its ingredients in such a manner as to form water. His celebrated experiment, by which he demonstrated the composition of water, will be mentioned particularly in another place.

Dr. Scheele chanced to take a very different view of the whole of this subject,—led to his singular notion of it by some preconceived and very strange theory of fire and heat. But Scheele found that this diminution of bulk was owing to a real abstraction of all the vital air which the atmospheric air contained. For when any of these phlogisticating proceffes of Dr. Priestley were performed in vital air, it was *totally* absorbed. The remainder therefore, when the experiment was made in common air, was considered by him as a primitive air, unchanged in its properties. He called it *faul air*, which may mean either *rotten* air, becaufe it is produced in vaft abundance by putrefying bodies, or fimply *foul air*, i. e. tainted occafionally, when the phlogifton is more than will faturate the vital air. When he burned phofphorus in atmofpheric air, he found the remaining noxious air free from all admixture.

Mr. Berthollet obtained this faul air, not only from animal and vegetable fubftances in the putrefactive fermentation, but alfo in great abundance, by pouring nitric acid on the frefh mufcular fibre. Fourcroy found that it was contained almoft pure in the fwimming bladders of carp, bream, and other fifhes. Other chemifts have found it in volatile alkali, as will be particularly noticed afterwards.

In all thefe methods of procuring it, it muft be cleared from fixed air by means of lime or cauftic ley.

This air, having been procured in fo many different ways, appeared in different lights to the difcoverers, as you have feen, and got different names,—phlogifticated,—foul,—mephitic,—choke-damp (in German), &c. Mr. Chaptal, and other chemifts of the firft rank, gave it ftill another, and perhaps more proper name, calling it *nitrogen.* This took its rife from a celebrated difcovery of Mr. Cavendifh, recorded in the Philofophical Tranfactions of 1783. A train of experiments, begun feveral years before, ended in his combining feven parts of oxygen and three of this air, by means of repeated difcharges of electricity. The mixture was wholly abforbed by a folution of potafh, and *formed perfect nitre.* Therefore, he concluded that this proportion of thefe two airs compofe that remarkable fubftance, nitrous acid. Hence thefe gentlemen called this gas nitrogen, with as much propriety as inflammable air was called hydrogen. But Mr. Lavoifier, after having repeated the experiment of Cavendifh with complete fuccefs, and after being convinced, by the comparifon of numberlefs facts, and particularly, after reflecting on all the proceffes for producing faltpetre, all of which employ animal and vegetable fubftances in a ftate of putrefcence, in the free air,—after this full conviction that this was really the compofition of nitrous acid, rejected the appellation of nitrogen, and called it Azotic Gas, from its deletereous quality. This denomination has taken place of all others ; and I fhall abide by it in the reft of the courfe, although I think that a better name might have been found than this, which really is no diftinction, becaufe all the gafes are azotic, that is, deadly to breathing animals, except thofe which contain a very great proportion of oxygen gas, or vital air. *(See Note 3. at the end of the Volume.)*

In the mean time, this difcovery by Mr. Cavendifh is one of the moft important in the whole fcience of chemiftry. I mean, the difcovery that Azote is the radical or characteriftic ingredient of the nitrous acid. I have given you only that fimple fact, by which the truth of this doctrine was eftablifhed by that gentleman. Numberlefs proofs, both by compofition and decompofition, will come before us, as we proceed in the examination of other bodies. I only remark at prefent, that the proportion of ingredients, in the experiment of Mr. Cavendifh, correfponds to the moft fuming ftate of the acid that can be prepared. It is even redundant in azote, which the nature of the experiment did not permit Mr. Cavendifh to obferve. The acids are fufceptible of different proportions of oxygen, as we fhall afterwards fee clearly ; and, in thefe different ftates, they are permanent, and have very diftinct properties and modes of action. None of them is fo remarkable in this refpect as the nitric acid ; and the knowledge of its diftinctions will lead us through the moft intricate paths of the inveftigation yet before us,—fo that it is with great juftice that I have faid that this difcovery by Mr. Cavendifh is one of the moft important in the fcience.

In nitric acid, which is its moft perfect ftate as an acid, the proportion of oxygen to azote is nearly that of 80 to 20. Nitrous acid, or *fpiritus nitri fumans Glauberi*, has that of 75 to 25. Nitrous gas, which efcapes from nitrous acid when inflammable fubftances are mixed with it, has the proportion 68 to 32 nearly. But this is fcarcely, if at all acid.   You will learn all this by degrees, when you previously know the nature of the substances mixed with the perfect acid.

With respect to the other properties of azote, I cannot say much at present, because you know as yet but few of the substances to which those properties relate.

It is rarer than common air, about $\frac{1}{18}$th.   The specific gravities of atmospheric air, oxygenous gas, and azotic gas, are 1,0000, 1,0625, and 0,9444 *.

Azotic gas does not mix with water.

It has no acidity ; it does not affect the ordinary test colours, nor contract any union with alkaline substances ; therefore it does not precipitate lime from lime-water.

Although it combines with oxygen by means of the electric spark, and composes nitrous acid, and although there is always this mixture in the air, we do not find nitrous acid there.   These gases are susceptible of simple commixtion, as water mixes with spirits, or one metal with another, yet they retain their properties unchanged, like the mixtures now mentioned.   It is not simply a high temperature that will cause them to combine, for they do not combine when forced through a red hot tube of glass. (*See Note 4. at the end of the Volume.*)

I have been led into this long digression by the experiments adduced in support of the third proposition of my theory of lime.   These gave me an opportunity of explaining to you, by means of the acquaintance which you have already acquired with several chemical substances, the properties of the three remarkable gases,—fixed air, vital air, and foul air; or, as I shall now call them, carbonic acid gas, oxy-

* A thousand cubic inches, or a box, whose length, breadth, and depth is 10 inches, contains, in the ordinary pressure of the atmosphere, and temperature 55° of Fahrenheit, very nearly the following quantities :

Common air	-	-	315 grains Troy.	One inch = 0,315		
Oxygenous gas	-		335	-	-	-	0,335
Azotic gas	-	-	297	-	-	-	0,297

EDITOR.

genous gas, and azotic gas. The knowledge of their leading properties will greatly expedite our progrefs through the reft of this
courfe. I now return to the theory of quicklime.

The experiments made to try the fourth and fifth propofitions
were found to agree equally well with the theory. Several of them
concurred, in the firft place, to fhew that the cauftic alkalis do not
contain any lime or limy matter combined with them. I fatisfied myfelf of this by three trials : 1/t, Evaporation to drynefs, and examination of the dry cauftic alkali. 2d, Saturation with fulphuric acid.
This, if there be any lime, fhould produce a felenite almoft infoluble
in water. I found none. I obferve fome experiments of others in
which a precipitate was obferved ; but, by the very account given, it
appears to have been vitriolated tartar, which a large quantity of hot
water would have diffolved. 3d, Expofition to the atmofphere, to
learn whether crufts of calcareous earth would be formed on its furface, like thofe on lime-water. But no fuch thing appeared. All
thefe trials verified the fourth propofition, fhewing that cauftic alkalis
contain no lime.

And in my experiments relative to the fifth propofition, I fucceeded
perfectly in changing the calcareous earth into lime, without expofing
it to the action of heat. The precipitate from a muriate of lime by
a truly cauftic alkali proved a perfect quicklime, diffolving completely in water, and making lime-water.

It now appears that by whatever way we contrive to feparate the
carbonic acid from calcareous earth, even without making ufe of fire,
the earth is thereby rendered active, and appears in the ftate of acrid
and foluble lime. We have, therefore, abundance of reafon to conclude, that the ftate of folubility and activity of this earth depend on
its being fimply reduced to a ftate of purity, or freed from the carbonic acid which it is combined with in its natural ftate. The fame

conclufion muft alfo be admitted with refpect to the volatile alkali, when we confider the proceffes by which it is made to appear in its cauftic or moft active form. The particular nature of this alkali enables us to deprive it of its carbonic acid, or make it free of it, by a greater variety of methods than can be followed with either of the fixed alkalis, or with quicklime. And, whatever way we take to free it from this acid, we are fure to find it cauftic, or in its liquid, or its highly acrid and volatile form. The common way of obtaining it is by decompofing fal ammoniac by quicklime. In the ammoniacal falt it has no carbonic acid. It is plain that it muft alfo be obtained in the fame ftate, for the fame reafon, if the fal ammoniac be decompofed by the action of a cauftic fixed alkali. Accordingly, in this way alfo are we fure to have the volatile alkali in a cauftic ftate. But what may appear the moft fatisfactory, is a way difcovered by Mr. Margraaf for obtaining a volatile alkali from an ammoniacal falt by heat alone,—namely, from the effential falt of urine, or microcofmic falt. This falt is an ammoniacal falt, at leaft in part, or contains the volatile alkali, which may be feparated, as from other ammoniacal falts, by either fixed alkali or by quicklime: And, in thefe cafes, we obtain the volatile alkali in the fame ftate as we would obtain it by the fame proceffes from other ammoniacal falts. But Mr. Margraaf difcovered that the fal microcofmi will yield its volatile alkali by the action of heat alone, becaufe the acid is of the moft fixed kind. He put fome of this falt into a retort, and fitted to it a receiver having a fmall perforation. There firft of all efcaped fome incondenfible matters with which we are now pretty well acquainted. Thefe were followed by the volatile alkali in a moft pungent ftate. Some of it condenfed with water in the receiver; and he was furprifed to find that it did not effervefce with acids.

There is, therefore, no doubt that if we could feparate the fixed alkali from an acid by an operation as fimple, we fhould have it alfo in a cauftic form; but I am not acquainted with any inftance of fuch feparation in chemiſtry. There are two which might at firſt fight appear of this kind. The firſt is the feparation of the fixed alkali from nitric acid by deflagration with charcoal ; and the fecond, its feparation from the vegetable acid merely by heat. But in neither of thefe cafes is the alkali obtained free from carbonic acid. On the contrary, it contains a good quantity of it, and is difpofed to effervefce ſtrongly with acids. And the reafon is, that in the firſt cafe the nitric acid and the charcoal, in acting upon one another, produce a great quantity of carbonic acid, of which a part, it feems, is attracted by the alkali. It is eafy to underſtand how this carbonic acid is produced. It is produced from the charcoal, and the vital air which the nitre affords in very great quantity when expofed to the action of red heat. And, in the fecond cafe, the acetous acid is not fimply feparated. It is deſtroyed by the fire, and converted into water, oil, charcoal, and a large quantity of carbonic acid, which, together with the charcoal, is found adhering to the fixed alkali *.

In every cafe, therefore, in which we contrive to have lime or the alkaline falts in a pure ſtate, or free from carbonic acid, we find them in their ſtate of greateſt activity, or caufticity as it was formerly termed. And if we examine the converfe of thefe experiments, and confider the different ways by which quicklime or the cauſtic alkalis may be rendered mild, we ſhall find that their return to a mild ſtate is always attended with the recovery of their carbonic acid, and is evidently a confequence of it.

One example of this is the reſtoration of the cauſtic volatile alkali to its milder ſtate, by an experiment which I made in the year 1757 or 1758, and communicated, with fome others, to my friend Dr. Hutchefon, of Trinity College, Dublin, and which has been publiſhed by Dr. Macbride in his Effays, page 50. A fmall quantity of cauſtic alkali is put into a phial. In another phial an effervefcing

* Nitre deflagrated with zinc gives cauſtic alkali. EDITOR.

mixture is put; and the fixed air expelled from it is made to pafs through a fyphon tube fixed into its mouth, and having the other leg inferted into the cauftic alkali. By this fimple apparatus, the fixed air is made to rife through the alkali, which rapidly abforbs part of it, and foon acquires fo much as to effervefce brifkly with acids. If a bladder, fitted up like a common medical injection bladder, with a pipe and a cork, be diftended with fixed air, and the pipe (having put in the cork) be now inferted into the phial containing the cauftic alkali, and if the pipe fit the mouth of this phial, fo as not to allow air to get paft it, then, on pulling out the cork, the compreffed fixed air will be forced in among the alkali, and will be gradually abforbed by it, and we fhall fee the bladder collapfe by degrees, till perhaps all be abforbed. The alkali will now effervefce with acids. The pure fixed alkalis are alfo eafily reftored to a mild effervefcent ftate by the fame procefs. They abforb the carbonic acid gas very quickly, even more quickly than milk of lime; not becaufe they have a ftronger attraction for it than lime has, (for the contrary is the fact) but becaufe we can employ a much ftronger folution of thefe alkalis than the folution we have of lime in limewater.

It is fcarcely neceffary to remark that the cauftic fixed alkalis become mild alfo by communication with the atmofphere for a fufficient length of time, and that when they become mild in this way, it is by attracting carbonic acid. This happens more quickly than ordinary, when folutions of them are boiled over an open fire; the reafon of which you will eafily perceive to be the quantity of carbonic acid which is inceffantly forming by the confumption of the fuel.

They are rendered mild alfo by adding to them a proper quantity of magnefia, that has not been calcined, (now named carbonat of magnefia) or a proper quantity of common mild volatile alkali (the carbonat of ammonia); the fixed alkalis having a ftronger attraction for the carbonic acid than either of these two alkaline fubftances; of which I affured myfelf by direct and conclufive experiments.

From the knowledge we now have of the effervefcence of alkaline falts with acids, it is eafy to explain the very remarkable manner in which the vegetable alkali, in its ordinary ftate, effervefces with fome of the weaker acids, of which you faw an example, when the acetite of potafh was defcribed, and the procefs for preparing it.

The fame phenomenon occurred to me alfo when I tried to make a borat of potafh by faturating the common alkalis with fedative falt.

Borax has fometimes been referred to the clafs of alkalis, on account of fome refemblance it bears to thefe falts : But it has been demonftrated by accurate experiments, that we fhould rather confider it as a neutral falt ; that it is compofed of an alkali, and of a particular faline fubftance, called the fedative falt, which adheres to the alkali in the fame manner as an acid, but can be feparated by the addition of any acid whatever, the added acid joining itfelf to the alkali in place of the fedative falt. As this conjunction of an acid with the alkali of borax happens without the leaft effervefcence, our principles lay us under a neceffity of allowing that alkali to be perfectly free of air, which muft proceed from its being incapable of union with fixed air, and with the fedative falt, at the fame time : Whence it follows, that were we to mix the fedative falt with an alkali faturated with air, the air would immediately be expelled, or the two falts, in joining, would produce an effervefcence. This I found to be really the cafe, upon making the trial, by mixing a fmall quantity of the fedative falt with an equal quantity of each of the three alkalis, rubbing the mixtures well in a mortar, and adding a little water. It is, however, proper in this place to obferve, that if the experiments be made in a different manner, they are attended with a fingular circumftance. If a fmall quantity of the fedative falt be thrown into a large proportion of a diffolved fixed alkali, the fedative falt gradually difappears, and is united to the alkali without any effervefcence ; but, if the addition be repeated feveral times, it will at laft be accompanied with a brifk effervefcence, which will become more and more remarkable, until the alkali be entirely faturated with the fedative falt.

This phenomenon may be explained by confidering the fixed alkalis as not perfectly faturated with air; and the fuppofition will appear very reafonable, when we recollect that thofe falts are never produced without a confiderable degree of heat, which may eafily be imagined to diffipate a fmall portion of fo volatile a body as air. Now, if a fmall quantity of the fedative falt be thrown into an alkaline liquor, as it is very flowly diffolved by water, its particles are very gradually mixed with the atoms of the alkali. They are moft ftrongly attracted by fuch of thofe atoms as are deftitute of air, and therefore join with them without producing an effervefcence; or, if they expel a fmall quantity of air from fome of the falt, this air is at the fame time abforbed by fuch of the contiguous particles of it as are deftitute of it ; and no effervefcence appears, until that part of the alkali which was in a cauftic form, or deftitute of air, be nearly faturated with the fedative falt. But if, on the other hand, a large proportion of the fedative falt be perfectly and fuddenly mixed with the alkali, the whole, or a large part of the air, is as fuddenly expelled.

In the fame manner, we may alfo explain a fimilar phenomenon, which often prefents itfelf in faturating an alkali with the different acids. The effervefcence is lefs confiderable in the firft additions of acid, and becomes more violent as the mixture approaches the point of faturation. This appears moft evidently in making the fal diureticus, or regenerated tartar. The particles of the vegetable acid here employed, being always diffufed through a large quantity of water, are more gradually applied to thofe of the alkali; and, during the firft additions, are chiefly united to thofe that are freeft of air.

But ftill this explication refted on a fuppofitio , and there was no neceffity for letting it reft there. It was eafy to try whether this fuppofition was agreeable to truth by an experiment.

I expofed a fmall quantity of a pure vegetable fixed alkali to the air, in a broad and fhallow veffel, for the fpace of two months, after which I found a number of folid cryftals, which refembled a neutral falt fo much, as to retain their form pretty well in the air, and to produce a confiderable degree of cold when diffolved in water. Their tafte was much milder than that of ordinary falt of tartar ; and yet they feemed to be compofed only of the alkali, and of a larger quantity of air than is ufually contained in that falt, and which had been attracted from the atmofphere ; for they ftill joined very readily with any acid, but with a more violent effervefcence than ordinary ; and they could not be mixed with the fmalleft portion of vinegar, or of the fedative falt, without emitting a fenfible quantity of air.

Thus, therefore, I learned that the fixed alkali, in the ordinary ftate of it, is never completely faturated with carbonic acid, or combined with the largeft quantity of it which it is capable of receiving. And further experiments on this and the other alkalis fhewed that they contained more or lefs of it according to the procefs by which they have been prepared, or the treatment they have undergone in preparing them ; but that they may be all completely faturated by different methods, efpecially by expofing them to the gas of fermenting liquors, or to the gas expelled from chalk by the fulphuric acid. As they are ufually prepared, I believe the mildeft or moft nearly faturated with fixed air, is that called black flux, which is tartar no farther burnt than to deftroy the acid, and intimately mixed with charry matter, from which it may be obtained pure by lixiviation. The chemifts of laft century, or rather the pharmacifts, mention a general clafs of alkaline falts by the name of falts of Tachenius, which were remarkably mild in a medical fenfe . By the account given of their preparation, they muft have been mild in the fenfe which I have affixed to the term. They were prepared from vegetables by a smothering heat, which did little more than char the plant, and must therefore have retained a great portion of carbonic acid. Pearl-ashes comes next to black flux in saturation with carbonic acid. White flux comes next to pearl-ashes, and should be followed by potash, which is often pretty acrid by long continuance in the melting heat. Of the volatile alkalis, the mildest is surely the volatile ammoniac prepared with chalk. Next to this is salt of hartshorn, tainted with much burnt oil ;—*spiritus cornu cervi* is more acrid,—so is *spiritus salis ammoniaci*, as prepared with fixed alkali.

The result of all this investigation is, that the alkaline substances, in their mild state, are to be considered as compound salts, consisting of the pure alkaline substances united to the carbonic acid.

As this is to be considered as a new acid, and must be added to our list of simple salts, we have a new column to be added to the six already placed in your table of compound salts. We must, therefore, see what is to be the arrangement of this column, having carbonic acid or fixed air at the head.

1*st*, then, It is evident from the whole tenor of the investigation, that pure quicklime must be placed first in the column of alkaline substances. The process for caustic fixed alkali, and the precipitation of lime from lime-water by a mild alkali in a mild state, prove it in respect of the fossil alkalis. That the attraction of lime for carbonic acid is greater than that of the volatile alkali, is also evident by the process for caustic volatile alkali without heat, and by the precipitation of lime from lime-water in a mild state by mild volatile alkali.

Lime must also come before the pure earth of magnesia, because crude magnesia precipitates lime in a mild state from lime-water.

2*d*, The fixed alkalis come next in order: For they will render a solution of volatile alkali caustic. Moreover, a caustic ley becomes mild, or will effervesce with acids, if we put into it crude magnesia, and digest the mixture for a little while.

3*dly*, Magnefia comes before the volatile alkali : For cauftic volatile alkali makes no change on crude magnefia ;—but calcined magnefia will effervefce with acids after being digefted with mild volatile alkali.

The column muft therefore ftand thus :

CARBONIC ACID.

Quicklime.
Fixed alkali.
Magnefia.
Volatile alkali.

It remains to determine the proportion of the attractions of quicklime and magnefia, and the alkalis for the other acids.

1*ft*, Fixed alkali has a ftronger attraction for the other acids than quicklime has. This appears from the procefs by which we make quicklime, by precipitating it from a folution in any acid by means of a cauftic fixed alkali. We have alfo feen, that fixed alkali, in its pure or cauftic ftate, precipitates magnefia from the acids in a non-effervefcent ftate ;—and we know that it difengages the volatile alkali without heat.

2*dly*, Quicklime muft have the next place : For we have feen that the precipitate by means of lime from a folution of magnefia was not a quicklime, but a pure non-effervefcing magnefia. Alfo we found that quicklime, whether ground with fal ammoniac in a dry powdery ftate, or put into a folution of it in water, inftantly detached the volatile alkali. It is true that a mild volatile alkali will precipitate calcareous earth from an acid. But the precipitate is a carbonat of lime, and the change has been effected by a double elective attraction, proving that the partiality of the volatile alkali for any other acid is greater than that of the quicklime. The precipitation of calcareous earth, in a mild ftate, by carbonat of magnefia, is explained in the fame way. The precipitate is a carbonat.

I cannot decide, by any experiment that has occurred to me, whether magnesia or volatile alkali has the strongest attraction. I am therefore obliged to clafs these two subftances in the same line of the column. And it ftands thus:

ACIDS.

Fixed Alkali.
Quicklime.
Magnefia—Volatile Alkali.

This undetermined fituation of magnefia and the volatile alkalis relates only to their mixture and action on each other in the ordinary low temperatures. For it is plain that if we employ great heats, the volatility of the ammonia difpofes it to quit the more fixed bafe to which it is joined. Magnefia, treated along with this compound, tends, by its attraction for the acid, to reprefs the volatility of the acid. This will evidently ftill further promote the feparation of the volatile alkali; and thus magnefia will certainly decompofe the ammoniacal falts with the affiftance of heat, and difengage the alkali, either cauftic or mild.

I cannot better conclude this inquiry, than by mentioning fome important confequences which have refulted from the difcovery of the carbonic acid. Befides the improvements in chemical fcience to which it has led us, it has alfo given rife to fome ufeful inventions, and has enabled us to decide queftions which occur in fome of the moft interefting of the arts. We know now, that the alkalis are compound falts, of immenfe demand in the arts of bleaching, dying, foap-boiling, glafs-making, &c. In their ordinary ftate, we have learned that they contain a very confiderable proportion of a fubftance which is of no value in thofe arts, that we know of as yet, and which therefore diminifhes the value of the commodity. Thefe artifts can now afcertain with a certain precifion, in what degree any parcel offered to them is of ufe, and can choofe or reject, or correct them as they fee occafion.

Thus, in agriculture, a queftion is now decided which has been often put to me,—whether frefh or acrid lime is better or worfe than fuch as has been long expofed to the air.   The effect of marles and of limeftone gravel, which are known to be the moft excellent of natural manures, decides this queftion.   The only ufeful purpofes of burning limeftone feems to be the reduction of its weight, and the eafy method of reducing it to powder, and thus fitting it for being uniformly fpread on the ground.

The value of marles is alfo eafily afcertained by thefe doctrines. Marles are valuable only in proportion to the calcareous matter they contain. Some do not hold $\frac{1}{20}$th of their weight; and it is a very rich marle that holds $\frac{1}{4}$th. Suppofe a marle to contain $\frac{1}{10}$th, a farmer will confult his intereft if he pays ten times the price for a cart of lime that he muft pay for marle at the fame diftance; for he will bring home the fame quantity of the ufeful material with $\frac{1}{10}$th of the labour and carriage.

Now of all the modes of trial of the quantity of calcareous matter in marle, the one beft fuited to the unlearned farmer is, to obferve how much fixed air it contains: And this he will learn certainly, within $\frac{1}{50}$th of the truth, by diffolving a little of the marle in an acid, the muriatic, for example, and obferving what portion of its weight it lofes.   Thus, if any little quantity, fuppofe half an ounce, lofes 40 grains, he may conclude that it contained 100 grains of crude calcareous earth, or 60 grains of lime.   Or he may reckon $2\frac{1}{2}$ grains of limeftone, or $1\frac{1}{2}$ grains of lime for every grain that his marle lofes by this operation.   But he muft take care to make the trial in a deep glafs, that fparks of water may not be loft, and computed for limeftone.   For the fame reafon, the acid muft be weak, that the effervefcence may be very gentle.   Vinegar, however, is too weak, and will not detach all the air: It becomes clammy alfo in the operation, which makes the froth remain on the furface for days, and this entangles much air that has been really difengaged.   Let the acid be fuch muriatic acid as is fold in the shops. Dilute it with twice or thrice as much water, and keep it for use. Weigh the marle exactly, and also the quantity of acid which is to be poured on it, taking care that it be more than enough for diffolving all the calcareous matter. When the folution is over, after repeated agitation without more effervescence, weigh the whole; the weight loft is two-fifths of the weight of the calcareous matter.

Should we attempt to meafure this by folution and precipitation, we are led into a number of examinations which will puzzle even an experienced chemift.

And having now finifhed the chemical hiftory of the calcareous earth and magnefia, I muft take notice of fome compounds produced by nature, in which the calcareous earth is combined with acids, and which are worthy of your attention.

CALCAREOUS EARTHS COMBINED WITH ACIDS.

Thefe compounds are, 1ft, Gypfum, or the fulphat of lime. 2dly, Fluor, or the fluat of lime, which is formed by a very peculiar acid not yet defcribed. 3dly, Phofphat of lime, which contains another peculiar acid not yet defcribed. 4thly, Borat of lime, but lately difcovered, which contains the fedative falt; or boracic acid, combined with lime and with magnefia.

1.—Gypfum.

GYPSUM, the moft abundant of thefe calcareous compounds, was formerly confidered as a peculiar fpecies of earth; but now we are better acquainted with its nature, and are well affured that it is a compound of the calcareous earth with the fulphuric acid; or is, in the new language, a SULPHAT OF LIME.

This compound is found in nature under feveral forms, or in feveral ftates. The greateft part is found in the form of ftony maffes,

which, however, are remarkably foft, fo as to be eafily fcraped with a
knife, or even with the nail. They are diftinguifhed from alkaline
earths and ftones by not effervefcing with acids, and from all other
ftony bodies, by making fulphat of potafh with the mild vegetable
fixed alkali, if boiled in water with this falt, or melted with it. When
this vitriolated tartar is feparated by repeated wafhing with hot water, the remaining powder is found to be a mild calcareous earth. The
fame materials alfo, with the addition of charcoal duft, form, by fufion, a
hepar fulphuris, or *fulphurat of potafh*. Thus, the ingredients of gypfum are made fufficiently evident. *(See Note 5. at the end of the Volume.)*

Gypfum is found in fome places in great abundance. In the
neighbourhood of Paris there are hills chiefly compofed of it. In
moft places where it occurs, it is intermixed with a marly clay, forming feparate maffes interfperfed through the ftratum: And it is alfo
fometimes found in veins. When pure, it is white and femi-tranfparent in the larger maffes, and perfectly tranfparent in its fmall particles.
The ftructure or aggregation of thefe maffes has fome variety. Often
fmall cryftalline grains are compacted together like fugar. This is
called *gypfum*, and fometimes *alabafter*. A fecond kind is the *fibraria*,
or fibrous gypfum, having a fomewhat fibrous ftructure, or being compofed of oblong cryftallized concretions, clofely compacted together,
which are moftly parallel among themfelves, but lie acrofs the mafs,
from the upper to the under furface. This fort has varieties according to
the fize and regularity of the concretions of which it is compofed. In
fome kinds, the concretions are larger, and more irregular in difpofition: In others, more flender and regular. A third fpecies is that
which is compofed of clear tranfparent plates, like the fineft glafs, lying parallel to one another, and in clofe cohefion through their whole
extent, but eafily feparable by fplitting them afunder with a knife. Thefe
plates are exceffively thin, or can be fubdivided, by fplitting into very
thin ones, and have an apparent flexibility, but no elafticity. This fpecies
is called *Glacies Mariæ*, and fometimes, but improperly, *Mufcovy glafs*,

or *Muscovy talc*. A fourth appearance of this substance is in the state of separate crystals, in the forms of which there is some variety. They are found, I believe, in those strata of clay which contain gypsum in some of the other states already described; and these separate crystals were especially named *selenites*, by naturalists. Fifthly and lastly, it is very often met with in waters, in a dissolved state; for it is evidently capable of dissolution in water, though only in very small quantity, like the artificial gypsum, requiring not less than 500 times its weight of hot water to dissolve it. But it can be dissolved more plentifully, if it be acidulated with some superfluous sulphuric acid. It occurs dissolved in the waters of many springs and wells, and is the most general taint of what are called hard waters; but most copiously in sea water. When water containing gypsum is slowly evaporated, the gypsum separates, or is deposited in a white sediment, which, by the microscope, is seen to consist of minute crystals like hairs.

The composition of gypsum was first clearly explained by Margraaf, in the Berlin Transactions. He first shewed it to be a compound of calcareous earth with vitriolic acid. And by thus knowing that it is a saline compound, we can more easily understand some of its properties.

When gypsum is exposed to a moderate heat, it loses its transparency and glittering appearance, and becomes a white opaque mass like chalk, in which the former appearance of its structure is destroyed. It also becomes very friable, rather more so than chalk, whatever may have been its former firmness. During this change, we hear a continual crackling. This must be considered as a sort of decrepitation. If powdered first, and then heated, it swells, seems as if set afloat, and is agitated like a boiling fluid. This is occasioned by the water being extricated from its crystals, which are thus destroyed. But after some time, it subsides into a dry powder. In close vessels it emits water, and some kinds of it a little sulphurous acid. After being heated in this manner, until the bottom of the

Q 2

veffel begins to grow red, if it be then cooled, it is difpofed to concrete with water in a very remarkable manner. When mixed haftily
with water, to the confiftence of very thick cream, it remains in
that ftate without any perceptible fubfidence. After a little while
it grows fenfibly warm; and, in a minute or two after, it is folid, and
a little enlarged in bulk. This concretion feems to be a hafty and
confufed cryftallization, or return of the gypfum to a cryftallized
ftate, or ftate of combination with water, which is the natural ftate
of it. In this ftate it is extremely porous and light, when thoroughly
dried, becaufe the water employed in diluting it is of more bulk
than the gypfum. It is ufed for cafting figures and impreffions of
every kind, and it gives them with wonderful fharpnefs. If one of
thefe cafts, of a medal, for example, be laid on the furface of melted
white wax with the impreffion uppermoft, it fills itfelf completely
with the wax, and has the appearance of a piece of fine fculpture in
marble, or of a porcelain caft.

If gypfum be expofed to a very violent heat, it is at laft brought
into fufion, without the feparation of any confiderable part of the
acid, provided, however, it be heated without touching the fuel.
If it be allowed to come into contact with the fuel, a very different
effect is produced. The inflammable matter of the fuel acts on the
acid, and volatilizes the greater part of it, in fuffocating vapours of
fulphurous acid and vapours of *hepar fulphuris.* Another part remains adherent to the earth in the form of imperfect fulphur, or
fulphurous acid, and forms with it a particular fpecies of phofphorus, of which hereafter. The other qualities of gypfum will be
eafily perceived from the nature of its two conftituent parts; and
it is needlefs to take notice of the action of alkalis and acids
upon it.

The ufes to which it is applied are, the cafting of figures in what
is called *plafter,* or *plafter of Paris,* and for making the moulds in
which they are caft; alfo for ornaments in *ftucco*: And at Paris it

is much ufed in the building of houfes. In Minorca, they have a coarfe and ftrong kind in plenty, which is employed in building. They form arches and floors of it, without needing timber to fup- port their work. It alfo has great effect as a fertilizer of land. Its effect is greateft upon the legumina and leguminous grains. Grain, however, was not fo much improved by it as by lime. Its effects laft two years, and are greateft upon ftrong and rich land. One of the beft ways of ufing it is to fcatter it upon wheat-ftubble, to pro- mote the growth of clover fown in the fpring, and harrowed in among the wheat. The gypfum muft be thrown on the ftubbles in autumn, or the following fpring. If done fooner, it would fill the wheat with weeds. The quantity need not exceed that of the grain fown. It is ufed in Switzerland, even at a very great ex- pence, both in a raw and calcined ftate; chiefly on the grafs lands, as a top dreffing. I do not find, however, that the trials made of it in this country have been fuccefsful. Yet it feems unwarrantable to doubt the numerous and circumftantial accounts given by the Eco- nomical Society of Berne, compofed of perhaps the moft intelligent practical farmers in Europe.

2.—Fluor, or Fluat of Lime.

Since the compounded nature of gypfum was difcovered by Mr. Margraaf, another kind of ftony matter, which was formerly known by the name of *fluor*, and *fluor mineralis*, and confidered as a parti- cular fpecies of earth, has been found to be another compound of the calcareous earth with an acid. We owe this difcovery to Mr. Scheele of Sweden, whom I have often had occafion to mention with praife; and in confequence of the light he has thrown upon the compound I fpeak of, it deferves now to be confidered as one of the moft remark- able objects of philofophical chemiftry.

The ftone or ftony fubftance I fpeak of, is called by the Germans

flus, or *flus fpat ;* in the books of natural history, *fluor* and *fluor
fpathofus.* It had thefe names from its effect in promoting the melting of ores and minerals in metallurgical operations.

Its appearance and more obvious qualities are thefe: It is a ftony
fubftance, which, fo far as I know, never compofes ftrata, but is always
found either in veins, or in fmall maffes. It has a clofe glaffy texture,
and receives a fine polifh, and generally is tranfparent, though often
tinged of a green colour, or purple, or yellow, or deep blue. There
is much of it in Derbyfhire ; and on account of thefe colours, it is
called in England, *blue John;* and formerly in apothecaries fhops, *falfe
emerald, falfe amethyft,* &c.

It is often found cryftallized in the cavities of veins ; and the moft
regular form of its cryftals is cubical.

In point of hardnefs, it holds a middle place between the calcareous
ftones and the ftones of the hard clafs ; being too hard to be eafily affected by the edge of a knife, but not too hard to be cut or fcraped
with hard fteel, and capable of being wrought by the turner with proper inftruments, and formed into very thin and delicate veffels.

It has a quality which is found alfo fometimes in the cryftals of calcareous fpar ; but this ftone has it more generally : I mean a power to
emit light, or a fubtile luminous volatile matter, when the ftone is
heated. After it is once red hot, it fhines no more. I faid " emits a
luminous matter,"—for the phofphorefcence of fluor is accompanied
by a pretty ftrong unpleafant fmell ; and there is a kind of it mentioned by Mr. Wedgewood junior, in his obfervations on phofphorefcent
bodies, which is foetid when rubbed, and this kind is much more luminous than the others. The light is firft green, then verges to a
purple or lilac. *(Phil. Tranf.* 1791.)

When we apply a violent heat to this fubftance, it melts moft perfectly ; and is very powerful in promoting the fufion of other earthy
fubftances,—efpecially if calcareous fpar and it be mixed together ;
then flowing almoft as thin as water. Whence it is much valued for the

fmelting of ores. If we melt any confiderable quantity of it in a crucible by itfelf, or without adding any thing to it, fuch is its power to
diffolve other earthy fubftances in the fire, or to promote their fufion,
that it will melt the bottom of the crucible and run out.

The moft furprifing part of this fubftance is the acid which it contains, and which can be expelled from it by the fulphuric acid. To
effect this feparation of the acid, the fluor muft be reduced to a very
fine powder, and an equal, double, or even triple, weight of the
ftrongeft fulphuric acid muft be poured on it in the retort. As foon
as the materials are mixed, they begin to act flowly on each other.
We muft apply heat to quicken their action, and join a receiver, previoufly warmed (to expel fome of its air) having water in it to promote the condenfation of the acid. While the water attracts and condenfes the vapours, a white and very tender fpongy earth is ufually
depofited on its furface, and hinders the further condenfation. We
muft agitate the veffels to break this cruft, and then the condenfation
is renewed. And thus, by frequent agitations, we condenfe the
whole acid. Such a quantity of earth is precipitated as to make the
water quite thick. The glaffes are found much corroded by this procefs. For a more particular account of it I refer you to Scheele's
Effays,—effay 1.

Mr. Scheele called this acid the *acid of fpar*, or *acid of fluor*. It is
now called FLUORIC ACID, and its compounds are called FLUATS. It is
very volatile, like the muriatic acid, and will not condenfe without
water. But it differs from the muriatic acid by many properties; as, 1*ft*,
By forming with calcareous earth a compound perfectly infoluble in
water. 2*dly*, It has the power to diffolve, and even volatilize, the
filicious earth, which is perfectly infoluble by other acids. And 3*dly*,
It has a greater attraction for lime than for alkalis; and it forms peculiar compounds with the alkalis, alkaline earths, and metallic fubftances.

Mr. Scheele learned by a number of experiments, that this fine
earth, which is mixed with the acid in the receiver, has all the qualities

which belong to the ſilicious when in fine powder. And a further inveſtigation by Mr. Scheele, and by other chemiſts, has proved that this ſilicious earth is generally a part of the materials of the glaſs veſ-ſels diſſolved and volatilized by the acid, but depoſited again in part, when the acid vapours unite with water. This was proved by diſtil-ling the ſpar in a retort of lead, and condenſing the vapours in a re-ceiver of lead, or of glaſs defended with wax. Thus we get a pure acid, not tainted with ſilicious earth. There are, however, ſome varie-ties of this ſpar unfit for this experiment, on account of their contain-ing ſome ſilicious earth in their compoſition, which is diſſolved and volatilized by the acid during the diſtillation. I could find no differ-ence between this curious compound, and the pureſt ſilicious earth that I could obtain from liquor ſilicum by precipitating it by the ſulphuric acid. I ſubjected theſe compounds to a great variety of trials. But whether the acid be tainted with ſilicious earth in this manner, or in conſequence of its being diſtilled in glaſs veſſels, there is a method by which the ſilicious earth can be ſeparated and the acid obtained pure, viz. by ſaturating the acid with volatile alkali, which precipitates the whole ſilicious earth from it, and forms with it an ammoniacal ſalt, purely ſaline. If a quantity of this ammoniacal ſalt be prepared, and then decompounded with ſulphuric acid in leaden veſſels, a pure ſpatoſe acid is thus obtained ; ſometimes, indeed, not quite pure, but tainted with a little muriatic acid. But Mr. Scheele has taught a method, by means of ſilver precipitated from nitrous acid, for ſeparating this alſo. The fixed alkali cannot be employed to pre-cipitate the ſilicious earth and to form a pure ſaline compound : It has an attraction itſelf for the ſilicious earth, and forms a triple com-pound, from which it is impoſſible to obtain by cryſtallization a pure neutral ſalt.

Mr. Scheele diſcovered many other qualities of the acid of fluor which it ſhews in mixture with metals, &c. for which I refer you to his eſſay.

Some of the French chemifts committed great miftakes and errors in their firft experiments and reafonings concerning fluor and its acid. You will fee Mr. Scheele's remarks on them in his effays. And Mr. Scheele himfelf was at firft in a miftake with refpect to the origin of this filicious earth, which this acid depofits in uniting with water.

Fluor is diffolved by nitric acid or muriatic acid. Sal tartari precipitates calcareous earth from this folution. But cauftic alkalis, or aërated volatile alkali, precipitate fluor in fine powder. Sulphuric acid precipitates gypfum. A gypfum is alfo formed by adding to the folution Epfom falt, or vitriolated tartar, or vitriolic ammoniac, which alfo acts by fublimation, in decompounding fluor by double elective attraction.

Phofphoric acid decompounds fluor by diftillation, and the refiduum is a fubftance perfectly the fame with bone afhes.

Diftilled vinegar and acid of tartar have no effect on fluor.

Cauftic fixed alkali melted with fluor, does not decompound it, but may be afterwards feparated by water, and it leaves the fluor unchanged. Mild fixed alkali, four parts, melted with powdered fluor, one part, decompounds it. The fixed alkali joins with the fluoric acid, and forms a falt not deliquefcent. The reft is mild calcareous earth. A folution of mild fixed alkali decompounds by digeftion fluor made of lime-water, and fluoric acid.

The acid of fluor may be obtained without employing any foffil acids. Melt the fluor with mild fixed alkali, and extract the compound falts from the mixed matter by means of water. To the folution add acetated lead. We obtain a precipitate, which, when expofed to a ftrong heat in a retort with charcoal duft, gives reduced lead and the fpatofe acid.

Fluor mineralis, if pure from quartz, can be diffolved completely by aqua regia. The fluor muft be in fine powder, and muft be digefted with a fufficient quantity of the aqua regia. Bergmann (*on Elective*

Attractions, p. 123.*)* propoſes this as a trial, to learn whether the fluor had contained any quartz.

The moſt remarkable property of the acid of fluor is its action on ſilicious earth, which is ſuſceptible of no union with any other acid, ſo far as has yet been diſcovered. Mr. Bergmann has given a curious example of this. He put into a phial ſome finely powdered quartz, and having filled it with diluted acid of fluor, he cloſed it up. After two years, he found intermixed with the flinty powder thirteen cryſtals as big as ſmall peaſe. They were of various forms; ſome were hexahedral pyramids; others were ſimilar pyramids on the end of hexahedral columns; moſt of them, however, were cubes, having the angles cut off. They had all the chemical qualities, and wanted little of the hardneſs of perfect quartz. Hence he was induced to believe that this acid had great influence in forming the hard-figured foſſils.

This action on ſilicious earth has been applied to a very curious purpoſe, namely, engraving, or more properly ſpeaking, etching on glaſs. A German nobleman, and Mr. de Puymarin of Thoulouſe, without knowing of each others labours, reflecting on Scheele's experiments, applied the acid to this uſe. The plate was covered with engravers varniſh, and traced with points in the uſual manner. A border of wax being alſo made as uſual round the plate, it muſt be covered with a mixture of equal parts of pounded fluor and vitriolic acid, and left there two or three days *.

---

* This action on glaſs has been known long before. In 1670, an artiſt at Nurenburgh, named Schankhard, practiſed it. Alſo in 1725, one Pauli in Dreſden, employed it for etching on glaſs. (See *Beckman Hiſtory of Inventions,* t. iii. *p.* 547. *Alſo Breſlaw Collection,* xxxi. 1725. *p.* 107.) Little ſeems to have been done with it, and it was forgotten till Scheele's experiments revived the art on principle.

The etching done in this way is far from being neat. As ſoon as the acid gets at the glaſs, it eats away ſideways below the varniſh. Alſo the lines are extremely ſhallow, and when viewed through a microſcope, ſhew us that the acid acts unequally on the different ingredients of the glaſs. The glaſs ſeems to exfoliate.

EDITOR.

3.—*Phosphat of Lime.*

The next fingular and remarkable compound of the calcareous earth with an acid, is the PHOSPHAT OF LIME. This was firft difcovered and analyfed by Mr. Scheele, in company with another Swedifh chemift, Mr. Gahn.

This compound confifts of lime, or calcareous earth, combined with a particular acid which was difcovered firft, or firft defcribed with precifion, by that excellent chemift Margraaf of Berlin. It is now called the phofphoric acid. He found it in the falt which firft cryftallizes from urine, when it is evaporated to about $\frac{1}{30}$. This is called *fal microcofmi, fal microcofmicum.* He alfo difcovered that the phofphorus of urine, of which we fhall fpeak when we defcribe the inflammable fubftances, is converted into this acid by inflammation. It was therefore called PHOSPHORIC ACID.

The phofphoric acid refembles the boracic, by enduring, in its pure ftate, very ftrong red heats, without being changed into vapour. It only melts very eafily into a tranfparent fubftance like glafs. This glafly-like matter diffolves, however, eafily in water, but cannot be cryftallized. It forms with the water a liquid acid, from which the water can be evaporated again; but near the end of the evaporation, the remainder of the water is ftrongly retained; and when forced off by heat, carries away a part of the acid, as you know happens with the boracic.

Mr. Margraaf difcovered a great many properties of this phofphoric acid, which are detailed in the Berlin memoirs. But he did not know that it exifted in any other natural productions excepting urine, and the feeds of fome plants in which he difcovered it.

More lately, however, fince the ftudy of the productions of nature has fo much engaged the attention of philofophers, the phofphoric acid has been found in other ftates or conditions. The firft example of this was difcovered by Mr. Scheele, and Mr. Gahn of Sweden, who were making a fet of experiments in company. They made fome on the earthy part of bones and horns, commonly called bone-afhes, that is, the white matter which remains from bones and horns, when all the inflammable matter has been completely burnt out of them. This white matter was formerly kept in our apothecaries fhops, under the name of *cornu cervi calcinatum*, and it was fuppofed to be an alkaline earth. I gave a few experiments on it in the effay of magnefia, which fhewed that it had very little of an alkaline quality. But the Swedifh chemifts, having examined it with different views, and in a quite different way, found it to be compounded of calcareous earth and phofphoric acid. Their firft procefs for decompounding it has been reduced to greater fimplicity by other chemifts. This fimpler procefs is given very diftinctly in the laft edition of the Edinburgh Pharmacopœia, as a ftep in the procefs for preparing the *foda phofphorata*, which is a phofphat of foda, recommended lately by Dr. Pearfon at London, as a pleafant and effectual laxative and purgative.

Since Scheele and Gahn publifhed their procefs for analyfing bone afhes, a fimilar compound, or phofphat of lime, has been found in vaft abundance among the foffil productions of nature. The firft example of this was communicated by a letter from Mr. Prouft, an able chemift, in the fervice of the King of Spain, to Mr. Darcet of the late French Academy of Sciences, and publifhed in the Journal de Phyfique June 1788, informing him that in Eftremadura it conftituted extenfive rocky ftrata. Mr. Prouft found it in a ftone fo abundant in the province of Eftremadura, that it is quarried and employed in building. In appearance, it refembles a ftone compofed of felt fpar; but it confifts of the calcareous earth faturated with phofphoric acid, or at leaft containing it in the fame proportion with the earth of bones, that is, nearly one-fifth. It is highly phofphorefcent when heated. It is alfo found in Saxony and in Bohemia. Mr. Werner difcovered it fimilar to the Spanifh in Saxony. At Schlackenwaldt, in Bohemia, it is found cryftallized; generally in fix-fided prifms, and alfo in the forms of tables and plates. It is feldom folitary, but generally mixed with fluor, fpar, litho-marga, fteatites, and feveral of the metals. It is rarely accompanied with quartz, as it is in Spain. The Germans fay that it contains 45 parts of phofphoric acid per cent, and they call the compound *apatit*. I fufpect that there is fomething like it in the north of Ireland, near the Giant's Caufeway. And, perhaps the fœtid marbles, and *lapis fuillus*, are of this nature.

The phofphat of lime is ufed in making veffels for the refinement of filver and gold, to be defcribed hereafter. From it too the phofphoric acid is now obtained. And lately, it has been recommended by a French practitioner as a remedy for the rickets. We fhall have a better opportunity for treating of its decompofition, when we confider it as the matrix of phofphorus. This interefting procefs has occafioned this compound to be tortured in every way that chemiftry can difcover.

4.—*Borat of Lime.*

There now remains to be mentioned but one more compound, formed by nature, of the calcareous earth with an acid: And it has been but lately difcovered, (in 1791.) It is a compound of this earth, and partly too of magnefia, with the boracic acid. It is found cryftallized. From its hardnefs, it was miftaken at firft for filicious matter. The above cryftals are found near Lunenburgh, in the Duchy of Brunfwick, in a vein of a mountain which abounds with gypfum. It is called cubical quartz. For an account of it fee the *Annales de Chymie, tome* ii. *p.* 101, by Mr. Weftrumb, and *p.* 132, by Mr. Heyer. It contains about two-thirds of its weight of fedative falt, about one-eighth of magnefia, one-tenth of lime, and fome other earths.

SPECIES III.—BARYTES.

Thus far we have been employed in confidering the common calcareous earth, or lime, and fome remarkable compounds which it forms in nature with different acids.

I am next to make known to you two other kinds of earthy or ftony bodies, which bear a great refemblance to the calcareous by fome of their properties, though they are quite different from it by others; and the differences are of great importance.

One of the earths I now mean had firft the name of TERRA PONDE-ROSA, on account of its being remarkably heavy. Now that name is changed to BARYTES, derived from the Greek, and alluding to the fame property.

When reduced to its pureft ftate, it has the acrimony and activity of lime, which it fhews by its acrid tafte and corrofive quality, like thofe of common lime. It likewife diffolves in water, fo as to form a lime-water, which, however, contains lefs of the earth diffolved in it than common lime-water does, viz. only $\frac{1}{1000}$th; that is, very little more than half a grain to the ounce.

In this pure and active ftate, it alfo decompounds the ammoniacal falts readily, and diffolves fulphur, by boiling in water, as common lime does.

It has a ftrong attraction for fixed air, or carbonic acid; and, when joined with it by art, forms a fpecies of chalk, quite infipid, and which effervefces with acids. The attraction of this earth for fixed air is fo ftrong, when it is pure, that it readily takes it from the alkaline falts, and renders them cauftic; and the air is with difficulty feparated from it by fire. It contains, however, much lefs fixed air than crude calcareous earth, and vaftly more water, almoft one-third of its weight.

When combined with fixed air, or in the ftate of a chalk, it can be

diffolved in fmall quantity by aërated water. When calcined, it diffolves fulphur.

In all thefe particulars, therefore, it agrees with the common calcareous earth.—It differs from the fame earth,

1. By its weight, or the weight of the compounds which it forms
with acids. They are all much heavier than the correfponding compounds formed by the calcareous earth.

2. It differs from the calcareous earth by having a ftronger attraction for acids in general, and efpecially for the vitriolic acid. It has
the power to feparate moft of the acids, but efpecially the vitriolic
acid, from the fixed alkalis, as well as from lime. This conftitutes a
very remarkable, and often a puzzling diftinction.

3. The compound which it forms with the vitriolic acid is much
more denfe and heavy than common gypfum, and it is *perfectly* infoluble in water.

4. With nitric or with muriatic acid it forms faline compounds, foluble in water, but not deliquefcent like the compounds of the fame
acid with lime. On the contrary, the nitrat and muriat of barytes are
eafily cryftallizable. The cryftals of the nitrat are of various flattifh
forms, as if they were flices of a very low pyramid. The cryftals of
the muriat are alfo flices of a very low pyramid; but their bafe is always an oblong rectangle, but generally having the four angles truncated, fo as to make it an octagon, having very unequal fides.

With the acetous acid barytes forms a deliquefcent faline compound, not cryftallizable; the reverfe in this refpect of the compound
formed by lime.

After the vitriolic acid, the acid of fugar has the ftrongeft attraction
for barytes, and, next to it, the acid of forrel. Thefe are very remarkable diftinctions.

The ftate in which barytes is moft commonly or abundantly found
in nature, is combined with the fulphuric acid, or under the form of
a fulphat of barytes. In this ftate it occurs moft frequently in the veins

of the mountains in whicch ores of metals are found ; for which reafon it is called by fome authors MARMOR METALLICUM. But in fome
rare veins it has alfo been found combined with the carbonic acid, or
in the ftate of carbonat of barytes. It is found in this ftate in the
mine called Anglefark, near Chorly, in Lancafhire. (*Vide Manchefter
Memoirs.*)

This earth has lately become an interefting object, fince it has been
found to have very uncommon and valuable powers in medicine. This
appears from Dr. Crawford's trials of it, in a number of cafes publifhed in the fecond volume of *Medical Communications* of a Society
at London. He gave it in cafes of fcrophula, or bad fores, and ob-
ftructed glands. Its fenfible effects were gently to increafe appetite,
and the difcharge by urine and perfpiration. Thefe effects were
in fome cafes impeded at firft by plethora, or an inflammatory
diathefis, which was removed by a vegetable diet. He has publifhed
all the cafes in which he has tried it, which is a very candid way of
communicating the knowledge of any new medicine. And it evidently has eminent healing powers. Much caution, however, is required in the ufe of it, as too large dofes of it might prove a poifon.
He, therefore, warns practitioners to diminifh the dofe, when it produces naufea or giddinefs. Mr. Watt poifoned dogs with half a
drachm of it. Profeffor Blummenbach, of Gottingen, found that the
warm blooded animals only are with certainty poifoned by it ; while
both the warm and cold blooded animals may take with fafety the
fimilar compounds of an earth which has been confounded with it,
and will be defcribed in the next place.

SPECIES IV.—STRONTITES.

The other alkaline earth, which I faid refembles the calcareous, by
feveral properties, has been known for fome time in this country as
cccurring in fome of our mines, and was fuppofed to be the barytes,

or a mixture of it with calcareous earth ; until lately that Dr. Hope of Glafgow employed himfelf in examining it, by a number of experiments, which he communicated to our Royal Society here. His experiments were very judicioufly planned, and the conclufions he drew from them are perfectly well fupported. I fhall now give a general abftract of them.

He finds reafon to conclude that it is a peculiar fpecies of alkaline earth, different from any defcribed before. The mine in which it is found intermixed with other fpars is in the weft of Scotland, near a village called Strontian : He therefore gives it the name of STRONTITES.

1. The carbonat of ftrontites of Dr. Hope is of a fpecific gravity from 3.650 to 3.726. The natural carbonat of barytes is 4.338. The carbonat of lime is about 2.700.

2. Its external characters are,—confiderable hardnefs, fibrous or cryftallized texture, muddy tranfparency, and colour inclining to yellow or green.

3. It is infipid, but has a little folubility in water. Four ounces of diftilled water being boiled with 10 grains of it in fine powder, diffolved $2\frac{1}{2}, = \frac{1}{768}$.

4. The gas extricated during its effervefcence with acids is carbonic acid gas ; and it lofes 30.2 per 100 during effervefcence.

5. The greateft heat of a common open fire is not fufficient to expel its air, but only makes it decrepitate a little, and become opaque by the lofs of fome water.

A violent heat in a fmith's forge, of 45 minutes, applied to a fmall mafs of ftrontites inclofed in a Sturbridge clay crucible, and which foftened the crucible, melted the outfide of the mafs into a green glafs ; while, within this vitrified cruft, the reft was white, opaque, and cauftic. When it is thus rendered cauftic, it lofes 38.79 per 100 of its weight. With water it now unites in the fame manner as quicklime, but more violently, and is flaked by the air in the fame manner. The vitrified part being dropped into muriatic acid, is flowly acted on. At length a jelly is formed, which becomes perfectly fluid by the addition of water; a minute portion of powdery matter, which probably came from the crucible, remained undiffolved.

Dr. Hope was not able to vitrify ftrontites with the flame of the blow-pipe. This makes it emit a brilliant light. *(Fourcroi.)*

The conftituent parts of natural carbonat of ftrontites, are by the above experiments, in 100 parts,

Earthy bafis,	-	-	61.21
Carbonic acid,	-	-	30.20
Water,	-	-	8.59
			100.00

6. Hot water diffolves a much larger quantity of pure or cauftic ftrontites than cold water, and in cooling depofites the fuperfluous quantity in cryftals. It is very remarkable by the great quantity of heat which is extricated from it in flaking. A few fmall bits thrown into a flafk of hot water, make it boil violently, and the ebullition may be kept up for a great length of time by frefh additions. Nine ounces and a half of water yielded by refrigeration rather more than an ounce of tranfparent cryftals. A hundred grains of thefe cryftals contain 68 of water. one part of which is eafily expelled from them by heat, at firft without fufion, but at laft a part of the water adhering more ftrongly, makes them undergo a watery fufion, which ceafes when the water is totally evaporated. Thefe cryftals muft be kept in phials very clofely ftopped, otherwife they attract carbonic acid, and fall down in powder. The great quantity of water rendered folid in thefe cryftals, and fo ftrongly united, accounts for the great heat produced by mixing calcined ftrontites with water.

One ounce of water at 60° Fahrenheit, diffolves flowly 8.5 grains of thefe cryftals. One ounce of water kept boiling diffolved no lefs than 218 grains. Thefe folutions have all the alkaline qualities of lime-water.

(N. B. Dr. Hope difcovered that barytes can alfo be cryftallized in the fame manner.)

7. Sulphuric acid forms with ftrontites a compound more difficultly foluble than gypfum. Four ounces of diftilled water, boiling, diffolved only one-half grain. The folution was rendered turbid by carbonat of potafh, by barytic water, and by muriat of barytes.

Sulphuric acid diffolves the fulphat, and dilution makes it feparate again.

8. Nitric acid diluted diffolves it totally, but does not act on it when not diluted. It rather precipitates a nitrat previoufly formed. The folution eafily yields cryftals, which are octohedral, or formed of two four-fided pyramids joined by their bafis. One ounce of water, at 60°, diffolves one ounce of thefe cryftals; and at 212°, one ounce feven drachms fourteen grains. In dry air, a part of their water is evaporated from them. In very moift air they deliquefce. They deflagrate with combuftibles, and give a bright red flame or light. Or if they be expofed to heat alone, they lofe the acid, and the pure or cauftic earth remains.

9. Muriatic acid muft alfo be diluted to diffolve ftrontites. And the folution gives long flender fix-fided cryftals, often difpofed in a radiated form. (N. B. By this mode of cryftallizing, this earth is diftinguifhable from others. By putting a little of the muriat on a plate of glafs, it will evaporate and cryftallize. The muriat of barytes cryftallizes into plates which are much lefs foluble in water.) Thefe cryftals are not pulverifed by the air, but in extremely damp air fhew a tendency to deliquefce. One ounce of diftilled water, at 60°, diffolves one ounce four drachms one fcruple of them. One ounce of water kept boiling, diffolves four ounces, or more. They contain 42 per cent. of water, and undergo the watery fufion. But the acid is not eafily feparated by heat. The muriatic acid may, however, be expelled from the ftrontites, by the heat of a blow-pipe applied to it in a platina fpoon.

10. Acetous acid diffolves ftrontites flowly. The folution tinges vegetable colours green; and by fpontaneous evaporation, gives minute undifcernable cryftals. One ounce of water diffolves, in a boiling heat. 196 grains of them, which remain diffolved when the water cools.

11. The oxalic, tartarous, and fluoric acids, form very infoluble compounds.

12. Phofphoric acid, applied in large quantity, diffolves it flowly; but if we attempt faturation, the whole compound precipitates. Four ounces of boiling water diffolved only one grain from 10 of this precipitate.

13. With fuccinic acid it forms cryftals which are durable in the air.

14. Acid of arfenic forms a folution which cannot be evaporated by heat, without undergoing a change in which the earth is more clofely combined with the acid, and alfo with a fmaller quantity of it than before; and this new ftate of combination renders the compound infoluble. One ounce of boiling water was able to diffolve little more than one grain.

15. With boracic acid, it is fenfibly alkaline, and water diffolves $\frac{1}{730}$.

16. With carbonic acid it prefents the fame phenomena as thofe which are exhibited by lime.

17. Strontites, and all its compounds, give a red colour to flame. The muriat does it moft, and is beft ufed for this purpofe by putting a cryftal of it on the wick of a candle. Muriat of lime alfo produces this effect in fome degree; but muriat of barytes gives a greenifh colour to flame. A certain portion of humidity is neceffary to enable thefe compounds of ftrontites to tinge the flame. Without it they have no effect.

18. Aërated alkalis precipitate ftrontites from its folutions; at firft in form of a gelatinous matter, but by adding more alkali, and agitation, an opaque white curdled precipitate is formed, which Dr. Hope calls

artificial carbonat. It lofes its air more eafily in the fire, and diffolves more quickly in acids than natural carbonat, but is in every other refpect the fame.

19. The pruffiat of potafh or of lime do not in the leaft precipitate ftrontites from its folutions in acids; in which property it differs remarkably from barytes.

20. With fulphur it forms a hepar, either in the dry or humid way, in the fame manner as lime.

21. Cryftals of cauftic ftrontites were diffolved, but very fparingly, by alcohol, which afterwards burned with a reddifh flame.

The following is the order of attractions of the acids for ftrontites: Sulphuric acid, oxalic, tartarous, fluoric, nitric, muriatic, fuccinic, phofphoric, acetous, arfenical, boracic, carbonic.

Strontites compared with other alkaline fubftances in its force of attraction for the different acids.

Sulphuric Acid.	Oxalic Acid.	Tartarous Acid.	Fluoric Acid.	Nitric Acid.
Barytes,	Barytes,	Lime,	Lime,	Barytes,
Strontites,	Lime,	Barytes,	Barytes,	Potafh,
Potafh,	Strontites,	Strontites,	Strontites,	Soda,
Soda,	Potafh,	Potafh,	Potafh,	Strontites,
Lime.	Soda.	Soda.	Soda.	Lime.

Muriatic Acid.	Arfenical Acid.	Boracic Acid.	Phofphoric Acid.	Carbonic Acid.
Barytes,	Lime,	Lime,	Lime,	Lime,
Potafh,	Barytes,	Barytes,	Barytes,	Barytes,
Soda,	Potafh,	Strontites,	Strontites,	Strontites,
Strontites,	Soda,	Potafh,	Potafh,	Potafh,
Lime.	Strontites.	Soda.	Soda.	Soda.

GENUS II.

PLASTIC EARTHS.

WE fhall now proceed to ftudy the fecond order of the earths,—— the PLASTIC, which are commonly called in our language CLAYS, or CLAYEY EARTHS.

The natural earths which are affembled under this divifion, all contain more or lefs of a particular kind of earth in their compofition, which gives them their plaftic qualities, or gives them, in different degrees, the qualities which belong efpecially to itfelf.

The earth I now mean is at prefent confidered by the chemifts as another of the pure elementary earths, and is called the argillaceous. It is diftinguifhed from the earths hitherto defcribed by thefe properties :

1*mo*, It does not effervefce with acids when they are fimply mixed with it.

2*do*, It is compofed of exceedingly fine impalpable particles : It therefore feels fat or fmooth between the fingers, like marrow ; and is not in the leaft gritty between the teeth.

3*tio*, When a dry mafs of it is applied to the tongue, it imbibes the fuperficial humidity of that organ fo ftrongly that it adheres to it ; and it gives a peculiar odour on all thefe occafions when humidity is applied to it.

4to, If it be mixed and well kneaded, and worked with a proper quantity of water, it forms a foft and plaftic mafs, not eafily diffufible or dilutible in more water, and which, if dried well, and afterwards burned with a ftrong heat, becomes very compact and hard, and impenetrable by water.

5to, It has fome notable qualities which it fhews in mixture with acids; the moft remarkable of which qualities fhall be mentioned prefently.

The plaftic earths, therefore, all contain more or lefs of this fimple earth; and they conftitute every where numerous ftrata. They alfo make a part of every ftrong and rich foil.

That property of plaftic earth, by which they become fo tough a pafte, and fo difficultly diffufible or penetrable by water, when wrought and comprefled with a proper quantity of that fluid, occafions their being employed for confining water in canals, and ponds, and refervoirs, and other works in which large quantities of water are to be confined, or preferved from being wafted by foaking through the foil. It alfo explains the bad effects of what is called *poaching* clayey grounds; that is, allowing cattle to tread on them much when they are wet or foft, as they are thus reduced to that plaftic ftate in which they do not tranfmit water eafily, but occafion it to ftagnate on their furface, and to rot or ficken the plants; and at the fame time they are fo denfe and vifcid, that the roots of plants cannot penetrate them without the greateft difficulty. The remedy for this is, to plough them when they are moderately dry, and when dry weather or froft is expected. If the clods once become dry, the firft rain will make them moulder down, and alternations of dry weather and fhowers will completely divide them.

Mr. Bergmann fays that a fine clay does very well for wafhing and cleanfing linens. Though it does not chemically combine with greafy filth, it adheres to it, and carries much off with it by rubbing. *Smectis*, or *fuller's earth*, is a marly clay, and is much employed in this way.

There is great variety of the earths which come under this division, and under the common denomination of clay. They are various by their colour, by the finenefs and fmoothnefs of their particles, by their degree of cohefion when we attempt to diffufe them in water, and the degree of toughnefs, or plaftic quality, which they afume when wrought or kneaded with a proper quantity of that fluid; and alfo by their qualities with refpect to heat, or the changes they undergo from the different degrees of its action on them. But all this variety is produced by the various admixture of filiceous earth, or magnefia, or calcareous earth, or iron, or inflammable and other matters, in various proportion with the argillaceous earth. It is fo liable to admixtures of this kind, that a pure argillaceous earth is one of the rareft productions of nature; and in the very few examples that have occurred, it was in fmall quantity.

The clays which contain it in the greateft quantity, or in a ftate lefs impure than ordinary, are either naturally white, or if they have a dark or dull colour, it proceeds from a fmall quantity of inflammable matter; and they become white when burnt in an open fire. Thofe that become red in the fire contain iron.

Clay, by being kept very long in a wet ftate, becomes evidently four to the tafte and fmell. I was affured of this fact by Mr. Wedgewood, who is perhaps the moft perfectly acquainted with all the qualities of the plaftic earths of any perfon in Europe, and the moft interefted to know them in their ftate of greateft purity. The fact is curious, and perhaps important to the chemical philofopher. Perhaps it receives fome explanation from another fact which he alfo told me, viz. That by long expofure to the air, the furface appears powdery; and when this duft is carefully fwept together, it is found tainted with a calx of iron. It is not unlikely that this arifes from a minute portion of pyrites, which is known to decompound by humidity, and to yield its acid. May not this alfo account for the hardening of clays by the fire, even in their pureſt ſtate?

The moſt uſeful qualities of clays in the arts are,—their diſpoſition to conſtitute, with a proper quantity of water, that ductile plaſtic maſs, which is eaſily formed on the potter's wheel, or otherwiſe, and which can be baked afterwards by fire to a ſtony hardneſs. In conſequence of which, they are employed for the manufacture of numerous veſſels, tiles, bricks, and many other uſeful productions of art. The beſt kinds of it acquire different degrees of compactneſs and hardneſs in the fire, according to their dryneſs in the firſt kneading, and the violence of the heat.

When clay has been made very ſoft, by kneading it with much water, in order to make it work eaſily on the potter's wheel, it muſt be very porous when burnt to a tile: For whatever quantity of water be employed, the piece does not contract much by drying; and by no means in proportion to the water in it. Such ware, therefore, muſt be very open and ſpongy, and may be uſed for filtres *.

In a moderate red heat they become hard, but porous, and are formed into tobacco pipes, &c. In a much ſtronger one they are rendered compact and hard like flint. But pure clays never melt in ordinary furnaces or fires. The more coarſe and impure clays have theſe qualities in inferior degrees. They do not grow ſo hard in moderate heat; and many of them cannot withſtand a ſtrong fire without melting, on account of the mixture of different earthy ſubſtances

---

* Accordingly, veſſels of this porous ware are uſed in the hot climates for cooling liquors; which ſervice they perform by the copious evaporation from the ſurface. Veſſels are alſo made in this country as porous as poſſible, and have their external ſurface turned into deep notches, or furrows, on the potter's wheel. They are filled with water; and ſmall ſeeds, ſuch as thoſe of creſſes and muſtard, are ſprinkled into thoſe furrows, where they receive enough of moiſture to make them grow, and cover the outſide with foliage. A ſallad is thus eaſily raiſed at ſea, and in winter in a warm room.

EDITOR.

and other matters which they contain, and which give them this disposition.

Experiments made with clays, in the way of mixture with other substances, have shewn that the argillaceous earth, although it does not effervesce with acids, can be combined with them; and that, when combined with the sulphuric acid, it constitutes ALUM.

This, however, is a late discovery, for although alum has long been in use, and well known to contain the sulphuric acid combined with an earth, the nature of this earth was not distinctly understood, until two of the academicians of Berlin, Pott and Margraaf, published their experiments on this subject, in which they made perfect alum by dissolving the argillaceous earth of clays in the sulphuric acid. Mr. Pott made the first step in this discovery, and Mr. Margraaf afterwards brought it to perfection.

It was ascertained by Mr. Margraaf's experiments that the argillaceous earth can be combined with the vitriolic acid in two different proportions, or so as to form two compounds considerably different. One of them contains a large proportion of the acid, and is very soluble in water, and deliquescent, and cannot be crystallized: This is called *alum liquor*. The other, viz. *alum*, contains a small proportion of acid to the earth, and is but moderately soluble in water, and easily crystallizes, and is not at all deliquescent, but rather calcines spontaneously.

The first of these two compounds is almost always formed when we attempt to make alum by combining its two ingredients, which may be done by boiling the best kinds of clay in strong sulphuric acid until the mixture is dry, and then adding water to dissolve the alum or aluminous compound which has been formed. We thus get an alum liquor, or acidulous liquid alum, from which it is necessary to abstract a part of the acid, by adding a small quantity of an alkaline salt, or by a proper degree of heat, before we can have alum that can be crystallized.

Mr. Beaumé of Paris has made an addition to thefe difcoveries of Margraaf. He difcovered that a third compound can be formed of the fulphuric acid and aluminous earth ;—a compound in which there is ftill lefs of the acid than in perfect alum ; and which is therefore not foluble in water. This compound is formed by boiling a folution of alum in water with fome of the pure argillaceous earth ; the confequence of which is, to change the whole of the alum into this third compound, which fubfides to the bottom, and forms a fediment which cannot be diffolved.

All thefe particulars have been difcovered by the experiments made to combine the fulphuric acid in different ways with the argillaceous earth.

It is not, however, in this way that alum is manufactured for the purpofes of different arts : It is obtained from materials which afford it much cheaper, by a procefs which was long practifed before it was well underftood. You will fee a full account of the proceffes by which alum is manufactured, and of the materials employed in making it, in Mr. Fourcroy's Elements, under the article ALUM ; and in Mr. Chaptal's, where he defcribes a new procefs of his own. Bergmann alfo has given an inftructive effay on the manufacture of alum.

Alum is generally manufactured from a foft laminated ftony matter found in ftrata, (and which bears fome refemblance to flate, but is fofter, and commonly dark grey or black), called *alum ore ; aluminous fchiftus,—Schale*. It is compofed of clay combined with fulphur, and very often fome bituminous matter. It hardly ever produces alum until after it has been burned with a very flow heat for a confiderable time, which confumes the bituminous matter, and changes the fulphur into fulphuric acid. It is accordingly burned by fetting fire to very large heaps of it, which are covered up in fuch a manner as to occafion a flow and long continued inflammation and very gentle heat. There are fome of thefe ores however, which, when formed in-

T 2

to heaps, take fire of themfelves, and burn flowly, without needing to
be fet on fire by means of fuel. And in fome volcanic countries,
they find materials which are already prepared for affording alum, in
confequence of their having been affected by the volcanic heat. In
fome few cafes alfo, the action of the armofphere alone, continued for
a length of time upon fuch materials, prepares them for affording
alum, although they never have been fet on fire. They are obferved
to be covered with a whitifh duft, which, when examined, is cryftallized alum. When this is wafhed off, more forms in a day or two.
Such matrices properly built up, fo that the rain may wafh it off into gutters which lead to a ciftern, afford a lixivium ready for boiling.
There is at Hurlet, in the neighbourhood of Glafgow, a ftratum of
fhale which has been left in the old workings of coal mines, which,
by the fingularity of its fituation, exhibits a wonderfully rich appearance. It is about ten inches thick, and, if cut out from a newly opened pit, does not yield an atom of alum : But lying in the open wafte
for more than a century, it has decompofed, and efflorefced. Moreover, there is above it a vaft thicknefs of fhale, which is too poor for
working; but in the procefs of time, has afforded, by filtration, enough
to enrich the ftratum below which efflorefces, and thus renews its
wafte. The refult of this has been, that each lamina of the fchiftus
has not only feparated from the next, but the interftice is filled with
cryftals continually increafing; fo that in fome places the ftratum of
10 inches has fwelled fo as completely to fill the fpace from which
the coal has been taken. I thought this fingular fact not unworthy
your notice.

In all thefe cafes, whether the alum ore has actually been fet on fire
and burned, or whether it be prepared by the long continued action
of the air alone, the change produced is a combination of the fulphur
with vital air, and confequently a change of it into fulphuric acid,
which acts on the clayey matter; after which, being fteeped in water, it affords an aluminous liquor, which requires the addition of

some alkali, fixed or volatile, to prepare it for affording good cryftals of alum by evaporation. The ufe of the alkali is partly to precipitate fome calx of iron, which is commonly diffolved in this liquor along with the alum, and partly, as is fuppofed, to faturate a portion of fuperfluous fulphuric acid. The cryftals are firft obtained of a moderate fize, but are afterwards united into large maffes by *roaching*.

Mr. Chaptal's procefs is very refined and artificial. He obferved that a gentle roafting of clay difpofes it to a more ready union with the acid, efpecially when this is in the form of fteam. He therefore makes up the clay into fmall balls, and roafts them in an oven or kiln. One effect of this muft be the driving out the water, and leaving the whole mafs very porous and acceffible to the fteam of the fulphurous acid. The balls are now fpread out on the floor and fhelves of a great room, whofe walls and cieling are covered with a fat luting impenetrable by the fumes. There is a furnace conftructed in a corner of the room, in which fulphur is made to burn in a manner which difperfes it through the room in fteams of the fuffocating acid: The balls are penetrated, and the clay unites with it. When judged to be fuperficially faturated with it, and alum formed in them, which requires fome days, the room is opened, and the balls taken out and expofed to the air under cover. The volatile acid becomes fulphuric by extracting oxygen from the air, and diffolves the clay, and the balls are then lixiviated. This manufacture is faid to be very profitable. (*Ann. de Chem.* vol. 3.)

The moft inftructive differtation that I have feen on this manufacture is, *Difquifitio Chemica de Confectione Aluminis, Auctore Guftavo Suedilio, Upfal,* 1767.

The appearance and more obvious qualities of alum are more or lefs known to you. It is expofed to fale in large cryftalline maffes containing much water. By the application of heat it melts, fwells, boils, and dries, and is then called *burnt alum,—alumen uftum,* which has often a good effect on foul ulcers. In a greater heat, it does not undergo real fusion, but part of the acid is diffipated. It diffolves in cold water, but flowly, and in fmall quantity. Hot water diffolves it much better *. Its tafte is fweetifh, four, and aftringent. It tinges fome vegetable blues red. Thefe laft qualities it derives from the acid, which is fo little neutralifed by the earth, or is prefent in the alum in fuch quantity, that it is lefs changed in its properties than in other cafes. It is alfo united with the earth by a weak attraction. This appears from the facility with which alum is decompofed by other fubftances ; for not only the fixed alkalis, but calcareous earth, the volatile alkali and magnefia, if added to a warm folution of alum, caufe effervefcence and precipitation. The effervefce..ce fhews that the argillaceous earth is not difpofed to unite with the air, or with very little of it, if the folution be warm. This earth, when properly feparated from the acid and neutral falt, is purer argillaceous earth than any clay ; and if it be precipitated from a cool folution, fo that the earth may unite with a portion of the air of the alkali, it has the plaftic qualities in the higheft degree of perfection.

The French chemifts have given a new name to this pure earth ; *alumine* in French, and *alumina* in Latin. I confefs I do not like this *alumina*. If a name is to be contrived, I would make it *argilla*.

Great quantities of alum are employed in the arts of dying, and in printing or painting of linens and cottons.

* It therefore readily cryftallizes by cooling. In repeating Mr. Lowitz's experiment with a mixture of alum and nitre, I obferved the whole alum cryftallized in one form (octaëdral) which I had never feen before, it being commonly a mixture of both kinds. The obfervation of Mr. Lowitz naturally made me think it probable, that not only the alum alone was made to cryftallize by touching the mixed folution with a cryftal of alum, but alfo that the form of the cryftal prefented would difpofe the whole to the fame aggregation. The notions which I had of cryftallization leads almoft neceffarily to this. I tried it, by prefenting to the fame aluminous folution a cubical cryftal, and the effect was as I expected. All were cubes or compofites of cubes. I have found the fame thing to happen in fome other falts, which often appear in truncated cryftals.

EDITOR.

To complete the natural hiftory of the argillaceous earth, we may remark, that it is found in many other natural productions befide the plaftic earths or clays; in the *alum ore*, for example, which is indurated clay with fulphur; in what is called *black halk*, in which there is fome inflammable matter which gives the blacknefs; in *flate*, which appears to have been formed from clay, indurated naturally by length of time and fubterranean heat: There is a very great variety of ftrata of this kind, which have different degrees of hardnefs and cohefion, and are more or lefs laminated like flate; all of which contain fome of the argillaceous earth mixed with others; and they appear evidently to have been clays, which have been indurated more or lefs by fome operations of nature, until many of them have acquired great degrees of hardnefs and durability. The general denomination for them all is Schistus. The hardeft of them are fo clofely concreted, that the ftrata they form, called Gneiss, are among the moft durable. The lefs hard conftitute different kinds of Slate, and the fofter ones Shale. In the neighbourhood of Geneva, and I fuppofe in France, fome are called Pierre de Corne.

The argillaceous earth is alfo a principal article in the compofition of fome of the hard ftones, as we fhall notice hereafter.

GENUS III.

HARD STONY BODIES.

IN this section I comprehend most of the stony substances called SILICEOUS, or FLINTY, by natural historians.

They are eminently hard, and they are unfusible by the most violent heats of common furnaces or fires. Their hardness is such, that they are not affected by the hardest steel, but on the contrary scratch it, and strike fire with it. And they also scratch or cut glass.

The greater number of the stony substances which belong to this division have also been called crystalline and vitrescent,—*crystalline*, on account of their being oftener found in the form of regular and transparent crystals than other kinds,—and *vitrescent*, as being the principal ingredients in glass, and more disposed than any other earthy or stony matter to produce good glass with proper additions.

All the stony bodies of this division contain a particular kind of earth, which I have had occasion several times to mention, but not yet to describe in this course. It predominates or bears a principal part in the composition of this order, and is at present considered as one more of the *simple elementary earths*. It has been commonly called the siliceous earth. The French chemists lately contrived the names for it, of SILICE in French,—and a barbarous term SILICA, in Latin.

We can easily extract it by a chemical process from the stones of of this divifion; and the nature of the procefs is fuch, that the earth is obtained in the form of a precipitate, which is an exceedingly tender, light, and fpongy earth.

The properties of it are thefe:

1*mo*, It is not diffolved or otherwife affected by any acid, except the fluoric, which, when applied warm to it, and efpecially in the form of vapour, not only diffolves this earth, but volatilizes it.

2*do*, It is not fufible in our moft violent fires.

3*tio*, When mixed in powder with half its weight of potafh or foda, or the carbonats of thofe alkalis, the mixture can be melted eafily in a ftrong heat, and forms a perfect glafs. It is even in fome meafure diffolved or combined with potafh, by boiling in a ftrong watery folution of that falt. This experiment will not fucceed with any filiceous fubftance reduced *mechanically* to a powder, however fubtile. The utmoft effects of mechanical divifion and trituration fall infinitely fhort of the fubtile divifion and attenuation which we obtain on many occafions by chemical folution and precipitation. And this earth, in order to make it combine with an alkali that is diffolved in water, muft either be in this ftate of the moft fubtile and tender powder, or it muft be previoufly combined with a fmall proportion of alkali in the dry way, or by fufion.

This experiment and others enabled me to underftand the nature of an earth which occurs in fome parts of Scotland, in the form of a fediment or mud, at the bottom of lakes of frefh water. There is a lake in Galloway in which it is found. Samples of it were brought to me a long time fince, to know if it was a marle: I quickly perceived that it was not a marle; but the properties of it were fuch, that for a confiderable time I was at a lofs to give it a name. At laft I found that the qualities of it were thofe of the filex, or filiceous earth, and that it was principally compofed of this earth. It may therefore be named *limus filiceus*, or *filex limofa*. It is undoubtedly depofited in thofe lakes, in confequence of the demolition and decompofition of ftones and rocks, which contain it in the higher parts of the country, from which thofe lakes are fupplied with water.

This completes the lift of the earths that are at prefent confidered as the moft remarkable and moft abundant *elementary earths*. They are fix in number, diftinctly known, namely, the *calcareous earth* or *lime, magnefia, barytes, ftrontites, alumina* or the *argillaceous earth, filica* or the *filiceous earth*.

We have been lately informed of a very few more lately difcovered, which appear to be fimple earths, and yet are different from any of thefe; but they are produced by nature in very fmall quantity, and have only been found in the compofition of fome particular and rare ftony concretions of a fmall fize; *nor have they yet been fufficiently examined*. I do not, therefore, think it proper at prefent to take up your attention with them. You will fee mention made of them in the new fyftems of mineralogy which I lately recommended to your notice.

To return to the confideration of the hard ftony bodies.—They appear to have their hardnefs more or lefs from the filiceous earth, or filica, in their compofition; but they contain it, however, in different proportions, or in different ftates of purity. I fhall firft enumerate thofe which contain it in largeft quantity, or in a ftate which approaches the neareft to purity. Thefe are,

1. CRYSTAL, or rock cryftal, which is tranfparent.

2. CHALCEDONY, which has an imperfect tranfparency refembling that of whey.

3. QUARTZ, which has a whitenefs like that of milk and water, and fhattered appearance, and breaks with an uneven furface, not a plated ftructure like fpar. It is abundantly produced by nature.

4. AGATE, in which the ftony matter is diverfified with ftreaks and fpots, whitifh, or of other colours.

5. FLINT, filex, which has an uniform dark colour like dark coloured horn, but becomes white in the fire.

6. An exceeding fine loose earth found at the bottom of some of our lakes in the mountainous parts of this country. It may be called *limus siliceus, or silica limosa.*

These varieties of hard stones contain the silica the most abundantly, or most approaching to a pure state.

But in many of the other stones of this order it is very impure, or mixed with a large proportion of other matter.

When iron is mixed with it in such quantity as to colour it strongly, and render it opaque, it forms JASPERS of various colours. These are distinguishable from flint by breaking without the smallest lustre, like dry clay, and void of all transparency. Or if the quantity of the iron is less, and the stone is semi-transparent, and of a red colour, it is named CARNELIAN, from the resemblance of its colour to that of raw flesh.

These are the principal appearances of the hard stony matter.

But it may also be proper to give a general view of the various situations and collections of the hard stones in nature.

1*mo*, They are found constituting numerous and extensive strata; and those which are the most abundant, are the strata of common sand and gravel, and sand-stone, and gravel-stone, which are very numerous and common, and of great extent. The origin of sand and gravel was formerly explained.

Some sands are white, or free from any colour, and are totally composed of small grains of crystal or quartz, such as the sand from Lynn in Norfolk. It is the fittest of all for fine glass.

The greatest number of sands, however, are variously coloured, by the admixture of other matter with the siliceous in the composition of many of the grains. And some are perfectly opaque, and dark coloured, and even black, from the large proportion of iron.

2*do*, GRAVEL is in many places of the same nature as common sand, and is found in the composition of strata, either by itself, or more commonly mixed with sand or clay in different proportions. It confifts principally of the fame kind of matter as the fands, only in larger grains or maffes, more irregular in their form and fize, and more coarfe and opaque by the more plentiful admixture of other earthy matter with the filiceous earth. Gravel is, however, of very different kinds in different places, and it muft be fo, as being formed of the hardeft fragments of the ftones of the country in which it is found.

3tio, SAND-STONE is the third kind of ftratified matter, which I faid is found in great abundance; and which, in moft places, is compofed of hard ftony materials. It has evidently been produced from common fand concreted together.

When the fand of which it is compofed does not cohere too ftrongly, this ftone is employed in buildings which are erected with hewn ftone, and it is then called FREE-STONE. All the free-ftone of this country is compofed of filiceous fand, and is therefore remarkably durable; and we have plenty of it. The fand of which it is compofed was firft collected and wafhed clean by water in the long lapfe of time, and was afterwards cemented together by fome operation of nature. The proofs of this are, an undulated appearance which often occurs in the furfaces of its ftrata, and which is exactly fimilar to the undulated furface formed on the fands of the fea-fhore, or of lakes, by the action of the waves. And we often find alfo in the ftrata of fand-ftone the relics of fea productions. There is, in particular, a very remarkable object found in many free-ftone quarries in this country. The ftone itfelf is obferved to take the form of a plaited (not twifted) rope, of confiderable thicknefs, and generally flattifh, the greateft diameter being horizontal. In the different plaits of it there is a fmall indentation, and the plaits are moft regularly difpofed. When broken acrofs, it has a fort of core, black and foft, and fibres are obferved to go out from this to the indentations on the furface. This object is fometimes of great extent, traceable through the ftratum in a horizontal direction many yards, varying in diameter from half an inch to five or fix inches, and fome-

times fending off branches.   An example of it was feen lately in a
quarry on the fea-fhore near Muffelburgh, which refembled the
trunk of a vaft tree fending out branches in all directions.

Sand-ftone is of very various hardnefs.   In fome the fand has but
a weak degree of cohefion.   In others the grains are fo clofely and
ftrongly coherent, that the ftone has the appearance of folid flint,
and cannot be wrought as a free-ftone.

4to, The ftrata of what I called gravel-ftone have been formed
in the fame manner as fand-ftone, only that gravel is intermixed
with the fand in the compofition of the ftone.   From its appearance
this ftone is named by the Englifh *pudding-ftone ;* and the name
has been adopted by foreigners.

Such are the ftrata, principally or totally compofed of the hard
ftony bodies.   There are alfo fome kinds of rock in which more or
lefs of them is contained, as granite, to be foon defcribed ; and the
more compounded kind of rock which abounds in this country, nam-
WHIN-STONE, fome kinds of which are a more coarfe and compound
granite.   The whin-ftone often contains nodules and pebbles of all
different fizes, fome of which are hard ftones. ·

The fchiftufes alfo, or indurated argillaceous ftrata, often contain
a large proportion of quartz irregularly intermixed through them.

The hard ftones conftitute therefore a great part of the materials
of many ftrata and rocks.   They are alfo found often in *veins,* or
otherwife interfperfed through rocky or ftratified matter.   The fpecies moft frequently found in this ftate is quartz ; but we alfo find
occafionally, cryftal, chalcedony, agate, jafper, and flint.

There are great quantities of flint in England, interfperfed in a
very irregular manner through the calcareous ftrata of chalk and
limeftone, traverfing and running through them in different directions.   As found in chalk, it is more efpecially named *flint ;* as occurring in limeftones, it is named *chert* by the Englifh.

From the appearance of this matter, as found in this ftate, there is reason to be convinced that it has been introduced in a very fluid state, or that it has been produced in these strata by the action of some very fluid matter which has penetrated them.   The proof of this is, our finding shells, lithophyta, and other marine productions, originally calcareous, but which are now completely penetrated by this sort of matter, and seemingly converted into flint or agate, without losing their external form.   The same fact is further established by the quantities of fossil wood penetrated and petrified with this matter, found in many parts of the world.   It is found in all states, from that of stony wood, still combustible, to that of pure flint.

It also deserves notice, that the nodules of flint found in chalk and lime, often in very strange forms, are not simply lying there, but seem forming, or else decaying, in that situation, being always surrounded by a crust, which changes gradually from a calcareous to a siliceous nature.

Another curious fact relating to this stony substance is, that masses of it are sometimes found which include water perfectly inclosed in the hard stone.   This has been observed in crystal, in common flint, and in agate.

In whichsoever of these states the hard stones are found, when masses of them occur that are not solid, but have vacuities within them, in these we find their matter crystallized, and very often into remarkably regular and transparent crystals.   Such are often found in hollows of veins, and of pebbles.   They are generally columns of six sides, terminated by a six-sided pyramid : Or sometimes they are pyramids alone.   All these crystals are chiefly found in cavities of veins, or of hard stones.

I have yet to mention one other state or condition in which the silica has been found, and that is, dissolved in water.   There is one very remarkable example of this in Iceland, particularly at the celebrated *Geyser*, where some hot springs contain so much as to form siliceous petrifactions.

These are the different forms and states in which we find the hard stony bodies. We are now to take notice of their chemical properties.

The purest kinds, or those which contain the most of the silica, when they are exposed to a strong heat, generally become opaque, white, and brittle. But they sustain the most violent heat without melting, or being softened by it. We can melt them, however, easily and perfectly, with fixed alkaline salts, with which they unite in the fire to form a perfect and workable glass. By workable glass, I mean a glass, which, while it is allowed to cool from its melted state, continues soft and ductile a considerable time, and passes through all the degrees of softness before it become hard and rigid. During these states of softness it has great ductility, and can be wrought or moulded to any form. Glass that is workable in this manner is one of the most valuable productions of chemistry.

All good glass contains siliceous matter and a fixed alkali as its only or principal ingredients. And sand, when it is composed of pure, or nearly pure siliceous matter, is preferred to the other forms of the hard earths, on account of its being easily mixed with the alkali and other materials, without requiring the expensive operations that would be necessary for reducing to powder, or to small grains, the hard masses of other siliceous matter.

In the composition of some kinds of glass, other materials are added to the sand and alkali; but the effect of these other materials is chiefly to increase transparency, or to affect the colour, or to diminish expence.

The siliceous matter and alkali must bear certain proportions to one another to make good glass. If the alkali be deficient in the due quantity or goodness, the siliceous matter is not completely dissolved, and the glass not sufficiently transparent. If the alkali is redundant, the glafs is liable to be corroded, or to have its furface tarnifhed by damp air. By increafing the alkali to twice the weight of the filiceous matter, we make a fort of glafs which diffolves completely in water. Thus is produced what is called the *liquor filicum*, which coagulates into a jelly by expofition, or by the admixture of acid, or by volatile alkali (according to Neumann). This is the chemical procefs by which we obtain the pure filiceous earth.

Although glafs turns out very faulty when it thus contains an over proportion of alkali, it is however found neceffary, in making good glafs, to ufe rather more alkali than what is barely fufficient to melt the fand; and then, by a continuance of the fire, to evaporate the fuperfluous quantity. Thus we are more certain of perfect folution or fufion of the filiceous matter.

It is alfo common to ufe a quantity of nitre in place of part of the fixed alkali, in making the finer kinds of glafs, to confume the inflammable matter; and borax likewife is frequently added in fmall experiments. A mixture of thefe falts with a very pure or chofen fand, or other filiceous matter, produces glafs of uncommon brightnefs and tranfparency, which can be coloured by addition of metallic fubftances, and gives thofe imitations of the gems which in this country are called *paftes*.

The author who has given the beft account of the art of making glafs of different kinds, efpecially the fine and coloured glaffes which are made in imitation of the gems, is Kunkel, in an edition he gave of *Neri's Art of making Glafs*.

In the finer kinds of glafs, therefore, the filiceous earth is combined chiefly with a pure fixed alkali, and brought into a vitrified ftate by means of that falt; for when nitre is ufed, it is changed into pure fixed alkali. The calces of lead, as they were named formerly, fuch as white-lead, and red-lead, are employed in the manufacture of the Englifh FLINT GLASS, fo remarkable for its pure tranfparency, which fits it for being cut into facets, and other ornamental forms, and gives it remarkable brilliancy. In making the coarfer and cheaper kinds, as bottle glafs, window glafs, and the like, they do not ufe a pure alkali, but take the afhes either of land vegetables, or kelp, or barilla, which are the afhes of marine plants, with the falts they contain. And thefe afhes are mixed with fand, or other filiceous matter, to make fuch kinds of glafs. In this cafe, the earthy part of the afhes, which in part confifts of alkaline earths, affifts the fixed alkali in bringing the filiceous matter into fufion, and enters into the compofition of the glafs. It is not nearly fo powerful a folvent, however, as the pure fixed alkali, and therefore we muft employ, in the compofition of fuch glafs, a fmaller proportion of filiceous matter than when the glafs is made with pure alkali. And it is alfo neceffary to frit the materials before they are melted in the pots in which the glafs is formed. This is expofing the materials in a furnace, generally adjoining to the working furnace, and deriving its heat from it. This is a reverberatory, where the materials are fpread out on its hearth, and kept roafting for many hours, or even days, with a red heat. This is neceffary, for two reafons: 1*ft*, and chiefly, to confume completely all remains of the inflammable matter of the afhes. 2*dly*, To expel from the fixed alkali and afhes fome part of their fixed air.

The earthy part of the afhes of vegetables always contains fome iron, which gives a green colour to fuch glafs. I muft obferve, that the coarfeft kind of glafs, on account of the large quantity of earthy matter which it contains, and the penury of falt, is fitted for giving what is called Reaumur's porcelain of glafs. This is a curious change produced on thofe hard glaffes, by long expofition to a red heat in contact with fand, gypfum, and other fubftances. The glafs gradually acquires a cryftalline ftructure, formed in fibres which are perpendicular to the external furface. The longer the piece is continued in that fituation, this fingular arrangement penetrates deeper, till the growth from each furface meets in the middle of the thicknefs. After this, if much longer continued, the fibrous ftructure becomes somewhat granular, and the grains in the middle become large and detached, and sometimes the piece separates in the middle of its thickness. When this change is completed, and no more than completed, the glass has greatly changed its nature with regard to the effects of heat. It is now infusible by the heat of a glass-house furnace, and will bear to be plunged into water when red hot. It is also extremely hard, and cannot be scratched by the hardest file. Mr. Reaumur discovered this by accident, and recommends this preparation for retorts, mortars, and other chemical vessels, where these qualities are of importance.

It is a curious circumstance in glass, especially the coarser kinds, that, when allowed to cool very slowly in great masses, it takes a crystalline opaque form. When broken, it would be taken for a natural fossil. The structure is fibrous, and the fibres all converge to centres, but are commonly interrupted by others converging to other centres. The whole is extremely like, and may be mistaken, even by a fossilist, for a zeolite. Sometimes we find, in large lumps of green glass, little round balls like pin heads, or pease, of the same dirty white colour. These, when examined by a microscope, are found to be crystallizations of the same kind,—the fibres all standing inwards, perpendicular to the spherical surface. Now, looking at the confused crystallization of the great masses just now mentioned, we see that they affect the same arrangement, and we can trace the circumference of each sphere, of which these radiated crystals are a portion. By the way, this appearance puts an end to the doctrine of those who ascribe all crystallization to watery solution. Here we see it produced altogether by heat. Indeed we know that many metals, and sulphur, always crystallize in cooling, and will congeal in no other way.

After this digression on the nature and manufacture of glass, which its importance will sufficiently justify, it is proper to consider a little more the union of the alkali and siliceous earths. I observed that they not only united by melting, but even by boiling the earth in a watery solution.

When the liquor filicum is mixed with an acid, the earth is precipitated in an impalpable powder, which will be completely rediffolved by adding more acid, and again precipitated by adding more alkali; and thus may be repeated as often as we pleafe. But there are differences in this refpect which are remarkable. The Chevalier Dolomieu obferved that when quartz was precipitated in this way, it would not be rediffolved by acid, if kept in the ftate of a precipitate even for a few hours. This, with fome other peculiarities of the quartz, attracted his attention. Quartz emits a very peculiar fmell when rubbed or ftruck by another bit of quarz, and emits much light. If the duft ftruck off be examined with a microfcope, particles of it are found fcorified, and fpungy. Quartz frothes when melted on charcoal. He therefore fufpected that it contained elaftic and inflammable matter.

He mixed 10 drachms of quartz with a folution of cauftic alkali in a retort, joined to a pneumatic apparatus. By gradually raifing the heat to the grateft intenfity that his retort would bear, he expelled from it, 1ft, 22 inches of azotic gas; 2d, 12 inches of hydrogenous gas; 3d, 22 inches of airs, of which 16 were carbonic acid, and the reft a mixture of azotic and hydrogenous gas. The refiduum was, in fome cafes, a fpongy glafs, and in others folid and green coloured.

He thinks that quartz, when diffolved in acids, has quitted the azote and hydrogen; and when precipitated by alkali, has regained it by decompofing the water. When deprived of thefe airs, it has an attraction for acids, but lofes it when reunited to the azote and hydrogen and carbone, as lime has its attractions diminifhed by its union with fixed air.

Dolomieu fubjoins to thefe experiments feveral ingenious obfervations on the filiceous earths, and fubftances which contain them, particularly the gems. Although he advances fome opinions which feem to me very unwarranted by experiment, his obfervations are curious and interefting. You will find an abftract of his differtation in the Journal de Phyfique, May 1772.

GENUS IV.

―――――――

FUSIBLE STONES.

THE ingenuity and induſtry of the modern chemiſts and natural hiſtorians, particularly of Margraaf, Woulfe, Bergmann, Scheele, Kirwan, and others, have diſcovered that the natural earthy or ſtony ſubſtances which I comprehend under the titles of FUSIBLE and FLEXIBLE. do not contain any ſimple elementary earth that is different from thoſe ſix we have already deſcribed.

The fuſible and flexible ſtones are all compounded of two or more of theſe ſix earths, intimately united and incorporated together.

You will find in Bergmann's Opuſcula, and in Mr. Kirwan's Mineralogy, an account of the proceſſes by which earths and ſtones may be analyſed, and the ſimple earths of which they are compoſed, ſeparated exactly from one another. We cannot deſcribe theſe proceſſes here. They are too complicated and tedious to be a fit ſubject for a lecture. A perſon who would make himſelf maſter of them muſt ſtudy them at home, and at leiſure.

Profeſſor Bergmann, in his *Outlines of Mineralogy*, and Mr. Kirwan, in his valuable work on the ſame ſubject, have formed their arrangement of the earthy and ſtony ſubſtances, and their diſtinctions of them, from the prevalence of the ſimple earths moſt abundant in the compofition of each ; and for this reafon they have no divifions or claffes which correfpond to thefe two laft of mine.

Their method, however, affembles together things fo very unlike to one another in every other refpect, that I have not chofen to follow it. The greater part of the gems, for example, the ruby, emerald, topaz, and others, are put into the fame clafs with the clays, on account of their having more of the argillaceous than of any other fimple earth in their compofition.

But a perfon who wifhes to have fome knowledge of the riches and variety of nature in this part of her works, ought furely not to confine his attention too much to that point. There are many other differential qualities of her productions, which are more eafily perceived, and more ftriking; and therefore more proper to be attended to in diftinguifhing and arranging them, and in giving general views of them. And this is the reafon that has induced me to prefer the method I follow in giving a general account of the earthy and ftony bodies. And I muft premife that there is fcarcely any method for thefe ftones that is not very imperfect. This refults from their compound nature. Accordingly, it is extremely difficult to clafs a fpecimen that comes in our way by means of the defcriptions given by the foffilifts. Nor are they agreed in their denominations: What one calls a granite, another calls a porphyry; and a third calls it a trap.

The natural ftones which I think it is proper to affemble under the title of *fufible*, are of fix kinds:

1. Feldt fpat, or felt fpar.

2. Porphyry.

3. The garnat.

4. The ftony matter called fchoerl or fcherle by the Germans and natural hiftorians, and cockle, by the Englifh miners.

5. The zeolite.

6. The lavas, bafaltes, pumice, and other fufible matters, which have evidently been thrown out of the bowels of the earth by volcanic

fires and explosions; or which appear to have been formed and accumulated by them under the surface.

The first of the fusible stones, therefore, is the FELDT SPAR, the appearance and qualities of which are these:

It is a stone generally less transparent than quartz: Some kinds are as white or free from colour. More commonly, however, it has a reddish tint or flesh colour. In hardness, it is nearly equal to the hard stones, and therefore strikes fire with steel. When it is broken we can always perceive by the reflection of the light from its surface, that it has a plated structure, and it has a disposition to be broken into rhombic fragments. These two qualities are never found in quartz.

FELDT SPARS are,

1*mo*, The most common, of a reddish colour.

2*do*, The white.

3*tio*, The crystallized.

4*to*, Crystallized and transparent,—*adularia*.

When feldt spar is exposed to the action of heat, it first becomes more brittle and pulverable; and afterwards, if the heat be increased to a violent degree, it melts into a viscid glass, which is white and semi-transparent; on account of which property, and some others, it is excellently fitted to be an ingredient in the composition of porcelain.

It has been analysed by Mr. Woulfe, Professor Bergmann and others, and the definition given of it by Bergmann, in consequence of this analysis, is,

" Siliceous earth united with argillaceous earth, and a small quan-
" tity of magnesia and sometimes barytes." The proportions have been found very different in different felt spars, or in the different analyses that have been made of them, of which you will find a comparison in Mr. Kirwan's Mineralogy. From the whole he concludes, that any compound of silica and alumina, in which the silica predo-

minates, and to which a fmaller proportion of lime and magnefia, or of lime, magnefia, and barytes accedes, fo as form a compound, fufible in a ftrong heat, may conftitute a felt fpar. The magnefia and alumina may be exactly feparated from one another by diffolving them in the muriatic acid, and precipitating by carbonat of ammonia. The whole of the alumina is precipitated, and the magnefia remains diffolved in the form of an acidulous carbonat.

Feldt fpar abounds in the compofition of the rock or ftone called granite, and is the diftinguifhing ingredient in that ftone. Granite has its name from being compofed of angular grains of various fizes, in the different varieties of this ftone; the greateft number of which grains are felt fpar; the reft quartz and mica, which we fhall foon defcribe.

The granites form fome of the higheft and moft extenfive chains of mountains in many parts of the globe, being in general very durable ftones, though fome of them are not fo remarkable for durability. And, as we never find any relics of marine productions in them, they are confidered by many as original or primeval ftones, and their formation as long prior to that of any ftratified matter. But my friend Dr. Hutton, having vifited fome parts of this country where granite is found, and examining with attention thofe places where the granite was bounded by the ftratified matter, or came in contact with it, he found the ftratified matter, which was of the hardeft and moft durable kind, fplit and broke in thofe places in a remarkable manner, and the granite conftituting or forming veins in it; that is to fay, that the granite had penetrated into the fiffures and rents of the ftratified matter, and had filled them up completely: fhewing therefore that the ftratified matter had, in this cafe, exifted firft, and that the granite was applied to it with violence and in a fluid ftate. Some of the hardeft granites that are known are rendered extremely friable by a very moderate heat, even that of a fmall bundle of ftraw burning in contact with them.

Feldt fpar, befides entering thus largely into the compofition of granite, occurs alfo fometimes by itfelf in veins.

The fecond fpecies of fufible ftone is PORPHYRY. It is of various colours, of which a dirty brownifh green is the moft common. It is alfo of a deep red, black, and rufty brown, and fometimes of a dark grey. All of them have fpots, and the ground is compact, and breaks with fomewhat lefs luftre than marble, but more than flint. The fpots in it are feldt fpar. Its hardnefs is greater than that of marbles, but inferior to that of hard ftones, and it is eafily fufible.

It forms large maffes and rocks in the places in which it is found; and fome kinds of it have beauty, by their colour and capability of polifh; but thefe are rare. The Upper Egypt alone fupplied Rome with fine porphyries.

The third kind of fufible ftony matter, the GARNAT, is very commonly cryftallized into roundifh polyhedral maffes of a fmall fize, and is generally of a deep red colour, and tranfparent. In fome of thefe ftones, the deep red colour is accompanied by great tranfparency and brightnefs; though in the greateft number it is obfcured, by innumerable flaws and cracks, or the admixture of other earthy matter. By chemical analyfis a confiderable quantity of iron can be extracted from the deep-coloured garnats, and to this metal the colour is imputed. But fome fpecimens of this ftone contain little or no iron, and fuch have very little of the red colour.

It is found, and appears to have been formed, in ftones that abound with mica or with feldt fpar. It has its name, *granatus*, from its refemblance to the pulpy feeds of the granate apple, or granate fruit.

The fourth of the fufible ftony fubftances is named SCHOERL, and is found in many compounded or mixed ftones which never are of the ftratified kind, but are fome of the fpecies of what I call rock.

The fchoerl appears on thefe ftones in the form of feparate grains or maffes, always cryftallized into extended columnar forms, or into fibrous or plated maffes. Sometimes it forms groups of ftraight fi-

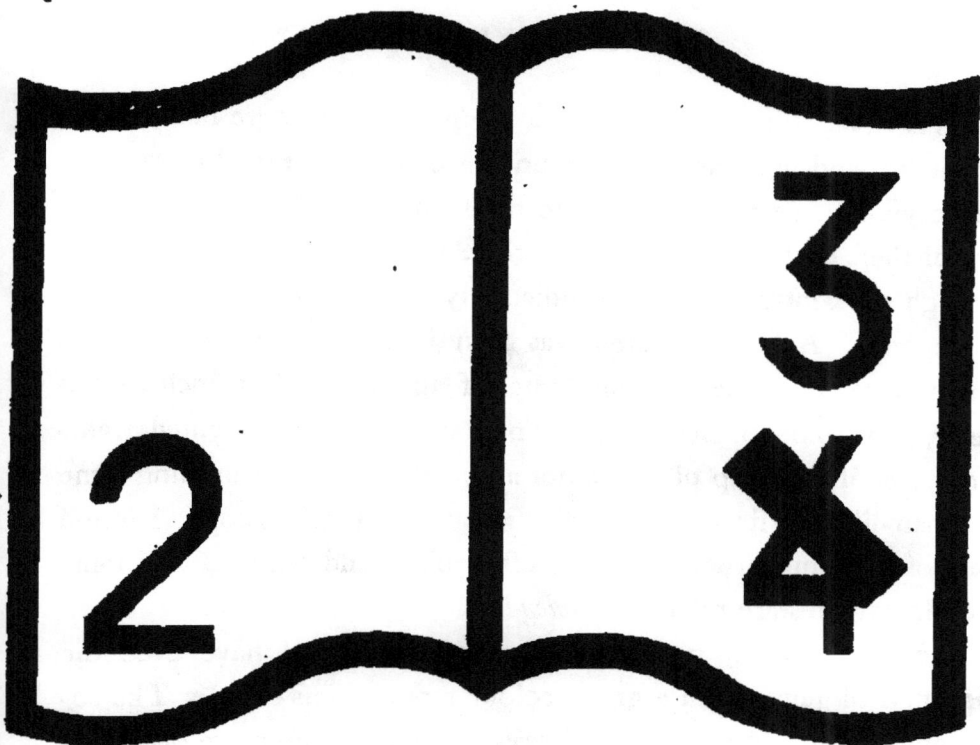

Pagination incorrecte — date incorrecte

NF Z 43-120-12

Pagination incohérente
Texte complet

brous cryſtals, like the briſtles of a bruſh, or even more ſlender than fine hair; and in ſome caſes is found in this form traverſing tranſparent cryſtals. Another ſingular form is the *croſs ſtone*, conſiſting of two of theſe cryſtals croſſing each other, not as adhering, but croſſing through the centre, ſome perpendicularly, and others in an angle of 60 degrees. A fine ſpecimen was ſeen in this country in cryſtals, which were hexagonal priſms, ſome of them almoſt an inch in diameter, and ſome inches long. They were deeply indented tranſverſely, as if made up of plates not all of one ſize. Examining ſome fine hair-like cryſtals in my poſſeſſion with a microſcope, I found them of the ſame ſtructure. It is often black and opaque, and ſome of theſe varieties are named *hornblend*.

Many kinds of it, however, are tranſparent, and have even the brightneſs of gems, with a green colour more or leſs deep. The colour and opacity in this ſtone always proceed from more or leſs of iron. The moſt remarkable of the ſchoerls is the *tourmalin*,—remarkable for its primitive and permanent electricity and electric polarity, perfectly analogous to magnetiſm.

All ſchoerls melt without difficulty in a ſtrong fire, and form a black or dark coloured vitrified maſs.

The fifth kind of fuſible ſtone in the enumeration I gave, was the ZEOLITE. It was firſt diſtinguiſhed and characteriſed by Cronſtedt. It occurs in general in the form of ſmall maſſes involved in rocky ſtones, or forming veins in ſuch ſtones. That part of the rock of Arthur's Seat which ſhews a tendency to the form of ſix-ſided columns, or priſms, contains ſmall veins of zeolite harder than common.

The zeolite is always cryſtallized, or ſhews ſome regular external form or internal arrangement of its matter. Some of theſe forms are the moſt wonderful and beautiful of all that have been obſerved in the mineral kingdom. Some are hollow, and are formed within into tranſparent or whitiſh cryſtals; ſometimes ſlender and long, like fine briftles, in diverging bunches; and fometimes even thefe brifly cryftals of the zeolite pervade chalcedony, and fometimes fmall filiceous cryftals like grains of corn are involved in them.

The chemical qualities of the zeolite are, to melt in the fire, or by the heat of the blow-pipe, with remarkable facility. And moft kinds of it fwell at firft into a fpongy mafs like froth, or fometimes as bo- rax does in its watery fufion. It is from this circumftance that it gets its name. Indeed, they contain a confiderable quantity of water in their compofition; but that cannot be the caufe of this fufion. But after this, by increafing the heat, the zeolite contracts again, and melts more perfectly, either into a clear and bright glafs, or fome kinds of it, which are more difficultly melted, form only a white opaque glafs, or white enamel.

When diluted nitric acid is applied to the zeolite, this ftone is dif- folved into a clear liquor, which after fome time becomes a jelly.

Thefe chemical properties of it were difcovered by Mr. Cronftedt. But my friend Dr. Hutton proceeded farther in this experiment; and by examining the folution, and the gelatinous matter formed in it, he difcovered that the zeolite is a compounded fubftance, and contains a fixed alkali combined with filiceous and with argillaceous earth. The proportion of the alkali to thefe earths is different in the feveral varieties of this ftone; and this is the reafon of the different degrees of fufibility which we find in them. They all contain a great quan- tity of water.

The laft kind of fufible earthy and ftony fubftances are thofe which are thrown out, or formed, by volcanic fires. Of thefe there is a confi- derable variety. We find among them what are called LAVAS and PUMICE, and what are called ASHES.

The lavas are very various. Some parts of them are black and per- fectly vitrified, refembling the darkeft and coarfeft kind of glafs. Generally, however, they are a mixture of earthy fubftances, which have not been brought to a ftate of perfect fluidity. Some of the materials only, were fluid when they came out of the volcano; and the whole had the confistence of thin mortar, or fometimes has been more fluid. When it cools, it has not an uniform or glaffy texture, but is compofed of particles and fragments of many different kinds, cemented together by the matter which was melted.

The fluidity of lava appears to me to depend on latent heat, combined with the materials by the long continued action of fubterranean fire. While they are ftill flowing, their heat is not very intenfe. *(Dolomieu Journ. de Phyf.* 1793). Yet they retain part of this heat a long time, with fome degree of fluidity or foftnefs. A lava, erupted in 1614, continued to move for ten years, and in that time flipped downwards about two miles. It was probably foft below; and as it lay on a flope, the indurated mafs above could flide a little way *.

Moft natural hiftorians are now of opinion that the rocky ftone, of which there are many varieties in this country under the name of WHINSTONE, but in other places called TOADSTONE, RAGSTONE, or ROWLEY-RAG, and alfo TRAP and BASALTES, and by other names, is of the fame nature with lava, and belongs to this divifion of volcanic matter,—a matter which has been melted by fubterranean fire. Dr. Hutton, however, makes a diftinction between many, or moft of thefe ftones or lavas. He confiders the term lava to be properly applicable to that melted matter only which has been thrown out by volcanoes, fo as to flow down their fides, or along the furface of the earth. But there cannot be a doubt, that the fame fires muft produce, at a great depth, large quantities of melted matter, which is never thrown out, but continues melted a long time, and is driven by the immenfe force of the explofive matter in lateral directions, forcing its way between the ftrata which are around, or penetrating into the rents and fiffures of thefe; and thus forming what we call dykes, or in other cafes, flat and extended maffes of unequal thicknefs, fuch as Salifbury Rock,

---

* The fpongy texture of the indurated matter muft greatly retard the emerfion of the latent hea., and confequently the congelation below. EDITOR.

and many others that are around it, and the toadstone of Derby-shire *.

Whinstone is called *basaltes*, when it is split into columns or prisms, most of them six-sided, and standing in close contact together, generally upright, though sometimes inclined to the horizon, and even bent.

There are famous examples of this in the Giants' Causeway in Ireland, and in the island of Staffa, and other islands and rocks on the west coast of Scotland. And there is a tendency to it on the south-east side of Arthur's Seat here, beside many examples on the continent.

This columnar appearance seems to arise from a kind of shrinking as the heat which gave it the imperfect fluidity gradually abates. I was informed by a gentleman who visited Iceland not long after the late eruption of Hecla, that the inhabitants shewed him extensive masses of lava of former eruptions, which were not thus shivered, when they had cooled so much that they could be approached, but that year after year, the regular columnar divisions appeared more evident: And at the time the gentleman saw them, they were very distinct basaltes. We observe an appearance very similar to this in common starch, which always separates by shrinking into pentagonal and hexagonal columns.

The opinion that this stone, and whinstone in general, has been formed from melted subterraneous matter, is founded on the similarity of its appearance to that of many lavas; and there is a similarity too in their constituent ingredients. And further, the lavas of some well known and now existing volcanoes, have been found in some parts of them formed into these pillars; as the Chevalier Dolomieu observed in the lavas of Ætna.

* A dyke of this kind pervades the coal strata in Newcastle Moor; and it has charred the coal on each side of it to the distance of several yards.

EDITOR.

In bafaltes and whinftone we commonly perceive many particles which have a plated ftructure or extended prifmatic figure, more or lefs refembling fchoerl. The colour of the ftone is commonly grey or black, or dark red ; but fometimes, though rarely, light grey, or whitifh.

There is a confiderable variety of the ftones or rocks which belong to this fpecies, or may be included in it ; and moft of them are not formed into columns, but are in general fplit into oblong angular maffes, which are often upright, as in Salifbury Rock; but in fome cafes inclined to the horizon. Some approach very nearly in nature to granite ; others as nearly to porphyry. The rock of the Calton Hill is a ftone of this kind, which has the nature of porphyry in fome parts of it. The rock of the Bafs approaches to granite.

They often contain nodules, or fmall maffes of ftony matter, different from that of the rock itfelf. Thefe fmall maffes are either filiceous, or calcareous, or barytic, or zeolite, or a collection of fome of thefe ftony bodies in the fame mafs. The greater number of agates, and efpecially the fmall ones, called Scotch pebbles, are fmall filiceous maffes, which have been formed in rocks of this kind ; and are found in the gravel and rubbifh produced from the decay of thefe rocks.

The toadftone, of Derbyfhire, is a whinftone containing nodules of calcareous matter.

It is always in thefe whinftone rocks too, and particularly the bafaltic kind, that zeolite is found, either in nodules or in veins.

Anciently, whinftone and other fufible ftones appear to have been employed, in this country, to conftruct walls or ramparts round little camps, or ftrong pofts, into which the inhabitants retired when invaded by an enemy. And they appear to have been cemented by fire,—the external parts, on both fides of the wall, being melted into an imperfect flag. There are ruins of fuch works in different parts of Scotland ; the conftruction of which muft have been long prior to any tradition that now remains, or to the use of lime. A description of some of them has been given in the Transactions of the Royal Society here, by my colleague, the Professor of Civil History.

Pumice is the next volcanic matter. It has the appearance of having undergone very perfect fusion. The most remarkable quality of it is its sponginess. It has nearly the contexture and appearance which glass would have, were it wrought up into a froth while it is cooling. The Chevalier Dolomieu, in his account of the *Isles Cyclopes*, a cluster of islands near Sicily, has given the natural history of pumice. And I have observed some facts and phenomena, occurring in some of the arts, which may suggest conjectures of the way in which it is formed, viz. what appears in flags of lead, and also flags of iron hastily cooled.

The last volcanic matter, called Ashes, is a sort of gravel and sandy dust, composed of vitrified and spongy particles of all sizes; and is undoubtedly a sort of rubbish, into which different materials are changed, when softened by the heat and exposed to the violent concussions and explosions which happen in the bowels of the mountains, or during eruptions. It is sometimes thrown up to a very great height in the air, and is then carried away by the winds, so as to fall, on some occasions, at a great distance from the volcano.

When this matter is amassed in great quantity, and compressed by incumbent weight, it concretes into a spongy soft stone, called *tufa* by the Italians. And other parts of it, which remain in the state of a sort of gravel, are excellently adapted for forming strong mortar with lime. They are brought from Italy for that purpose, and are called *puzzolana*. They are used for water buildings.

GENUS V.

════════

FLEXIBLE STONES.

THE laſt ſection of the earthy bodies only now remains to be deſcribed,—that of the FLEXIBLE STONY BODIES. They are diſtinguiſhed by being flexible, or eaſily diviſible into parts which have great flexibility.

Theſe are concretions which are in general ſo ſoft as to be eaſily ſcraped or cut with a knife, and do not ſuffer any change of their hardneſs or ſtructure from the heat of a moderate fire, but in a ſtrong one many of them are eaſily melted.

Two of the claſſes formed by Cronſtedt, his *Micaceæ* and *Aſbeſtinæ*, are comprehended in this ſection of flexible earths. And it may with propriety be diſtinguiſhed into three ſub-ſections, the *plated*, the *fibrous*, and the *membranous* or *cellular* kinds.

The firſt of theſe ſubdiviſions, therefore,—the plated flexible ſtony bodies, contains the micaceæ of Cronſtedt, or the *talcky* foſſils of the Engliſh naturaliſts; ſo called from the principal ſpecies long known by the name of mica or talc.

It is a ſtone which feels unctuous and ſlippery between the fingers, and ſplits eaſily with a knife, like the foliaceous gypſum, into innumerable and exceſſive thin leaves, not only very flexible, but claſtic;—employed for confining the objects in microſcopical obſervations. It is eaſily diſtinguiſhed from the foliaceous gypſum by its elaſticity, and by the effects of heat.

This ſtone is generally tranſparent in the thin plates, and often colourleſs. It is more commonly tinged, and ſometimes quite opaque. The colour is very various,—duſky, greeniſh, ſilvery, golden, and many other glittering appearances, which often deceive thoſe who are unacquainted with it. Theſe appearances do not proceed from metal, but from the diſpoſition of the plates.

Mica is found very plentiful in the compoſition of ſome kinds of rock, eſpecially ſome of thoſe which contain a large proportion of the fuſible earths ;—in granite eſpecially, and in the ſchiſtuſes which contain garnets or granates.

But it is mixed beſides with ſome of the ſtrata which have been formed from the materials of demoliſhed rocks, as ſand, and ſandſtone, and gravel, and ſome clays.

The largeſt ſpecimens of tranſparent mica come from Ruſſia, where they have it in ſuch plenty and perfection, that in ſome parts of the country it is uſed for windows and lanthorns.

The flexibility of this foſſil explains the flexibility of a ſtone which was exhibited here in 1791, and is now in the cabinet of Lord Gardenſtone. It is about half an inch thick, and more than a foot ſquare. It is flexible to a moderate degree, but without elaſticity. It is compoſed of ſmall thin plates of mica, all parallel to one another, and intermixed with thin diſcontinuous plates of quartz. The great number and ſmallneſs of the plates, and the preſence of the quartz, prevent this ſtone from being elaſtic.

The ſecond diviſion of the flexible earths,—the fibrous,—comprehends foſſils which have the ſame chemical qualities with mica, and differ from it chiefly in their ſtructure, which is fibrous inſtead of being plated. There are many varieties of theſe, diverſified by the fineneſs and flexibility of the fibres, and by the cloſeneſs or looſeneſs of their connection. Thoſe reckoned the moſt curious and perfect, called amianthus, may be teaſed into a matter like cotton, which, being mixed with a ſmall quantity of flax, may be ſpun into yarn,.

and wove into a cloth, which does not fuffer from ordinary fire. The only ufe that has been made of this fibrous ftone, is one for which modern cuftoms do not find room. The ancients made a fort of cloth of it, on which they laid dead bodies on the funeral pile, to preferve the afhes of the body by themfelves, and unmixed with thofe of the fuel. Some have attempted to form it into paper; but it makes very bad paper.

The other foffils which belong to this fubdivifion, are formed into membranes like leather, or elaftic fpongy maffes like flefh or cork. Thefe are named *mountain-leather*, *mountain-flefh*, and *mountain-cork*, by the miners.

All thefe fibrous or membranous flexible ftones are found in veins, or cavities of veins, and all contain the filica in largeft quantity, but intimately combined with magnefia, alumina, lime, and a fmall portion of iron. You will find the proportions ftated in Mr. Kirwan's Mineralogy, and in Profeffor Bergmann's Effays.

Magnefia is always a conftituent part of the flexible ftones, and is generally their moft abundant ingredient after the filica.

APPENDIX.

PRECIOUS STONES.

AND now, gentlemen, I have taken fome notice of the different kinds of earthy and ftony bodies that appear to be moft remarkably diftinguifhed from one another. Among all thefe, however, I have as yet made no mention of the PRECIOUS STONES, which may perhaps be thought a material omiffion.

The reafon is, that this title does not affemble together a fet of ftones that are fimilar to one another by their nature or chemical qualities. It is a general name given to all ftones remarkable for more or lefs tranfparency and brightnefs, or for fome colour, or mixture of colours, or fmoothnefs of furface, which has pleafed the tafte of mankind. And the ftones in which thefe qualities are found happen to be very different with refpect to their nature and to the materials of which they are compofed. I fhall here give a fhort enumeration of them.

I may begin by making a divifion of them into STONES and GEMS. Under the firft title I comprehend thofe produced by nature in large maffes, or in confiderable quantity. Under the fecond, thofe which occur of fmall fizes only, and much more rarely, on which account they are far more coftly than the others.

To the firft divifion belong,

1. Marbles.
2. Serpentines.

3. Alabafters.
4. Some Porphyries.
5. Some Granites.
6. The Englifh Pudding-ftone.
7. Some Spars.
8. Some Jafpers.
9. Some of the filiceous petrifactions of wood.
10. Some kinds of the Lapis Nephriticus.

1. *Of Marbles.*

The marbles are all limeftones, or are compofed of calcareous matter, and are different from the more common limeftone only by agreeable colours, or mixtures of colours, and a more perfect compactnefs and greater hardnefs, which fits them for being more highly polifhed.

Many marbles appear to have been formed from fmall fragments of fhells, ground down by agitation and attrition into fand, which has afterwards been cemented by fome natural operation. In many others we fee diftinctly the relics of the fhells or lithophyta entire, or in large fragments. Marbles in which fhells are feen in this manner are named by the Italians *Limacelli*.

When marbles are clouded or fpotted with different colours, this is produced, either,

1*mo,* By thefe relics of fhells and corals appearing in them ; or more generally,

2*do,* By an unequal and irregular intermixture of iron, or inflammable matter through the ftone, fo as to form ftreaks and fpots which pleafe the eye, by the contraft of their colours, and the variety of their fhades and diftribution ; or,

3*tio,* The variety of colours in fome marbles have been formed by an operation they have undergone. We can plainly perceive that the ftone has been fractured into a number of pieces, which have been fet afloat in a liquid marble paste of a different colour, and are now cemented together by it.

Such marbles are named *Breccia* by the Italians.

All these have more or less beauty. But we may reckon among the most valued marbles, the pure white statuary, and the perfect black. Such must naturally be rare. It cannot often happen that a block, of size sufficient for a statue or a groupe, is entirely free from spots or clouds, or streaks of a different colour or tint.

2. *Of Serpentines.*

The most obvious distinction of serpentines from marbles is by their colour, in which a deep and dusky green prevails, and this is a colour but very rare in marbles. As to the nature of serpentines, some are partly calcareous, but all contain magnesia and iron. A species of porphyry is improperly called serpentine.

3. *Of the Alabasters.*

This name is applied by the lapidaries to white stones that are softer than marbles, but capable of being well polished. They are generally composed of the natural sulphat of lime. But often calcareous stalactites and spars are also named alabasters by the stone-cutters.

4. *Of the Porphyries.*

Of the porphyries there are only a few species reckoned precious or beautiful. One is dark green, with light green or white spots. Another is dark red or purplish, with whitish spots. And there are some the ground of which is almost black. They are harder than the marbles, but softer than the hard stones. I rank the porphyries among the fusible stones, to which division they certainly belong by

their properties and conftituent ingredients. And many of our whin-
ftones, and many lavas in the volcanic countries, are coarfe porphyries.

5. Of Granites.

Some of the granites are alfo confidered as ftones of value. The
nature of this ftone was formerly defcribed. The only certain fpecies
of it reckoned valuable are the Egyptian, the Ruffian, and the
Scotch.

6. Of Pudding-Stone.

Many pieces of the pudding-ftone or gravel-ftone found in England
are alfo reckoned beautiful and precious.

7. Of Spars.

The fpars reckoned precious or beautiful are fome few calcareous
fpars, but chiefly *fluors* and *lapis Lazuli.* Fluors, admired for their
beauty, are found plentifully in Derbyfhire. *Labrador-ftone* is a
feldt fpar, or is nearly allied to it by its conftituent parts. It derives
its beauty from a certain difpofition of its tranfparent elements, which,
like thofe of pearl and mother of pearl, exhibit different lights as they
are viewed in different directions. It frequently appears as if the furface was undulated, although it be perfectly flat. The opal has its
beauty from the fame caufe. We fee appearances of the fame kind in
fome woods, fuch as mahogany, fatin-wood, &c. and in thefe the microfcope difcovers the circumftance on which it depends. We even
fee a little of it in fome flates. When examined in their natural plates,
they are really undulated; but when the furface is made flat and
polifhed, the undulated appearance is not entirely removed.

8. *Of Jaspers.*

The name of jafper is fometimes applied to fome marbles, but improperly. When ufed with propriety, it means hard ftones, opaque and agreeably coloured by admixture of other matter with the filiceous, efpecially iron. The moft noted jafpers that are beautiful are Sicilian, yellow, red, and white; green jafper fpotted red; Heliotropium.

9. *Of Siliceous Petrifactions of Wood.*

Some of the filiceous petrifactions of wood, when they happen to have the denfity of agate or jafper, and are agreeably coloured, are alfo wrought by the lapidaries, as precious ftones; and fnuff boxes or other toys made of them.

10. *Of the Lapis Nephriticus.*

Laftly, fome kinds of the *lapis nephriticus* are in requeft, on account of their agreeable colour, and the polifh or fmooth furface that can be given to them. The Turks, in particular, are fond of this ftone to make the handles of their fcimitars.

Having now faid enough of the firft fection of precious ftones, thofe, to wit, produced by nature in larger maffes, we proceed to the fecond; in which I include the gems, or thofe produced in fmaller maffes, and more rarely.

The title of GEM was applied not long ago to a much greater number of ftones than it is by late authors. I fhall follow the old method, according to which the ftones called GEMS were divided into SEMI-PELLUCID and PELLUCID GEMS.

1. *Semi-Pellucid Gems.*

The greater number of thefe belong to the order of hard ftony fubftances, and are chiefly compofed of the filiceous earth.

They are, in general, formed into fmall maffes in fome kinds of rock, or fill up fmall veins in it ; or they are found among the gravel and earth into which fuch rocks are refolved, by the influence of time and the weather.

The filiceous matter, of which they are principally compofed, has generally fome degree of tranfparency; but this tranfparency is diminifh-ed, and in fome cafes a great degree of opacity induced, by the admixture of other matter. Often this diverfity of materials is fo difpofed as to produce clouds, or fpots, or ftreaks of different colour and appearance, which add to the beauty of thefe ftones.

Among the moft notable kinds of thefe ftones, we may reckon,

1*mo*, Agates.—Great numbers of them are found in fome parts of this country, of a fmall fize, but fome of them very pretty when polifh-ed,—called *Scotch Pebbles*. I have already pointed out the origin of them. The internal arrangement of their matter is remarkable, and will be explained by Dr. Hutton, in a work he is now publifhing on the formation of ftones.

2*do*, Chalcedony, of the colour of the whey of milk. It is much efteemed by feal-cutters, for a kind of *toughnefs*, as they term it, which makes it work without tearing up by their powders. This muft arife from its having no grain or fymmetrical arrangement of parts.

3*tio*, Carnelian, alfo valued by feal-cutters, as being very workable. The natural formation of carnelian is one of the moft curious objects that mineralogy prefents to our eye. It has certainly been fluid as chalcedony, and the fluidity has been imperfect or clammy, and has been a very flow exudation from the rock or matrix to which it adheres. It is found hanging in tears from the fides of the cavities, and thefe appear to have hardened superficially as it sweated out; so that the long string is all incrusted with little excrescences, which have been, in succession, the end of the drop. These icicles often dip into a mass of chalcedony, whose surface is perfectly smooth and flat, as if it had been fluid in a dish, and the icicles can be traced through it, (it being in a small degree transparent) to the bottom, without mixing with this mass. The curious, and not inelegant form of the exudation, may be precisely imitated, by putting a lump of hard pitch on a shelf, in warm weather. It will soften and drop down from the shelf in strings, which are figured in the very same manner.

4*to*, Onyx, fit for cameos, by reason of its strongly contrasted coats. Sardonyx is either a mixture of chalcedony and carnelian, or is a chalcedony spotted with small red points.

5*to*, Mocoe, finely diversified by tree-like figures, elegantly ramified.

6*to*, Opal, of varying colours like mother of pearl.

7*mo*, Turcoise.

2. *Pellucid Gems.*

Most of these also belong to the order of the hard stony substances, and some of them are the hardest substances produced by nature. A few belong to the fusible class. They are always found in very small masses, crystallized into regular figures, more or less peculiar to each kind; and when they are reckoned gems, have extraordinary transparency and brightness. They appear to be originally formed by nature in some kinds of rock, or in the veins of rock, and some are actually cut out of such veins. But most of them are picked up from among gravel or earth formed by the decay of such rocks which contain them.

The more common kinds of them, which bear but a moderate

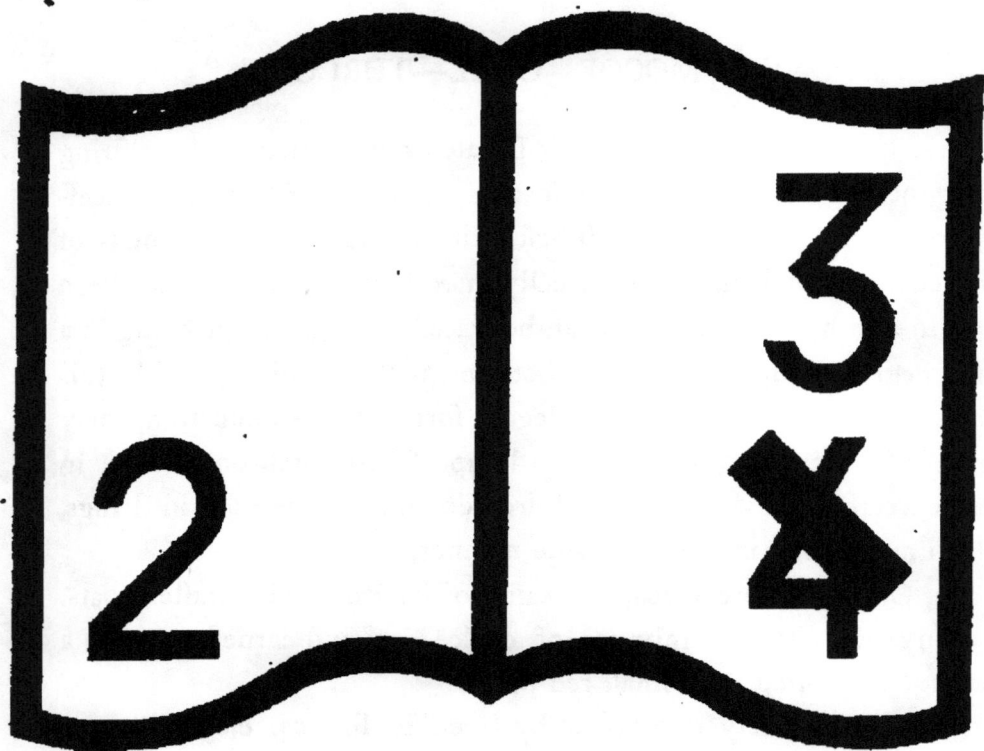

Pagination incohérente
Texte complet

price, are filiceous cryftals, compofed of the filiceous earth, with very little admixture of any other.

But thofe which have the greateft hardnefs and brilliancy, and which are named by the jewellers *true* or *oriental gems*, have been found lefs fimple in their compofition. A number of accurate analyfes of them, by Achard of Berlin, Profeffor Bergmann, Mr. Kirwan, and others, have fhewn that they contain the argillaceous earth in greateft quantity, intimately combined with a fmaller proportion of the filiceous, and a ftill fmaller of the calcareous, and of iron, to which metal they owe their beautiful colours. And it is only thefe tranfparent ftones, which contain more of the argillaceous than of any other earth, which are diftinguifhed by the title of gems by Profeffor Bergmann. I fee no reafon however for reftricting the title fo much.

We may here remark, that thofe gems which contain a larger proportion of the argillaceous than of the filiceous earth, have not the fame chemical qualities as the other hard ftones in which the filiceous earth prevails above all others. They not only refift the moft violent heat without melting, when expofed to it without addition, but it is extremely difficult to melt them by means of alkaline falts, which eafily diffolve the filiceous cryftals into a glafs. You may fee, in Bergmann's Opufcula, the other particulars by which they differ from the filiceous cryftals. Some of the filiceous cryftals are tinged, as well as thofe reckoned true gems, with blue, or purple, or yellow, or other colours. But thefe are much rarer than the colourlefs cryftal, and, when they have brightnefs, they pafs for gems; though, on account of their inferior hardnefs, they are not fo highly valued, nor indeed have they fo much brilliancy.

There are alfo filiceous cryftals which have a dufky or brown colour, and yet have great tranfparency and brightnefs. Such are thofe found in the mountains of Arran, and fome other mountains of this country. And fome of thefe brown cryftals, by being heated equally and cautioufly, to a degree that is fhort of ignition, lofe the brown colour, and retain a yellow one, which may be loft alfo if the ftone be heated too much. I knew a perfon who made profit by collecting thefe brown cryftals, and converting them into yellow ones. Thefe convertible brown cryftals are found in fome mountains to the north-weft of Aberdeen.

The pellucid gems may therefore be thus enumerated :

1*mo*, SILICEOUS CRYSTALS, colourlefs or coloured.

2*do*, The EMERALD, of a bright green colour, of which there are many varieties. It is much allied to fchoerl ; and fome of the pale emeralds named *aqua marines* are true fchoerls.

3*tio*, The TOURMALIN, though it be not a beautiful ftone, being of a very dufky green, is however at prefent reckoned a gem. Its re-markable property is to be electrified by gentle heat alone, without being rubbed. It is alfo a fchoerl.

4*to*, GARNATS that have tranfparency and brightnefs are ufed as gems. They have all of them a rich red-colour, but differ very much in the deepnefs of this colour ; when it is very deep the ftone has lefs brightnefs. The very bright garnats in which the colour is lefs in-tenfe are named *fcarlats*.

5*to*, The AMETHYST, of a bright and beautiful purple.

6*to*, The SAPPHIR, of a perfect and rich blue.

7*mo*, The TOPAZ, of a beautiful and brilliant yellow.

8*vo*, The RUBY, which has a red colour of great richnefs and bright-nefs.

9*no*, The DIAMOND the moft valuable of all. It is fometimes flight-ly coloured, and even dufky and opaque ; but the moft colourlefs and brilliant are the moft highly valued. The diamond is remarkable for four particulars :

1. It is one of the rareft productions of nature. We are not cer-tain that it is to be found any where except in the peninfula of India, and in the Brazils. In both places it is always found in detached fo-

litary cryftals, which are octaëdral, or often prifmatic, terminated by a pyramid. Thefe have fometimes fmaller cryftals as it were growing out of them; but we have never met with the diamond in groups adhering to a fhapelefs bafe, as is the cafe in all other cryftals. They have never been met with in the cavities of other ftones; nor do we know any kind of rock or vein which may be faid to be the matrix of diamond. All that have been hitherto feen are found in loofe gravelly earth at no very great depth. This is wafhed and picked for the diamonds it may chance to contain. There are to be met with, in the cabinets of the curious, a very fmall number of diamonds terminated at both ends with pyramids. This is, indeed, a very rare cryftal of any kind. From the circumftance of one end wanting a pyramid, we are entitled to fuppofe, that, like other cryftals, they have originally adhered to fome matrix. But, as has been faid, they have never been met with, except among what may be called the rubbifh of fome former condition of things. Several fpecimens have been found wholly or partly incrufted with an opaque whitifh matter of great hardnefs, but yet fo foft as to have all its afperities rounded by the attrition which the pebble has received. The jewellers call this mother of diamond, and, I think, with fome propriety,—for in fome that I have feen, the tranfition from diamond to this fubftance is not abrupt, but gradual, as in the cafe of the onyx. It was in the courfe of this century that diamonds were difcovered in Brazil; and their price is now greatly fallen.

2. Diamond is no lefs eminently diftinguifhed by its extraordinary brilliancy and great refractive and difperfive power,—far exceeding all other natural cryftals. Mr. Zeiher, a German artift, has approached very near to the brilliancy, and to the refractive and difperfive power of the diamond, in fome of his compofitions for imitating precious ftones. But thefe fall infinitely fhort of the diamond in its next remarkable property, viz.

3. Its fuperior hardnefs. In this it exceeds all fubftances that are

known. It is therefore employed for fawing and boring the hardeft ftones, and for engraving feals. For this purpofe it is reduced to powder.

This powder is ufed in the fame manner as fand, emery, or other cutting powders, are employed by the lapidaries, viz. by moiftening it with empyreumatic oil, and applying it to their wheel or drill; or the metal point ufed by the feal engravers, like a crayon. The metal employed for this purpofe is the pureft and fofteft iron. This takes faft hold of the diamond powder,—it being preffed into it by the force exerted in the operation : And the little particle never quits its place to roll about, but is carried along by the tool, acting like the tooth of a file, and tearing up whatever it touches. There is fomething curious in the way in which diamond acts in cutting common glafs. The glazier's diamond is by no means fharp pointed. I have feen them as blunt and round and fmooth as the head of a large pin ; yet this, with a moderate preffure, caufes the plate to fhiver under it wherever it is drawn. This is by no means a crack ; but the glafs being made fomewhat weaker there, fplits immediately, when gently patted on the other fide with a hard body.

4. The diamond is ftill more diftinguifhed from the reft of the gems by its chemical properties, or the effects produced on it by heat and mixture with other bodies. The utmoft violence of heat, even that of a burning mirror, does not induce the leaft appearance of fufion. Its afperities are not perceptibly rounded, as has been obferved in the ruby. The only effect of fimple heat is, to diffipate fome foulneffes which fometimes taint its purity, or water, as it is called by the jewellers ; but it makes no change in its texture, if the diamond has been free from previous cracks or flaws.

Nor does the utmoft violence of fimple heat volatilize the diamond. If it be protected from the action of the air, it fuffers no diminution of its weight by the longeft continuance in the fire. But if it be fubjected to intenfe heat, and to a current of free air, it will be entirely diffipated, and this diffipation is accompanied by the emiffion of a dazzling white light. This was long thought to be a phofphorefcence, as in the cafe of many other cryftals and fpars; and the diffipation was thought to be an exfoliation and difperfion, like the decrepitation of falts. But if this were the cafe, the difperfed duft might be found. As this has never happened, chemifts began to conclude that the emiffion of light was an inflammation, and that the diamond was wafted by combuftion like a piece of charcoal. The great rarity of diamonds, and the inability of moft fpeculative chemifts to afford the expence of the neceffary trials, has retarded the decifion of this queftion. But it is now paft a doubt that the diamond is a pure inflammable fubftance. I fhall therefore defer the confideration of its combuftion and its chemical relations till I come to the confideration of that clafs of inflammables in which it muft be placed.

Of the Corune, or Corundum.

After this notice of the gems, it may be proper to mention a fingular foffil, which, though not a gem, (for it is totally deftitute of beauty) is yet allied to fome of the gems by its conftituent parts, and by fome of its properties. This is the CORUNDUM, or, as it is called by the Britifh lapidaries, the *adamantine fpar*. The firft fpecimens of it came from China to Britain and it was afterwards found that it was to be got in India, not far from Bombay; and our naturalifts have now found it in feveral granites of France and Spain. It is of a grey colour like emery. The entire pieces are opaque, but very thin plates of it have confiderable tranfparency. What comes from China has very evident hexagonal cryftals; but all other fpecimens that I have feen are of an indeterminate granulated ftructure. What comes from Bombay is more unlike to emery, and confiderably whiter than all others; and it is this which is called corundum by the natives.

The remarkable quality of corundum, and for which it is chiefly valued, is its extreme hardnefs.   It fcratches every fubftance but diamond ; and is therefore of great value to the lapidaries and feal engravers of this country, who employ it by the name of adamantine fpar.   It is but a little harder than the ruby, the fapphir, or the oriental topaz.   It is far fuperior to emery, particularly for grinding on the wheel, to which it adheres like diamond duft.   One part of it does as much work as four of emery, and does it in half the time.

The corundum unites with cauftic foda, but with great difficulty. When the liquor filicum formed with it is decompofed by acids, it is found to confift of one-third of a peculiar earth, and two-thirds of clay.

———————

Many of the precious ftones are imitated by art.   The marbles, and porphyries, and jafpers, by means of paftes compofed of lime and gypfum, with other materials added to increafe the hardnefs and fmoothnefs, or to give the colours.   The art of making thefe imitations has been carried to a great degree of perfection.   Large columns, pilafters, pannels, tables, mouldings, feftoons, and other embellifhments of the infide of buildings, have been executed very like to the more beautiful and coftly kinds of marble and porphyry.   And thefe imitations have the advantage of being procured at very little expence compared with the originals.   But they are only fit for the infide of houfes : The materials do not withftand the weather.

The pellucid gems are imitated with glafs, made with great care and in fmall quantities for this purpofe alone ; the very pureft materials being employed, and certain additions being made to improve the tranfparency and brightnefs of it, fuch as borax, and fome of the metallic fubftances ; fome of which alfo give to the glafs the different colours of the gems.   The moft celebrated authors who have written on this art are Neri, Merret, and Kunkel, whofe works appeared firft in the German language, but have been collected together and tranflated into French, with the title of *Art de la Verrerie*,—the art of glafs-making.

Thefe artificial gems, or paftes as they are called by our jewellers, are very good imitations in fome refpects, and reprefent the beauty and brightnefs of the natural productions very nearly, but they cannot be compared with them in hardnefs and durability.

Setting afide, however, the attempt to imitate the natural gems, we may venture to fay that art has in fome refpects far excelled nature in the productions it has obtained by working on the earthy fubftances. We have examples of this in the manufacture of the common kinds of glafs, and in that of porcelain, and of the other beautiful and ufeful kinds of pottery.

The common kinds of glafs muft no doubt be allowed to be far inferior in brightnefs, as well as in hardnefs and durability, to fome of the tranfparent ftony bodies produced by nature. But the inferior hardnefs of glafs is a great advantage in fome refpects, by rendering it more eafily cut, engraved, and polifhed, and thereby making it a fitter fubject for manufactures. And the facility with which we can procure large maffes and quantities of it, and the foftnefs and ductility which it receives from heat, in confequence of which it is fo eafily formed into any fhapes we defire, render it a moft ufeful production, and raife its intrinfic value far above that of the gems, which are produced by nature in fuch very fmall quantities, and are fo difficultly wrought into any form, that they are rarely applied to any other purpofe than the gratification of weak-minded vanity.

Of Porcelain.

Porcelain is another production of art which has been invented upon principle, by making chemical experiments upon the earthy and ftony fubftances, and which deferves to be admired for the elegance and ufefulnefs of its productions.

The art of making porcelain was long confined to the eastern parts of Asia. It is now underflood and practiced in several parts of Europe, especially Dresden, Paris, and some other places in France, and in several towns in England. It is a branch of the art of pottery, brought to an extraordinary degree of perfection, so as to produce vessels and other pieces of work, not only very elegant, and capable of high decoration, but which have the qualities the most desirable in vessels of earthen ware intended for the purposes of common life. The qualities I mean are, 1st, Great compactness of texture, by which it is both stronger, and made impenetrable and untainted by any thing put into it; whereas all the soft pottery is comparatively much weaker, and is moreover bibulous, and indelibly impregnated with almost every thing that we put into it; 2dly, A smoothness of surface which gives the vessels beauty, and makes them easily cleaned; 3dly, A texture which allows considerable alterations of heat and cold suddenly applied without cracking, as glass does.

We cannot tell by what steps this art arose to perfection so early in China and Japan, from whence the art was imported into Europe about fifty years ago. Du Halde has given a faithful account of the Chinese manner, as far as he underflood it; but unluckily Du Halde was but a sorry chemist.

Perhaps the circumstance which gave origin to it in those countries was the abundance of proper materials, which are but rarely found in Europe. The materials of which the best kinds of porcelain are composed are two: The first is a clay of the best and whitest kind, called by the Chinese *kaolin*; the other is a stone, which they are under a necessity of reducing to a fine powder, to prepare it for mixing with the clay in different proportions. This they call *petuntse*. This stone is no other than either pure feldt spar, or, in want of it, a species of granite abounding with feldt spar, and quite free from any admixture of iron, or other dark coloured matter, which might spoil the whiteness of the porcelain. The effect of this stone, when added in moderate quantity to the clay, is to make it affume the greater degree of compactnefs in a moderate fire, and to give it femitranfparency, without communicating too great a degree of fufibility. The Chinefe call it the *flefh*, and the clay the *bones*. A granite, quite free from iron or other colouring matter, is rare in Europe; but there are places where fuch are found, particularly in Cornwall, where there is a great variety of granite called *moor-ftone*. And a company fome time ago obtained a patent for the eftablifhment of a porcelain manufactory of the beft quality, in England, with intention to ufe that Cornifh moor-ftone.

Though thefe two ingredients are in general the conftituent parts of the beft porcelain, it has not been always neceffary to mix them artificially together. It has happened both in China and in England that nature has in fome places performed this part of the work, or furnifhed materials which do not need admixture. In fome parts of Cornwall, where the granite or moor-ftone abounds, they find a white clay mixed with the fandy rubbifh of the moor-ftone in fuch proportions that, without any other trouble than that of feparating it by water from the coarfer particles, it makes excellent porcelain.

A confiderable mafs of Chinefe porcelain clay was brought to this country fome time ago, and worked up by our beft artifts. It makes perfect porcelain, but of that kind called *ftone china*, viz. lefs tranfparent than common. To give it the degree of femitranfparency expected in common porcelain, it is neceffary to add fome of the *petuntfé*. But it may alfo be ufed alone, and then forms *ftone china*, or *Nankeen china*. This kind of clay is manifeftly produced by the gradual decay of the granite into rubbifh and powder, during which decay it appears to lofe fome degree of its fufibility. We have proof of its origin by microfcopic examination, by which we find in it the mica and quartz, and other materials of the granite. Magnefia, or the earthy bodies which contain it, are alfo excellent ingredients in the compofition of porcelain. There is fome of it contained in feldt fpar and the granite, &c. Some of the porcelain manufactories in England employ the *soap-stone*, or *steatites*, in their compofitions with good effect. The good qualities of this earth, are, 1*st*, To increafe or improve the whitenefs of the porcelain; 2*dly*, To preferve it from foftening too much, or melting by heat.

The clay being worked into the intended form, is dried in the air, and then put into the kiln, and baked with a moderate heat, and for a fhort time. This brings it into the ftate of our foft potteries. It is then glazed and painted. The glazing is made by dipping it into a cream of powdered flint, with the addition of a little borax or barilla and other colourlefs materials, to make it take a thin fufion. Before dipping the piece, the blue colours have commonly been put on, becaufe this colour is too apt to diffufe or run. The other colours are penciled on after it has been dipped. Many of the more ordinary European porcelains are covered with an opaque white varnifh or enamel; but this is to cover ill-coloured clay, and may be known by its thicknefs and foftnefs. The fine porcelain glaze is perfectly tranfparent, and as thin as a film of water. The piece is now baked again with a much greater heat, and for a much longer time; the particles now take a new cohefion and arrangement; and the piece becomes hard and compact, breaking with a dull luftre and fomewhat of a granulous texture: But in the very fineft porcelain no grain can be perceived, even with a microfcope; and the fracture has fcarcely any luftre, and refembles the fracture of white wax.

Thefe are, therefore, the materials of the beft kind of porcelain. But that we may not leave any part of this fubject untouched, it is neceffary to mention, that very near imitations of this kind of pottery may be made with materials different from thefe; and many of the manufactories which have been eftablifhed in Europe make ufe of thofe other materials, and produce only imitations of porcelain. They muft all employ a fine and white clay for one ingredient in the compofition. In this they neceffarily agree: But they have gone different ways in fearch of their fufible ingredient, which gives to the clay the proper degree of compactnefs and femitranfparency. As a granite or feldt fpar fufficiently free from irony particles occurs only in very few places of Europe, they have endeavoured to fupply its place by other fubftances. Some have employed different kinds of gypfum, which, being fufible in a violent heat, anfwers in fome meafure the purpofe of feldt fpar. Others employ fome of the alkaline earths, which, though not fufible themfelves, produce fufible mixtures with clay and flinty earth. But greater numbers have had recourfe to the ingredients of glafs made into what is called frit, ground to a fine powder. The quantity of alkali in this compofition is not fufficient to make it melt into a tranfparent and fluid glafs, but makes it approach a little to that ftate, or makes it femitranfparent. But thefe imitations are not porcelains. They are glaffes or enamels not completely vitrified; and a greater heat will melt them. Even in inferior heats they go out of fhape in the kiln by their own weight. Such ware cracks by any fudden change of heat. It is alfo foft externally, and foon fcratched and fullied, becaufe the ware will not ftand the heat neceffary for fufing a hard varnifh. Mr. Reaumur, of the French Academy, who firft examined the Chinefe materials, was alfo the firft who fuggefted this fubftitute of frit in place of petuntfé where it could not be procured; and upon the fame principle he attempted to change common glafs into porcelain. He was led into this project alfo by accidental experience, as has been already mentioned.

Another author, in purfuit of the fame object, fometime ago was induced to make a very great number of experiments, which have proved a confiderable addition to our knowledge on the fubject of earths. The author I now mean is the late Mr. Pott of Berlin, who publifhed thefe experiments in two fmall volumes, with the title of *Lithogeognofia.* But the nature of many of the earths and ftones was not fo well underftood at that time as it is now; and he confidered many compounded fubftances as fimple earths. His experiments, however, excited a curiofity in other chemifts to examine fome of the purer earthy bodies in this manner.

And there have since appeared, in the *Memoirs of Berlin*, papers, containing experiments of this kind, by Mr. Margraaf, and by Mr. Achard : And a multitude of experiments have been made in France, by Monf. d'Arcet, and by feveral others. Such experiments are interefting, as having a tendency to improve or illuftrate the arts of pottery, glafs-making, and metallurgy.

But no perfon has made, perhaps, fo many experiments on the mixture of different earths for the compofition of earthen ware and porcelain as Mr. Wedgewood. His object was to difcover paftes, or compofitions, which fhould have, not only the denfity and hard-nefs of porcelain, but different colours ; and he has fucceeded in finding fome of very agreeable colours,—fuch as what he calls his artificial jafpers, with which he executes beautiful pieces of work-manfhip, with figures in baffo relievo, of the fineft white porcelain, on a coloured ground. In this way he has imitated a famous vafe, that is reckoned one of the moft extraordinary examples of the ex-cellence of ancient workmanfhip,—the Barbarini vafe, now in the poffeffion of the Duke of Portland. Mr. Wedgewood viewed this piece of antique pottery with the higheft admiration, declaring that the artift deferved a thoufand guineas for his work. The work in relievo is executed with the moft perfect fharpnefs and precifion, and is an enamel altogether opaque ; fo that where it is not fo thick as the fineft poft paper, the colour of the ground does not appear through it,—an excellence which all his labours had not yet attained.

The true porcelains are now made in feveral parts of Europe in great perfection,—equalling the Afiatic in the beauty of the material, and even exceeding it in its moft valuable qualities, of ftrength and its abi-lity to withftand great and fudden heats without fplitting or foftening, and going out of fhape. Saxony produced the firft of fuperior qua-lity, but it is now excelled by the porcelain of Seve in France. The manufacture is daily extending, by the difcovery of more copious ma-terials ; but ftill, the Oriental is far cheaper than any European porce-lain of equal goodnefs.

APPENDIX.

OBSERVATIONS BY DOLOMIEU ON QUARTZ AND PRECIOUS STONES.

(Obſerv. de Phyſ. May 1792.*)*

1. QUARTZ, or cryſtal, phoſphoreſces when heated or ſtruck.

2. It alſo, when ſtruck, emits a peculiar odour. Theſe phenomena indicate inflammable matter.

3. It decrepitates when ſuddenly heated, and the more phoſphoreſcent it is it decrepitates the more.

4. It boils and bubbles very much when melted with the heat of vital air acting on charcoal; and is formed into globules, white or opaque, with air bubbles.

5. When melted with fixed alkalis, there is a great efferveſcence or boiling, although the alkali be cauſtic. (The author has alſo obſerved a flame come from this mixture; and ſo has alſo Mr. Pelletier.)

6. If there be enough of the fixed alkali, the compound diſſolves in water and forms the liquor ſilicum, which contains the ſiliceous earth, certainly different from what it was before. For, if we precipitate it with an acid, and inſtantly add ſuperfluous acid, it is rediſſolved by that acid.

7. The author had long ſuſpected that quartz contained ſome condenſed gas or elaſtic aëreal matter, and he at laſt made ſome experiments along with Mr. Pelletier. They put into an earthen retort of

twelve cubic inches capacity, ten drachms of levigated quartz, and two
ounces of cauſtic potaſh, freſh made and quite dry, and ſet the retort in
a reverberatory furnace, with a receiver and apparatus for elaſtic fluids,
which were confined in this proceſs by water.   1ſt, There came ſome
air, which he expected to be atmoſpherical air, from the cavity
of the retort, but after two or three cubical inches were come, all the
reſt extinguiſhed flame, and appeared to be azote, to the quantity of
22 cubic inches.   2d, A little ceſſation happened in the emiſſion of
elaſtic matter, and a tendency appeared to the abſorption or intropulſion of the water.   The fire was therefore increaſed ; and ſoon after,
that is, when the bottom of the retort began to be red, an elaſtic fluid
was again emitted, along with a good quantity of white watery vapours, and a white ſmoke that, while it riſes in bubbles through the
water, is not totally diſſolved or combined with it. This white ſmoke
diſappears after it has riſen out of the water into the bell.   During
this ſecond production of elaſtic matter, 12 cubic inches of it are obtained ; and there is a limit here too, marked by a ceſſation, attended
with ſtill greater danger of intropulſion of the water than in the firſt
inſtance. ·It muſt be avoided, either by letting in ſome air, or by increaſing the intenſity of the fire with bellows.   The nature of this
ſecond product is different from that of the former :   It is all inflammable air which explodes with atmoſpherical air, excepting a ſmall
portion of it which is mixed with air and gas azote.   3d, The third
product, which requires a very ſtrong heat to make it come, amounts
to 20 or 22 cubic inches :   About 16 of theſe are fixed air, which is
gradually abſorbed by the water,—the remaining five or ſix are a
mixture of inflammable and azotic gas, in which the laſt prevails.
After this, whatever heat be applied, no more elaſtic matter can be
obtained.   (The author obſerved ſigns as if ſome of this matter were
attracted and abſorbed by the water.)   They repeated this experiment
with levigated cryſtal of Madagaſcar, and cauſtic fixed alkali, which
was prepared by Mr. Pelletier, with all precautions to have it pure

and untainted with inflammable matter. And the procefs was twice performed with this cryftal : The firft time the heat was not raifed fo high as to bring over the third product of gafes, but the fecond was completed. The refiduum in the retorts, after the firft and laft experiment, was a white, vitrified, opaque, blown-up matter ; that of the fecond experiment was perfectly vitrified, and proved a greenifh glafs. But all thefe three were very diffolvable and even deliquefcent ; and were diffolved by water into a liquor filicum, which by reft depofited a fuliginous-like matter.

8. The author is of opinion that the inflammable and azotic gas come from the quartz, for thefe reafons : 1ft, The alkali is not decompounded or deftroyed, but can be feparated again entire from the filica by acids. But the quartz is certainly very much changed, for it is become foluble in all the acids, even the acetous, provided thefe are added in a fufficient quantity as foon as it is precipitated from the liquor filicum. The fulphuric acid fometimes rediffolves it fo quickly that the precipitation or feparation of it from the alkali is not perceptible ; and the fame thing happens with any other acid, if the folution of the liquor filicum be largely diluted. That there is a real folution here by the acid, becomes evident when we add enough of alkali to faturate the acid, for then the filiceous earth is precipitated again, efpecially when the alkali we ufe is a mild alkali, and moft certainly if the volatile alkali. For if we ufe a cauftic alkali, it is neceffary to take care that no more be added than juft enough. A little fuperfluity of it rediffolves the filica very quickly again ; and thus it can be rediffolved by acid or alkali as often as we pleafe.

The author believes that the inflammable and phlogifticated airs come from the quartz ; and that the boiling and blowing of it up into a fpongy glafs, when it is melted by the heat of vital air, is occafioned by the extrication of thofe airs from it ; that the fire or light, and the fmell of burnt air, produced by ftriking pieces of quartz againft one another, are to be imputed to the bafes of thofe airs in its compofition; and that the difference between quartz in its natural ftate, and quartz frefh precipitated from liquor filicum, is fomewhat analogous to that of quicklime and calcareous earth ; the frefh precipitated filica, in confequence of its purity and fimplicity, having an attraction for acids, which quartz has not. And as the quicklime recovers its fixed air, if left expofed, the author is of opinion that the precipitated filica can recover from water the principles which it loft when it was melted with the alkali ; for the precipitated filica continues foluble in acids for a very fhort time only after it is precipitated. If we obferve it while we are precipitating it, we can fee intermixed with it, and form- ing around the particles of it, numerous little bubbles of aëreal matter, which rife afterwards to the top of the water. And heat promotes and increafes this appearance, which is foon over, and then the filica is infoluble in acids, as much as quartz in its natural ftate. And if we now wafh and dry it, and melt it again with an alkali, it gives the fame effervefcence that it gave at the firft. It muft be acknowledged that inflammable air is not abforbed by newly precipitated filica ; but the author fuppofes that it is only the bafis of it which is prefent in quartz ; and that bafis, when once combined with caloric, is not very eafily feparated from it.

If thefe airs are fuppofed to come from the alkali, how can we ac- count for the effervefcence or intumefcence of pure quartz and cryftal when melted with the heat of vital air, or for the light and the fmell of burnt matter, which are produced by ftriking pieces of quartz againft one another ? The Chevalier de Lemanon, performing this ex- periment over white paper, and examining afterwards with a micro- fcope the minute duft or particles which had been rubbed off from the quartz, found among them, particles which had been fcorified, and which, when rubbed or bruifed on the paper, marked it like charcoal. (*Journ. de Phyf.* July 1785.)

The matter we find thus combined with the filica in quartz is proba- bly the matter which, when combined with water at a great depth be-

low the furface of the earth, enables it to diffolve quartz, and to tranf-
port it and to cryftallize it, which matter cannot probably remain
combined with water when it comes to the furface and is expofed to
light. Bergmann was deceived when he imagined fixed air combined with water to be a folvent of quartz: The experiments of other
chemifts have not fupported this opinion. I have not been able to
make it diffolve filica, even when frefh precipitated from liquor filicum.
Siliceous cryftals are often formed in nature on the furface of calcareous ones, without the fmalleft appearance of diffolution or corrofion
in thefe laft, which would certainly happen were the folvent of the
filiceous matter *carbonic acid water.* Mr. Morveau alfo miftakes
this water. (*Encyclopedie Methodique*). For, in an experiment in which
filiceous cryftals were formed by fhutting up in a veffel aërated water
and filiceous earth, there was iron alfo, which became corroded and
rufted, and the cryftals were found among the ruft of the iron. He
had four glaffes, containing aërated water and filex. Into one was
put filex; into another limeftone; into a third, argilla; and into the
fourth, iron. After nine months, no change appeared, except in
the laft, in which both the pieces of quartz and the iron were evidently corroded, and fmall cryftals found among the ruft of the iron. Dolomieu therefore concludes that the hydrogen, feparated by the action
of the diffolving iron on a fmall part of the water, produced, in conjunction with the reft of the water, a folvent which acted on the
quartz. Dolomieu learned, however, by experiments, that iron has
no effect on the liquor filicum, or the liquor on the iron. A bit of
polifhed iron preferves its brightnefs unimpaired in this fluid. That
the folvent of filica in nature is an inflammable fubftance, is alfo rendered probable by the dufky colour of fome cryftals and of flint,
which dark or dufky appearance is diffipated by fire. He adds one
more argument to fupport his opinion o the compounded nature of
quartz and cryftal. This is drawn from its inactivity, or want of attraction for moft other fubftances.

He then enters on the confideration of the gems, which have ftill more inactivity, and are infoluble even by alkali, though they are penetrable by it to a certain degree. He appears inclined to fuppofe that thefe are ftill more completely faturated than quartz, with the matter which he has difcovered in its compofition. And fome of the properties of the diamond may give room for fuppofing that it is the moft completely faturated of them all, and therefore the moft difficultly penetrable and diffolvable by the moft active folvents.

He confiders the argil, which is the moft abundant ingredient in many of the gems, as diffolvable by the fame folvent which is neceffary to quartz. The argillaceous earth fhews, by its odour when moiftened, and by many other particulars, that it has a difpofition to unite with inflammable matter.

He afterwards confiders the different gems, as differing not only by the number and proportion of earths they contain, but alfo by the clofe coalition of thofe earths, which has been produced by the action of their common folvent,—and fuppofes that the hardeft and brighteft of them retain a larger proportion of the remainder of this common folvent, or are more completely faturated with it, than the reft are.

But I refer you to the conclufion of his paper, which, I confefs, is not fo clear and diftinct in the preceding parts.

CLASS III.

─────────────

INFLAMMABLE SUBSTANCES.

THE third clafs of the objects of chemiftry, in the plan which I
have adopted, is that of the Inflammable or Combustible
Substances.

By the *inflammation* of a body, is meant a rapid deftruction and
change, which it fuffers when expofed to the action of heat and air at
the fame time ; which change is attended with the emiffion of a great
quantity of heat and light, and ends in a total lofs or privation of the
quality of inflammability.

When the general effects of heat were formerly explained to you,
fome notice was taken of the phenomena of inflammation, and of the
general nature of this clafs of bodies. And the opinions which formerly prevailed, as well as thofe which now prevail concerning the
nature of it, were briefly ftated, and have fince been more fully explained to you occafionally.

The opinion that is now the moft generally approved, had its rife
from the numerous inveftigations and experiments that have been made
fince the laft twenty or thirty years on the nature and properties of
the different elaftic fluids which are found in nature, or may be produced by art.

After I had difcovered the particular nature of the carbonic acid, and had fhewn that fome of it is produced by the action of air and burning fuel on one another, and alfo by the breathing of animals, I fuppofed that it was formed by the union of common air, with a quantity of the phlogifton of the chemifts, the exiftence of which was not doubted at that time. And I fuppofed that atmofpherical air had a ftrong tendency to unite with this principle, and to feparate it on many occafions from other bodies.

This opinion of a tendency in the atmofpherical air to unite itfelf with the fuppofed phlogifton, was afterwards adopted by Dr. Prieft-ley and others. But the Doctor did not admit the carbonic acid or fixed air to be produced in the manner I had fuppofed. He examin-ed with more care than I had done the change which the air under-goes in contributing to the inflammation of burning bodies and the breathing of animals, and thus difcovered the diftinction between car-bonic acid gas and azotic gas. This diftinction had been in fact clearly pointed out before, by my colleague Dr. Rutherford, in his in-augural differtation, printed in June 1772.

When atmofpheric air is completely vitiated by the breathing of animals, or the burning of fuel, we find in fuch vitiated air a much greater quantity of the azotic gas than of the carbonic. Dr. Prieftley was of opinion that the azotic gas was formed by the combination of the atmofpherical air with the phlogifton; he therefore named the the azotic gas *phlogifticated* air. The carbonic acid gas, he fuppofed, had exifted before, but was concealed in the atmofpherical air, or was intimately combined with it until phlogifton was added, which, uniting with the air, made it feparate from the carbonic acid gas. Or, as he expreffed it, the carbonic gas, or fixed air, was precipitated, or extri-cated, in confequence of the phlogiftication of the atmofpherical air.

So far had Dr. Prieftley proceeded, when the late Dr. Crawford of London employed his attention upon this fubject. By taking hold of the difcoveries of others, and making ufe too of experiments made.

by himfelf, he formed a new theory of inflammation, which he publifhed in his work on Animal Heat.

He made a number of experiments to learn what capacity for heat
different fubftances have when compared with one another. Thefe
experiments were made in the manner which I pointed out, by applying different bodies one to another unequally heated. A part of
the heat of one is communicated to the other, until they come to an
equilibrium or equal temperature of heat. While this happens, the
alteration of temperature in the one body is very different from that of
the other, although the heat which the one receives be precifely the
fame quantity which the other lofes.

This fhews, therefore, that different kinds of matter have different capacities for heat; that fome are more heated by the fame
quantity of heat than others, or that a fmaller quantity of the matter of heat is fufficient for raifing their temperature or thermometrical
heat by the fame number of degrees, (for this is the mark and the
only meafure of capacity).

Dr. Crawford, making a great number of experiments in this way,
with different materials, thought that he difcovered that thofe which
he fuppofed to contain the phlogifton in their compofition had lefs
capacity for heat than others, or required lefs of the matter of heat
to raife their temperature, and that in proportion as they contained the
more of the imaginary phlogifton, they had the lefs capacity for heat.

He therefore began to think that atmofpherical air, while it received this phlogifton from burning fuel, muft have its capacity
for heat diminifhed, and muft throw fome of its heat into the contiguous bodies; and that the increafe of heat which appears during
inflammation, might be this very heat extricated and expelled from
the air, and not from the burning body.

This led him to examine, by actual experiments, the capacity for
heat of the air in its different ftates. And he thought that he difcovered fuch a very great difference between the capacity of atmofpherical

or refpirable air, and that of the azotic and carbonic gafes, that he was able to fhew by calculations that this is fufficient to account for all the heat that appears during the inflammation of fuel.

Thefe are the general outlines of his theory of inflammation, which he has applied alfo to explain the heat maintained in the bodies of animals. For particulars I muft refer you to the firft and fecond editions of his treatife on this fubject, which contains experiments made with amazing labour and much ingenuity.

While Dr. Crawford was thus employed, another perfon in a diftant part of the world, had already formed for himfelf a very different theory of inflammation. This perfon was the late Mr. Scheele of Sweden, whom I have had frequent occafion to mention already in this courfe, as an eminent chemift and philofopher.

He engaged himfelf in an inquiry which had for its object the nature of inflammation, and how heat and light are produced by it : And he thought that he had difcovered how they are produced. He thought that he had reafon to conclude that heat and light are compounded fubftances, and that he could actually produce them, by combining together their conftituent ingredients; and that he could decompound them, by feparating thefe ingredients from one another.

Affiduoufly occupied in this refearch, he was one of the firft difcoverers of vital air, or oxygen gas. Dr. Prieftley alfo difcovered it about the fame time, having obtained it from nitric acid, and from other things, in fome of his numerous experiments. Mr. Lavoifier alfo difcovered it foon after, while he was employed in inveftigating the action of air on the metals.

But Scheele was the firft perfon who, from a number of ingenioufly contrived experiments, concluded by very fair reafoning, that atmofpherical air is a mixed fluid, compofed of about two parts of azotic gas, and one part of vital air or oxygen gas, along with a very fmall admixture of carbonic acid. The greater part of the carbonic acid gas

found in the air which has contributed to the burning of fuel, he fuppofed to be extricated from the fuel, moft kinds of which he fuppofed
contained this acid, or the bafis of it. But he obferved that there
are fome inflammable fubftances which do not contain any of this
bafis, and therefore do not communicate carbonic acid to the air
which contributes to their inflammation. Such are fulphur, phofphorus, and fome of the metallic fubftances.

Scheele, after making many experiments with atmofpherical air,
repeated them alfo with the oxygen gas, by burning fome of the moft
inflammable fubftances in limited quantities of it, which he had confined in clofe veffels; and he found that in fome of thefe experiments
the *whole* of it was expended, or difappeared during the bright and
violent inflammation which it occafioned. This he learned by opening the phials under water, after the experiment. The water was
pufhed in by the preffure of the atmofphere, and filled them quite
full. He therefore concluded that the oxygen gas had penetrated
through the glafs of the phial, and had efcaped in the forms of heat
and light, into which forms it had been changed by uniting with the
phlogifton of the inflammable matter. The heat, he fuppofed, was
produced by a leffer, and the light by a greater proportion of the phlogifton combined with it.

We muft not form a light opinion of Scheele's acutenefs and judgment from the extravagance, and I may almoft call it abfurdity, of
fuch a theory, as it appears to us at prefent. Scheele was one of the
moft judicious, as well as ingenious chemifts that ever lived; but in the
active time of his life, the exiftence of phlogifton was univerfally believed. It was fo firmly eftablifhed in the imagination of every chemift,
that it prefented itfelf to their minds on every occafion. It muft be
confeffed, however, that he may be blamed for a grofs overfight or
neglect, in omitting to weigh the phial and its contents before and
after the inflammation. Had he weighed it accurately, he would
have learned that the oxygen gas had not flown away through the

fides of the phial, but was ftill there, having only loft its elaftic aëreal form. Had he examined the weight of the acid into which the inflammable body was changed, he would have found that this acid matter contained the oxygen in a condenfed ftate ; the weight of it being equal to that of the inflammable body and of the oxygen gas taken together.

When he performed the fame experiment with atmofpherical air, no more than one-third or one-fourth of it was expended, or had difappeared. The remainder was become totally unfit to contribute to the inflammation of burning bodies, or to fupport the life of animals by refpiration. And when he examined it further, he found it to be either pure mephitic gas, in fome cafes, or a mixture of this gas with carbonic acid gas in others. He then, by the ufe of lime and water, feparated the carbonic acid gas when it was prefent, and he added to the mephitic gas as much pure oxygen as made up the whole to the original quantity of the atmofpherical air with which he had begun the experiment. He found this mixture to be exactly fimilar, in its powers and properties, to good atmofpherical air. He exhaufted it of its oxygen, by burning bodies in the fame manner as before. And he again renewed its powers in the fame way feveral times over ; and, as the quantity of the azotic gas always continued the fame, he concluded that atmofpherical air is effectual in promoting inflammation and fupporting the life of animals, only in confequence of its containing near a third part of its bulk of oxygen gas in its compofition ; and that it is only this part of the atmofpherical air that is capable of uniting with the phlogifton, and of being converted along with it into heat and light.

The moft furprifing and ingenious part of his treatife was the apparent facility with which he thought he could explain the various proceffes by which oxygen gas could be obtained in its feparate ftate. They are all proceffes in which fubftances, which were fuppofed by the chemifts to have a ftrong attraction for the phlogifton, were expofed to the action of heat. And he fuppofed that they decompounded the heat applied to them,—that they attracted the phlogifton, and difengaged the oxygen from it. And no perfon difcovered fo many ways to obtain the oxygen gas in a feparate ftate, or to extract it from fo many different fubftances, as Mr. Scheele did.

You muft perceive that all the attempts to explain combuftion that have been mentioned, agree in fuppofing or affuming the exiftence of a common or general principle of inflammability contained in all inflammable fubftances. It is further fuppofed that this is a moft fubtile kind of matter, and that it is feparated from fuch bodies with great rapidity during their inflammation, appearing then in its feparated ftate, according to the firft opinion of the chemifts, in the form of heat and light; or, according to the opinion which I had formed, and to that of Prieftley, Cavendifh, Crawford, Kirwan, and others, uniting, at leaft in part, with common air, and forming carbonic acid and azotic gas; or, thirdly, according to Scheele, uniting itfelf with the vital air of the atmofphere, and forming with it the heat and the light.

I muft now make you acquainted with fome other opinions which have been more lately imagined, and which are of a quite oppofite and contrary nature to all thofe I have yet explained. The principal author of the firft of thefe new opinions is Mr. Lavoifier, whom I have frequently had occafion to name as the author of many excellent experiments upon elaftic fluids, as the fubjects of chemical inveftigation.

Mr. Lavoifier was induced to form this opinion chiefly by a number of facts obferved in the inflammation of bodies. Some of thefe have been already touched on occafionally, but muft now be more particularly infifted on.

You will find the detail of thefe facts in a feries of differtations, publifhed in the *Memoirs of the Academy of Sciences*, particularly in the volumes 1781, 1782, 1785;—in his *Opufcles Chymiques* publifhed in 1777, and in his *Elements of Chemiftry*.

He remarks, that although it has been hitherto fuppofed by the chemifts that a fubtile kind of matter flies off from bodies, or is feparated from them, during their inflammation, no perfon has been able, either to exhibit this common fubftance by itfelf, or to fhew that the body, which was fuppofed to fuftain this lofs of matter, fuffered any diminution of its weight. Juft the reverfe appears, in the greateft number of experiments, when the inflammable body is of fuch a nature, that we can collect accurately together all the inflammable matter that is left after the inflammation is over. In fuch cafes, we always find that this matter exceeds in weight the inflammable body from which it was produced. This fact alone had long occafioned many to doubt of the exiftence of a principle of inflammability.

But further, among the numerous experiments which have been made of late upon different kinds of aëreal fluids, feveral have been made by Mr. Lavoifier, in which the inflammable bodies were expofed to the action of *meafured quantities* of air, in clofe veffels. They were burnt, in part, and nothing was loft or gained by the whole apparatus. This was weighed with moft fcrupulous accuracy before and after the inflammation. But the inflammable body was found to have gained a quantity of weight proportioned to the quantity that had been burnt. Moreover, the air was found to be diminifhed both in bulk and in weight. That a quantity had been abforbed by the burning body, or had fomehow difappeared, was evinced by opening the veffels under water. The water rufhed in, and occupied the room of the abforbed air. The fpecific gravity of the remainder being examined and compared with the diminution of room, it appeared that the air remaining alfo weighed lefs than the air before the inflammation; and, laftly, it was found that the lofs of weight in the air was exactly equal to the augmentation of weight in the remains of the inflammable body.

The moft fimple, elegant, and unexceptionable experiment to this purpofe, is that of F. Beccaria, of Turin. Two fmall glafs matraffes

were joined hermetically by the necks. One of them contained a
small quantity of an inflammable body which emitted no vapour in
burning. The rest of the space in both veſſels was filled with vital
air; and the veſſels were then ſealed up and carefully weighed. This
apparatus was accurately poiſed on an axis, ſo as to vibrate like a common balance which is in equilibrio. A burning-glaſs was now employed to kindle the body, and to keep up the combuſtion as long as poſſible. It was obſerved, that as ſoon as the combuſtion had proceeded a
very little way, that end of the balance which contained the burning
body began to preponderate. When the burning could be maintained
no longer by the action of the burning-glaſs, the balance remained in
a very oblique poſition, ſhewing a great addition of weight on the ſide
of the burning body. But, as the heat may be ſuppoſed to have expanded that arm of the balance, the whole was allowed to grow as
cold as at the firſt. It required about 13 grains to be laid on the other
end to reſtore the equilibrium.

Here, therefore, is an evident transference of matter from one end
of the apparatus to the other. For, when the apparatus was again
weighed, it was found of the ſame weight as at firſt. The veſſels
were now opened, and air ruſhed in: It was again weighed, and had
gained five grains. The aſhes or remains of the body were now
carefully collected and weighed. They were found ſeven grains
heavier than before.

Nothing can be conceived more convincing and unexceptionable
than this experiment, as a proof that, in the inflammation of this
body, it had united to itſelf part of the air contained in the two veſſels. In other experiments made by Mr. Lavoiſier and his copart-
ners, contrived for aſcertaining the preciſe quantities of air conſumed
or combined, and the weight gained, it was found that the one was
preciſely equal to the other.

It was chiefly on theſe facts that Mr. Lavoiſier founded his new
theory of inflammation and combuſtion. He was of opinion that
there was no ſuch thing as a principle of inflammability, the phlogiſ-

ton, affumed by the chemifts, nor any feparation of a fubtile principle from bodies, in the act of their inflammation. The very reverfe of this happens, fays he; the inflammable body fuftains no lofs, but receives a confiderable addition of matter, which is now ftrongly combined with it, and, during its combination, produces a total change in its nature and qualities, making it appear a fubftance of a quite different kind from what it was before.

The matter thus combined with the inflammable body is fuppofed by Mr. Lavoifier to be the bafis, or ponderable part of vital air. He confiders this air as a compound of this matter, and of the matter of heat, or *calorique*, which calorique is fo combined with the other matter as to give it the form of an elaftic fluid, not condenfable by cold, like the vapour of water, but requiring the application of fome *proper* fubftance, for which it has a ftronger attraction than for calorique. An inflammable body is a proper fubftance; but a certain high temperature is neceffary for enabling them to act on each other. The bafis of vital air then combines with the inflammable body; and the calorique is allowed to efcape, in the fame manner that fixed air is allowed to efcape, when a mild alkali combines with an acid. The heat thus let go is fufficient to enable the adjoining particles of the inflammable body to act upon, and decompofe more vital air, and befides, to heat all furrounding bodies.

This opinion of inflammation, and of the change which inflammable bodies undergo, was held by Mr. Lavoifier as proved by the moft convincing experiments. For 1ft, It is proved that vital air is abforbed during its action on inflammable fubftances; 2dly, Many of thofe fubftances which have been burnt, or have been expofed to the action of air and heat, fo as to fuffer a change fimilar to inflammation, can afterwards be made to afford, by means of heat, very confiderable quantities of vital air. For the proof of this, Mr. Lavoifier refers to thofe very experiments in which Dr. Scheele fuppofed that fuch bodies decompofed the heat applied to them. Lavoifier could not conceive that heat could be decompofed in any of our experiments ; and maintains that it acts fimply by expelling the vital air from fuch bodies in which it is contained, by furnifhing what is to be its latent heat when it is in its elaftic gafeous form.

Mr. Lavoifier further fays, that it has long been remarked, with refpect to inflammable fubftances, that the incombuftible matter into which they are changed during inflammation, is, in the greateft number of cafes, either an evident acid, or has the qualities and appearances of a matter which contains a quantity of acid combined with it, and which it had an opportunity of getting during the combuftion. This is eminently the cafe in the combuftion of fulphur, of phofphorus, of charcoal,—from which we obtain the fulphuric, the phofphoric, and the carbonic acids. An acid is obtainable from the afhes or calxes of fome metals deftroyed by fire and air; and all thefe calxes are fimilar to what the metals are changed into by actually combining them with a due portion of acknowledged acids.

Mr. Lavoifier, therefore, induced by thefe general facts, fuppofed that vital air is the general principle of acidity. Although it has not the properties of an acid itfelf, it forms acids of different kinds, by combination with inflammable bodies. Combined with fulphur, it forms the vitriolic acid,—with charcoal, it forms the carbonic acid, &c. &c. He therefore gave it the name of the oxygenous or acidifying principle. *(See Note 6. at the end of the Volume.)*

As to the heat and light which are emitted from thefe bodies in fuch quantity during their inflammation, or as Mr. Lavoifier views it, during their combination with the bafis of vital air, he fuppofes that it is extricated chiefly, or rather folely, I think, from this air; which, in its aëreal ftate, contains it in great quantity, in confequence both of what is neceffary, as latent heat, for its aëreal form, and alfo becaufe in that form it has a very great capacity for heat, requiring much of it to elevate its temperature any number of degrees.

This theory of Lavoifier is bold and ingenious : And, affuredly, it applies with great facility to explain very many of the facts which belong to this fubject.   We muft certainly admit as a thing proved by his experiments, that when bodies are inflamed, a great quantity of vital air is combined with them, and increafes their weight.   But there are many chemifts, and chemical philofophers, who, although they admit this as a fact fufficiently proved, are not yet fatisfied that nothing elfe happens in inflammation.   They ftill fufpect, or fuppofe, that the burning body fuftains the lofs of fome fubtile and active principle, (fuppofe it heat and light), and that it is the lofs of this principle which difpofes them to attract the air, and unite with it fo ftrongly as they are known to do.   For my own part, I was much difpofed to this opinion, on the general tenor of chemical combinations.   The uninflammable matter produced by combuftion is generally a much more active fubftance, or has an attraction for a greater number of bodies, than the inflammable fubftance had from which it came.   This is manifeft in fulphur and fulphuric acid, and many other inftances.   This difpofed me to confider thefe fubftances as in a ftate of greater fimplicity, when they were fo much more active on other fubftances.   But, when I confidered that inflammation cannot now be viewed as a decompofition alone, it being now proved that the inflammable body is, in fact, combined with a great quantity of vital air, we cannot fay that it is reduced to a ftate of greater fimplicity than before inflammation : For, admit that it has loft one principle, it muft be acknowledged that it has gained another, and therefore the obferved increafe of activity does not entitle us to fay that it is rendered more fimple. It is a new fubject, and has new relations ; and we really do not know whether thefe are or are not more numerous and clofe than before *.

* Dr. Black ufed to remark in fome of his courfes, that Mr. Lavoifier's fyftem did not explain the remarkable effect of light on bodies, and that Dr. Lubbock had given fome

The difficulties, therefore, and objections againft this theory, are now become fo few and of little weight, and the experiments which fupport it are fo numerous, direct, and conclufive, that it is gaining the afcendancy over all the others, and becoming the moft general opinion among the chemifts.

OBSERVATIONS BY THE EDITOR.

There were, however, fome points that prefented great difficulties, and almoft put a bar in the way to the confident adoption of the theory, in the extent in which it was propofed. For it muft be remarked, that Mr. Lavoifier's theory goes much farther than the mere explanation of the phenomenon of combuftion. He ftates the bafis of vital air as the principle of acidity,—therefore the combination of this principle with an inflammable body is equivalent, chemically fpeaking, with the burning of that body. The theory, therefore, embraces almoft the whole of chemiftry;—and combuftion, the moft remarkable phenomenon of material nature, and almoft characteriftic of chemiftry, is now but a fubordinate fact,—a particular mode of oxydation. But there were feveral effects of the vitriolic and muriatic acids which could not be explained by the theory in this its fimple form. Moft fortunately, fome experiments were made by Mr. Cavendifh at the very time while this theory was in its cradle, which opened a way out of all the difficulties which then embarraffed it. The difcovery of the compofition of water by Mr. Cavendifh in 1781, and fully demonftrated by him in June 1783, was carried

---

ufeful hints on this fubject, in his differtation *de principio forbili*, which alfo profeffes to be a theory of combuftion and acidification. But I do not fee that Lavoifier is any how bound to explain this phenomenon. Berthollet and others affect to affign its mode of action, which is always accompanied with feparation of oxygen. And they very frequently explain phenomena by fhewing that they are really inftances of this expulfion by means of light.

EDITOR.

to Paris by Mr. Blagden, fecretary of the Royal Society, and by him communicated to Mr. Lavoifier, who immediately repeated the experiments, and with great addrefs and ingenuity, applied the difcovery to his theory ; and not only got over the difficulties now mentioned, but by inverting the experiment, and refolving the water into its conftituent parts, he gave his principles an influence almoft unbounded, explaining almoft all the phenomena of active nature. It is here, much more than in the firft conception of the theory of combuftion, that the penetration, the inventive genius, and the found judgment of Mr. Lavoifier are moft confpicuous. The precife logic, to which he endeavoured always to adhere, would have preferved Lavoifier from many errors, into which *his followers*, in all parts of Europe, have frequently fallen,—mifled by precipitant and overweening notions of their own knowledge. The compofition and decompofition of water affords a mode of explanation fufceptible of fo many forms, according to the fancy and *the wifhes* of the employer, that there is fcarecely a phenomenon of which a fpecious explanation may not be given in more ways than one.

Dr. Black fays moft juftly, therefore, that fcience has caufe greatly to deplore the death of that eminent philofopher. He always exprefled a high opinion of Mr. Lavoifier's genius and found fenfe, but was much difpleafed with the authoritative manner in which the junto of chemifts at Paris announced every thing, treating all doubt or hefitation about the juftnefs of their opinions as marks of the want of common fenfe.

But, perhaps, Dr. Black was not a competent judge of the matter. In the courfe of his own difcoveries, he was fatisfied with the juftnefs of his view of the fubject ; and he found himfelf able to communicate his knowledge to his ftudents by means of very plain arguments, and the moft familiar and fimple experiments. He defpifed the parade of multiplying experiments and argument ; but he employed, with becoming acknowledgment of his obligation, the experiments furnifhed him by Mr. Watt and other friends, in further confirmation of his doctrines. Having fufficiently inftructed his ftudents, he had no farther care, and was contented with that reputation which he enjoyed without ftruggle, and which he was confcious of deferving.

But Mr. Lavoifier was in a very different fituation. He faw that he was about to operate a complete revolution through the whole extent of chemical fcience. He could not but forefee doubt and oppofition on all hands. Confident of victory (after his happy employment of Mr. Cavendifh's difcoveries), the profpect was very flattering. I may perhaps add to this the genius and character of his nation. This is fcarcely left in my choice,—for, almoft at the firft, the doctrines of Lavoifier were preached by the affociated chemifts as the fyftem of *French chemiftry*. Mr. Fourcroy, Monge, De Morveau, and others, repeatedly give it this name, with fome exultation. It was propagated as a public concern; and even propagated in the way in which that nation always choofes to act,—by addrefs, and with authority. Every thing pertaining to the fyftem was treated in council, and all the leading experiments were documented by committees of the academy of fciences. To accomplifh this purpofe more effectually, they publifhed the *Annales de Chymie* in concert, and they formed a new language, with the pretext indeed of improving fcience, but, in reality, that every thing might be forgotten which did not originate in France. A Swifs gentleman, affectionately attached to Dr. Black, was in Paris at the time, viz. 1787, and wrote to him in thefe words: " *L'objet qui occupe les chy-* " *miftes furtout à préfent, c'eft la nouvelle Nomenclature. Il paroit qu'on* " *veut par là donner le coup de grace au pauvre phlogiftique; quant a* " *l'air fixe il faut qu'elle devienne l'acide carbonique,*" &c. The writer had furely caught the patriotic flame, otherwife he would have recollected that it could not amufe his friend to learn that his difcovery, which had led the way, muft vanifh with the reft. The plan was the fame with that of Fabre d'Eglantine with his new calendar;

and the principle was that of Rabaud,—" *il faut tout detruire,—oui,—* " *tout detruire,—parce qu'il faut tout recréer* *."

Dr. Black difliked this way of proceeding, fo unlike fcience and philofophy. He difliked the avowed principle of the nomenclature, thinking it more likely to corrupt fcience than to promote it; and he began to write fome obfervations on it, but he foon defifted.

Some time after this, he had more reafon to be difpleafed, and even to be offended. Mr. Lavoifier faw that his theory of combuftion depended on the doctrine of latent heat, and was extremely anxious to obtain Dr. Black's acquiefcence. In the courfe of 1789, Dr. Black received two letters from the Marquis de Condorcet, full of refpect for his " *illuftre confrére,*" (Dr. Black having not long before been elected affocié etranger de l'Academie des Sciences). In October 1789, Mr. Lavoifier wrote to him in thefe words : " C'eft un des " plus zélés admirateurs de la profondeur de votre genie, et des im- " portantes revolutions que vos decouvertes ont occafionné dans la " chymie, qui profite de l'occafion du voyage de Mr. B. à Edin- " bourg," &c.

Learning, by the return of this gentleman, that Dr. Black thought well of his theory, and had introduced it into his lectures, he wrote to him again in July 14th 1790, as follows :

---

* It is not undeferving of remark, that not only does this principle or aim of the new nomenclature greatly refemble that of the new calendar, and the new meafures of France, but that alfo feveral of this chemical convention were alfo affiftants, officially, to Fabre d'Eglantine in his project. La Place was in a high department of public bufinefs. Monge was a minifter of ftate, and ere this, had figned the death-warrant of his fovereign. Meunier was a general officer. Morveau was a commiffary of the convention ; and perfe- cuted with the moft cruel virulence the noblefle of his province, who had twice paid his debts, and given him 24,000 livres to enable him to profecute his chemical inquiries. He was the chief agent in framing the nomenclature. Haffenfratz, the publifher of the nomenclature, and of the fymbols which he had contrived, was alfo high in office, and moft active in all the projects of Robefpierre. It is not, therefore, on light grounds that I have affigned the fame motive to the nomenclature and to the calendar.

EDITOR.

" J' apprends avec une joie inexprimable que vous voulez bien at-
" tacher quelque merite aux idées que j'ai profeffé le premier contre
" la doctrine du phlogiftique.   Plus confiant dans vos idées que dans
" les miennes propres, accoutumé a vous regarder comme mon maitre,
" j'etois en defiance contre moi même (credat Judæus Apella) tant
" que je me fuis ecarté, fans votre aveu, de la route que vous avez fi
" glorieufement fuivie.   Votre approbation, Monfieur, diffipe mes
" inquietudes, et me donne un nouveau courage.   Je ne ferai content
" jufqù a ce que les circonftances me permettent de vous aller porter
" moi même le temoignage de mon admiration, et de me ranger au
" nombre des vos difciples.   La revolution qui s'opere en France de-
" vant naturellement rendre inutile une partie de ceux attachés à
" l'ancien adminiftration, il eft poffible que je jouiffe du plaifir de la
" liberté, et le premier ufage que j'en ferai fera de voyager, et fur-
" tout en Angleterre, et à Edinbourg, pour vous y voir, pour vous en-
" tendre, et profiter de vos leçons et de vos confeils."

Dr. Black wrote him a very plain, candid, and unadorned letter in
anfwer, expreffing his acquiefcence in his fyftem.   Mr. Lavoifier
anfwers this by praifing in the higheft terms the elegance of the ftyle,
the profoundnefs of the philofophy, &c. &c. and begs leave to infert
the letter in the *Annales de Chymie.* Dr. Black, who had been in very
poor fpirits when he wrote that letter, and was much diffatisfied with
its feeblenefs, was difgufted with what he now conceived to be art-
ful flattery, and refufed to grant the requeft.   Yet his letter appeared
in that work before his refufal could reach Paris.

This wheedling, in order to fcrew out of Dr. Black an acquiefcence,
on which he put a high value for the influence which it would have
on the minds of others, was furely unworthy of Lavoifier. Dr. Black
was not only difgufted with the flattery, but ferioufly offended with
its infincerity; and with a fort of infult on his common fenfe, by the
fuppofition that he could be fo wheedled, by a man whofe publications
never expreffed the fmalleft deference for his opinions. For, by this

time, Dr. Black had read Mr. Lavoisier's *Elements of Chemistry*, and the various dissertations by him and Mr. De la Place, published in the *Memoirs of the Academy*. His name is not once mentioned, even in the dissertations on the measures of heat, where his doctrine of latent heat is delivered and employed as the result of Mr. Lavoisier's own meditations. Nor is he named in those passages of the earlier dissertations, where the characters and properties of fixed air, and of the mild and caustic alkalis, are treated of. All appears to be the train of Mr. Lavoisier's own thoughts, for which he was indebted to no man. Such inconsistency with the deference expressed in the above cited letters, provoked Dr. Black to such a degree, that he resumed his critique on the nomenclature, and began to express his dissatisfaction with some parts of the theory, and his utter disapprobation of the unscientific and bullying manner in which the French chemists were trying to force their system on the world. But, by this time, his health was become so delicate, that the least intensity of study not only fatigued him, but made him seriously ill, and forced him to give it up. I saw him but seldom at this time, being then in very bad health myself; but had this information from Dr. Hutton, who shared all his thoughts. It was at this time that he gave up his intention of making a considerable change in the arrangement of his lectures, and that he expressed himself, as I have related, at the end of the introduction to the particular doctrines of chemistry. But still, notwithstanding the contempt which he expressed for the folly of a man who had tried, by fulsome and insincere flattery, to obtain what he had given him unasked, by teaching all his doctrines, Dr. Black considered the death of Lavoisier as a great loss to the science *. He expected much from his penetration

---

* This ornament of France fell a sacrifice to the ambition of the very men whom he had associated with him in his labours and honours. They were all persons in office, or national representatives, and, in that character, gave their consent (to say the least of it) to his sentence of death. But he was rich, and loyal,—they were——and——

and found fenfe ; and he confidered him as the only perfon who could keep his followers right, by checking their precipitant manner of proceeding.

Profeffor Lichtenberg, of Gottingen, a man of extenfive and accurate knowledge in every department of natural fcience, gives an entertaining and inftructive account of the introduction of thefe doctrines into Germany. It is to be found in his preface to the edition 1794, of *Erxleben's Introduction to Natural Philofophy ;* as alfo in the *Literary Magazine* of Gotha. Great hefitation, doubt, and objections, were to be expected in Germany, the native foil of chemiftry, and the refort of all who wifhed to perfect themfelves in mineralogy. The new doctrines were even received with averfion and difguft. This, he fays, was chiefly owing to the character of the nation from whence they came. The Germans, who had been accuftomed to confider themfelves as the chemical teachers of Europe, could not bear to hear the opinions of their mafter, Stahl, treated with contempt ; to be told by Frenchmen, living among them for inftruction, that the principles of Stahl were fuch as no man could embrace who had a fpark of common fenfe ; to be told, in letters from France, that the principle of Stahl was a *mera qualitas ;* a *mera contemplatio,* a fancy of the brain, which difgraced any man who entertained it for a minute ; and to have it added, with faucy politenefs, *dulci requiefcat in pace !* But what moft provoked them, was the pitiful triumphs of victory in which the French chemifts indulged themfelves. He fays, that when the affociation had finifhed their experiments on the compofition and decompofition of water, which filled up all the gaps of the fyftem, they had a folemn meeting in Paris, in which Madame Lavoifier, in the habit of a prieftefs, burned on an altar Stahl's *Chemiæ dogmaticæ et Experimentalis Fundamenta,* folemn mufic playing a *requiem ;* and he remarks, that if Newton had been capable of fuch a childifh triumph over the vortices of Des Cartes, he could never be fuppofed the man who wrote the *Principia.* I might add, that if Newton or Black had fo exulted over

Des Cartes and Meyer, their countrymen would have concluded that they were out of their senses. But at Paris every thing becomes a mode, and must be *fêté*. Dr. Black's nice sense of propriety made the intriguing conduct and arrogant assumption of all merit by the French chemists extremely offensive to him, and has probably made him so minutely careful to place in full view all the labours and discoveries of the British and Swedish chemists, particularly those of Cavendish and Scheele, which supplied the great facts on which the ingenious doctrine of Lavoisier is established.—I flatter myself that this statement of facts, and these reflections, will not be thought improper or unimportant.

We shall now proceed to take a nearer view of the different kinds of the inflammable substances.

The most remarkable inflammable substances may be arranged under seven titles, which are these:

1. Inflammable air.
2. Phosphorus.
3. Sulphur.
4. Charcoal.
5. Spirit of wine.
6. Oils.
7. Bitumens.

Of these, the first four may be called SIMPLE INFLAMMABLES, because we have not not been able to resolve them into substances more simple. Ardent spirits, oils, and bitumens, are very easily resolvable, and are vastly complex.

I.—INFLAMMABLE AIR.

The first in this order, the *gas pingue* of Van Helmont, the substance which has been long known by the name of INFLAMMABLE AIR,

is certainly the moſt ſubtile and moſt highly inflammable of all the bodies that belong to this claſs. The peculiar phenomena and conſequences
of its inflammation, as well as what was obſerved in combining it with
other bodies, cauſed it to be conſidered by ſome authors, of the firſt
rank among chemiſts, as the true phlogiſton of the chemiſts, or as
totally made up of their principle of inflammability. This opinion of
it was formed by Mr. Cavendiſh, and by Mr. Kirwan, who publiſhed
his thoughts on this ſubject in the Philoſophical Tranſactions for the
year 1782; and, after that time, in a ſeparate volume. And Dr.
Prieſtley had nearly the ſame opinion of it. But, ſince that time,
Mr. Kirwan has abandoned the ſuppoſition of a phlogiſton in the inflammable ſubſtances.

This inflammable ſubſtance has been known a long time. Indeed,
it muſt have been known as early as any conſiderable progreſs was
made in the knowledge of nature and chemiſtry. Van Helmont
calls it *gas pingue.* It is not always preciſely the ſame. There are
great varieties of it, occaſioned by impurities or admixture, and ſome
of theſe are found in almoſt all that is gotten already formed by nature; but it may be alſo produced, or extricated from different ſub
ſtances artificially, by a variety of chemical operations.

It is frequently met with in mines, eſpecially in thoſe of coal, and
renders the working of them extremely dangerous. It is called by
the miners the *fire-damp,* or *wild-fire.* It is alſo extricated or produced from animal and vegetable ſubſtances, when theſe are decompounded and deſtroyed by fire or putrefaction. And hence it is that
there is a ſmall quantity of it mixed with the black mud of putrid
ditches and marſhes in ſummer, which mud is compoſed of the putrid remains of vegetables and animals. If a ſtick be thruſt down to
the bottom of ſuch ditches, and the mud be ſtirred, a number of air-
bubbles riſe to the ſurface of the water, and a candle being held near
it at the ſame time, the air will take fire, and give a momentary flaſh.
Or it may be collected in inverted glaſſes filled with the water, and may

be afterwards fired. This was firft difcovered by the celebrated Dr. Franklin *. The knowledge of that kind of it which occurs in coal mines muft have been as early as the art of mining coal.

In thofe coal mines which are infefted with it, it is obferved to iffue from crevices of the ftrata in the fubterranean chambers of the mine, in form of a vapour or aëreal fluid, which, mixing with the air of the mine, is faid to produce fome degree of mifty appearance in it. And, when it is not very abundant, it does not communicate any un-wholefome quality to that air.

The greateft danger attending it proceeds from its high degree of inflammability; the fmalleft flame, as that of a candle, being fufficient for firing the largeft quantities of it, with a violent explofion. And, unfortunately, in thofe places, no work can be done without artificial light. In fome mines they are obliged to work by the dull light pro-duced by a piece of flint rubbing againft the circumference of a fteel wheel, which is jagged like a file. When it happens unfortunately that confiderable quantities take fire, the inflammation of it is as rapid and violent as that of gunpowder; and it produces an explofion almoft like thunder, and which is attended with moft dreadful effects. In fome cafes, the whole works of the mine are demolifhed, and numbers of miners killed, and fometimes blown up, with heavy machinery, to a confiderable height in the atmofphere. Thefe fhocking effects are produced by the great expanfion which the flame occafions in the air and vapour that is mixed with it. The inflammable air itfelf does not expand in the act of inflammation, but, on the contrary, collapfes into very fmall bulk; but it is then mixed with the vapour of water, and with azotic gas, both of which it expands, and forces them along

---

* Van Helmont mentions this, and adds, " Stercoraceus flatus, per flammam can-
" delæ tranfmiffus, tranfvolando accenditur, ac flammam diverficolorem, inftar iridis
" exprimit." (De Flatibus, § 49.)

It is even produced by fome living plants. The *Dictamnus Fraixnella* emits it from its flowers in fuch abundance in a calm evening, that it may be fet on fire by a candle, nay, take fire of itfelf.

the long and narrow chambers of the mine, and sweeps every thing along with it, just as the firing of gunpowder accelerates a bullet along the barrel of a musket.

In some mines, in which the sources of this vapour are but scanty, and in which it requires a long time before a considerable quantity is collected, they preserve themselves from danger by firing it frequently, and therefore, by small quantity at a time. They observe the places in which it collects, which are always over their heads, or in the hollows of the roof of their subterraneous workings, it being the rarest or lightest of all fluids that we know. When pure, it is but one-fifteenth of the weight of an equal bulk of common air. At stated times they set fire to it with a candle fastened to the end of a long stick, or tied to the middle of a long string, the two ends of which are held and drawn by two men at a distance from the spot.

But there are some coal mines in which the sources of it are too abundant to be easily managed in this manner. In these they have recourse to another method by which it is constantly carried off and destroyed, namely, by having wooden trunks or pipes conducted along the roof of the workings, (so the chambers of a mine are named) with branches carefully leading to all the places where this air most copiously gathers. All these trunks meet at the bottom of a shaft, and from thence a great trunk is carried up, to the surface, in a corner of the shaft, where it enters into a small chamber having a tall chimney. A fire is kept in this chamber, and air is supplied to it only from this trunk. The warmth of the chamber and chimney produces a current, and thus the air is collected from all parts of the mine. When the inflammable air is very copious, it is said to burn at the top of the trunk, and produce heat enough, without any more fuel, for maintaining a continual flame and current.

There are several places on the Continent, particularly in Italy, where a vapour of this kind breaks out at the surface of the earth, and is liable to take fire, producing a lambent flame. In England there is an example or two, recorded in the Philofophical Tranfactions. In Perfia, there is a fmall diftrict, where the inhabitants collect it into one place by means of covered gutters, and there fetting fire to it, ufe it for dreffing their victuals, and even for lighting their huts. I doubt much that thefe are exaggerated accounts; for this extremely rare fluid gives fo little heat, that its flame fcarcely will fcorch the hand; and the light is proportionably feeble. When it rifes through the water of wells in bubbles, they burft at the furface; and when they are very copious, fo that while one is burning another rifes clofe by it, then a candle being applied, flame catches at the furface, and continues with a crackling noife. In the burning well at Chittagong, in India, it takes fire of itfelf after having been extinguifhed by dafh-ing pailfuls of water in it.

Befides thefe examples, where it is formed by nature, I faid that it is alfo very frequently produced from different fubftances artificially, by a variety of chemical operations.

The examples of this are too many to admit of my enumerating them all here *. I fhall for the prefent mention only one fimple pro-cefs by which this gas is produced in a high degree of purity. We need only to mix fome of the ftrong fulphuric acid with about eight times its weight of water, and throw into the mixture fome very clean iron filings, equal in weight to half of the fulphuric acid. The iron is diffolved with ebullition and heat, and a great quantity of this aëreal

---

* We muft, however, carefully diftinguifh the inflammable air of which Dr. Black is now fpeaking, from a vaft multitude of inflammable vapours which nature and art produce. Oils, vinous fpirits, and many other inflammable fubftances, can be changed into vapour by heat, and the vapours are inflammable, of courfe. But we are now fpeak-ing of a *peculiar fubftance*, which we have never been able to decompound, and therefore affume as a fimple fubftance. This, when pure, is always the fame, in whatever way we procure it;—the other inflammable and incondenfible airs are not fimple, but yield, by burning, carbonic acid gas.

EDITOR.

inflammable fubftance arifes from the folution, and may be collected in veffels filled with water, and inverted into a veffel of water.

This has been long known to the chemifts, and it was ufual with them, when they had occafion to diffolve iron in the fulphuric acid, to amufe themfelves by firing this vapour, to make it give explofions, or burn in different ways.

But no other experiments were made to inveftigate its nature in other refpects, until after I had made the experiments on quicklime, and on fixed air, which I have already defcribed to you.

I had then the curiofity to try whether this gas was attracted by alkaline fubftances, in the fame manner as fixed air is attracted by them; and I fatisfied myfelf that it was not attracted by them. I was alfo accuftomed to exhibit to my pupils the different manner in which it burns when pure or when mixed with air. When it iffues pure, in a continued ftream, from a pipe, a candle applied to the jet will kindle it, and it will continue to burn quietly as faft as it iffues; —but not till it is out of the pipe, becaufe it burns only where it is in contact with vital air. But, if the veffel from which it iffues contain vital or atmofpheric air alfo, mixed all over with the inflammable air, then a candle applied to the mouth will fire the whole in an inftant, with an explofion that will burft the veffel, if not very ftrong.

The Honourable Mr. Cavendifh, fome time after, (anno 1766) publifhed in the Philofophical Tranfactions, experiments on this and other kinds of air; by fome of which he afcertained with great ingenuity and exactnefs, the denfity of different airs or gafes, compared with that of common air. He found that the gas we are now fpeaking of has a furprifing degree of rarity. It weighs lefs than the onetenth part of the weight of an equal bulk of common air. Therefore, if you would keep it in an open veffel for any immediate experiment, you muft keep the mouth of the veffel down, and the bottom uppermoft. If you fire it in this fituation in a tall glafs, and immediately turn it up, the inflammable air rifes in a beautiful burning column.

As foon as this difcovery of its great levity was publifhed, it pointed out an obvious confequence, which immediately occurred to me,—that if a quantity of gas could be confined in a veffel, or other containing matter that was exceedingly thin and light, the gas and veffel together might form a mafs lighter than an equal bulk of common air, and which would rife in the atmofphere, as cork does in water. I therefore thought of providing a veffel or envelope for this purpofe; and that which firft occurred to me was either the allantois or the amnion of a calf, which I procured, but not getting it ready foon enough to fhew the experiment at the time I intended it, I did not exhibit it, but mentioned it in my lectures as a thing which might be found practicable, though I did not fee that it could be applied to any ufe. I did not imagine that the fame idea would have been improved to fuch a degree as it was afterwards in France. All the world has been amufed with their air-balloons, which they foon made of fuch a monftrous fize as to lift very heavy weights to a great height in the air *. Of thefe balloons two kinds have been made ufe of in their experiments.

Mr. Mongolfier, a paper-maker at Lyons, firft thought of making a balloon fo light as to float in the air, merely by burning fhreds of paper or ftraw under the mouth of a great globe of paper or thin linen. Having fucceeded, he made another fo large as even to carry up a confiderable weight; and, in order to keep it afloat, he hung under it a choffer, in which the fuel was renewed by the perfons who were carried up by it. He made them of 40 feet diameter, and 70 feet high, which carried fix or feven perfons.

When accounts of this contrivance reached Paris, the inhabitants of

---

* It is worthy of notice, that Dr. Mayhow, in his differtation *de fpiritu nitro-aëreo*, defcribes experiments which have certainly been made with oxygen gas. He alfo gives very plain hints of balloons filled with a fubftance much lighter than air, which he fays was known to him.

EDITOR.

which are keen for amufement, Mr. Charles, an intelligent chemift, immediately recollected the immenfe fuperiority of inflammable gas for a project of this kind, and readily found men of fortune and plea-fure to contribute to the great expence of fuch entertainment, which in feveral cafes amounted to more than 500l. His balloons were made of a thin but ftrong filk, made air-tight by varnifh, and they were filled with gas produced from iron diffolved in vitriolic acid,—but fo far from pure, that it was not more than feven times lighter than common air. Indeed, this is as much as fhould be reckoned on, becaufe the iron employed is generally rubbifh, rufty, and even caft iron, which produce much fixed air. His balloons were much fmaller than Mongolfier's, but rofe to a very great height, fometimes almoft three miles.

Since, in order merely to float, the weight of the balloon and air muft not exceed that of common air, there is a minimum of fize, de-pending on the weight of a fquare foot of the covering. The fmalleft balloon of thin poft paper, that will merely float, is feven inches in diameter. Strong poft paper muft not be lefs than thirteen inches,—oiled filk will float, if two feet ten inches,—and oiled linen, if four feet fix inches. An oiled linen balloon, of nineteen feet diameter, will lift a weight of 250 pounds a mile high; and other fizes will lift nearly in the proportion of the cubes of the diameters, when they are large.

Thefe experiments gave rife to a number of projects for perform-ing voyages with fuch machines; though it muft have appeared evi-dent to any perfon who underftood their nature, and confidered it well, that they were totally unfit to be applied to any fuch pur-pofe.

That air balloons cannot be applicable to the purpofe of making voyages, or of traverfing the air in any direction that we pleafe, is evident, by thefe reafons:

1ft, We cannot find a power that will be fufficient for moving such bulky masses through the air, and that can be lifted up by them. The force of one man, or of any number of men which the balloon can lift, is very far from sufficient for moving it with the requisite velocity for performing voyages, even in a perfectly calm air. And if there be the least of a contrary wind, they could not move it an inch in the proposed direction. All this is true, even supposing that the whole force of those men could be employed or exerted in order to move it. But how are we to employ or exert this whole force, when the balloon is suspended in mid air? Had the men some fixed immoveable body, towards which they could draw the balloon with ropes, or from which they could push it with poles, their whole force might be exerted. But they have nothing to push it from, except the empty air, (as it is styled by the poets) a fluid so rare, that it gives very little resistance. And the force of the men would be mostly expended in moving their own limbs and the instruments, whether oars or wings, with which they should attempt to beat or to push that air. When a boat is impelled through the waters by men and oars, the oars are applied to a medium, which, though fluid and yielding, yet gives incomparably more resistance and power to the stroke of the oar, than air. It is more than 800 times as dense as air. And besides, the boat is formed for moving through the water with the least resistance possible. The two machines cannot be compared together.

Some again have thought of the example of ships, which can be made to go in different directions with the same wind, by setting their sails obliquely to the wind in different positions. And they thought that some sort of sails might be applicable to a balloon. But here again the case is totally different. A ship has a hold of the water by her bottom and hull while she is impelled by the wind: And the form of the vessel disposes it to glide easily through the water with the prow foremost, but to be difficultly moved with the broad side. In consequence of this, it is made to sail in many different directions with the fame wind. The balloon has no hold of any thing whatever, but the air or the wind itfelf. It is, therefore, carried along by that air, as a feather would be, and has as little power to refift the action of the air on it as a feather has.

2dly, Befide all thefe, there is ftill another and an infuperable reafon againft the poffibility of ufing balloons to make voyages, and to command their motion and direction. They muft neceffarily be made of very thin materials, and of the moft flimfy conftruction, that they may be fufficiently light. Now, fuppofing we could apply a power to them that could move them with fufficient velocity, they could not fupport the impulfe and refiftance of the air againft them while they moved through it. They would immediately be torn in pieces.

In the common manner of ufing them, when no attempt is made to direct or modify their motion, they are fecure from this accident, being then carried along by the wind as a feather would be; fo that although they are moved fometimes with great velocity, they are not expofed to any violent impulfe of the air on any one fide of them more than another. They move as faft as the air itfelf, and the perfons who are in them feel as if they were becalmed.

3dly, They cannot remain fufpended in the atmofphere for any confiderable time.

4thly, Another inconvenience, and even danger, to which balloons are expofed, is a whirlwind, which they fometimes raife, and by which fome of them have been agitated, with great danger to the aëronauts.

The only ufe to which they have been found applicable, is one which occurred to me, and which I have occafionally mentioned ever fince they were contrived, viz. for reconnoitering the pofition and ftrength of an enemy's encampment, and pofts, in the art of war. It has been common to afcend fteeples and rifing grounds for this

INFLAMMABLE AIR.

purpofe; but in calm weather, a balloon, fecured with a rope, can eafily be made to rife to many times the height of a fteeple.

Let us now return to the examination of inflammable air.

Another remarkable property of inflammable air, is its high degree of inflammability *.

The flame of this gas, in its unconfined ftate in common air, is extremely weak, as is reafonable to expect from a fluid fo very rare, that a cubic foot weighs only 37 grains. The beft way of obferving it in this fituation, is to blow up a foap bubble with it, and fire it with a candle. A more fhowy way is to fill with it a tall glafs, having a movable bottom. It muft be held with the bottom uppermoft, otherwife the gas will foon difperfe by its great levity. Holding the glafs upright, and a candle a little way above it, (about twice the height of the glafs) remove the bottom fuddenly, and the gas, pufhed up by the air below, will rife and meet with the candle, and form a beautiful column of lambent flame, not hot enough to finge the fineft down. It burns with a brighter flame at the mouth of a pipe, becaufe it is denfer there, efpecially if ftrongly preffed out, by fqueezing the bladder which contains it.

* It catches fire by the fmalleft fpark. The moft trifling electric fpark is fufficient. But it feems as if the mere elevation of its temperature is not enough; for it is very difficult to fire it by paffing it through a red hot tube, or by blowing it againft a lump of ignited, but uninflamed matter. Even in the cafes where we fucceed, I am doubtful whether it is not by a fpark of inflammable matter in the act of compofition, that it is kindled. For I obferve, that when the pureft that I could obtain, by diffolving a metal in muriatic acid, is kindled at the end of a tube, from which it iffues in a ftream, the dull flame with which it burns has in the very middle a continual train of brighter ruddy fparks, which not only rufh ftraight forward, but frequently fplit and dart fideways, with a momentary brilliancy, like the fparks of brandifhing iron. I fufpect that thefe are fcorifications and explofions. Thefe may be effected by the contact of a red hot tube, and will fire the gas. We know, without being able to explain it, that the action of the electrical fpark is of the fame kind. It feems to be only a comparative fecurity which the light from flint and fteel gives to the miners. Yet I confefs that I have not been able to fire it in this way. EDITOR.

It burns with more vivacity in vital air, but the difference is not fo remarkable as one fhould expect. Yet, on reflection, we muft be fenfible that fo fmall a quantity of inflammable matter muft be completely and inftantaneoufly inflamed, even in common air, with which it readily mixes.

The examination of the phenomena which accompany the inflammation of it in vital air has been productive of fuch extenfive confequences, and has fo enlarged our knowledge of the chemical operations of nature, that I think it neceffary to give you a hiftorical account of it, and of the contributions which different authors have made by its means to the general ftock of knowledge.

Dr. Prieftley's examination of the gafeous fluids had exhibited a great number of concrete fubftances brought into an aëreal form, contrary to all our former conceptions of things. But this particular gas occafioned a difcovery ftill more unlooked for. A fubftance which had been confidered as an element, not only in the very dawn of natural philofophy, but which had maintained the character undifputed by the moft acute and penetrating chemifts of this moft inquifitive age, is now found to be a compounded fubftance, and its ingredients fairly put into our hands.

Dr. Prieftley was occupied with the examination of inflammable air, and tried the effect of almoft every fubftance on it : He alfo tried the effect of the electrical fhock and fpark. As he expected, he found that it was expanded by it, but could not be inflamed by it in clofe veffels, unlefs mixed with common air. In this ftate it fired with a violent explofion. He was particularly furprifed at the great diminution of bulk,—finding that a mixture of one part of inflammable air, and two of common air, might be made to contract into half the bulk, and that it was now phlogifticated air. Having already difcovered the vital air, he fired a mixture of thefe, and found that when two parts of inflammable air and one of vital air were exploded together, it collapfed into almoft nothing, or nearly the whole difappeared. Mr.

Warltire, who affifted in thefe experiments, obferved that the infide of the veffel in which the deflagration had been made, was always moiftened with dew. Dr. Prieftley naturally afcribed this to moifture, which probably adhered to the airs employed, as they were always produced in proceffes in which water in fome form or other was prefent. Thefe experiments were made about the year 1782.

My friend Mr. Watt had taken great intereft in thefe experiments of Dr. Prieftley's, and communicated his opinion concerning them to Mr. De Luc, in a letter dated April 1783. This letter is, in part, a tranfcript of one written fome months before to Dr. Prieftley, with a defire that it fhould be communicated to the Royal Society. In this he declares his opinion, that the water obferved in thefe experiments arofe from the combination of the two airs; and fays, that water is the compound of dephlogifticated or vital air, and inflammable air, deprived of their latent heats; and that dephlogifticated air is water deprived of its phlogifton (i. e. of the inflammable air) in an aëreal form, that is, faturated with the matter of light and heat. Dr. Prieftley did not communicate this to the Society, becaufe (he fays) fome experiments which he had made fince he faw Mr. Watt, were directly contrary to this opinion.

Dr. Prieftley's experiments excited the attention of the Honourable Mr. H. Cavendifh, and recalled to his mind his own obfervation of the moifture in the veffels in which he had exploded thefe two airs. Thefe experiments had been begun in the fummer 1781, and were continued from time to time, along with thofe by which he had difcovered the compofition of the nitrous acid. He immediately fet about repeating the explofion of dephlogifticated and inflammable airs by the electrical fpark; and in May 1783, he found that when fix parts by weight of pure dephlogifticated air were exploded with one of inflammable air, they difappeared entirely, and that the refult was a quantity of pure water, equal in weight to the airs employed. The utmoft care had been taken to free the airs made ufe of, by making them pafs through the dry muriat of lime. The veffel burft in feveral of his experiments, becaufe, in the inftant of explofion, the vapour of the produced water was expanded by the heat extricated from the airs. Much of this heat, to be fure, was expended in giving thefe the vaporous form, or fupplying it with latent heat. But the veffel was inftantaneoufly heated, fhewing that the heat contained in the two airs more than fufficed for this purpofe. Thefe experiments were publifhed in the Philofophical Tranfactions for 1784.

Such curious experiments, and fo interefting a refult, could not remain a fecret, had fuch a thing been intended. But there was no fuch intention. Mr. Blagden, fecretary of the Royal Society, went to Paris in June 1783, and communicated thefe experiments of Mr. Cavendifh to Mr. Lavoifier, and his affociates, De la Place, Meunier, Monge, &c. knowing that they were much interefted in their refult, which was fo intimately connected with the new theory which Mr. Lavoifier was then eftablifhing.

Accordingly, Mr. Lavoifier, who faw the immenfe confequence of this difcovery to his theory, immediately fet about repeating the experiment of compofing water by the combination of the two airs ; and in September 1783, with the affiftance of Mr. Meunier, effected the compofition in a way that admitted no doubt. Inftead of effecting it by the explofion of a few grains of gas, which is all that a manageable veffel of glafs can contain, he did it by admitting two fine ftreams, one of each gas, into a balloon, through two tubes leading from large magazines of gas, and having their points fo near to each other that the ftreams mixed immediately. The gafes were fupplied *in the due proportion* by regulated preffures of water on the gafes in the magazines. While the pipes were thus delivering the due proportion of gas, it was fired by an electric fpark, and the flame continued as long as the gafes continued to be fupplied. In this manner, very great quantities of gas were inflamed, fo that the unavoidable errors in the ultimate meafurement bore a very fmall proportion to the whole. At the end of the operation, there was commonly a remainder of carbonic acid and of azote, from which it is almoft impoffible to free the gafes completely.

The refult of this capital experiment was perfectly conformable to that of Mr. Cavendifh. When the exact proportion of gafes was attained, the refult was water flightly acidulated. This proportion was fourteen parts by weight of inflammable air, and eighty-fix of vital air.

Mr. Monge made a fimilar experiment at Mezieres, in June 1783, with the fame refult, and (he fays) without having heard of the experiments of Mr. Cavendifh, or thofe of Lavoifier and Meunier. The vital air employed weighed five ounces five drachms twelve grains; and it left thirty-five grains of water in the muriat of lime through which it paffed. The inflammable air weighed fix drachms thirty grains, and left forty-four grains of water. Therefore the quantities which really burned were—

	oz.	dr.	gr.
Vital air　-　-　-	5	4	49
Inflammable air　-　-		5	58
	6	2	35
Mixture remaining unburnt　-		6	24
	——	——	——
Quantity of gafes compounded　-	5	4	11
Water produced　-　-	5	4	41

This excefs of thirty grains muft be afcribed to errors in the eftimation and weighing of the different articles. The water was not perfectly pure, but contained five grains of nitrous acid in each ounce.

Some time after, in 1798, Mr. Seguin again repeated this experiment, expending 25582 cubic inches, or nearly two hogfheads of inflammable air, and 12457 of vital air. The firſt weighed 1039½ grains, and the ſecond 6210, amounting to 7249½ grains; and the water obtained amounted to 7245 grains, which is nearly 5944 grains Engliſh troy, about three-fourths of an Engliſh wine pint. No greater loſs than four grains, in an experiment of this kind, is an exactneſs of which one has no idea. I doubt not, however, as this relation is a formal report from the Academy of Sciences, but that the experiment was very accurately performed, and the reſult extremely ſatisfactory.

Another experiment ſtill was made by Le Fevre de Geneau, in which 35085 inches of oxygenous gas, and 74967 of inflammable gas were burned, weighing two pounds three ounces and ſixty-four grains, (French *poids de marc*), and two pounds three ounces and thirty-three grains of water were obtained, containing twenty-ſeven grains of nitrous acid. Still more trials were made by Von Hauch, employing 3000 and 5000 inches, alſo with 1600 and 3000. The aim of ſo many laborious and expenſive trials was, to hit the proportion of gaſes ſo exactly that the water ſhould be pure;—but it was always contaminated with nitrous acid. This valuable information, however, was obtained from it,—that it was indifferent what acids had been employed for obtaining either of the gaſes. *Nitrous acid only* was found in the water.

A very candid and intelligent account is given of all the diſcoveries relating to the inflammation of theſe two gaſes in the tranſlator's preface to the ſecond edition of Fourcroy's Chemiſtry.

Mr. Cavendiſh, the original author of theſe experiments, and of the doctrine deduced from them, concluded that water is a compounded ſubſtance, and that its conſtituent parts are vital air and phlogiſton, viz. inflammable air. For at that time Mr. Cavendiſh was of opinion that inflammable air is the true phlogiſton of the chemiſts.

But, although Mr. Cavendiſh is the undoubted author of this decifive experiment, and, with Mr. Watt, is alſo the author of the important doctrine of the compoſition of water, Mr. Lavoiſier has the ſtill greater merit of ſeeing this propoſition *in all its importance.* *This* incited him to undertake theſe laborious and expenſive experiments, which confirmed thoſe of Mr. Cavendiſh beyond a doubt ; and he had alſo the ſagacity to perceive *immediately*, that by means of this propoſition, he ſhould extricate his great ſyſtem from difficulties and objections which I think would otherwiſe have been unſurmountable, and even to convert them into ſtrong arguments in its favour, and make them the means of extending it to chemical facts, and to the great operations of nature, which ſeem otherwiſe inexplicable.

Thus excited, Mr. Lavoiſier was not contented with having demonſtrated that the exploſion of vital and inflammable airs produce pure water, but, in September 1783, with the aſſiſtance of Mr. de la Place, he confirmed this demonſtration of the compoſition of water by decompoſing it, and producing its ingredients in a ſeparate ſtate. I ſhall therefore relate to you Mr. Lavoiſier's experiments on this ſubject, and the manner in which he reaſoned from them.

His firſt experiment conſiſted in ſimply putting ſome clean filings of the pureſt iron into diſtilled water, which filled a jar ſtanding in the ſame water. After ſtanding ſome days, a quantity of pure inflammable air is found collected in the top of the jar, and the iron is found corroded and black,—and when carefully dried, weighs more than before, and if expoſed to heat in a retort, affords vital air. The iron, ſays he, attracts the oxygen of the water, and the inflammable air is ſet at liberty.

The ſecond experiment is more remarkable. A ſmall glaſs retort, having a very long neck, and holding an ounce or two of water, is ſo placed that this neck paſſes through a choffer of live coal, by which it is maintained red hot, while the water in the retort is made to boil gently. The vapour is collected and condenſed in a proper pneumatic apparatus. It is found to be pure water, and precifely equal in weight to the water boiled off.

But, having put into the neck of the retort 28 grains of pure charcoal, and repeated the diftillation, he found that the charcoal had vanifhed, and that the water collected in the receiver was $86\frac{4}{10}$ grains lefs than the water which had quitted the retort, which was $113\frac{7}{10}$ grains. But he found in the pneumatic veffel connected with the retort and receiver, 100 grains of carbonic acid, and 12 of inflammable air. Here, therefore, it appears that the carbone had attracted the oxygen of the water, and thus decompofed it. The quantity of the carbonic acid was determined by pafling the whole elaftic matter obtained through milk of lime, which abforbed it. Now 100 grains of carbonic acid had been long before proved, by his experiments, to contain 72 grains of oxygen. This quantity of oxygen, combined with $13\frac{7}{10}$ grains of inflammable air, would compofe $85\frac{7}{10}$ grains of water. There appeared, therefore, a deficiency of $1\frac{7}{10}$ grains of inflammable air.

In a third experiment, he placed, inftead of the charcoal, 274 grains of fine foft iron wire, loofely coiled. At the end of the operation, he found the iron changed into black iron fcales, fuch as are found to fly from iron in forging, and 85 grains heavier than before. In the receiver he found 15 grains of pure inflammable air; and there was a lofs of 100 grains of water. Now 85 grains of vital air, when united by inflammation with 15 grains of inflammable air, fhould compofe very nearly 100 grains of water.

The conclufion from thefe experiments is fo evident that it is needlefs to go through it minutely, I have not ufed the precife numbers which occurred in Mr. Lavoifier's experiments, but fuch numbers *in the very fame proportions* as enable you to make the calculations without any trouble. Mr. Lavoifier and his affociates give the name HYDROGEN to the ponderable bafis of inflammable air, (which they call HYDROGENOUS GAS), becaufe, when combined with oxygen, it compofes water.

Notwithftanding this body of evidence, fome chemifts, unable to relinquifh their habits of explaining every thing by phlogifton, have made many obfervations and objections to thefe experiments; and when they could not gainfay the facts, they endeavoured to explain them by different fuppofitions. But I confefs that I think their explanations infinitely embarraffed and hypothetical. There has appeared fince, long after the date of thefe experiments, a proof of a very different kind.

Meffrs. Van Trooftwyck and Dieman, of Haerlem, produced electric fparks under water by the difcharge of a fquare foot of coated glafs between two balls of gold. At every fpark a fmall bubble of air was formed between the balls, which afcended through the water, and occupied the upper part of the tube that contained it. The air thus collecting, caufed the furface of the water gradually to fubfide. At laft, it funk fo as to be below the uppermoft of the two gold balls. The next fpark, therefore, was not under water, but in the air. The very firft fpark that was made in thefe circumftances, exploded the whole air which had been collected. When this did not burft the tube, (which it fometimes did), the water immediately filled it to the very top, the whole air having vanifhed. Here is a decompofition, and fubfequent recompofition of water, which I fee no way of gainfaying *.

* If further proof be ftill wanted, the wonderful effects of galvanifm fupply them in abundance. Mr. Woolafton alfo has greatly improved the method of the Dutch philofophers, by availing himfelf of the well known property of fine points, by which they promote the transference of electricity; (a property happily explained by Mr. Cavendifh, in the 61ft volume of the Philofophical Tranfactions). By employing a very fmall fhred of fine gold-leaf, all coated with fealing-wax, except the very extremity formed by breaking it acrofs, he fo conftipated the ftream by which the electric fluid is fuppofed to flow in or out, that a very middling machine, without coated glafs, produces a continual decompofition.

EDITOR.

By the help of this difcovery of the compofition of water, Mr. Lavoifier eafily explains the production of inflammable air during the folution of iron in diluted fulphuric acid, which you may remember I
told you was the beft procefs for obtaining it pure.   While the fulphuric acid diffolves the iron, and divides it into parts inconceivably
minute, it puts it into a condition for more powerfully attracting the
oxygenous principle of the water.   It attracts fome of it, and thus
leaves difengaged hydrogen, which affumes the form of inflammable
air *.   Lavoifier remarks that we do not obtain inflammable air, but
fulphur, or fulphurous acid, when the fulphuric acid is concentrated.
It muft be largely diluted.   Here the diffolving iron, having plenty
of water to act upon, attracts oxygen more eafily from it than from
the acid.   It is certain that it unites with oxygen on this occafion,
for if we feparate it from the acid, we find it in fact combined with
oxygen, and can obtain this from it again.

Thus, you will now underftand the opinion which at prefent prevails concerning the nature of water, and of this highly inflammable
fubftance, and the confequences of inflaming it.

We may now further remark with regard to inflammable air, that it
is at prefent confidered as one of the fimple or elementary bodies in nature.   I mean, however, the *bafis* of it, called the *hydrogen* by the French
chemifts ; for the inflammable air itfelf, namely, *hydrogen gas*, is confidered as a compound of that bafis, and the matter of heat.   What appearance and properties that bafis would have, were it deprived of its latent heat and elaftic form, and quite feparated from all other matter, we

---

* I do not fee how its attraction for oxygen is increafed.   The iron is certainly, by
this doctrine, combined with oxygen derived from the fulphuric acid,—a compound of
oxygen and fulphur.   How can this increafe its attraction for oxygen already combined
with hydrogen ?   The firft action of the iron is generally fuppofed by the French chemifts to be exerted on the water ; and they fuppofe that the fulphuric acid acts only on
the compound already formed of iron and oxygen.   This may perhaps affift, by removing
the iron already faturated with the oxygen taken from the water ; but there is ftill a difficulty,—to be noticed afterwards.   EDITOR.

cannot tell; but it is fuppofed to be an elementary principle in the compofition of a great number of natural bodies; particularly, it is an ingredient in all animal and vegetable fubftances. The attention of chemifts was much directed to this object by fome moft ingenious experiments and reafonings of Mr. Lavoifier and Mr. Berthollet; and it is now pretty generally received as a thing fully demonftrated, " that nothing " is to be found in the bodies of plants and animals, except the four " kinds of gas which we have difcovered, namely, hydrogenous, car- " bonic, oxygenous, and azotic; and a fmall quantity of earthy and fa- " line matter." And thefe fubftances, hydrogen, carbon, oxygen, and azote, are fuppofed to exift in the plant or animal in various ftates of compofition, forming folids and fluids, in which thefe fimple fubftances do not exhibit their peculiar properties, by reafon of their compofition. The chemifts of the new fchool further hold, that the union by which thefe fubftances exift as ingredients in the diftinctive folids and fluids of organifed bodies, is but flight; fo that it readily gives way to changes of temperature, and to the vital functions of the plant or animal; and that this is the caufe of thofe fermentations and corruptions which we obferve in them all. By thefe mutual actions of the fenfible ingredients, thefe ultimate fimples change their partners, (fo to fpeak), and become ingredients of new compounds, uniting by pairs or triplets, in confequence of a change produced in their former attractions,—a change occafioned by a change of temperature, or by the living powers of the plant or animal.

When we maturely reflect on the fubject, we fee that the opinion has a great degree of probability, and that an inconceivable variety of appearances may fairly refult from this feemingly very fimple conftitution of things. The theory here aimed at is moft magnificent and comprehenfive, embracing almoft all the chemical phenomena of nature; and it even promifes fome introduction to the knowledge of thofe myfterious functions of vegetable and animal life, by which plants and animals affimilate, or convert into their own peculiar fubftances, the various materials which ferve them for food and nourifhment, fo that even air, and light, and heat, become part of their compofition. But it is at the fame time very evident, that the moft fcrupulous caution is neceffary in all our difquifitions on this fubject, and the utmoft moderation in our theories. The combinations of pairs and triplets, in a collection of five ingredients, are fo numerous, that it is in our power to bring out any ultimate compound that we pleafe, by properly felecting the order of fucceffion of their mutual actions.

I fhall give you one example of this manner of proceeding, which feems to meet with general approbation, and will give you a pretty clear notion of the kind of reafoning employed in the French fchool. This is the explanation given of the compofition and formation of the volatile alkali. *(See Note 7. at the end of the Volume.)*

All the volatile alkali that we know is produced from animal and vegetable fubftances, or fubftances derived from thefe, by the action of heat, or by putrefaction. Yet we cannot difcover it to be prefent in the fubftances from which it is thus obtained *(ex. gr.* in filk, animal jelly, &c.) before the action of great heat, or before putrefaction.

We have long had reafon to believe that volatile alkali is a compound fubftance, and contains inflammable matter. The deflagration of nitrous ammoniac, and of all the ammoniacal falts with nitre, put this paft doubt. Putrid fteams, when copioufly produced, are generally alkaline; they are always inflammable; and it is not till the putrefaction, accompanied by fuch fteams, has proceeded a certain length, that the alkaline fmell is perceived. It becomes gradually more fenfible and lefs fœtid.

Mr. Berthollet obferved that the mufcular fibre, when frefh, gave out a great quantity of azotic gas, if digefted with nitric acid; but if fo treated when in a putrid ftate, it gave none, but gave inflammable air and volatile alkali. He was induced by this to think that the ammonia, which appears in this procefs of putrefaction, arifes from a combination of the azotic gas and the inflammable air, in the inftant of their extrication from the compounds from which they were difengaged. He fuppofed that ammonia is compofed of inflammable air and azotic gas, or of the hydrogenous and azotic gafes.

This opinion appeared to Mr. Berthollet almoft demonftrated, by the effect produced on volatile alkali by the muriatic acid furcharged with oxygen. Although the liquid alkali was perfectly cauftic, the mixture produced a confiderable effervefcence. But the gas which efcaped was not carbonic acid, but pure azotic gas: This, being a fimple fubftance, muft have exifted in the materials. The acid contains none; it therefore made a part of the alkali. In the mean time, the acid became ordinary muriatic acid,—its redundant oxygen had difappeared,— it had combined with the hydrogen of the alkali, and formed water.

This opinion explains, and is confirmed by many obfervations of different chemifts before this conjecture of Berthollet; and, in the firft place, fome very remarkable experiments by Dr. Prieftley, in which he fhewed that inflammable air had the properties of other inflammable bodies, but with circumftances that were characteriftic. Thus, we know that minium, or red lead, is converted into lead by making it red hot, in contact with oils and other inflammable liquids and folids. In like manner, it is converted into lead, if heated by a burning-glafs, when expofed in a veffel filled with inflammable air. In this experiment, the infide of the veffel is covered with dew, which trickles down the fides, and proves to be pure water. The inflammable air is almoft completely abforbed.

But it is alfo converted into lead, if treated in the fame manner, in a veffel filled with pure cauftic volatile alkali. But in this experiment, there is a great remainder of unabforbed gas. Dr. Prieftley expected to find this nearly pure vital air,—the lead having (according to his theory) abforbed all the phlogifton of the alkali. He was aftonifhed to find it, on the contrary, highly phlogifticated; that is, nearly pure azotic gas.

Mr. Berthollet explains thefe two experiments in a very fatisfactory manner. Minium contains (as we fhall learn in due time) a quantity of oxygen. This, uniting with the hydrogen in the firft experiment, produces the water which Dr. Prieftley obferved, and the hydrogen difappeared. But, in the fecond experiment, the azote, which is a fimple fubftance, and muft have exifted in fome of the ingredients, but cannot be demonftrated in minium carefully prepared, muft have come from the ammonia, and muft have been one of its ingredients. The other is, in all probability, hydrogen; for it is a fimple fubftance, and it is yielded by putrid mufcle when digefted with nitric acid; which putrid mufcle alfo yields volatile alkali. It yields volatile alkali only when it ceafes to yield difengaged azote, the putrefactive procefs having combined it with the hydrogen.

This is the general theory, founded on a very few experiments indeed, but thefe abundantly fimple. They are not, perhaps, decifive; but this theory gathers ftrength by attending to a number of more complex facts, and taking along with us the two difcoveries of Mr. Cavendifh,—the compofition of water, and that of the nitrous acid, as propofitions fully demonftrated.

1. Our newfpapers inform us that the French chemifts procured faltpetre for the army, by blowing alkaline gas, and even putrid fteams, through red hot fubftances which readily yield oxygen. We know that fuch fteams yield both inflammable air and azotic gas. The laft of thefe feizes part of the oxygen prefented to it, and forms nitrous acid, while another part combines with the inflammable air, and compofes water, which dilutes the acid. It feems to be for fuch reafons that putrefcent fubftances are ufeful in nitre beds, and that the nitre firft obtained is frequently nitrous ammoniac.

We often find the older chemifts expreffing their furprife at the ftrong fmell of volatile alkali from the mixture of fubftances which contain none. Thus, iron or copper filings, when diffolved in ftrong nitrous acid, often emit the fmell of volatile alkali, inftead of the offenfive fmell generally emitted from this mixture. The metal may be fuppofed to decompofe the water; and the hydrogen, uniting with the azote (now redundant in the nitrous gas in confequence of the diffolution of fome metal by the acid) will form ammonia.

Dr. Auftin, who had early formed the fame opinion of volatile al-kali, narrates, in the Philofophical Tranfactions for 1788, feveral facts of the fame kind. The following is one of the moft fimple, and is very conftant:—Put a fmall quantity of iron filings, very pure and clean, into a phial containing azotic gas, having previoufly wetted the filings with pure water. The gas will be quickly abforbed, and figns of am-monia will appear: The fmell becomes very diftinct,—teft-paper be-comes green,—paper dipped in the folution of copper in muriatic acid, will become blue;—thefe are all marks of ammonia. All this will happen, but more flowly, in common air, which contains azote in abundance. The procefs is obvious: We know that the iron de-compofes the water which wets it,—the difengaged hydrogen combines with the azote. For the fame reafons, we fmell volatile alkali in a mixture of iron filings, fulphur, and water. It would feem, however, that in all thefe experiments one or both of the gafes muft be catched by the other in its nafcent ftate, in the very act of its extrication. Dr. Auftin could not combine them when already in the form of a gas. (See Note 8. at the end of the Volume.)

On the other hand, facts have frequently occurred which are beft explained by the decompofition of volatile alkali. Mr. Milner, by blowing akaline gas through red hot manganefe, which yields oxygen in great abundance, produced nitrous acid, as the French did. Alfo, the muriatic acid, taken in a ftate in which it is overcharged with oxygen, being blown through, or made to mix with alkaline gas, (ammonia) produces water and azotic gas. The oxygen feizes the hydrogen of the alkali, and forms water; and the azote is difengaged. In general, when ammonia is forced to bear a red heat, in contact with fub-ftances which readily yield oxygen, we obtain azote and water. If they ftrongly attract azote, we obtain inflammable air.

Such are the facts which feem to eftablith Mr. Berthollet's opinion of the compofition of volatile alkali. He imagines it to confift of one part of hydrogen, and four parts by weight of azotic gas nearly.

The union of inflammable with vital air came in our way in confidering the effect of heat on inflammable air. Its importance has occafioned me to take up a good deal of time with it. I now proceed to confider how inflammable air is affected by mixture with other fubftances.

It has little difpofition to mix with pure water. It does, however, mix in a fmall proportion, and gives it a very naufeous fmell. It is this that offends fo much when water is thrown on red hot coals. In its pure gafeous ftate, it is by no means offenfive, although rather unpleafant. I fpeak of the pure hydrogen gas, obtained from iron or zinc, diffolving in the mineral acid. This is diftinguifhable, by its great levity, from that obtained from vegetable and animal fubftances. When its weight is more than one-thirteenth of that of common air, it is impure, and has various fmells (all bad), according to the fources from which it was obtained, or the mixture it holds in folution. All of them are confiderably heavier than the one now under confideration. They generally emit a brighter flame. The inflammable air from marfhes is ufually called the *heavy inflammable air*. Another produced from charcoal has fome remarkable properties, particularly its operation on living animals who refpire it. · It is called *hydrocarbonat*, and will be noticed very foon. It is reafonable to doubt whether thefe inflammable gafes have any thing in common with hydrogen befides their inflammability. Yet all compofe water by deflagration with oxygen. Therefore they contain fome common fubftance.

Hydrogenous gas lofes much of its inflammability by frequent agitation with water, and gives it a bad fmell.

It has little action on the alkaline falts. Mr. Lavoifier was once difpofed to confider it as the alkaline principle ; and I obferve that Chaptal, and others, have ftill a leaning to this opinion. This gas combined with quicklime is fuppofed to form potafh, and foda when mixed with magnefia *. But I fee very little to fupport this notion.

* See a letter from Van Mons in the *Annales de Chymie*, of which an extract is given

Its action on the acid falts is much more diftinct. If a little ftrong fulphuric acid be heated in the bottom of a very narrow and tall glafs, which fwells out at the other end like a pear,—and if, in this ftate, inflammable air be blown through it by a long pipe from a bladder, they unite very readily; the hydrogen from the gas unites with the oxygen of the acid,—and they form water. The acid, deprived of its due proportion of oxygen, becomes redundant in fulphur, which is manifeft by its fuffocating fumes; and even fulphur in powder will fometimes collect in the wide part of the glafs. When the acid is concentrated as much as poffible, and made to boil before the inflammable air pafs through it, it acts in another way. Inftead of abftracting the oxygen, and combining with it, it combines with fome of the fulphur, and comes off in form of an abominably fœtid gas, which will be confidered afterwards by the name of *hepatic gas*, or *hydrofulphuret*, only noticing at prefent that it preferves in fome meafure the chemical relations of an acid. When paffed through an alkaline folution, it forms a fort of neutral falt, having a difgufting fœtor and tafte. But,

Nitric acid alfo abforbs this gas very readily. We can fee at once what muft be the effect of an acid which yields oxygen fo readily,— it becomes ruddy and fuming, becaufe water is formed of part of its oxygen, and the nitrogen or azote is now predominant.

This mixture gives me the firft opportunity of confirming the difcovery of the compofition of nitrous acid by Mr. Cavendifh, which I mentioned when giving you an account of the gafes difcovered by different chemifts, after the publication of my differtation on fixed air and quicklime. I confidered it as abundantly proved by that experiment of Mr. Cavendifh. But it has been proved fince that time by many experiments, in which the two ingredients oxygen and azote are feparated. None exhibit this more clearly than the mixture of nitrous acid, or nitrous falts, with inflammable fubftances. Inflammable air is the firft and moft fimple of them, at leaft in its chemical relations.

When a long continued ſtream of inflammable air is made to paſs through nitric acid, we obtain (in a pneumatic apparatus) the gas which Prieſtley calls *nitrous air*, which is colourleſs at laſt, though ruddy in the beginning, and has no acidity, nor changes the colour of teſt paper, and is ſcarcely abſorbed by water. This gas, when mixed with vital air, (oxygenous gas), collapſes with it into nitrous acid. Here then is a proof that vital air is one of its component parts. The inflammable air had united with a part of the oxygen that is more eaſily detached from the perfect acid, and the gas which came from it is therefore deficient in it; ſo deficient as not to be acid. But when oxygen is preſented to it, it is combined, and we have again nitrous acid. In the mean time, the nitric or perfect acid is become fuming, and alſo weaker, or diluted, by the water that is formed by the inflammable and vital air.

It is to be particularly remarked in this experiment, that although the gas produced is clear, and nearly colourleſs, and vital air is the ſame, yet theſe two, on mixing, form, for a few ſeconds, a thick ruddy cloud; and a very ſenſible heat is produced by their condenſation into nitrous acid *.

If the experiment be made with common air, in the place of oxygenous gas, we ſtill produce nitrous acid, and have the ruddy cloud; but the two airs do not vaniſh, but leave a conſiderable reſiduum of azotic gas. The reaſon of this is now obvious. The nitrous air combines only with the oxygen of the atmoſpheric air, and leaves the azote. It was no part of the nitric acid employed in the experiment. We now alſo ſee the cauſe or origin of thoſe red fumes which appear in moſt experiments with the nitrous acid. They are formed during the combination of nitrous air with the vital portion of common air.

This experiment is therefore inſtructive; but it is extremely

* Theſe ruddy fumes are moſt evidently veſicular, and ſome of the veſicles very large.
EDITOR.

tedious, becaufe the quantity of matter in a reafonable bulk of inflammable air is fo very minute. We fhall have much better examples as we proceed.

Its relations to the other acids, to the compound falts, and to the earths, have been but little examined as yet. We are now to be occupied with the other inflammable fubftances; and fhall confider the relation of it to each. Some are very remakable.

II.—PHOSPHORUS.

The next fort of inflammable fubftances we propofed to defcribe are thofe named PHOSPHORI. The principal fpecies of thefe is commonly named the PHOSPHORUS OF URINE, on account of its having been prepared formerly from urine.

When pure and newly prepared, it is femitranfparent; and by its degree of cohefion bears fome refemblance to white bees wax. It makes nearly the fame refiftance to a knife in cutting it; and it melts with a heat even inferior to that of melting wax, coinciding rather with the heat of the human body.

It may be melted with this heat fafely, if immerfed in water, and covered by it from the contact of air; or even without water, provided the glafs veffel containing it be of a fmall fize, and clofed up, to prevent the renewal of the air in it.

In a fmall retort, with a fuitable receiver clofely luted to it, this phofphorus can both be melted, and by an increafe of the heat to the loweft degree of ignition, can be converted into vapour, and thus made to pafs over into the receiver, which being kept cool, the phofphorus quickly congeals in it, and is in the fame ftate as before. It is fometimes fubjected to this fort of diftillation in order to purify it.

But, in all fuch operations with it, we muft be very careful that no frefh air be admitted into the veffels until they be perfectly cold again, and, even then, the moment the receiver is feparated from the retort, it ought to be filled with cold water.

This precaution is neceffary, on account of the extraordinary propenfity of this fubftance to be inflamed by the action of frefh air, efpecially if it be in the leaft warm, or expofed to the air, with an extenfive furface of communication with it. And, when it takes fire, it burns with amazing rapidity and violence. It is therefore proper to be cautious in handling it, as the warmth of the fingers may fet it on fire.

The low degree of heat at which it takes fire fhews the extraordinary inflammability of this fubftance, and is the foundation of many of its remarkable properties, and of fome of the tricks that are played with it, fuch as fetting paper on fire by rubbing the paper *. We may kindle tow, which conceals a little bit of it, and is loofely wrapped round a phial, by pouring hot water into the phial, or by mixing in it two cold liquors which grow hot by mixing. Another trick is to light a candle by touching a glafs of cold water. The glafs has a minute bit of phofphorus ftuck on its edge. The fhowman pours cold water into it, blows out the candle, and while the wick is ftill hot, he touches the phofphorus with it, which inftantly takes fire. With this phofphorus alfo are made what are called phofphoric matches. Thefe philofophical toys are flender wax tapers, having an atom of phofphorus at the end; and each muft be kept in a glafs tube hermetically fealed. It is warmed by putting that end into the mouth,—then broken, and the taper fuddenly

* This is done by previoufly drawing a ftrong line with a piece of phofphorus on a piece of ftiff and rough paper, fuch as cartridge paper, laid upon fome very cold body during the friction. The warmth of the fingers being fufficient to fet phofphorus on fire while it is fo hard rubbed, it muft be wrapped two or three times round in a piece of wet paper. In order to fhew the trick, the paper is folded upon the ftroke, and brifkly rubbed very hard one part upon the other. The phofphorus that was left upon it is generally fufficient to fet it on fire.

drawn out. It generally takes fire. They are coftly, offenfive by their vile fmell, and very childifh.

During the rapid inflammation of the phofphorus, it is quickly changed into a faline fubftance, which is no longer inflammable in the fame degree, and which is thrown out of the flame or burning vapour in the form of a thick white fmoke. A part of this fmoke is difperfed in the air, and unavoidably loft ; but a confiderable part of it is immediately condenfed on the furface of the veffel, in the form of a faline cruft, which has a ftrong attraction for the humidity of the air, and is very foon liquefied by it, forming with it an acid liquor.

Although this acid matter, when firft produced, is no longer inflammable in the fame degree as the phofphorus, it ftill retains a fmall degree of inflammability. This we learn when we expofe it to a ftronger heat, fuch as that of red hot iron. Then, after glowing a little, it becomes, at laft, a white or tranfparent faline fubftance, which has loft all remainder of inflammability, and is a perfect acid.

The properties of this acid were firft inveftigated by Mr. Margraaf of Berlin, and publifhed in the Tranfactions of the Royal Academy there, and fince, in his *Opufcula*, which have been tranflated from the German into the French language. The compounds which it forms with different fubftances are alfo defcribed by M. Fourcroy.

Mr. Margraaf made it return to the ftate of phofphorus, by mixing it well with charcoal powder, and expofing this mixture in earthen retorts to a violent heat in the way of diftillation, with the proper precautions for preventing the phofphorus from taking fire in the receivers.

His opinion of the converfion of the phofphorus into an acid by inflammation was the fame with that of other chemifts at that time ; namely, that it happened in confequence of abftracting the phlogifton ; and yet he particularly remarks, that the quantity of the acid which he collected by burning it gradually, or by small bits in succession, greatly exceeded the quantity of the phosphorus.

This fact has been more exactly ascertained since that time by the experiments of Mr. Lavoisier, and others. And many instructive discoveries have been made by means of the inflammation of phosphorus, and its conversion into an acid. It is peculiarly fitted for being useful in investigations relating to combustion. Sulphur and inflammable air are equally simple and effective. But the compounds which they make by inflammation are so volatile, or offensive, or troublesome, that they are almost unmanageable. Here it is quite otherwise. This pointed it out to Scheele and Lavoisier as the fittest substance for ascertaining the change produced on air by combustion. No volatile matter from the phosphorus will in any way taint it. It is now well known, that when it is burned in a limited quantity of atmospherical air, the vital part of the air, or the oxygen gas, is expended and disappears, without any formation of carbonic acid. And when burned in pure oxygen gas, it burns with amazing brightness and violence, and the whole of the gas is absorbed and disappears.

The burning of phosphorus in oxygen gas was first tried by Mr. Scheele ; but the same experiment was afterwards repeated by Mr. Lavoisier, with larger quantities of the phosphorus, and with the most careful attention to every circumstance which could affect the conclusion to be drawn from it.

Scheele put his bit of phosphorus into a phial filled with oxygen gas, closed the mouth of the phial with a cork, and then warming the phial, he thus kindled the phosphorus. When the inflammation was finished, and the phial was cool again, he plunged it into water, and drew out the cork under water. The water was suddenly pressed into it by the atmosphere, and filled the phial perfectly full, or very near it. He therefore concluded that the oxygen gas had united with the phlogiston of the phosphorus, so as to form with it heat and light, which had passed through the glass.

But Mr. Lavoifier, performing a fimilar experiment, took the pre-caution to weigh the glafs with what it contained immediately before the inflammation of the phofphorus. Weighing it again, after the inflammation, and before it was opened, he found it was precifely of the fame weight. Afterwards, examining as exactly as poffible the weight of the acid into which the phofphorus was changed, he found that it exceeded the weight of the phofphorus confumed, by a quantity exactly equal to the weight of the gas which had difappeared. It was therefore evident that the gas was now changed, by the lofs of its latent heat, into a denfe matter, which made up the greater part of the weight of the acid. One part of phofphorus abforbs in this way a little more than $1\frac{1}{2}$ parts, by weight of the oxygen gas. The acid is made up of the phofphorus and oxygen, in the proportion of 100 to 154 [*].

The change of phofphorus into an acid is, therefore, frequently named the OXYDATION of phofphorus; and other fimilar changes of other inflammable fubftances are alfo called OXYDATIONS [†].

[*] This experiment muft be confidered as one of the moft convincing proofs of the doctrine of the French chemifts. Whatever notion we have of the phofphoric acid, it is plain that the phofphorus itfelf enters into its compofition; for the phofphorus difap-pears, and we have the acid in its ftead. It can never be faid, therefore, that phofpho-rus confifts of the phofphoric acid and phlogifton; for phofphoric acid confifts of phofphorus, and fomething befides: Thefe are irreconcilable. EDITOR.

[†] It is rather an unlucky term, chofen in the heat of difcovery, and before Mr. Caven-difh had difcovered the compofition of water. Had this been known, it is not likely that a perfon of Mr. Lavoifier's general knowledge and good fenfe would have included the formation of water in the lift of acidifications. Moft of his followers were lefs fenfible of the violence done to common language. Perhaps they even liked a diction which will caufe the uninitiated to ftare. We fee this very plainly in the phrafeology of fome of them, who are fond of ufing the term combuftion in cafes where a plain man can fee nothing like it. What he would call combuftion they call oxydation; and what he would call oxydation they call combuftion. The formation of water is an *oxydation of hydrogen ;* but the formation of nitrous acid is the *combuftion of azote.* There is little fcience in this, but abundance of vanity. EDITOR.

Since thefe difcoveries were made, Mr. Pelletier, another French chemift, who has diftinguifhed himfelf by many inftructive experiments relating to phofphorus and the procefs for preparing it, contrived a procefs for oxydating phofphorus, which appears to be very effectual, and very well fitted for preventing the lofs of any part of the acid.

He melts the phofphorus under water, in a cylindrical glafs veffel, and then throws into it oxygen gas, by very fmall quantities at a time, through a glafs tube, connected with a bladder, containing the gas. Every little addition of the oxygen gas kindles that part of the phofphorus with which it comes in contact, and is abforbed by it. That part of the phofphorus is thus changed into acid, and diffolved by the water: And thus the whole of it can be thus oxydated, or changed into acid, which is diffolved by the water.

Another procefs or experiment, founded on the fame difcoveries, may alfo be mentioned here. It is an experiment by which we learn the proportion of oxygen gas contained in atmofpherical air. *The eudiometer of phofphorus*, or method for learning exactly, by the ufe of phofphorus, the proportion of oxygen gas contained in atmofpherical air, contrived by Seguin and Lavoifier. *Annales de Chymie*, *tom.* ix. *p.* 301.

Get a glafs tube about one inch in diameter, clofed above and widened below, and about feven or eight inches long. Fill it with quickfilver, and invert it into a ciftern of the fame fluid. Throw up or introduce into it a little bit of phofphorus, and approach a burning coal to the upper end of the tube, to warm and melt the phofphorus by blowing on the coal. Then, having meafured a proper quantity of the air to be tried, throw it up by fmall portions at once, each of which will caufe the phofphorus to burn, and will thus confume or faturate fome of it, and in fo doing will be faturated, and changed into acid, in as far as it confifts of oxygenous gas. When it is all thrown up, warm the tube again, to be fure that the oxygen gas fhall be completely ex-

pended or faturated. Laftly, let it cool, and transfer the air into another tube which is graduated for meafuring exactly the bulk of the
remaining air. This will fhew the quantity of the oxygen gas which it
originally contained. A glafs funnel, having the extremity of its tube
clofed up, may ferve very well in place of the above apparatus.

Thefe are the moft remarkable particulars of the properties of phofphorus with refpect to heat, and the manner in which it is inflamed
and changed into phofphoric acid.

It has alfo a quality, upon which have been founded many of the
curious experiments and furprifing tricks that have been performed
with it. This quality of it appears when it is expofed to the air in
an extended furface, but in degrees of heat inferior to that neceffary
for its bright inflammation. It then emits a pale light, vifible in the
dark only. We do this very effectually by drawing lines with phofphorus on ftrong white paper.

When we view the paper in day light, the light of the phofphorus
is not perceived, in confequence of its weaknefs, but a white fmoke is
feen to arife from it. A fmall folid mafs of phofphorus continues to
emit this pale light and fmoke a furprifingly long time. Mr. Boyle
relates that a bit of phofphorus, weighing only three grains, continued
to emit this weak light fifteen days and nights before it was exhaufted.

From the confequences obferved, when a bit of the phofphorus is
thus expofed on a plate of glafs a little inclined, we find reafon to be
fatisfied, that this luminous ftate of it is only a ftate of very flow inflammation. It is gradually converted into an acid, which attracting
humidity from the air, forms an acid liquor, much the fame with that
formed by the bright and violent inflammation of the phofphorus;
only that it is not quite fo perfect an acid. It is ftill fomewhat deficient of the proper quantity of oxygen : This trickles down along
the glafs, and may be thus collected.

One of the proceffes for converting phofphorus into an acid is

founded on this property. The phofphorus, formed into fmall cylindrical pieces, is put into a glafs funnel, the throat of which is flightly
ftopped with a pebble or a bit of glafs, to prevent the defcent of the
phofphorus into the pipe. A phial is then placed under the funnel,
and the whole apparatus fet in a moderately cool place. Thus the
acid of the phofphorus, in proportion as it is formed, drops into the
phial. But we muft take care that the place in which the apparatus
ftands be fufficiently cool. A fmall increafe of the heat of the air has
great effect in accelerating the procefs; and if it be much accelerated,
the phofphorus becomes warm by the very oxydation, and there is
danger of its taking fire, and burning with violence. Scheele difcovered that even the weakeft lumination of the phofphorus is attended with the production of fome perceptible heat, difcoverable by
a thermometer. But when the procefs I fpeak of is rightly conducted, this heat being weak and flowly produced, it is carried off by the
furrounding air, fo that it never is accumulated in fuch quantity as to
difpofe the phofphorus to be rapidly inflamed.

We come now to confider phofphorus in mixture with other bodies.
Being a fubftance of recent difcovery, its chemical relations have not
yet been thoroughly examined. Many of thofe we know are very
remarkable. Mr. Margraaf, to whom we are indebted for the knowledge of the chemical nature of this fubftance, has made thefe combinations with his ufual judgment and accuracy.

A drachm of phofphorus was diftilled by him with an ounce and a
half of fulphuric acid, giving at laft a violent heat. A few grains of
phofphorus remained, mixed with a fpongy white mafs, which deli-
quefced in the air. The liquor which diftilled was thickifh, and a
little milky. Both liquors contained a mixture of fulphuric and phofphoric acids.

A drachm of phofphorus was diftilled with an ounce of ftrong nitric
acid. As foon as it was dropped into the acid, blood red fumes were
difengaged, which made him fit on the receiver in a hurry. The folu-

tion went on with great vehemence and heat, and in a fhort time the greateft part of the nitrous acid came over, without applying any fire to the retort; at laft the phofphorus yet remaining took fire, and burft the retort with great noife.

The union of the oxygen and azote in the nitric acid is fo weak, that fcarcely any body which attracts oxygen can be prefented to the acid which will not overcome their union and decompofe the acid. The azote, or rather nitrous gas, efcapes and produces thofe red fumes, by uniting with the atmofpheric air. The latent heat of the oxygen, as it is contained in the nitric acid, greatly exceeding what is neceffary for the acid of phofphorus, emerges and produces heat enough to in-- flame the remaining phofphorus.

But, by another way of conducting the procefs, it is very manage-able. Put ftrong nitric acid into a tubulated retort fitted with a re-ceiver. Have ready a quantity of phofphorus cut into very fmall pieces; drop in one, and immediately ftop the hole. The phofphorus will diffolve very quickly, and will produce heat. When all is cool, drop in another bit, and continue this till the laft bit diffolves and feems to faturate the acid. Now apply heat, and the nitric acid will diftil over in fiery fumes, and leave nearly pure phofphoric acid in the retort.

This mixture alfo gives us a pretty enough phenomenon. Put about half an ounce of ftrong nitric acid into a fmall and thin flafk with a narrow mouth. Drop in a bit of phofphorus like a large pea. There is an effervefcence, or rather an explofion, during which a flender column of flame is darted out of the glafs along with the vapours and drops of acid. Care muft be taken fo to regulate the quantity to the fize of the orifice, that the veffel do not burft, and to have it of fuch a fhape that few drops of acid may come out.

This is the next opportunity given us for examining the compofition of nitrous acid; and it is incomparably better than the laft with inflam-mable air. Accordingly, all the phenomena are the fame in kind, but much more remarkable. The production of the phosphoric acid, in the same quantity precisely as if it had been produced by inflaming the phosphorus, shews evidently that oxygen constitutes a great portion of nitric acid. The gas which escapes is the same as when inflammable air is employed, provided that the combination be slowly effected. If rapidly, much phosphorus is volatilized, and taints the gas, sometimes even making it inflammable. The fumes are blood red, while they mix with common air or with oxygen; and nitrous acid is produced by the union. · If all access of air be prevented, the gas is colourless, and without acidity. By comparing the gain to the phosphorus, with the oxygen necessary for forming nitric acid with the gas, Lavoisier found that the proportion of oxygen and azote in the gas, is that of 68 to 32 nearly. It requires about 24 parts (in bulk) of common air to saturate 11 parts of this gas, and form with it nitrous acid. In this mixture, the whole nitrous gas disappears, and a fourth part of the common air, so that 35 cubic inches, after mixture, will only occupy 18 inches. This is azotic gas.

That the red vapours are true nitrous acid, appears from an experiment of Dr. Priestley. Hang in the glass a piece of sal ammoniac. When the red fumes are over, a snowy powder settles all over the vessel, which is imperfect nitrous ammoniac.

The nitrous gas may be decomposed, and the composition of nitrous acid completely demonstrated, by filling a jar with nitrous gas, and putting into it a quantity of *hepar sulphuris*. This will immediately attract the oxygen remaining in the gas, and leave the other ingredient alone. This is found, by this experiment, to be pure azotic gas.

We may surely now assume the composition of the nitrous acid as a thing as firmly established as any doctrine in chemistry. You will recollect, that, in order to answer an immediate purpose, I mentioned the precious experiment of Mr. Cavendish, in which he found, that when seven parts of vital air were deflagrated with three of phlogisticated air, or azotic gas, the whole collapsed into pure nitrous acid.

But I did not at that time affume this conftitution of the acid as a fixed point, becaufe I was not then in a condition to fhew you how the acid might be feparated or refolved into thofe its conftituent parts. I was unable' to do this, becaufe you were unacquainted with the fubftances whofe properties and manner of action were to operate this decompofition. Enough of thefe have now occurred; and I have given you moft diftinct examples of the fact. More will yet occur as we proceed; and fome of them perhaps ftill more perfpicuous. But we have enough; and it will be agreeable now to meet with the others, as phenomena which are explained by this principle now in our poffeffion.

The muriatic acid did not appear to Mr. Margraaf to act at all on the phofphorus, which did not diffolve. At the end, indeed, of the diftillation, the phofphorus alfo came over, but without any remarkable change.

The pure or cauftic fixed alkalis may be combined with phofphorus, or can be made to diffolve it, as they diffolve fulphur, and indeed all the inflammable fubftances in the humid way. In confequence of this combination, an inflammable gas, extremely fœtid, fmelling like rotten fifh, is produced, which takes fire the moment it comes in contact with atmofpherical air; and which therefore expofes the veffels to the rifk of being burft by its explofions.

This gas is undoubtedly produced from a fmall portion of the water decompounded by the diffolved phofphorus. It is the hydrogen in its form of gas; and as faft as it is produced, it diffolves and volatilizes a fmall portion of the phofphorus itfelf, and from this receives the quality of fpontaneous inflammability *. This fmall portion of

---

* Is the feries of operations here ftated very fatisfactorily eftablifhed? I cannot help confidering this experiment, and thofe which are analogous to it in the treatment of fulphur with alkaline fubftances, as examples of that gratuitous employment of the decompofition of water, which the followers of Lavoifier indulge in with fo little circumfpection or moderation. All depends on the *order* in which the different actions fucceed

the phofphorus, however, is very flightly united with the hydrogen, for it is depofited or feparated, if the gas be kept for fome time confined with water. It gradually depofits the phofphorus on the fides of the veffel, lofes its fpontaneous inflammability, and is changed into common inflammable air or hydrogen. The fame gas is produced by quicklime. *(See Note 9. at the end of the Volume.)*

By thefe experiments, you will perceive that phofphorus is one of the moft eafily inflammable bodies that we know ; that it has a ftrong propenfity to take fire, and to be inflamed,—or, in other words, that it has a ftrong attraction for oxygen. A very ingenious and inftruc· tive experiment has lately been founded on this ftrong attraction, or has been contrived in confequence of it. It has been a great defideratum among the chemifts to decompound the carbonic acid ; that is, to contrive fome way by which the carbone or coal may be feparated from the oxygenous gas or vital air, which is now held to be its other ingredient. This has been attempted with very flattering appearances of fuccefs by Mr. Smithfon Tenant. You have an account of it in the Philofophical Tranfactions, vol. 81. part ii. I fhall give you a brief account of it.

Mr. Tenant put a fmall bit of phofphorus into a glafs tube that was fhut at one end, and then put over it marble reduced to a fine powder. He fhut the other end, but loofely, that the common air expanded by heat, might efcape, and yet fo that the free circulation of air, which might kindle the phofphorus, fhould not take place. He heated this apparatus red hot for a few minutes. When all was cold, he broke the tube, and found therein a black powder, which confifted of coal, phofphorated lime, and phofphorus mixed with quicklime.

each other. I think that phofphorus and fulphur, and perhaps charcoal, afford a probable opportunity of fettling the point, by means of a previous determination of their fimple affinities. This is not in our power, perhaps, in the more complicated fubftances of animals and vegetables.

EDITOR.

When the phosphorated lime was separated by solution in an acid, and filtration, and the phosphorus by sublimation, the coaly matter that remained did not appear to differ from vegetable coal in any respect. Mr. Tenant explains the experiment by saying that the coal is separated from the oxygen, although it has a stronger attraction for it than the phosphorus has, in consequence of the sum of the attractions of the phosphorus for oxygen, and the phosphoric acid for quick-lime.

Dr. Pearson observing that the compounds of phosphoric acid and the fixed alkalis could not be made to yield phosphorus by treating them in contact with charcoal in a red heat, while the calcareous phosphat yields it with great readiness, thought that the fossil alkali would be a better intermedium than calcareous earth, for operating the decomposition effected by Mr. Tenant. He therefore employed, with perfect success, a mixture of two parts of phosphorus, and eight of carbonat of soda, cleared of its water of crystallization. The process was similar to Mr. Tenant's, and very easy. The rationale of these processes is by no means obvious; but the production of carbone, and this in due quantity, completes the theory of carbonic acid. A letter of Dr. Pearson to Hassenfratz, and some observations by Fourcroy, which accompany it in the 18th volume of the *Annales de Chymie*, deserve perusal.

Phosphorus will also combine, though loosely, with the caustic volatile alkali, and produces a gas with an abominable smell. This gas also takes fire in the air, and, which is remarkable, deposits the greatest part of the phosphorus in the flash. Indeed, in all these gases, containing phosphorus volatilized by hydrogen, the union seems exceedingly slight. They all decompose by long keeping, even in corked phials. Westrumb says that they deposit the phosphorus, even in phials hermetically sealed. That the phosphorus is not inflamed in these experiments, but deposited, appears singular at first view, seeing that it is so very inflammable. But when we consider the extreme rarity of the gas, and reflect that the inflammation even of this rare gas, which takes place by the mere contact of air, is probably the low inflammation, in which phofphorus itfelf only fhines without ordinary combuftion, we fhall be fenfible that the complete inflammation of the phofphorus is not to be expected.

Thefe are fome of the moft remarkable qualities and relations of this inflammable fubftance. It is alfo capable of being combined with fome of the other inflammables *, and with the metals; and forms fingular compounds, which fhall be noticed hereafter.

The acid which it affords is alfo now become one of the important objects of chemiftry. It is fixed and vitrifiable. With the foffil alkali it forms a falt, now ufed in medicine, and recommended by Dr. Pearfon by the name of SODA PHOSPHORATA. A procefs for preparing this neutral falt is given in a new edition of the Edinburgh Pharmacopœia †.

---

* Indeed all the inflammable fubftances are fufceptible of a perfect admixture; and this has, in many cafes, fome appearance of a chemical combination,—the compounds having certain general properties which do not belong to the ingredients, at leaft in the fame degree. They are, in general, more fufible, volatile, and inflammable, than fhould be expected from the ingredients: And they are more difpofed to union with another inflammable fubftance. EDITOR.

† *Procefs for Soda Phofphorata, communicated by Dr. Pearfon.*

Take bone afhes: Thofe to be had at the hartfhorn manufactures are commonly ufed, and ground to a coarfe powder, in the ftate they are in for manure, and as fold for about a fhilling a bufhel—ten pounds.

Pour on them, in an earthen, or iron pan, oil of vitriol, of fpecific gravity about 1800 —fix pounds.

Stir the mixture well, and add to it gradually rain or river water—nine pounds.

Stir the whole well together, and fet the mixture to digeft in a fand heat of about 130°, for two or three days; then add to this mixture nine pounds more of very hot water, and pour it on a filtre of coarfe linen cloth, upon which pour boiling water till it paffes through with little acid tafte. Let the filtrated liquors, all mixed together, ftand for the fediment to fall; decant the clear liquor, and evaporate to about nine pints.

When the acid is completely freed from phofphorus, and may therefore have the diftinctive name PHOSPHORIC ACID, it forms with the alkalis, falts, which cryftallize with great difficulty. But, when it retains fome phofphorefcence and volatility, in which ftate we may call it PHOSPHOROUS ACID, thefe compounds cryftallize very well.

With calcareous earth it forms a fubftance infoluble in water, refembling, in all refpects, the earth of bones.

Filtrate, to feparate the felenite precipitated in boiling, and evaporate again to feven pints; cool the liquor, and feparate more felenite. Heat the liquor in an earthen veffel, and add pure cryftallized foffil alkali, diffolved in 1½ its weight of water, until the effervefcence ceafes. Filtre the faline liquor hot, into a fhallow veffel, and let it cryftallize three or four days. Decant the remaining liquor from the cryftals; and if it be acid, neutralife it again with the folution of foffil alkali, and evaporate again and cryftallize; repeating thefe operations, until a liquor remain which will not give any more cryftals, either by evaporation or addition of more alkali.

Shorter Procefs, but with more wafte of phofphoric acid.

Add as much water to a mixture of fulphuric acid and bone afhes as will reduce it to a thin pafte, which muft be put into a coarfe thick hempen bag, tied up clofe at the mouth. Then prefs and moiften it with water repeatedly, until the whole, or greateft part of the acid liquor is extracted. The turbid acid liquor muft be purified from the felenite, by decantation and filtration; and if the quantity of this liquor be more than fix pints, reduce it to that quantity by evaporation, and faturate, as before defcribed, with a folution of foffil alkali.

The quantity of phofphorated foda fhould be at leaft equal to the weight of the alkali, and about five-fixths of that of the bone afhes; fo that the above quantities fhould yield above 8½ pounds of the falt.

This falt is liable to be contaminated with Glauber's falt, or fuperabundant alkali, which are diftinguifhable in it by the tafte. The tafte of pure foda phofphorata is a flight tafte of marine falt. The bone afhes fometimes contain more, fometimes lefs acid, and the foffil alkali is often impure.

The fureft way to fucceed, is to be careful that the fulphuric acid be rather too little for decompounding the whole of the bone afhes, than that there fhould be too much of it. In the above procefs this is attended to.

Origin and Preparation of Phofphorus.

The firft procefs by which the phofphorus was obtained in its pure ftate appears to have been an accidental difcovery in 1669, (*Leibnitz de Inve. t. Phofph. Mifcell. Berol. 1.*) by Brandt, a merchant in Hamburg, who was employed in experiments on urine, in the hopes of making gold; and it remained in his hands, or in the hands of a few others, who acquired the knowledge of it, but concealed it from the public for a long time after. It was brought to England by a Dr. Kraft. At laft fome hints relating to it were publifhed in the Philofophical Tranfactions, No. 196, by Mr. Boyle, who had practifed Kraft's procefs with fome degree of fuccefs. Meantime, Kunkel, a chemift in Drefden, by dint of labour, difcovered an effectual procefs, which he publifhed, and claimed the invention. A full defcription of a better procefs appeared afterwards in the Memoirs of the Royal Academy of Sciences at Paris for the year 1737, in confequence of the purchafe of the fecret from a chemift who was poffeffed of it. All this is detailed very diftinctly by Mr. Macquer, in his Dictionary of Chemiftry. The urine was evaporated to a dry extract or coaly matter; then fome of the falts were feparated from this extract by water; and the remaining matter was diftilled with a violent heat in earthen retorts.

Although fome phofphorus may be produced by this procefs, it is far from being a good one. Some parts of it are actually very detrimental, or diminifh the quantity of phofphorus which might be obtained from the materials; and the procefs is exceedingly troublefome and hazardous. The beft management of the procefs, and the manner in which the phofphorus is actually produced by it, were never underftood until Mr. Margraaf applied himfelf to the ftudy of it, and publifhed his experiments in the Berlin Tranfactions. By them it appears that the production of phofphorus from

urine depends upon a falt contained in the urine, and which had
been obferved in it before, but not fufficiently examined.   Dr. Boerhaave takes notice of it, and calls it the effential falt of urine.   Mr.
Margraaf collected a quantity of this falt and examined it.   He found
it ammoniacal, or containing a volatile alkali in its compofition.   It
is a mixed neutral, compofed of a peculiar acid, combined partly with
each of the fixed alkalis, and partly with ammonia.   With charcoal
duft this falt gave a larger yield of phofphorus than ever had been
got from any other materials.   And the remains of the urine from
which this falt had been extracted gave hardly any fenfible quantity
of phofphorus.   Mr. Margraaf alfo fhewed, by a train of experiments, that the phofphorus is changed by inflammation into an acid,
now named the phofphoric acid; which acid can again be changed
into phofphorus.   This cleared up the whole matter, and gave reafon to be affured that the falt juft now mentioned contains this very
acid.

In confequence of thefe difcoveries, Mr. Margraaf contrived fome
improvements of the procefs for preparing phofphorus from urine;
by which improvements the operation was facilitated, and the yield
of phofphorus much increafed.

This procefs of Margraaf's continued to be the beft, until Mr. Scheele
of Sweden, in company with Mr. Gahn, another eminent chemift,
taught us one ftill better.   The two gentlemen juft mentioned have
faved us the trouble of evaporating urine, by difcovering that the
acid of phofphorus is contained in the bones of animals; and when
combined with the calcareous earth, forms their bafis, or their moft
folid and fixed matter,—that is, the white earthy-like matter which
remains of them after all their inflammable matter is confumed; in
fhort, what is ufually called *the earth of bones*.   The procefs thefe
gentlemen propofed for extracting the acid from this matter was a
little complicated.   A more fimple one has been difcovered fince that
time.   We need only to mix with the bone afhes fome fulphuric

acid and water, &c.—(*Vide first part of the process for soda phosphorata, in the Pharmacopœia Edinensis*).

But the phosphoric acid may be extracted by other processes, as 1*mo*, By applying the mild volatile alkali, or carbonat of ammonia, to the bone ash reduced to fine powder. (N. B. The double exchange will probably happen best in a cold place).

2*do*, By applying sulphat of ammonia, dissolved in water, to the bone ashes.

Since these discoveries have been made, the production of this inflammable substance is a much easier operation than it was formerly. Mr. Pelletier, at Paris, has sometimes made 60 ounces of it at once ; and in one year he made upwards of 300 pounds of it. (*Journ. de Phys.* 1785. Jul.) The only nicety is in the choice of the vessels, and management of the condensation, which must be so contrived as to allow a very great quantity of gases produced during the operation to escape. These gases seem to arise from the water of the acid, and its action on the charcoal and the phosphorus nearly formed. It acts on these substances in the form of vapour, therefore in the most extensive surface possible, and furnished with great store of heat. With the charcoal it forms inflammable air of the heavy kind, as we shall learn afterwards ; and with the phosphorus it forms a gas which takes fire when it comes in contact with the external air. This immediately kindles the inflammable air which issues along with it, and the whole produces a bright flame. The gases are not formed in an equable manner, but by sudden paroxysms, or explosions, which frequently burst the vessels. Mr. Hellot found it necessary to have a hole in the receiver, lightly stopped with a wooden peg, which was blown out when the vapours became too elastic. Other chemists had a tube, which communicated with the receiver, and had its mouth immersed. A small quantity of generated phosphorus was saved in this way.

Having thus learned how phosphorus may be obtained pure, we are curious to learn how its acid comes into the compofition of animal bodies; for phofphorus muft furely be added to the lift of fimples which the French chemifts allot to the formation of animal and vegetable fubftances. There is reafon to think that it is introduced into them from the vegetable fubftances by which animals are ultimately nourifhed. Margraaf obtained phofphorus from the charcoal of wheat, rye, and other nutritive vegetable fubftances. Whence thefe receive it is a queftion not fo eafily anfwered. It may be produced in them by the powers of vegetation. For we know that there is in vegetables a power to combine their elements together in different ways, fo as to generate productions which would not otherwife be formed. Oil, fugar, &c. are manifeftly formed in the veffels, or organs of plants, of materials which before were of a quite different nature *.

But a quantity of this acid has alfo been found in fome foffil fubftances; for example in the green ore of lead. There is alfo a bed of ftone in Spain, in the province of Eftramadura, and diftrict of Truxillo, near Lagrofana, which contains this acid united to calcareous earth, as copioufly as the earth of bones does. This was difcovered by Mr. Bowle, and communicated to the French Academy by Mr. Prouft. It forms very extenfive ftrata, refting on quartz; is of a white colour, and fibrous texture, not fo hard as to ftrike fire with fteel, but hard enough for building. It is fo employed for the houfes and inclofures. Thrown into the fire, it burns faintly, with a

---

* We muft, as yet, fuppofe that the phofphorus exifts in the foil, and is taken up by the vegetative functions of the plant; for we have no authority to fay that it is a fubftance compofed of more fimple ingredients. But further, as thofe plants now mentioned will grow and thrive in pounded glafs moiftened with diftilled water, enjoying at the fame time the benefit of air and light, we muft, if they yield this falt when fo raifed, look for phofphorus in thefe fources. This leads to very nice fpeculations. I found it, in very great quantity, in a piece of fine coral rock from Port Royal, in Jamaica; and am inclined to think that all corallines contain it, and that, by a proper treatment, they may be rendered true phofphori, like the Bononian ftone, without the fulphur employed in that preparation.   EDITOR.

pretty, green light, which it keeps for some time after being taken out of the fire. A fossil of the same kind was discovered by Werner at Lunenberg in Saxony, and at Slackenwald in Bohemia, crystallized in hexahedral prisms, and in plates, generally mixed with fluor, spar, lithomarga, and steatites, but rarely with quartz. The pure specimens are called APATIT by the German fossilists; and they say that it contains $\frac{9}{10}$ths of phosphoric acid.

I think that a stone, forming thin white strata, near the Giant's Causeway in Ireland, is of this nature; and perhaps the *lapis suillus*, and other foetid marbles are so *.

III.—SULPHUR.

This is the next inflammable body in the order in which I proposed to consider them. Its common appearance is too familiar to need any description. When held in the warm hand, we feel it crackle, and even hear it. This is a real splitting; and the same thing is observed in the large crystals of saltpetre, and some other salts. It smells when rubbed, and becomes highly electric.

Several of its properties have already been occasionally noticed. It has long been an object of great attention by the chemists, and has acted the chief part in all their explanations of phenomena. It was considered as the first principle of metallization, and of every species of combustion. It has now sunk to the condition of a simple ordinary chemical substance, of vast extent indeed, existing in an endless variety of bodies, but deprived of all its former pre-eminence.

* It may be advisable to observe with some care the vegetation of the Spanish district now mentioned. If this stratum of phosphat crops out on the surface, it is not improbable that its rubbish will affect the vegetation. Experiments should be tried with plants raised on powder of this stone, especially of such plants as are superficial. It is most probable that plants so raised will take up some phosphoric acid, or phosphorus. If they do not, the origin of it in plants and animals is not a little mysterious; and speculations on this subject seem to open a door to much research, which may greatly affect our current theories. EDITOR.

Sulphur melts and evaporates without change, in very moderate heats; but with some peculiarities that are worth mentioning. When heated to 170° of Fahrenheit's thermometer, it begins to evaporate, and we feel a very disagreeable suffocating smell. This, long continued in open air, produces a considerable change in it, which we shall not consider at present. In close vessels, there is no such change; but at 185° or 190° it begins to melt, and before 220° it is fluid. If the heat be quickly increased, it loses its fluidity, and becomes firm, and of a deeper colour. It regains its fluidity, if we reduce the temperature; and this may be repeated at pleasure in close vessels, if the changes of heat be not too slow, otherwise it begins to evaporate so copiously, that we cannot apply heat in sufficient plenty suddenly to raise its temperature. These peculiarities were first observed by the eminent naturalist Fontana.

If, after being quite melted, we let it cool, it congeals in a crystalline form, but so confusedly, that we cannot easily define the shape of the crystals, further than that they are slender interlaced fibres. If a great mass be kept fluid below, while it fixes at the surface, the crystallization there is much more distinct. If melted sulphur be poured into water, the congealed mass has a considerable pliancy; but this does not last.

When heated in open air above 300°, it takes fire and burns with a very weak blue flame. By this inflammation it is totally changed, and from being a solid, mild, tasteless substance, now becomes a fluid immensely corrosive, and in the highest degree acid, being now what you are well acquainted with by the name of *vitriolic* or *sulphuric acid*.

When this acid is treated in a red heat, and in contact with an inflammable body, it is changed into sulphur; and any inflammable body whatever will answer, if it can be made to bear the heat. Even inflammable air will do.

Hence it was inferred by Dr. Stahl, that there was a principle common to all inflammables, and the cause of their inflammability,—the

PHLOGISTON. Hence it was inferred that fulphur confifted of the peculiar acid now fpoken of, and phlogifton; and that this phlogifton was expelled during inflammation, but reftored to the acid by any inflammable body, in confequence of a greater attraction for the acid.

But no perfon ever faw this phlogifton in a feparate ftate, or could prove that there was fuch a thing in fulphur. Mr. Lavoifier firft obliged the chemifts to attend to a thing which they had either overlooked, or accounted for in an unwarrantable manner,—namely, that the weight of the acid is vaftly greater than that of the brimftone from which it is produced. From eight grains of fulphur we can produce twenty-fix of vitriolic acid,—part of this indeed is water, collected from the atmofphere, but by no means the whole. Moreover, he found that in burning, the fulphur abforbed and attached to itfelf a great quantity of the pure part of the air; and at laft, he found, by careful meafurement, that it abforbed a quantity exactly equal to the augmentation of its weight by inflammation; and he gave good reafons,—reafons which have at laft been acquiefced in, that this air is the caufe of the acidity of vitriolic acid,—he therefore called it *oxygen*.

Sulphur, therefore, is now confidered as a fimple fubftance, and fulphuric acid is a compound of fulphur and oxygen; and in this light I fhall continue to confider it.

At the temperature of 140°, or 150°, fulphur begins to attract oxygen fenfibly, and if the heat be increafed to 180°, or 190°, the combination becomes pretty rapid, accompanied by a faint light; and if this be long continued, the fulphur will be converted into acid, in the form of fuffocating fteams. But the heat is infenfible,—at leaft, it is fo weak, that the inflammability of the fulphur in gunpowder will be completely deftroyed without fetting fire to the charcoal, and the powder is now quite ufelefs. This is fimilar to what happens in phofphorus and many other inflammable fubftances. It is an imperfect, or rather a a flow inflammation, which will be noticed more particularly in the fequel.

When fulphur is evaporated in clofe veffels, it fublimes into a fine duft or powder, called *flowers of fulphur,*—they are fine cryftals, generally a little acid, by reafon of the air which filled the veffels. This acidity may be removed by wafhing the flowers in water.

I come now to confider the properties of fulphur difcoverable by mixing it with other bodies.

Sulphur unites readily with the fixed alkalis, as you have already feen, efpecially when they are cauftic; and this combination is effected both in the dry and the humid way. It combines alfo with all the alkaline earths, and the compounds are called SULPHURETS. It can alfo be combined with volatile alkali. It then forms a volatile hepar, called alfo *volatile tincture of fulphur.* This combination of fulphur with volatile alkali is had by mixing together two parts of the muriat of ammonia, two parts of lime, and one part of fulphur, and diftilling the mixture brifkly. The fulphuret does not begin to form till towards the end of the operation; the lime immediately detaching the ammonia in its pure ftate. This volatile hepar is mixed, but not compounded with the pure volatile alkali. It may be feparated, and will even cryftallize.

The propenfity which fulphur has when thus combined with alkaline fubftances, and diffolved in water, to abforb oxygen more quickly and eafily than when it is pure, was formerly noticed. In its pure or feparate ftate, or even when combined with alkaline fubftances, *via ficca,* it does not attract oxygen, except when it is heated to fuch a degree as to make it melt and begin to evaporate. Then indeed its union with oxygen begins to take place, or it enters into a ftate of imperfect inflammation. But, when united with alkaline falts or earths, and at the fame time diffolved in water, it gradually attracts oxygen from the air to which it is expofed, and is thus changed into fulphuric acid, without needing the affiftance of heat. You doubtlefs recollect examples of this when I was confidering the different kinds of gas, and the experiments by which Scheele and Lavoisier demonstrated the composition of atmospheric air.

There cannot be a doubt but that this increase of its disposition to attract oxygen proceeds from the change induced upon its cohesive attraction by the alkali.

Many of the foreign chemists explain this in another way. Sulphur, say they, attracts oxygenous gas so weakly in low heats, that it cannot take it from the air in that form; but being mixed with alkali, this prevents the first approach to saturation of the sulphur with oxygen, by its attraction for the nascent sulphuric acid, with which it instantly unites and forms a sulphat. This is carried off by the water, and leaves the rest of the sulphur equally ready for combination.

The attraction of sulphur for alkaline substances is not very strong. It may be separated by any acid, even the carbonic; and in the moment of this separation, the mixture emits a gas or elastic fluid. If the acid be poured on a dry hepar sulphuris in powder, the copious escape of this hepatic gas occasions a brisk effervescence, although the hepar has been prepared with a caustic alkali.

This hepatic gas has very singular properties.

1mo, It has an intolerably offensive smell, which is commonly said to resemble the smell of rotten eggs. Indeed, the vapour coming from an egg when boiled, whether fresh or rotten, contains this gas, and blackens silver, &c. like it. If a bit of hepar sulphuris, a boiled egg, or a piece of newly manufactured horn, be shut up in a cupboard, all the silver in it will be stained almost black in a few days,—a few hours will stain it brown.

2do, It can be condensed into a sort of oily-like matter by intense cold, as appears from experiments of Mr. Monge, narrated by Fourcroy in his *Preliminary Discourse, page* xxxi.—*Kerr's edition,* 1788.

3tio, In its elastic or vaporous state, it is highly inflammable, and burns something like hydrogen gas, but not so rapidly; and while it burns, a small quantity of sulphur is separated from it. The sulphur alfo undergoes a flight degree of inflammation, and is in part changed into fulphurous acid. This gas is much heavier than the hydrogenous. It unites with alkalis, in the fame manner as the hepar from which it was produced; and is feparated without decompofition by an acid, provided that acid do not too readily part with fome oxygen.

The nature of it was not underftood, until the late improvements were made in our fcience by the ftudy of gafes or elaftic fluids. It is now confidered as an hydrogen gas, holding a fmall quantity of ful-phur combined with it. *(See Note 10. at the end of the Volume.)*.

The inferior degree of rarity, or of elafticity and volatility, which it has, when compared with pure hydrogen gas, proceeds from the ful-phur which it contains. The fulphur, by its attraction for the hy-drogen gas, diminifhes its volatility, and therefore renders it a denfer or heavier fluid, and difpofes it alfo to be condenfed by a great degree of cold; and, probably from the fame caufe, it derives an aptitude to be condenfed or abforbed by water, to which it communicates its own deteftable odour. Hence the fmell of fulphurous mineral waters, and the depofition of fulphur obferved at many fuch fprings.

If fuch water, however, be expofed to the air, or put into bottles, without ufing every poffible precaution to prevent the efcape of the elaftic matter, a part of the gas evaporates, or is loft. And it leaves behind in the water a part of the fulphur which was united with it. This renders the water milky, and is depofited by it. Hence the fulphur depofited by fome famous mineral fprings. This happens in confequence of the great volatility of the hydrogen gas, which makes it evaporate, carrying away with it a part only of the lefs volatile fulphur *.

Such water is alfo made to depofit the fulphur immediately, by

---

* Is it not more probable that this decompofition of the gas, and precipitation of the fulphur, is effected by the oxygen in the air? This was Scheele's eudiometer, and is per-haps fuperior to any other.

EDITOR.

adding to it a fmall quantity of very ftrong nitrous acid. And the fulphurous acid produces the fame effect. But neither the fulphuric acid, nor the muriatic in its ordinary ftate, have any power to precipitate the fulphur; and even the nitric has but little of this power.

It is fuppofed that the ftrong nitrous acid and the fulphurous acid act by a loofe and feparable oxygen, which unites with the hydrogen, and changes it into water. The fulphur, when alone, not being foluble in water, muft therefore make its appearance in an undiffolved ftate. But I cannot underftand why the fulphuric and the nitric do not produce the fame effect. *(See Note* 11. *at the end of the Volume.)*

Several of thefe experiments were firft made by Profeffor Bergmann, and alfo by Scheele. Bergmann was the firft who explained by them the nature of many of the mineral waters called *fulphurous waters.*

They have the odour, and the other properties, more or lefs, of an artificial compound of water with this gas. This will be exhibited again hereafter, when we defcribe the mineral waters.

The authors who have thrown the moft light on the formation of this gas, are a little fociety of chemical philofophers in Holland, Meffrs. Von Deiman, Trooftwyck, and others, who have publifhed a volume or two of their experiments on different fubjects. Their doctrine on this fubject may alfo be feen in the 14th volume of the *Annales de Chymie*, at page 294.

According to thefe gentlemen, a fmall portion of the diffolved fulphur acts on a fmall portion of the water, fo as to take its oxygen from it. The feparated hydrogen of that fmall portion of the water does not immediately affume the form of a gas, but is joined to the remaining unchanged fulphuret, attaching itfelf to it by an attraction for both the fulphur and the fixed alkali. But when we add an acid to faturate the alkali, the feparated hydrogen has nothing left with it but the fulphur, the attraction of which is not fufficient for entirely repreffing its volatility. It therefore affumes the form of gas, aided in this by the heat produced or extricated by the action of the acid and alkali on one another. And while it assumes the form of gas, it volatilizes a part of the sulphur, which is combined with the gas thus produced *.

This hydrogenous sulphuret has already come before us, while we were considering sulphur in its connection with the alkaline substances.

We come next in order to consider its relation to the acids. Its relation to the sulphuric acid is already known.

We may foresee what will be the effect of mixing sulphur with the nitric acid. It must be very similar to that of a mixture of phosphorus with it. Accordingly, if we pour on a quantity of sulphur twenty times its weight of nitric acid, and distil it, we shall obtain sulphuric acid double the weight of the sulphur. This arises from the very loose and separable state of the oxygen in nitric acid. For the same reason, this change may be produced by the oxygenated marine acid.

You all know that melted nitre forms the sulphuric acid with rapidity and vehemence, by deflagrating it with sulphur. For we have in the crucible a sal polychrest, or sulphat of potash.

Phosphorus can be combined with sulphur in the way of fusion and sublimation, in a retort and receiver. Margraaf formed a compound of this kind, in which the attraction of both these inflammable substances for oxygen appears to have been [more powerful] than when in their separate state. The compound took fire, and burned with

* This explanation seems both gratuitous and embarrassed. We assume, without warrant, that water is decomposed, and is afterwards recompounded. In the *Ann. de Chymie, vol.* 14. p. 311. the hepar sulphuris is supposed to decompound the water, not of itself, but by help of an acid, whose caloric forms a hepatic gas with the hydrogen and part of the sulphur. But here the decomposition of the oxygen becomes still more mysterious, and we cannot see why the hydrogen, which at first detached the caloric from an acid, should afterwards let it go, and again unite with the acid and compose water.

EDITOR.

violence, whenever it was expofed to the air *. The chemical combination of thefe highly inflammable fubftances produced this effect
probably by a diminution of their cohefive attraction.

Natural Hiftory of Sulphur.

Sulphur is to be found in all the kingdoms, as they are called, of
nature. It is formed by the decompofition both of animals and vegetables. The fœtor of animal excrements is found to arife from hepatic gas. A confiderable quantity of fulphur was collected in cleaning
an old common fewer at Paris. (See Chaptal, tom. i. p. 91.) The fame
gas is found in all putrid collections of decaying vegetables. The pit
in which flax is fteeped, year after year, impregnates the clay around
it to fuch a degree that it burns with a blue fuffocating fmoke. Nay,
fulphur is found in the juices of fome plants, particularly the rumices
or docks.

But it is moft abundant in the mineral kingdom. And there it is
generally in a ftate of combination, chiefly with metals. The ores
of moft metals contain it. But the ore or mineral fubftance, moft remarkable for the quantity of fulphur which it contains, is the Py-
RITES, in which the fulphur is combined with iron or copper,—chiefly,
however, and ofteneft, with iron. This compound is of metallic
opacity and luftre, and a pale braffy colour. It is found in entire
veins, or fometimes in feparate maffes. It generally is fo found

---

* I rather think that Margraaf's account of it indicates lefs inflammability. It inflamed with fome difficulty, he fays, by friction, giving a yellow light. A dry heat,
equal to that of boiling water, caufed it to burn with violence, giving a ftrong fmell of
hepar fulphuris. It fwelled in water, imbibing much of it, and rendered the water acid
and fulphurous. The water into which it dropped in the diftillation effervefced violently with alkalis.

This is a very curious and fomewhat puzzling experiment. Whence did the acid
come, and what was it? Did the compound decompofe the water, and become oxydated? EDITOR.

among coal. Relics of animal and vegetable fubftances are frequently found completely penetrated with it, and, as it were, changed into pyrites. In the ifland of Sheppey, and other parts of Kent, fuch fpecimens are very abundant; particularly vaft quantities of the *cornu ammonis*, fome of them of enormous fize. What is curious, the fhell only is penetrated with pyrites, and its hollows are filled with fandftone or other common matters. Pyrites is alfo found very frequently cryftallized. The cubical brafs-like cryftals in common flate are the pureft fpecimens of iron pyrites that I know. But there is a vaft variety of other forms, many of them very beautiful, adhering to almoft all ores and fpars in metallic countries. Thefe cryftals are commonly the moft fhowy parts of a foffilift's cabinet.

It is only from the richeft pyrites that fulphur is now extracted. Such as give lefs than one-fifth of their weight are not worth the labour and expence; but are long roafted in open heaps, to drive away the fulphur, and reduce the pyrites into a ftate fit for fmelting for its metal. In fome of thefe heaps, cavities are left in proper places, into which the fulphur, which evaporates or melts from the reft of the heap, is directed and there collected.

Very rich pyrites are treated for their fulphur by a rude kind of diftillation. The retorts are nothing but conical earthen pipes, about four feet long, ten inches wide at one end, and two at the other, and open at both ends. The furnace is an oblong fpace between two walls three feet afunder, having feveral holes in the roof for the fmoke, and alfo a row of holes in the floor, to admit air to maintain the fire. The conical pipes lie acrofs this gallery, the wall on one fide having holes ten inches wide, and the oppofite wall having as many two inches wide. Thus the pipe fits both holes, and is made tight with lute. It lies with a fmall defcent towards the narrow end, which projects three or four inches, and a rude receiver is there joined to it. The lumps of pyrites are put in at the wide end, and when the pipe is full, that end is fhut up by a round tile luted into it. The narrow end of the pipe is also plugged with a bit of tile shaped like a star. This allows the melted or vaporized sulphur to get out, but keeps in the lumps of pyrites. The space round the pipes being filled with fuel, the fire is kindled, and kept at a moderate heat, by opening or shutting the holes in the roof. The sulphur melts and evaporates, and runs into the vessels under the small end of the pipes; and when no more runs, the tile is taken out of the wide end, and the effete pyrites are raked out, and fresh put in immediately; so that the work is not interrupted. The pyrites taken out still retains sulphur. This is driven away by roasting them in a heap, and then the remainder is smelted for its metal.

But the greatest part of the sulphur used in Europe is obtained much more easily. It is found in vast abundance, ready formed, in the volcanic countries, particularly at Solfatara in Italy. It is undoubtedly produced in those countries in the same way that we produce it by art. The subterranean fires are occasioned chiefly by the decomposition of the matrices of sulphur; and they act on them in the same way as the fires in our laboratories, causing the sulphur to fly off in vapours through every fissure and cavity. It sublimes into those cavities, and collects in vast quantities, so as even to break down by its own weight. The vapours even penetrate upwards to the very surface, impregnating even the soil, so that the very earth taken up with the spade will yield sulphur. In the cavities, and adhering to the stones and rubbish, it is often found pure and crystallized,—sometimes transparent. This is called *sulphur vivum.* But it is generally foul, like what is obtained from the pyrites.

Crude sulphur is contaminated by earthy matters diffused in it, and also by metals which are in some measure united to it. It is purified by melting, and keeping for some time fluid; by which means the impurities subside, and the pure sulphur is raked off, and poured into moulds which form it into the little rolls in which we commonly see it.

It is alfo purified more completely by fublimation. For this purpofe, a furnace, holding a pot, in which the fulphur is melted, is
built at one end of a room, which is fhut on all fides, and has the walls,
ceiling, and floor, covered with lead or a proper cement. The fulphur rifes from the pot in fteams, which are condenfed on the walls,
ceiling, and floor, in the form of fine duft, called FLOWERS OF
BRIMSTONE. Thefe are fometimes a little acid, owing to the air
which originally filled the room, the oxygen of which decompofes
fome of the fulphur. They muft be cleared from it by wafhing
before the fulphur can be ufed for all purpofes of medicine.

The medicinal preparations of fulphur are not many. It is combined with quicklime, in the moift way, forming a calcareous hepar,
which we have paid fufficient attention to already : But this is only a
previous ftep for the preparation of LAC SULPHURIS, by precipitating
the fulphur with any acid. The muriatic is the one prefcribed in the
London Pharmacopœia for this purpofe.

Sulphur is alfo combined with fome oils, to compofe BALSAMS OF
SULPHUR,—naufeous medicines, having a moft offenfive fœtor, occafioned by emanation of an inflammable gas.

The chief preparation from fulphur, and it is a very important one,
is its acid. This can be obtained only by inflammation, for we need not
fpeak of the procefs of diftilling a great quantity of nitric acid from
fulphur. This, indeed, gives the acid, but in a way vaftly too expenfive. In the ordinary way of burning fulphur, the acid is obtained
in a volatile fuffocating form. It is an acid vapour, holding, but very
flightly, a confiderable portion of fulphur in folution, and a great
quantity of water, collected from the atmofphere. It is what we
named *fulphurous acid*. That this is its nature and compofition, is
proved by diftilling vitriolic acid, into which a bit of fulphur, or almoft any inflammable fubftance, has been put. If we expofe fome of
this volatile acid to vital air, under a glafs, it will abforb it in a minute, gain weight, and lofe entirely its volatility and fuffocating fumes,

becoming *fulphuric acid*. This is not altogether by faturating the redundant fulphur, for fome of it is precipitated.

Sulphurous acid is fcarcely acid to the tafte; alfo its chemical relations as an acid are exceedingly weak, fo that almoft any acid will detach it from an alkali. It has diftinguifhing properties; particularly that of quickly difcharging vegetable colours. The fuffocating vapour of burning brimftone will make a red rofe white in a few minutes. It is employed for whitening woollen, and filk, and linen goods. This volatile acid could not be obtained in a tangible form, were it not for the moifture diffufed through the air; and it is this alone which makes it liquid in the old procefs. This procefs was to burn fulphur with a fmall flame, under a large glafs bell, kept cool by a wet cloth. The acid formed on its infide, and thence it dropped into a funnel. This produced the *fpiritus fulphuris per campanam*, now called fulphurous acid, to denote its redundancy in fulphur.

To obtain fulphuric acid free from thefe fuffocating fumes, we need only expofe the fulphurous acid to the air. The redundant fulphur partly attaches to itfelf more oxygen, and partly is loft, perhaps by precipitation in the air.

The manufacture of fulphuric acid, in the ftate which it is required in chemiftry and the arts, is a very difficult procefs; and a very lucrative manufacture to thofe who are poffeffed of the fecret. It is, as yet, in very few hands. This much is known of it, that the fulphur is kept burning in a pot, fet on a fmall furnace, built at one end of a chamber which is clofed on all fides, and lined and floored with lead. Water is kept on the floor, to aid the condenfation of the acid vapours. Nitre is employed, mixed with the fulphur, before inflammation. This maintains the combuftion, which otherwife would foon ceafe, becaufe the fulphur quickly abforbs the oxygen of the air that fills the chamber in which it is burned, and its vapours condenfed. The remaining azote, and the acid vapours already produced, foon put an end to the inflammation: The nitre maintains it by fupplying oxygen. But nitre is a very coftly article, and no more can be afforded than what juft keeps up the inflammation. Even with this help, a great many nice contrivances are practifed in the great manufactures of Great Britain, (which alone, as yet, poffeffes the fecret) to keep up the inflammation; and, after all, the acid is loaded with much redundant fulphur, and is very weak, becaufe much water was employed for condenfing the vapours. It muft therefore be concentrated. This would be a fimple operation with fulphuric acid, which is much more fixed than water. We needed only to evaporate the fuperfluous water. But here it is difficult, partly becaufe the volatile fteams contain fulphur, which would be loft, and partly becaufe, in a very diluted acid, the water will carry off much acid. It is concentrated by a diftillation in great retorts, feveral times repeated, in a way which would be tedious to defcribe, but which a little reflection on the properties of the fubftances will fuggeft to you. Attempts have been made to fupply oxygen in this part of the procefs by means of manganefe, a mineral of low price, which contains much of it very loofely combined. The acid muft be dephlegmated till it acquire a fpecific gravity almoft twice as great as that of water. It muft be at leaft 1,6 or 1,7 before it is fit for diffolving indigo for the dyers.

When fully faturated with oxygen, it feems to contain feven-tenths of fulphur, and three-tenths of oxygen.

IV.—CHARCOAL.

Charcoal is the fourth kind of inflammable matter which I enumerated, and which deferves a feparate confideration.

What is commonly called charcoal is not produced by nature, though there are native inflammable fubftances which approach to it nearly by their qualities and conftituent parts. But they are known by other names. What is commonly meant by this term is artificial, and produced from wood, by burning it, or rather fcorching it with a fmothered fire, until it be red hot, and then ftopping the further progreſs of combuſtion, by covering it ſo cloſely as to exclude complete-ly the further action of the air *.

This proceſs is practiſed, as I ſaid, with wood only, when we wiſh to prepare what is commonly called *charcoal,* which is employed in refining iron, preparing ſteel, &c.   But a ſimilar product may be ob-tained, by ſimilar treatment, from all vegetable and animal ſubſtances, and from the bitumens; all of which, if heated gradually until all their volatile matter is expelled, and with proper precaution to pre-vent inflammation, will afford charcoals.   Such is the charcoal from pit-coal, called COAKS.   Small pieces of charcoal may be finely pre-pared for experiments, by plunging a piece of hard wood into red hot lead, and keeping it there till all ebullition or eruption of vapours of any kind is over; or by putting it into a crucible with good ſand, and luting on a cover, to prevent all action of the air, and then keep-ing this for an hour or more in a red heat.   Dr. Prieſtley obſerves, that charcoal which is made with extreme ſlowneſs retains much in-flammable matter, which a haſty operation diſſipates in vapour; and that, after it has ſo retained it, no heat can afterwards diſſipate it in cloſe veſſels.

All theſe ſubſtances from which charcoal can be prepared contain a conſiderable proportion of inflammable matter.   Their other conſti-tuent parts are a ſmall portion of earth and ſalts, and a large quantity of water.   By the action of the ſmothered heat, the water is diſſipated, partly alone, in watery ſteams, partly combined with other matters, in oily or ſooty vapours, which are ſtill very inflammable.   But a parti-cular kind of the inflammable matter remains united with the earth and the more fixed ſalts, and with theſe conſtitutes the charcoal that is at preſent to occupy our attention.

Charcoal is always black and opaque, and, if produced from a ſolid ſubſtance, ſuch as wood, bone, or the like, generally retains the exter-

* Vide Les Arts et Metiers, CHARDON.

nal form, and fome appearance of the organic ftructure of the mafs from which it was produced.

Charcoal thus formed is diftinguifhed among the inflammables by feveral remarkable and peculiar qualities:

1ft, When we expofe it to the action of heat alone, and take care that all accefs of frefh air to it be effectually prevented, it appears to be perfectly fixed and unalterable by heat. The utmoft violence of heat applied in this manner has no power either to melt it or to volatilize it. Dr. Prieftley, it is true, thought that he had converted charcoal totally into a fpecies of inflammable air by the action of heat alone. He made the experiment by placing the charcoal in a vacuum of an air-pump, and directing the focus of a burning glafs on it. But there is great reafon to fufpect that this effect was not produced by the action of heat alone, but by that of heat and watery vapour applied at the fame time to charcoal. That watery vapour, affifted with a red heat, very quickly confumes and volatilizes charcoal, is now very certain, from the experiments of Mr. Lavoifier, Dr. Prieftley, and others; viz. thofe already mentioned, which were made by pufhing the fteam of the water through a red hot tube, in which different fubftances were expofed to its action. When the experiment was made with charcoal, it was attended by the production of hydrogen gas and carbonic acid. *This* hydrogen gas from charcoal contains fome charcoal diffolved in it, and being thereby heavier than pure hydrogen gas, it is named by fome authors *heavy inflammable air*, by others, *carbonated hydrogen*.

Mr. Lavoifier is of opinion that the charcoal is only diffolved on this occafion, but not, properly fpeaking, decompounded. He thinks that it is the watery vapour which is decompounded, and fupplies the hydrogen gas, the charcoal joining with the oxygenous principle, or bafis of vital air, which the water contains, and being thereby diffolved and converted into carbonic acid. We have not, therefore, as yet, any proof that charcoal can be volatilized by the action of heat

alóne, unlefs we reckon on fome light black powder obtained in diftillations of fulphuric acid from inflammable fubftances.

As charcoal fuffers no change from the moft violent action of heat
alone, fo it does equally refift the powers of the air and humidity,
when thefe are not aided by heat.   It is a common practice to fcorch
the ends of ftakes which are to be driven into the ground, that they
may be lefs liable to rot and decay; and this is done with confiderable fuccefs.   They become thereby a great deal more durable *.
But when wood is perfectly charred, we have no experience of any
end to its duration.   It is totally exempted from any change or decay
to which fo many other fubftances are liable by the long continued
action of the elements on them.

The knowledge of this incorruptibility of charcoal has given occafion to a moft ufeful contrivance, which has been lately thought of,
for preferving water uncorrupted at fea.   Water at fea corrupts, in
confequence of its drawing a tincture, or diffolved matter, from the
wood of the cafk, which diffolved matter is a very corruptible fubftance.   But it has been lately contrived to burn or fear the internal
furface of the cafk with flame or hot irons, until that internal furface
is changed into charcoal, after which it no longer yields any corruptible matter to the water.

Charcoal equally refifts the moft powerful and deftructive folvents..
Thofe which tear into atoms the metals themfelves have not the leaft
effect on it fo long as the charcoal remains cold.   It is only by being
heated that it becomes liable to the action of various bodies.   In its
hot ftate, all thofe fubftances which are difpofed to act remarkably on

---

* About forty years ago a number of pointed oak ftakes were difcovered in the bed
of the Thames, in the very fpot where Tacitus fays that the Britons fixed a vaft number of fuch ftakes, to prevent Julius Cæfar from paffing his army over by that ford.
They were all charred to a confiderable depth, and retained their form completely; and
were fo firm at the heart, that a vaft number of knife-handles were manufactured from
them, and fold as antiques, at a high price.   EDITOR.

the inflammable bodies in general shew that they have the same power on charcoal. Such are the sulphuric acid, the nitric acid, nitre, the phosphoric acid, calces of metals, and the vapour of water. Hepar sulphuris has also the power to dissolve charcoal.

When a pound of sulphuric acid was distilled with half an ounce of charcoal, the charcoal was totally dissolved, and it changed a part of the acid into sulphurous acid. The liquor became of a sea-green colour, which disappears when the liquor is cooled, and return when it is again heated. A greater portion of charcoal produces sulphur, by distillation,—the operation is extremely troublesome,—the production of sulphur very partial,—and a great part of the charcoal is volatilized, in the form of an impalpable powder, which collects in the neck of the retort, and often chokes it up. The distillation may be repeated again and again with the volatilized acid and charcoal, and always produces the most intolerable suffocating fumes; but the colour at length disappears. It also deserves remark, that both the sulphur and the volatilized charcoal appear before any considerable quantity of the acid has come over.

Nitric acid also acts remarkably on dry charcoal, when very strong. When the mixture takes place properly, it bursts out into a flame. In all cases, the charcoal decomposes much of the acid, by depriving it of oxygen, and we obtain nitrous gas and carbonic acid.

We have already seen the effect of the acid of phosphorus on charcoal. It is decompounded by the charcoal, and we obtain phosphorus and carbonic acid; which last must be allowed to escape.

It dissolves in hepar sulphuris both in the humid and dry way, rendering it of a much deeper colour. In the dry way, charcoal combined in a very great proportion, produces very singular effects. To produce this combination, however, in the best manner, the sulphur must be taken red hot, in its nascent state, as in the process for producing fulphur *de novo*, by treating charcoal in a red heat with a vitriolic falt. In this compound, the united attractions of both the fulphur and the charcoal for oxygen produces fuch a rapid combination and heat, that the mixture takes fire when expofed to the air. Such mixtures are therefore called *pyrophori*. I fhall conclude this article with a particular account of them.

As heat difpofes charcoal to be attacked and decompounded by thefe fubftances, fo does it alfo prepare and difpofe it to be acted on by the air. You all know that the confequence of heating it red hot in the open air is an immediate beginning of its inflammation, during which it is gradually confumed, and a great deal of heat is produced. The apparent quantity of the uninflammable matter into which the charcoal is changed, or which remains in its place when the inflammation is completed, is a very fmall quantity indeed, when compared with that of the charcoal before it was inflamed. It is only one-ninth of its weight. But this appearance is a deceitful one. There is a much greater quantity of uninflammable matter produced, but it affumes an elaftic aëreal form, and is diffufed in the atmofphere imperceptible to the fight. It is fixed air, or the carbonic acid gas, the total weight of which far exceeds the weight of the charcoal that has been confumed. We can collect and condenfe this gas, fo as to reduce it to a more perceptible form, by the attraction of cauftic alkalis or quicklime. And there are other ways by which we can render it perceptible, and eftimate the quantity of it. When this is done with exactnefs, the weight of it is always found equal to the joint weights of the confumed charcoal and that of the oxygen gas expended in confuming it, and changed along with it into carbonic acid gas. Mr. Lavoifier, by a very careful meafurement, found that 100 parts, by weight, of carbonic acid, contains nearly 72 parts of oxygen, and 28 of charry matter.

We have an example of this in the deflagration of charcoal with nitre. You remember what a blowing noife, and fucceffion of explofions, accompanies that deflagration. Curiofity to learn what happens on this occafion, and what was the caufe of the aftonifhing force of gunpowder, fuggefted a contrivance to fire fome of it, or a mixture of charcoal and nitre, in a ftrong piftol barrel, the end of which was foldered clofe, and the touch-hole fcrewed up. And the quantity of materials fired at once being but very fmall, there was no report or explofion ; the ftrength of the machine being fufficient to confine the flame and elaftic matter. The only external fign of the firing of the powder or compofition was, that the barrel became hot. It was allowed to cool, and then opened under water, by unfcrewing the touch-hole. A quantity of elaftic aëreal matter came out, which was furprifingly great when compared with the quantity of charcoal and nitre which had been fired. This experiment was made by Mr. Robins, Engineer to the Honourable Eaft-India Company.

But others having repeated it fince the nature of the aëreal fluids became a fubject of inquiry, they found that the elaftic aëreal matter produced by charcoal and nitre is a mixture of carbonic acid gas and of azotic gas.

Knowing, as we now do, that nitrous acid is compofed of feven parts of oxygen, and three parts of azote nearly, we are enabled to explain very clearly the production of azotic gas during the deflagration of charcoal with nitre. And the other experiments which have been defcribed to you explain the production of carbonic acid at the fame time. It appears, that the charcoal when heated, acts by a powerful attraction for the oxygen, which abounds in the nitric acid. By uniting, they form the carbonic acid. The other principle of the nitric acid, the azotic gas, is thus fet at liberty, and being added to the fixed air, makes up that great quantity of elaftic aëreal matter which is difengaged from charcoal and nitre.

There is another production fometimes obtained from this deflagration, viz. a very fmall quantity of volatile alkali. This is explained by another late difcovery which I already mentioned, concerning

the component principles of the volatile alkali, which is now sup-
pofed to be a compound of hydrogen and azote. This being admitted, we can account for the formation of a fmall quantity of it from
fome of the mixtures of charcoal and nitre. Some part of the water
or humidity inherent in charcoal, and in nitre cryftals, is decompounded during the heat of deflagration, by the attraction of the charcoal
for the oxygen of the water. Thus the other principle of the water,
the hydrogen, is let loofe, and joining with a part of the azote, forms
the volatile alkali. This, however, is in this cafe an accidental production, in fmall quantity only, and not always perceptible. A chemift
in Birmingham, whofe name I do not recollect at prefent, forms volatile alkali in abundance, by paffing common air over a mixture of
charcoal and vitriolic falts with an earthy bafis (fuch as alum or Epfom falt) made red hot.

After this explication of the deflagration of nitre with charcoal, I
may fo far take notice of the deflagration of the fame falt with fulphur, as to remark that the products of that deflagration are accounted for by the fame principles. The products of the deflagration of
nitre and fulphur, are fulphuric acid, formed by the union of the fulphur with the oxygen of the nitric acid, and azotic gas, formed of the
azote of the nitric acid and latent heat. The fulphuric acid, the moment it is formed, joins itfelf to the alkali of the nitre, fo that very
little of it is volatilized.

By the fame ineftimable difcovery of Mr. Cavendifh, by which we
firft learned the conftitution of nitrous acid, we can alfo account for
what happens in the operation by which the largeft poffible quantity
of vital air may be extracted from nitre. In that operation, the nitre
is expofed to the action of heat in a fmall retort of earthen ware, or
of glafs well coated with clay and fand, and which has a long neck.
Sometimes an iron retort or gun-barrel is ufed, but it is not fo proper.
The retort is expofed to the immediate contact or action of the burning fuel, but heated gradually in the beginning. As foon as the re-

tort and nitre begin to be ignited, the vital air begins to come forth. At the firſt it is very pure, provided the nitre be quite free from duſt or admixture of vegetable or animal matter. If ſuch matter be preſent in it, though in exceedingly ſmall quantity, ſome carbonic acid gas is formed at firſt, and comes out mixed with the firſt portions of vital air and ſome azote. After this, or from the beginning, if the nitre was quite pure, we obtain a great quantity of good vital air, which, if not quite pure, contains a ſmall admixture of azotic gas. But, towards the end of the operation, when the heat muſt be increaſed to a higher degree, the vital air comes over mixed with a more conſiderable portion of azotic gas. The reaſon of this difference in the purity of the vital air is, that in the beginning, the nitric acid gives out only a part of the oxygen which it contains, (nearly $\frac{70}{100}$) and it is changed into nitrous acid, retaining perhaps $\frac{65}{100}$, and in this ſtate continues adhering to the alkalis. In the nitrous acid, therefore, there is a deficient proportion of oxygen, and a ſuperfluous quantity of azote. When the proceſs is continued, and the heat increaſed to produce a more complete decompoſition of the acid, the azote muſt neceſſarily make its appearance in a proportion continually increaſing.

All theſe facts, and many more, to be mentioned as we proceed, and which are ſo well explained by Mr. Cavendiſh's diſcovery, confirm the ſolidity of that diſcovery, making it one of the moſt important which the modern chemiſtry has produced.

One of the moſt remarkable properties of charcoal, when recently tak.n from the fire, is an attraction for a certain quantity of humidity, and for various odorous and colouring matters of different fluids, containing animal or vegetable ſubſtances, ſubject to fermentation or corruption,—as alſo for the acetous acid.

We have proofs of its attraction for humidity in many curious experiments of Mr. Scheele and Dr. Prieſtley. Although indeſtructible by heat in cloſe veſſels without addition, yet, if moiſtened, it will yield carbonic acid and hydrogenous gas. This may be repeated by an-

other moiftening; and fo on, till it is all expended in thefe productions.   This is evidently owing to its ftrong attraction for oxygen, in
which it exceeds all fubftances yet examined.   It decompofes the
water,—combining with the oxygen, and thus forming carbonic acid,
and thus alfo leaving the hydrogen at liberty.

Its action on odorous effluvia is no lefs remarkable.   If laid (frefh
made) on filk or linen gummed or oiled for umbrellas, a preparation
which continues to exhale a heavy fickening fmell for many years, it
will remove it in a few hours.   It fweetens bilge water, and all kind
of corruption that is accompanied with emiffion of hepatic ammonia.
It clears faline folutions of their colouring matter and rank fmells,
caufing them to cryftallize in fnow-white purity; and is much ufed
for this purpofe in pharmacy; as in the preparation of the *terra foliata
tartari*, which was formerly a tedious procefs, and confidered as a teft
of pharmaceutical dexterity.   It removes in an inftant the heavy flavour of corn fpirits haftily diftilled.   It clears foul camphor in the
fublimation from all fuliginous taints.   It fweetens water which has
grown putrid by long keeping.   It even fweetens meat which has
already putrified to a very great degree.   Mr. Cappe at Lille has publifhed valuable experiments on this fubject,—as has alfo Mr. Lowitz
an eminent chemift at Peterfburgh in Ruffia.   Charcoal is therefore
an excellent dentifrice, as very well adapted to the mechanical operation of cleanfing the teeth, and ftill more as the moft powerful corrector of all putrefcence, which is the chief caufe of all diforders of the
teeth and gums.

The acting principle in thefe effects is not yet diftinctly underftood.
As they are generally accompanied by an immediate and great increafe
of the offenfive fmells, we are led to afcribe its efficacy to its attraction for oxygen, by which moft of thofe gafes are fet at liberty.

Powdered charcoal clears water impregnated with carbonic acid fo
completely that it no more renders lime-water milky.

Charcoal is found to act powerfully in relieving from the pain of heartburn.

In confequence of its ftrong attraction for pure acetous acid, it becomes a powerful agent for concentrating it by diftillation. We are indebted for this, as well as for the full confirmation of the laft mentioned chemical property of charcoal, to Mr. Lowitz. After having concentrated this acid as much as poffible by freezing, he mixed it with a great proportion of charcoal frefh made, and diftilled it till the charcoal was feemingly dry ; then, changing his receiver, he obtained from this charcoal acetous acid, in the utmoft ftate of concentration and purity, and which cryftallized in a cold little below that of freezing water. This is fomewhat of an anomalous fact, becaufe charcoal exhibits no remarkable attraction for acetous acid in a lefs concentrated ftate.

I have already obferved, that charcoal attracts oxygen more powerfully than any other fubftance does that we are acquainted with. We cannot decompofe carbonic acid by any fingle elective attraction. Charcoal is employed for feparating it from all other bodies ; and, in confequence of this power, charcoal is the great inftrument in all metallurgic operations; and in all operations by which we reftore to bodies their quality of inflammability. For inflammability is deftroyed only by attracting oxygen from the atmofphere, and becoming faturated with it. I have already mentioned the ingenious procefs by which Mr. Tenant effected the decompofition of the carbonic acid, by means of phofphorus and an alkaline fubftance.

From all that has been faid of charcoal, you perceive that there is a principle common to every combuftible fubftance procured by charring combuftible bodies, and that this principle is exceedingly fubtile, fince it is found to compofe fo pure a fluid as carbonic acid, or fixed air. You now fee, that although the black fubftance obtained by the fmothered burning of many bodies be folely the production of art, yet the common principle, the *carbon*, fo called to diftinguifh it from the groffer body in which it is found, is one of the moft copious and uniform productions of nature. The calcareous, and other abforbent earths, muft now be added to the numerous claffes of bodies that contain it; and we muft, in fhort, confider every thing as an ore or matrix of carbon, which contains fixed air, or which produces fixed air by union with oxygen.

It is remarkable that a principle fo abundant in nature fhould never be feen in its native form, pure and unmixed. But this arifes from its activity and difpofition to combine with almoft every fubftance in nature. The wonderful changes of external appearance which may be induced by fuch combinations, are now fo familiar to you, that although you may be furprifed, you will not be difpofed to doubt, when I venture to fay that in all probability the native, unmixed, form of carbon, is what is known to the world under the name of the DIAMOND! Surely nothing can be more unlike than the moft brilliant, the moft tranfparent of all bodies, to a fubftance effentially black, and completely impervious to light. You will be eager to know the evidence on which I venture fuch an unlooked-for opinion.

When confidering the filiceous earths, I obferved that quartz, cryftal, and others abounding in them, were diftinguifhed from the reft by feveral peculiarities. A remarkable fmell when they are rubbed together; the light which their friction produces; and feveral appearances of elaftic matter iffuing from them in their union with fixed alkalis; and their refufal to unite with other filiceous earths, were confidered by many chemifts, and particularly by the Chevalier Dolomieu, as indications of inflammable matter in them.

The diamond, befides thefe qualities, is farther diftinguifhed by being totally diffipated by a violent heat; but this only when expofed to the joint action of heat and air. Till the neceffity of this combination was difcovered, the diffipation of diamond was confidered as analogous to the decrepitation of falt, and of many foffils, which may be difperfed by heat in fmall fragments, which in fome cafes are even a fine powder; but thefe fragments could always be collected, whereas the diamond difappears altogether. It has long been fufpected, therefore, to be inflammable.

Accordingly, feveral chemifts have lately examined it with the exprefs view of afcertaining this point. Crucibles of fine porcelain, having ground ftoppers of the fame fubftance, were employed; and it was found that only a part of the diamond could be evaporated in this way; fo much the more, as the veffels contained more air. When covered with the powder of charcoal in the crucible, no change whatever was produced by the moft intenfe heats. When fo heated in free air, it gradually confumed, and was all the while of a dazzling brightnefs, much more brilliant than the capfule on which it was lying. Although there may be fome impropriety in faying that diamond is inflammable, like oils, yet it appears to be combuftible like charcoal. Count Sternbergh, a gentleman in Bohemia, burnt a diamond in oxygen gas, by fixing to the point of it a bit of iron wire, which he made red hot, and then plunged the whole into a veffel containing vital air. The wire burned with great fplendour and production of heat, and this being communicated to the diamond, it took fire and burned in the fame manner. The experiment was repeated in a glafs veffel, whofe infide was moiftened with lime-water. It very foon became dim, and was totally obfcured, fo that the progrefs of the combuftion could not be obferved. But this gave ftrong indication of the prefence of carbon in the diamond.

Mr. Smithfon Tenant mixed $2\frac{1}{2}$ grains of diamond duft with one-fourth ounce of nitre cleared of the water of cryftallization, and expofed the mixture in a tube or retort of gold. A good deal of nitric acid was difengaged before the nitre began to act on the diamond. By this circumftance, the carbonic gas, produced by the action of the oxygen in the inftant of decompofition, was abforbed by the fixed alkali, for which it has a ftrong attraction, and it was all retained. This was obtained by diffolving in water, and adding muriat of lime, which formed a digeftive falt, and precipitated a crude calcareous earth, and thus gave Mr. Tenant hold of the carbonic acid. He found this calcareous earth to contain $9\frac{1}{7}$ grains of carbonic acid. By Mr. Lavoifier's repeated examination of the proportion of ingredients in this acid, he found that it held $2\frac{1}{7}$ grains of carbon, which is precifely the weight of the diamond employed.

By this ingenious experiment, which has been repeated both in London and Paris, we feem entitled to conclude that diamond is carbone in a cryftalline form.

Another experiment leading to the fame conclufion, has lately been made by Mr. Morveau (Guyton de). He expofed to an hour's intenfe heat, a diamond inclofed in a tube of iron, put into a crucible, and furrounded with a mixture of filiceous and argillaceous matter, which had been burnt in a burnt clay crucible. This crucible was inclofed in another, which was coated with a fimilar compofition. The diamond vanifhed entirely, though the tube had no aperture, and the infide of the tube was converted into perfect fteel. You will afterwards learn that this change is produced by the union of iron and carbon. You fee the propriety of Mr. Morveau's precaution to cut off all communication from the fuel to the tube.

The diamond burns at a much lower heat when mixed with nitre, than by any other treatment. When merely heated in contact with atmofpheric air, a very ftrong red heat is neceffary to induce that bright fhining, which is the indication of its uniting with oxygen, or of burning. When this is done in pure oxygen, the combuftion begins a little fooner,—but when mixed with nitre, the combination begins with the loweft red heat. Mr. Tenant found that common charcoal may be rendered extremely hard, fo as to fcratch and work upon tempered fteel, by long continued ignition in clofe veffels ; and that by this treatment its tendency to combuftion is fo much repreffed, that it does not begin to burn till red hot. There are fome very remarkable examples of this kind which have lately come to my knowledge. In the dreadful eruption of lava in Iceland, trees growing in crevices of the rocks have been buried under the melted lava, and there charred and covered with melted matter, which has not ceafed to be red hot for more than two years. Fragments of thefe are fome- times brought into view by the convulfions and fhatterings of fubfe- quent eruptions, and are called *fvaërt fouterbrant*. Some pieces of them in my poffeffion require a full red heat for their combuftion, and they are totally confumed. I have obferved the fame thing in a native coak, lately difcovered near Newcaftle, which has been produced by the protrufion of a mafs of melted whinftone, which now forms a dyke in the ftrata, and it has charred above two fathoms of the feams on each fide.

Appendix.—Pyrophori.

I conclude this article with the account of a very curious clafs of chemical preparations, which derive their diftinguifhing proper- ties from charcoal. They are called PYROPHORI, becaufe they are always in a difpofition to take fire and burn whenever they are ex- pofed to the free air. They are as commonly known by the name of *phofphori,—light bearers*, (but with lefs propriety), becaufe they have this refemblance to Brandt's phofphorus of urine. They are diftin- guifhed, however, by the name of *Homberg's phofphorus*, or *phofphorus of alum.*

This is generally a blackifh or dark coloured powder, like char- coal, or the half burnt afhes of combuftible bodies. It is kept in bot- tles well corked; and if a quantity be poured out into the air, efpe- cially if the air be a little damp and warm, it grows hot, fmokes, and prefently takes fire, burning like as much charcoal, with a difagree- able fmell.

This phofphorus of Homberg is prepared by firft roafting a quantity of meal or flour of any kind, till it is almoft charred, taking care that it be not completely fo. This charry powder is mixed with a quantity of alum, deprived of its water of cryftallization; and the mixture is put into a phial, fo as nearly to fill it. The phial is fet among fand in a crucible. The fand muft furround the phial as far up as it is filled with the powder. The whole is kept in a mode-rately hot fire, till all fumes have ceafed, and a very faint flame, like that of fulphur, has continued for fome time. It is now removed, and the mouth of the phial ftopped with a cork as foon as it is cool enough to allow it.

When a little of this powder is fhaken out of the phial, and formed into a heap, it grows hot, fmokes, fmelling ftrongly of hepar fulphu-ris, and then takes fire, if well prepared. If much over or under cal-cined, it is apt to fail; and it is generally in its beft ftate when it has fomething of a dirty green colour mixed with it, as if it contained powder of fulphur.

Subfequent experiments and obfervations have fhewn that any falt which contains the fulphuric acid will do as well as alum for the preparation of this pyrophorus. Scheele obtained the fineft from vitriolated tartar, or fulphat of potafh. Nay, the natural produc-tions, gypfum, fluor, and others, formed by the fame acid, will an-fwer. We know that all fuch compounds, when treated with com-buftible bodies in a ftrong heat, produce fulphur or hepar fulphuris. Thefe pyrophori may therefore be called *fulphureous pyrophori*, or perhaps *hepatic pyrophori*. Mr. Scheele has obferved that the fteams of hepar fulphuris generate much heat when they abforb vital air. It is very probable that the copious abforption in the prefent cafe, (for it is very copious), may produce enough to kindle the mixture of naf-cent fulphur and charcoal.

This fubftance has not yet been examined with the attention which it deferves; and we are not yet well affured of the procedure of na-ture in its inflammation. Mr. Lavoifier has beftowed particular attention on compounds nearly allied to it, in his first essays, published in 1777, and also in his subsequent papers.

There are many other substances which take fire of themselves, by the heat produced in the mutual action of their ingredients. Of this we have already seen an example in the mixture of nitric acid with essential oils; also with phosphorus. We know that the same thing happens in many cases, where vegetable or animal substances are heaped together in a moist state. They ferment, grow hot, and frequently burn, or at least are reduced to ashes. We shall have occasion to take notice of some remarkable examples of this kind as we proceed in our examination of inflammable bodies.

At present, I propose to bestow a little attention on another class of bodies known by the name of PHOSPHORI, and to which that name does more peculiarly belong. These bodies do not take fire, that is, do not suffer a decomposition of their parts, accompanied with the production of heat and light. The bodies I now mean to consider only shine in the dark, but without any sensible heat. And they acquire this faculty, not immediately by exposing them to the action of the air, but to that of a strong light.

Of these there is a class very nearly allied to the substance I have last described, by its composition and chemical properties. The most remarkable of them, and the first observed, is that called the *Bolognan phosphorus*,—the *Bolognan stone*, because first found in the neighbourhood of Bologna, about the year 1630, by one Cascarioli, (a shoemaker I believe) who observed, that when taken from the light into a dark place, it continued to shine for some time. As soon as this was publicly known, several philosophers of that time examined the appearances with more care. It was found that two or three seconds exposition to the light was sufficient to make it shine, and that longer exposure did not increase its luminousness. It shone for four or five minutes, and some fine pieces shone for a quarter of an hour. The light of the sun was the most effectual. After this the clear light of day. Nay, the light of a torch made it fhine. Moon light had no effect. The light emitted was enough to make the fmalleft print legible, when held very near.

Thefe facts attracted much attention, and were very interefting to the philofophers of laft century, who were at that time much divided in their opinions concerning the nature of light. Some imagined, with the vulgar, and with the followers of Newton, that light was a material emanation from the luminous body; while others, with Des Cartes at their head, imagined that light, or vifion, was the effect of a tremulous motion of an elaftic fluid, as found, or hearing, is the effect of the tremulous motion of elaftic air. The firft clafs drew ftrong arguments in fupport of their opinion from the phenomena of this ftone, for it feems to imbibe the light to which it is expofed, and afterwards to give it out again.

It is fomewhat fingular that fo curious a phenomenon did not engage the chemifts of that age in a very minute examination of the conftitution or ingredients of the Bolognan ftone, and caufe them to fpeculate about the way in which it produced thofe effects. This inquiry was more ftrongly fuggefted to them by the methods which muft be taken to make the ftone a powerful phofphorus. It muft be beaten to fine powder, and then made into a pafte with water and oil, and then calcined. But it lay a half century neglected; till, in 1675, Baldwin, a German chemift, obferved that the refiduum from the diftillation of the nitrat of lime imbibed the light and emitted it in the fame manner; and in 1693, Homberg difcovered the fame thing in the muriat of lime; and in 1730, Dufay found that many other fubftances, when calcined in the fame way, had the fame property. A feries of curious obfervations of this kind by Beccaria (Giacomo Bartholomeo) greatly augmented this lift. But although thefe difcoveries naturally occafioned fome chemical examination of the Bolognan ftone and calcined phofphori, no addition of knowledge was obtained, till Mr. Margraaf examined them with his accuftomed fkill, in 1749.

He found that the Bolognan stone was a sort of gypsum, or a heavy spar, containing the sulphuric acid united to calcareous earth. Subsequent observations, however, prove it to be a barytes united to that acid. He found that all heavy spars could be made phosphorescent by proper calcination. Leibnitz made the same observation in the first volume of the *Miscell. Berol.* with respect to fluor. Margraaf gives the manner of preparing those substances in the best manner, viz. by pulverizing, and then making the powder into small cakes with a little gum water, then drying them slowly, and calcining them amidst the coals in a reverberatory furnace. So prepared they shine like glowing embers. He obtained phosphori from all earths which contained the sulphuric acid; and the earth of alum was the best, and some calcareous ones scarcely inferior. Some time after this, Mr. Canton of London produced a shining preparation of the same kind, by calcining oyster shells half an hour, and then reducing them to powder, and mixing them with one-third of sulphur. The mixture was rammed into a small crucible, and calcined for an hour in a clear fire. The mass, when broken, was found unequally luminous. The best parts were therefore selected, and were very brilliant. It is to be remarked, that different parts emitted light differing considerably in colour. Figures may be drawn on paper with white of egg, which will take hold of the powder and exhibit the figure in the dark. Canton's phosphorus equals any of the natural stones in brightness, and in the quickness in which it is saturated with light, so as to attain its greatest possible brilliancy. Indeed, in this respect, it may be thought superior; for Mr. Margraaf says, that two or three minutes exposure is necessary for his preparations. When this light emitted by these phosphori has expended itself, or is no more sensible, heating the body renews it for a little time; and when this has ceased, a greater heat causes another emission. But, in all these cases, a greater heat sooner exhausts the power of shining. In short, in this experiment, heat operates in the same way as in expelling dampness, or any other volatile

matter.   Alfo a great heat applied to the phofphorus, while it is expofed to the light, hinders it from acquiring the luminous faculty.
All thefe circumftances concur in ftrengthening the opinion, that this
compofition imbibes, and then emits fomething material, which is
light, or the caufe of light.   Beccaria faid that the Bolognan phofphorus moft certainly did fo, becaufe, when illuminated by coloured
lights, it emitted only the colour which had illuminated it.   This
would indeed be almoft a demonftration of the doctrine of imbibition,
and of the materiality of light ; but the experiments of Mr. Wilfon
with Canton's phofphorus, render this doubtful.   Certain parts emitted a particular colour, whatever had been the colour of the illuminat-
ing light.

When we reflect on the compofition of this curious fubftance, we
cannot but afcribe its fhining to the fame caufe which produces the
inflammation of the pyrophorus of Homberg.   They confift of the
fame ingredients, united by the fame procefs, and they continue to
emit the fame hepatic odour, though long kept, if kept from damp air.
Accordingly, Mr. Macquer confiders the fhining of thefe phofphori
as inftances of a low and imperfect combuftion, fimilar to what we
obferve in the phofphorus of urine, which begins to fhine when expofed to the air, and by this fhining, is flowly converted into phofphoric acid, ftill mixed however with a very great proportion of un-
burnt phofphorus in a volatile ftate.   It is well known too, that fulphur, fat oils, fuet, and many other fubftances, nay, even gunpowder
when heated to a degree far below ignition, emit a faint lambent
flame and fuffocating fumes, which flowly confume them to a certain
degree, and that this happens, without their becoming of themfelves
fources of heat fufficient for the continuation even of this low degree
of inflammation.   If an iron bar be heated red hot at one end, and
allowed to become very hot alfo at the other end, and if we then
make a line along it with a bit of tallow, or white bees wax, or fulphur, or rather make a fet of detached fpots, we fhall fee them fhine

with various brightnefs according to their diftance from the hotteft part of the bar, and the brighteft are fooneft exhaufted of this faculty. Macquer, therefore, afcribed thefe and all other phofphorefcences to the fame caufe, and fuppofed that they were cafes of flow and imperfect combuftion. This opinion is very plaufible, and is well fupported by the emiffion of the hepatic fmell, and fome other appearances. Many kinds of fpar are very luminous when thus heated. The Derbyfhire fluor is remarkably fo. Corrofive fublimate fhines the brighteft of all. Next to this is pure chalk, crude magnefia, earth of alum, vitriolated tartar, French chalk, &c. Mr. Wilfon of London obferved fimilar appearances in many other bodies,—indeed in moft bodies of a white colour. Even paper had this property.

But there are circumftances attending the phofphorefcence at prefent under confideration, which will by no means agree with this explanation. All the emiffions of light juft now fpoken of require the application of heat, and the contact of vital air, and confume this air in the fame manner as common combuftion does. But the Bolognan ftone, and the phofphori of this clafs, require nothing but expofition to the light, and do not require the renewal of the air. Mr. Canton put fome of his phofphorus into fmall glafs balls, and fcaled them hermetically, and found them as good as ever after twelve months fubjection to every trial. He found this to be the beft way of keeping his powdered phofphorus. After twelve months, two feconds expofition to the fun made it fhine, fo as to make the figures on his watch dial perfectly vifible. An electrical flafh, or even the light of a fingle candle, fufficed for caufing it to fhine. Long expofure to dry air impairs it, and damp air does this immediately. In fhort, thefe phofphori only emit the light they have received, or they are rendered luminous by the action of light alone. The firft appears the moft probable, becaufe a greater heat occafions a brighter light and a more rapid exhauftion of all that will be diffipated by that heat. This is very like the expulfion of fomething received. Another circumftance confirms this. No length of keeping in the dark will hinder the emiffion of the light which had been imbibed, and would have been expelled by that heat. A ball kept fifteen months in the dark, after having been illuminated by the moon only, and which did not fhine immediately after, on account of the weaknefs of the light, was diftinctly feen by plunging it into boiling hot water.

Mr. Canton found that humidity inclofed with the phofphorus, that it might ftick to the glafs, impaired it much, and deftroyed it. Alcohol did not hurt it fo much, and æther not at all. Æther, therefore, with a minute portion of refin or gum, would make the powder adhere as Mr. Canton wifhed. The phofphorus might be attached in this way to a plate of glafs, and another plate laid on it, and the edges fecured with varnifh and paper pafted round.

Thefe circumftances evidently diftinguifh this phenomenon from ordinary combuftion. But it were worth while to examine by fome proper train of experiments, whether the fhining is effected by the abforption of vital air, and a fubfequent decompofition by the action of frefh light. As humidity fpoils the phofphorus, perhaps a permanent compound is now formed which frefh light cannot decompound again. We might perhaps learn this by examining the change produced after a long time on the air which was fhut up with the phofphorus in a damp ftate. It would be proper to try this with two balls, one of them to be expofed day and night, and the other kept in the dark.

Mr. Wilfon thinks that this phenomenon is not merely an imbibition and fubfequent emiffion of light, but rather that light acts on fome matter in the body, in fuch a way as to caufe it to emit light. Thus only, he thinks, can we account for white light being emitted by a part which was illuminated only by a red or blue light.

The foregoing obfervations on the phofphori, properly fo called, are, in my opinion, as interefting to the chemift as to the optician, and they are very interefting to both. To us, they are important,

being nearly connected with the whole doctrine of combuftion,—
a doctrine ftill full of difficulties, notwithftanding the very great difcoveries which have been made. The feparability of light and heat by
a plate of glafs, in the valuable obfervation of Scheele, and their feeming feparability in the prefent inftances, are undoubtedly facts of
great moment in philofophical chemiftry, efpecially when confidered
along with Herfchel's obfervations. The obfervation alfo of Mr.
Goettling, that phofphorus fhines bright in azote and in volatile alkali,
and the fhining of vegetable and animal fubftances, (particularly fea
fifh) in a certain ftage of putrefcence, and their fhining in *vacuo* and
other fituations incompatible with combuftion, merit a much more
careful attention than has yet been given.

The four inflammable bodies which have now been confidered
appear to us in the character of fimple fubftances. I do not mean
that they are elements, but only that we are not authorifed, by any
obfervation or experiment, to fay that they are compounded of other
more fimple fubftances known to us in a feparate ftate. We
have never decompounded them, nor formed them by the union of
known fubftances. We have only been able to extricate them from
more complex bodies, and to fhew that they are ingredients in their
compofition.

This fimplicity, and the manifold relations of hydrogen, phofphorus, fulphur, and charcoal, have been of great fervice to us, by giving
us a conception of the remarkable phenomenon of combuftion, which
is incomparably more agreeable to our general knowledge of chemical
facts than the ingenious doctrine of Dr. Stahl, and is really fupported by
proofs which feem incontrovertible. Even though imperfect, this new
doctrine furnifhes us with a fact, formerly unnoticed, which accompanies all combuftion, namely, the combination of the body called combuftible with vital air. We fhall find that this fact, when traced through
all other cafes of inflammation, will give us almoft all the knowledge
of chemical changes that we poffefs, and enable us to explain a vaft

number of the moſt complicated operations of nature, inaſmuch as all explanation of phenomena conſiſts in ſhewing that they are particular caſes of general laws or facts already known.

We proceed now to conſider inflammable ſubſtances of a more complex nature : In doing this, I ſhall ſtill endeavour to introduce them to your acquaintance in the order of their ſimplicity, as far as I can perceive any gradation or order in this reſpect. With theſe views I begin with *ſpirit of wine.*

V.—ARDENT SPIRITS.

VINOUS SPIRIT is produced from ſome vegetable ſubſtances by fermentation, and ſubſequent diſtillation ; and the vegetable materials which by theſe proceſſes yield the moſt of it are,

1ſt, The ſweet juices of vegetables, in their naturally diluted ſtate.

2d, The ſugar, or ſweet matter extracted from vegetables, when properly diluted with water.

3d, Grain, or other farinaceous parts of vegetables, malted and diluted in water.

4th, Grain, or other farinacea, diſſolved or diluted in water, without being malted.

All theſe are capable of the vinous fermentation, by which the vinous ſpirit is formed ; but not all with equal facility and perfection. They ferment the more eaſily and perfectly, and yield the more ſpirit, nearly in the order in which I have now enumerated them.

The nature of fermentation will be conſidered when we ſhall treat of the vegetable ſubſtances in general. It is ſufficient at preſent to know that the vegetable matter which undergoes fermentation, or a part of that matter, is changed by it into vinous ſpirit.

This ſpirit is at firſt diluted and combined in the fermented liquor with a large quantity of water, a portion of vegetable acid, ſome mucilaginous and colouring matter, and a small quantity of a subtile and volatile oil.

The spirit is separated more or less perfectly from these substances by distillation, it being more volatile than most of them, especially the acid, mucilaginous, and colouring matter. The water is but imperfectly separated at first, on account of the small difference of volatility between it and the spirit.

To reduce the spirit to a state of purity, we must perform several other operations; such as distilling it again, once or twice, with a gentle heat, which is called *rectifying*. By this we separate the greater part of the water which had come over in the first distillation.

But, even after these rectifications, it still retains a confiderable portion of water, and along with it another ingredient, which being very volatile, always arises with the spirit, though distilled ever so often. This is the volatile oil which I mentioned as being intimately blended with the spirit at first. The quantity of it and the flavour are different, according to the nature of the particular vinous liquor which we have chosen, and also according to the manner in which the fermentation has been conducted. It is this oily ingredient in vinous spirits which produces a diversity among them in point of flavour. When it is separated from them as much as possible they are all alike, or are distinguished with difficulty. It is this oil which makes the spirit obtained from all sorts of grain in particular so nauseous at first, when compared with some others. This oil becomes most obvious in them when they are hastily diluted with water, being then more perceptible by its disagreeable smell and flavour, and often by some degree of milkiness, which appears when the spirit and water are mixing together, and for some time after.

In order to free the spirit from this oily principle, and from a portion of the water which it retains too strongly to admit of its being separated by distillation alone, we must have recourse to an elective attraction; and the usual method is to employ the common vegetable fixed alkali in its ordinary ſtate, or in the ſtate of pearl aſhes.   About one pound weight is added to every gallon of the ſpirit, and allowed to remain with it 24 hours.   It is diſſolved by the watery part of the fluid; and forms a liquor which remains at the bottom, and cannot be mixed with the ſpirit above it.   The ſpirit may therefore be poured off from it into another veſſel, in which we may add to it half a pound more of the ſame fixed alkali perfectly dry.   This may not perhaps become perfectly liquid, but it may attract as much humidity as will make it ſoft, and will half diſſolve it.   After waiting 24 hours more, the ſpirit may be poured from this alſo into a clean veſſel, and we may make a third addition to it of the dry ſalt.   By theſe repeated operations, we bring it to that ſtate of ſtrength in which it no longer imparts any watery humidity to the dry alkali.

Thus we at laſt ſeparate the whole of the water which the alkali can attract from it.   But, at the ſame time, the ſpirit receives a diſagreeable taſte, and a yellowiſh colour.   This happens in conſequence of its diſſolving a ſmall portion of the alkali, which, acting on the volatile oily matter, produces the yellow colour.   The quantity of alkali thus diſſolved, is, however, but ſmall, though it is ſufficient to give the bad taſte and flavour.   But in order to ſeparate this alſo, as well as the oily matter with which it is combined, another operation muſt be performed, which is, to diſtil the ſpirit with a gentle heat, until a ſmall quantity of it only remains in the ſtill.   The alkali and oily matter will be found remaining with this ſmall quantity in the ſtill; and the ſpirit that has paſſed over in the diſtillation will thus be brought to the higheſt degree of purity and perfection to which it can be reduced by theſe operations.

The artiſts named rectifiers and compounders, who employ themſelves in purifying coarſe ſpirits, and changing their flavour, uſe alſo, in ſome of their diſtillations, a ſmall quantity of the ſulphuric acid, and ſometimes the nitric.   Theſe act on the ſmall portion of water, and of the oil which may ſtill remain; and they diminiſh the volatility of both thefe ingredients; and by their action on the fpirit itfelf, they communicate more or lefs of an agreeable flavour.

When fpirit of wine is thus highly purified, it is lighter than water, in the proportion of 82, or 83, or 84, to 100, and in this ftate it is called ALCOHOL. But I have fometimes brought it up to an higher degree of ftrength, fo that, in the heat of 60ᵛ of Fahrenheit, 80 parts of it by weight were exactly equal in meafure to 100 by weight of water. It was brought to this ftrength by diftilling it with the addition of dry muriat of lime, which has a very ftrong attraction for water, and retains it powerfully in the diftilling veffel. This intermedium for rectifying fpirits has alfo the great advantage of not acting fenfibly on the vinous fpirit, and therefore imparts nothing of that difagreeable foapy tafte and flavour which are often produced by the fixed alkalis.

Vinous fpirits make an article of commerce that is very extenfive. As they derive all their value from the alcohol which they contain, a method for accurately determining how much alcohol, and how much water, there is in any fpirit, is a very valuable acquifition. This is beft difcovered by the fpecific gravity. A cubic foot of water, of the temperature 55°, weighs 1000 ounces precifely. The like meafure of the pureft alcohol that I have been able to prepare weighs 800; and ardent fpirits approach to this levity, in proportion as they contain alcohol. A mixture of equal meafures of water, and of an alcohol which weighs 820 ounces, forms what is called PROOF SPIRITS. Its fpecific gravity is 0,925, or the weight of a cubic foot is nearly 925 ounces *.

---

* There is confiderable uncertainty in this. *Proof fpirits* is that ftrength by which the liquor pays the excife duty. The ftatute by which this is impofed declares that an Englifh wine gallon, which is 231 cubic inches, or $\frac{231}{1728}$ of a cubic foot, fhall weigh feven pounds and 12 ounces, avoirdupoife weight. This gives 927$\frac{6}{10}$ ounces for the weight of a cubic foot. The hydrometers ufed by the officers fuppofe it ftill weaker. (See *Encycl. Britannica*, SPIRITS.)   EDITOR.

When a vinous fpirit is by fuch operations brought to an high degree of purity, it is an exceedingly fluid, penetrating, fragrant, and highly inflammable liquor, named by the chemifts ALCOHOL ; a word introduced into chemical language by the Arabians, and which means, I believe, fomething very fubtile and elaborately prepared.

It is an inflammable fubftance, diftinguifhed from the reft by its fingular qualities.

In the firft place, it has an extraordinary difpofition to retain the form of fluid. The moft intenfe cold that has yet been obferved in nature, or produced by art, is not fufficient to congeal it.

Another remarkable property of alcohol is, a difpofition to be much expanded by heat and contracted by cold. It is therefore often ufed in the conftruction of thermometers.

But thermometers made of it can only be employed in meafuring intenfe colds, and the heats of the atmofphere and of animal bodies. In a heat equal to 174° of Fahrenheit, which is far below the boiling point of water, alcohol begins to be changed into elaftic vapour, which, if allowed to increafe in quantity, would burft the thermometer. Its expanfion is not in proportion to its increafe of temperature.

We may therefore reckon as a third of the remarkable qualities of alcohol, its volatility. In the vacuum of an air-pump it would boil and produce elaftic vapour in lower heats than the ordinary heats of the atmofphere. And in an open veffel, even under the preffure of the atmofphere, it evaporates fpontaneoufly much fafter than water. If a perfon dip one finger into alcohol, and another into water of the fame temperature, and then expofe them to a dry air, the finger which was dipped into the fpirit will immediately feel vaftly colder than the other, owing to the more quick evaporation, and the rapid abforption of heat. Its latent heat is fcarcely inferior to that of water, if it does not exceed it.

A fourth well known quality of it is an high degree of inflammability. The vapour of it, in whatever manner produced, is eminently inflammable. Whenever it is touched or approached by flame it immediately takes fire, and burns with a blue tranfparent flame, which has not the fmalleft appearance of fmoke; and the whole of the alcohol is gradually confumed in this manner, with as little appearance of its leaving any earthy or other fixed incombuftible matter behind. This, as I formerly obferved, induced Dr. Boerhaave to confider alcohol as the pure pabulum of fire, or as a matter which was totally fpent and confumed in producing heat and light. And he infinuates a fufpicion that the other inflammable bodies may have their inflammability from alcohol prefent in their compofition.

But this notion of its being totally fpent and confumed by inflammation, and converted into heat and light, was a great miftake. The fact is, that it is converted into a great quantity of water.

Newman obferves, that if we burn the pureft alcohol in a deep veffel, fuch as a large brafs mortar, and keep the fides of it cool by furrounding it with cold water, a quantity of water will remain equal to one-third of the fpirit; and if we fufpend another mortar inverted over the firft, a quantity will remain equal to one-half; and that by other contrivances which ftill more effectually condenfed the watery vapour, he could obtain a much greater quantity.

I have long been of opinion that a quantity of water might be collected from burning alcohol equal in weight to the alcohol itfelf, or even exceeding it. And this has been found to be fact by fome experiments ingenioufly contrived and carefully executed by Mr. Lavoifier, in which he employed very effectual means for condenfing the watery vapour which arifes from its flame. He caufed it to burn very flowly, under a tall chimney of thin plate furrounded by cold water. By a medium of repeated trials, he obtained nine ounces of water from eight ounces of alcohol burnt in dry air; yet the air unavoidably carried off with it a very fenfible portion of moifture. (*See his Effays*).

Some part of this water was undoubtedly prefent in the alcohol before it was inflamed ; for it is extremely difficult, and perhaps impracticable, to feparate from it completely the whole of the water in which it is originally diluted. But as we get more of the water than there was of the alcohol, it is evident that the whole of this water could not be contained in it. Mr. Lavoifier, therefore, has accounted for the origin of it in another way, by fuppofing that alcohol contains a great quantity of hydrogen in its compofition, and that the atmofpheric oxygen combines with this by inflammation, and thus produces water. This fuppofition is founded on very fatisfactory experiments, the firft of which were made by Dr. Prieftley, who publifhed an account of them without attempting to build any theory upon them. And Mr. Lavoifier immediately afterwards repeated them with the greateft accuracy.

They were made by forcing the vapour of alcohol to pafs through a red hot tube of metal or earthen ware. The greateft part of the alcohol was thus changed into hydrogen gas or inflammable air, the bulk of which, on account of the rarity of that fort of gas, was aftonifhingly great, when compared with the bulk of the alcohol ; for in the alcohol, it is not in the form of hydrogen gas, which is a compound of hydrogen and latent heat. It is in the denfe form of hydrogen, deftitute of that latent heat, and this, together with a fmall portion of carbon, makes up almoft the whole of the alcohol ; for it appears to contain, befides thefe two, a very fmall quantity only of fome other principles ; fuch as a very little oxygen, and perhaps of azote. The prefence of the carbon is known by a fmall quantity of very light charcoal which it leaves in the red hot tube, and a part of which is diffolved in the hydrogen gas. And the prefence of the oxygen is proved by a fmall quantity of carbonic acid gas, which is found mixed with the inflammable air.

Mr. Lavoifier, therefore, is of opinion that the greater part of the water which arifes in vapour from burning alcohol is formed by the union of the hydrogen with the oxygen of the atmofphere, which contributes to its inflammation. This opinion appears to be very well founded.

Befides thefe ultimate principles which have been found in the compofition of alcohol, we can extract from it, by fome lefs deftructive proceffes, a fmall quantity of vegetable acid, fimilar to vinegar or the acetous acid, and we can even convert a great part of it into that acid, by diluting it largely with water, and making it undergo a particular fermentation, to be defcribed hereafter. But this is not furprifing, as the acetous acid itfelf is now known to be compofed of oxygen, combined in a particular and loofe manner with carbon, and with a fmall portion of hydrogen. The fmall quantity of acetous acid, which is often concealed in alcohol, is loft however when the alcohol is con-fumed by inflammation. In paffing through the outfide of the flame where the heat is produced, it is burnt and deftroyed, together with the carbon, and is changed by the oxygen of the atmofpherical air into carbonic acid gas, and a fmall quantity of water. (See Note 12. at the end of the Volume.)

This may fuffice for a general account of the nature of alcohol, and of the manner in which it is affected by heat. We muft next con-fider its properties in mixture with other bodies.—

One of thefe properties, which appears remarkable when we con-fider this fluid as an inflammable body, is its mixing fo readily and completely with water in any proportion, which no other inflammable fubftance will do. It even fhews a confiderable attraction for water. When I have prepared alcohol of an extraordinary ftrength, I found it was difficult to preferve it in that ftate. It attracted humidity even through the corks of the bottles. This fhews a very ftrong attraction between this fluid and water. And this attraction further appears by the readinefs with which thefe two fluids will depofit other fubftances that they may unite together. Moſt of the compound ſalts may be precipitated more or leſs from water by the admixture of alcohol. And alcohol, which can diſſolve a variety of oils and reſinous ſub-ſtances not ſoluble in water, deſerts or depoſits theſe to unite with water.

The union of alcohol with water, in equal weights, produces eight or ten degrees increaſe of temperature; and the bulk of the mixture is leſs than that of the ingredients by one part in thirty-four.

It is this attraction which renders it ſo difficult to make alcohol very ſtrong by diſtillation alone, although the alcohol in its ſeparate ſtate is much more volatile than water.

Among the ſaline ſubſtances, there are a number that act one way or other upon alcohol. We have already noticed one property of the com-mon vegetable fixed alkali with reſpect to this fluid, that of attracting the water from it when weak, and therefore aſſiſting us to make it ſtrong. And I obſerved, that if much alkali be employed in this way, a ſmall part is diſſolved, and gives the alcohol a yellow colour and diſagreeable taſte, which can only be removed by diſtilling it ſlowly, until a ſmall quantity only remains in the ſtill. I muſt now add, that alcohol thus tinctured with fixed alkali has been eſteemed a more powerful ſolvent of ſome ſubjects than a purer alcohol would be; and hence the chemiſts ſome time ago took much pains to learn the beſt manner of preparing it, or the way to have it as ſtrong of the alkali as poſſible,—and they called it *tartarized ſpirit of wine*. You will find that Dr. Boerhaave gave much attention to the combination of ſpirit of wine with alkali, and conſidered the preparing of a good tartarized ſpirit of wine as a nice and difficult operation. He recom-mends or enjoins attention to two particulars: 1ſt, To uſe the ſtrongeſt or pureſt alcohol; 2dly, To uſe alkali of tartar well calcined, and put into the alcohol perfectly dry and hot. If there be the leaſt moiſture in the ſalt, or water in the alcohol, it will be impoſſible to diſſolve the proper quantity of alkali. But I muſt add, that Dr. Boerhaave met with ſo much difficulty, in conſequence of his uſing the alkali combined with carbonic acid, or in its ordinary ftate, as we find it in pearl-afhes or alkali of tartar; that ftate of alkaline falts being fuppofed at that time to be their pureft ftate, which it is not in reality, the cauftic ftate of alkalis being the pureft. If we take an alkali that is perfectly cauftic, or totally deprived of its carbonic acid, we can diffolve as much of it as we pleafe in fpirit of wine, weak or ftrong, or though the alkali itfelf be not very dry. This property I difcovered in the cauftic alkali; and it is a confequence of its being more foluble, and its having a greater attraction for other bodies than mild alkali has. It unites with alcohol, as we have feen it unite with other inflammable fubftances.

Thus we have an eafy method for making a tartarized fpirit of wine, as it was called, as ftrong of the alkali as we pleafe, which Dr. Boerhaave thought to be fuch a difficult bufinefs.

As the effects produced with alcohol by a fixed alkali, perfectly pure, are different from thofe produced by the fame alkali in its ordinary ftate, fo, on the other hand, are they very different, if we take the fame alkali perfectly faturated with carbonic acid. If alcohol be fuddenly poured upon a fpirit of fal ammoniac, formed by diffolving as much as poffible of the cryftallized volatile alkali in water, the alcohol feparates the falt from the water, forming a thick, and fometimes firm coagulum, called the *offa Helmontii alba*. It is a cryftalline fponge, containing ardent fpirits.

We fhall now turn our attention to the mixing of this fluid with the different acids, taking them in their ufual order: The fulphuric therefore in the firft place.—

This acid is known to act powerfully on the inflammable bodies in general; and it accordingly unites with alcohol rapidly and violently. It will be proper to pour in the acid at one fide of the retort, by little at a time, that it may flide down under the alcohol; and after each addition, to agitate the mixture with a circular motion. Thus we temper the very great heat produced by the mixture. Equal weights may be thus mixed by cautious agitation with interruptions; but in whatever way we proceed, a violent commotion is excited, and a heat which the hand could not bear. A thin glafs muft therefore be ufed.; for if we were to proceed fo flowly that the heat fhould never be confiderable, we fhould lofe much of the valuable product of the operation. Each of the firft additions of fpirits produces a puff of ebullition, but this becomes moderate by the time that two-thirds of the fpirit have been mixed, and the mixture now requires fomewhat greater heat to make it boil.

This mixture has not a little engaged the attention of the chemifts, on account of fome remarkable productions which are got from it when it is diftilled. A condenfing apparatus muft be immediately fitted to the retort, and this muft be immediately fet upon hot fand for diftillation. This may be carried on at firft with a pretty brifk heat. But this muft be quickly diminifhed, when a certain fign (to be mentioned prefently) appears, or when the liquor in the receiver is reduced to nearly one-half of the alcohol.

The diftillation produces as follows:

1mo, There is condenfed a clear liquor, of a penetrating diffufive aromatic odour, the quantity of which is equal to half the fpirit of wine employed.

2do, Sulphurous acid and *oleum vini dulce* then come over. But to have thefe without danger, the heat muft be gentle, and long continued ; if otherwife, the matter boils over, and the hot froth cracks the top of the retort.

3tio, A thick bituminous matter or coal is left in the retort.

But the fragrant fluid which comes firft, called ÆTHER, is the defirable product of this diftillation, and the one on account of which it is commonly performed. It owes its fragrancy to a fubtile and volatile oily fluid, which makes up the greater part of it, and which was formerly called the *vitriolic æther*, but now the *fulphuric*. As it is the principal product of the operation, I shall point out the best method of managing the process to obtain it in quantity, and perfect.

First, of the different proportions in which the acid and alcohol may be mixed together. The proportion of equal weights is the best for producing the greatest yield of æther from the same quantity of the materials. If more alcohol be taken, a great part of it rises unchanged. If a larger proportion of acid is used, the mixture soon becomes black and thick, and forms sulphurous acid.

The chemists are indebted to Mr. Beaumè for having investigated the best manner of conduct ng this operation. He tried many different proportions, and found that the one we have taken was the best. You may see an account of his experiments in his little volume on æthers.

In the conduct of the distillation, it was formerly the practice to apply a gentle heat, and distil slowly from the beginning to the end. It is, however, quite unnecessary to be so cautious at first ; we may distil briskly in the beginning, and until the æther has distilled over. But then, indeed, it is absolutely necessary to diminish the heat greatly ; and the best way to do this is by removing the retort from the hot sand, and thus putting a sudden stop to the distillation. If we neglect to do this, the sulphurous acid begins to be produced so suddenly and abundantly, that it makes the matter in the retort boil over in the form of a black foam, blows up the vessels, and poisons the whole air of the house with an insupportable and suffocating stench.

The signs by which we may know when it is time to stop, that we may avoid this accident, are these—

It is time to stop when a quantity has distilled equal to one-half of the alcohol, or a little more. This critical period of the operation may be also perceived by the appearance of a whitish vapour, like a mist, appearing in the retort. And the bubbles formed in the retort by the boiling of the liquor are more numerous, and remain longer

before they burſt.   The odour of the vapours perceived at the luting
is alſo leſs fragrant than at the beginning.

Another reaſon for ſtopping is, that after having performed one
diſtillation, we may perform a ſecond with the ſame acid, by adding
to the retort a quantity of freſh alcohol, equal in meaſure to the fluid
that was diſtilled off, and then, proceeding as before, we obtain a ſecond product, which is exactly ſimilar to that of the firſt diſtillation.
And in the ſame manner, by another freſh doſe of alcohol, and a third
diſtillation, we get a third product as good as the two firſt ; and this
repeatedly a number of times.   But after a certain number, ſuppoſe
ſix or eight, the acid becomes too weak, and has much leſs effect on
the alcohol.   In practiſing theſe diſtillations, a little of the acid always
paſſes over along with the laſt of the æther, and is intimately blended
with it ; and as a ſmall quantity of ſpirit of wine comes over in the
beginning unchanged, it is always neceſſary to rectify the æther, or
re-diſtil it, to have it quite pure.

This ſecond diſtillation is very ſimple, requiring only a very gentle
heat, the veſſels to be well luted, and the operation performed with
day-light.   A ſmall quantity of cauſtic ley is put into the liquor in
the retort, to abſorb the volatilized acid, and we muſt ſtop when two-
thirds have come over, reſerving the reſt, which ſtill contains æther,
to be mixed with the materials for another proceſs.   Some prefer the
cucurbit for this operation, but the retort anſwers perfectly well.
(See Note 13. at the end of the Volume.)

It is time now to examine this æther, and attend to. the properties
of it which have attracted notice.—

Firſt, it is called an oily liquor, being a liquor which does not mix
with water, except in ſmall proportion.   It was once repreſented as not
mixable at all with water ; but Monſ. Lauragais has ſhewn that this
was a miſtake, and that water will diſſolve one-tenth part of its bulk
of æther, but no more.   I knew this, and had been long in the uſe of
mixing it with water, to give as a medicine.

Along with this oily nature, it is the lighteft of all fluids, and of great volatility. Its fpecific gravity is about 0,735. A little of it poured out very quickly, evaporates, and fpreads its flavour generally through the whole houfe in which the bottle is opened. A few drops from the height of the arm will feldom reach the floor. As it evaporates very faft in the fpontaneous way, fo alfo it very foon arrives at a boiling heat. It boils at about 100° of Fahrenheit, even under the preffure of the air. And when we examine its proper boiling point, by removing the preffure of the air, we find it as much lower than thofe of other fluids in the fame circumftances, and far below the freezing point of water. *(See Note* 14. *at the end of the Volume.)*

Cold is produced alfo by the fpontaneous evaporation of æther. Beaumè made Reaumur's thermometer defcend below froft with cloths wet with æther, and wrapped round the phial.

Another quality for which æther is eminent, is inflammability. It is very liable to catch fire by the mere approach of a candle while we pour it from one veffel into another. Therefore, as it cannot always be fully condenfed, the operation for æther ought always to be performed in day light.

It is an ufual experiment by itinerant fhowmen to throw a lump of fugar, foaked in æther, into a glafs of warm water. When a candle is applied to the furface of the water, it catches fire, and burns in a very amufing fluttering manner. The heat of the water expels the æther in a ftream of bubbles, which take fire at the top of the water. The glafs muft be deep and narrow, that the fucceffion of bubbles may efcape near enough to fet each other on fire.

Æther burns with a remarkably bright flame, frequently emitting fparks more brilliant than the reft, and produces a fenfible foot.

The remaining properties of this fluid are chiefly thofe of a folvent of many fubftances. It diffolves a number of refins, gums, &c. as we fhall learn afterwards more particularly.

The medical qualities of this fingular fubftance are alfo eminent, and deferve our notice. Internally taken in water, in the quantity of 10, 20, 30, or 40 drops, it is a powerful antifpafmodic. But its effects as an external application are the moft remarkable. Applied to the forehead in the palm of the hand, it performs all the wonders of Dr. Ward's volatile effence, in refolving fpafms and removing nervous pains in a moment, as it were by charm. Toothache and headache commonly yield to it. I am inclined to think that it acts in fuch cafes by a fort of revulfion. It brings on, in a moment, heat and inflammation upon the fkin, which, to fome, become infupportable. But it goes off immediately when the hand is removed. It brings on this fuperficial inflammation more quickly than any blifter, finapifm, or fuch application ; and it is much more under command,— for as foon as the hand is removed from the part, the heat and uneafinefs abate, and foon go off entirely. I am therefore perfuaded that it is very proper in may cafes in which it has not been thought of, as in pleuritic ftitches, rheumatic pains, and other fuch cafes, in which bliftering and cupping are of fervice. There are many fuch cafes, in which it is expedient to apply remedies of quick operation. Rectified æther fhould be ufed.

Of the effects of mixing alcohol with the nitric acid, chemiftry furnifhes many examples, which are remarkable, both for the appearances which they exhibit, and the information that we derive from them.

This acid, when obtained from nitre by the procefs formerly defcribed, and which has the name of *Glauber's fmoking fpirit of nitre*, and was thought the ftrongeft and pureft, is in fact the weakeft as an acid, and impure. It is a mixture of two acids, now diftinguifhed by the name of NITRIC and NITROUS. The Latin names exprefs their diftinction more precifely, *nitrofum* denoting an abundance of that which diftinguifhes it as nitrous, viz. the fiery colour, and copious deep blood-coloured fumes, and offenfive fuffocating fmell. The *nitric*, on the other hand, is colourlefs, and emits no fenfible fumes. Yet, in all the diftinguifhing properties of an acid, it exceeds the other, having a much stronger attraction for water, alkali, and every thing that is diffolved by it ; nay, even for that which feemed to characterife the excellency of the other,—I mean for inflammable fubftances. It acts on all fubftances much more violently than the nitrous, diffolves more alkali, earth, or metal, or even inflammable fubftance ; and its virtues are permanent, whereas the red fumes of the nitrous acid wafte by expofure, and the acid becomes unable to diffolve the fame quantity of alkali as before. It was long fufpected, therefore, that Glauber's fpirit of nitre was a compound, and that it contained inflammable matter. combined with it, arifing from impurities in the nitre employed in the procefs for obtaining it. This fufpicion was confirmed by ob- ferving that fmall additions of the more inflammable fubftances to the colourlefs acid immediately produced that fiery colour, and thofe red fumes, for which Glauber's fpirit of nitre is remarkable.

I was the firft, I believe, who entertained a notion fomewhat dif- tinct on this fubject. Finding that Glauber's fpirit of nitre, when diftilled to about two-thirds of its bulk, had loft entirely its fuming quality, and that the liquor which came over poffeffed it more emi- nently, while the firft was ftronger as an acid, I was led to confider the original liquor as a mixture of two feparable fubftances, of which that was the compound which exhibited the weakeft action on the fubftances which are diffolved by both. This I conceived to be conformable to the general facts in chemiftry.

I was therefore difpofed to confider Glauber's fpirit of nitre as a compound of the acid of nitre and inflammable matter, or perhaps of the principle of inflammability. With this view of the fubject, I tried to form anew this fuming fpirit of nitre, by adding to the pale fpirit of nitre a fmall quantity of alcohol, which I confidered as an inflam- mable fubftance fufficiently fimple not to contaminate the new com- pound with unfuitable ingredients : And I fucceeded quite to my wifh. A very minute portion of alcohol being added to a quantity of the pale and very ftrong acid of nitre, from which I had feparated the highly fuming acid by diftillation, immediately imparted to it the fiery colour and blood red fumes, and made it in all refpects fimilar to Glauber's fuming fpirit of nitre, having all its properties. This was, I think, in the year 1760.

This experiment being very inftructive, it will not be amifs that I tell you the method of making it in the neateft and moft perfpicuous manner. Take a glafs tube, about one-tenth of an inch in diameter, and draw one end of it more flender. Having put the acid into a folution glafs, dip the fmall end of the tube five or fix inches into fpirit of wine, and then, clofing the top of it with the finger, take up the fpirit with it, and dip it to the bottom of the nitrous acid. Then, taking off the finger, let fome (about half an inch) of the fpirit run out of the pipe into the acid, and then ftop the pipe again. You will fee the union take place immediately, and fmall bubbles form on the mixture, few of which will reach the top ;—the whole will acquire an orange colour. Purfuing this method, you will fee the progreffive alteration very diftinctly. One of my ftudents, at Glafgow, afked me, after lecture, whether inflammable air would anfwer the fame purpofe, feeing that fome imagined it to be the principle of inflammability. The thought pleafed me, and I tried it with perfect fuccefs. It produced the fame effects, but required an immenfe quantity, which did not furprife me, by reafon of its great rarity.

But this mixture of nitric acid and alcohol merits further attention. I found that, by gradually adding more alcohol, the volatility of the acid, or a difpofition to emit red fumes, increafed, and the attraction for water, and its acidity, diminifhed ; and that, by proceeding in this manner, the acid may be totally diffipated in thofe offenfive fumes, leaving only acidulated water, having no inflammability. nor the fmell of nitrous acid, but rather that of vinegar, which it alfo refembles in tafte and in its mixture with other fubftances.

This procefs is accompanied with great heats, notwithftanding the copious eruption of thefe fumes; and it is very hazardous, becaufe every addition produces a great and fudden increafe of heat, which contributes to increafe the explofive power of the mixture, and will throw it about the room, while the veffels run the rifk of fplitting by the fudden changes of temperature.   To fucceed to the degree I have mentioned requires a very long time, mixing very fmall portions of alcohol at once, and keeping both ingredients in veffels furrounded with ice and water.   It will often be obferved, that, after alcohol, amounting to one-fourth of the weight of the acid, has been added, the explofions are colourlefs and tranfparent, and are accompanied with a fragrant fmell, refembling that of vitriolic æther.   This indicated the production of fomething different from the naufeous blood red fumes, and gave the chemifts hopes of obtaining an ætherous fluid, different from the vitriolic.

They were better conducted to this by another mixture, well known in pharmacy,—the preparation for obtaining what is called DULCIFIED SPIRIT OF NITRE.   This is prepared by cautioufly mixing with rectified fpirit of wine, one-fifth or one-fourth of its weight of ftrong nitric acid, and diftilling the mixture.   It produces a liquor with no remarkable acidity, having a fragrant fmell, much refembling that of apples.   This encouraged to a farther pufhing the addition of acid to the alcohol.   Their mutual action was incomparably more quiet and manageable.   For, as it already appears that a very fmall quantity of alcohol gives this ungovernable volatility to a great deal of the acid, it is plain that, fince a fmall quantity of acid is not converted at once into incoercible fteams by mixing with a great proportion of alcohol, repeated additions of fuch fmall quantities, after the heat and ebullition produced by the preceding addition have ceafed, muft in all probability be quiet and fafe.   So the cafe turns out; but ftill, although the fucceffive mixtures go on without much trouble, it is found that when fomething more than one-third of acid (by weight) has been added, the mixture begins to explode and become troublefome; and, if one-half be added, it is almoſt unmanageable. Mr. Navier, a French phyſician, was the firſt, I believe, who publiſhed a practicable and ſuccefsful procefs, founded, I prefume, on ſimilar obſervations. He preſcribes the cautious and gradual mixture of one-half of acid with the alcohol, in a very ſtrong glafs veſſel, which he immediately corks up, and fecures the cork with leather, tied hard over it, and fecured by packthread ; and the bottle is kept in cold water. Thus are the elaſtic ſteams prevented from forming, by the great preſſure produced by thoſe already generated. The fluids gradually act on each other, and an æther is produced, which, like the vitriolic, floats a-top. As all this goes on under a very great preſſure, it is plain that if we pull out the cork, or even untie the packthread, the elaſtic exploſion will take place in an inſtant, through the whole liquor, and it would be thrown out. Mr. Navier, therefore, directs the cork to be pierced with a pin, and the vapour allowed to efcape. After this, the æther may be feparated by a funnel. It amounted to one-third, or one-half of the alcohol.

This procefs was ſtill hazardous, for the bottle often burſt. Mr. Beaumé, of the French Academy of Sciences, improved this procefs, by carefully inveſtigating the beſt proportion and manipulation. He found that two parts of acid to three of ſpirits gave the greateſt produce of æther from the fame alcohol, and directed both ingredients to be ufed in the coldeſt ſtate, by keeping each in melting ice, or water and ice, and by fetting the corked-up bottle in the fame fituation. This proportion of ingredients fecures us againſt the chance of exploſions wholly ungovernable, and the low temperature greatly moderates the action that is unavoidable. He alfo directs us to give the liquor in the bottle a briſk whirling motion immediately before pouring in any more acid. This prevents any accumulation in a particular ſpot.

By this procefs we obtain, in three or four hours, a conſiderable quantity of æther, which is obferved to form in little drops all over the liquor, and rife gently to the top. By allowing the bottle to remain undiſturbed for eight or ten days we obtain about half the weight of the alcohol, after which no more is produced in these cir-cumstance:.

I am by no means certain that this process will give the greatest product. I suspect that the external pressure really prevents the che-mical union, in the same manner as it certainly prevents it in the boiling of water. That it is prevented in this instance is evident, be-cause heat is not absorbed unless the pressure be removed. Therefore the heat is not combined when vapour is not produced. The like may happen here. I am justified in this, from observing that more nitric æther is obtained by other artificial processes, in which this pres-sure does not take place. I shall mention one, which I practised before I heard of those of Navier and Beaumé, and is extremely simple and easy.

Into a strong phial, having a ground stopper, I first pour four ounces of strong pale nitric acid. I then add three ounces of water, pouring it in so gently that it swims on the surface of the acid. I then pour in, after the same manner, six ounces of alcohol. I put in the stopper slightly, and I set the phial in a tub of water and ice. The acid mixes slowly with the water, and, in a diluted state, comes in contact with the alcohol, on which it immediately acts, and æther is produced slowly and quietly. The liquor gets a dim ap-pearance, because imperceptible bubbles are formed, which get to the top, and having collected to a certain degree, they lift the stopper and escape. After eight or ten days, I find upwards of three ounces of nitric æther, though I am certain, by the smell, that much escapes with the vapour. This is, however, a certain, easy, and safe pro-cess, though it is slow and imperfect.

More artificial processes have been followed by several eminent chemists.

1. Mr. Woulfe's, in his general manner of managing all distilla-tions where fumes of difficult condensation are produced. He uses a succession of receivers, which are tubulated. A tube goes from

the top of the firſt into the ſecond, down to the bottom, and another from the top of the ſecond into the third, and ſo on. By this contrivance, what is not condenſed in the firſt receiver is condenſed in the liquor of the ſecond, &c. A mixture of equal parts of ſtrong acid and alcohol is thus diſtilled with a very moderate heat, and a good produce of æther is obtained.

2. Nitric or nitrous acid is made to act, in the very inſtant of its formation, on alcohol. Nitre is put into a tubulated retort, to which is fitted a receiver. Vitriolic acid is poured on this, and immediately after, ſpirits of wine. The nitrous acid is diſengaged, which riſes through the vitriolic, and acts on the alcohol ; and æther is produced, whoſe ſteams are condenſed in the receiver.

3. Inſtead of putting the alcohol upon the vitriolic acid and nitre, it is put into a glaſs veſſel, which communicates, by means of a bent tube, with a large receiver luted to the retort containing the nitre and vitriolic acid. Heat being applied to the retort, the nitrous acid is diſengaged, and part of it is condenſed in the receiver, and part paſſes on to the bottle containing the alcohol, on which the vapours act ; and æther is produced without any troubleſome exploſions.

Mr. Chaptal, who prepares a great deal in the way of commerce, uſes two receivers in ſucceſſion, the firſt being ſet in water, and the ſecond covered with wet cloths, and it has a tube proceeding from the top, which is bent downward and immerſed in a bottle of water. He ſays that the proceſs is eaſy and ſure, affording very pure æther, and in good quantity.

But all the proceſſes, and indeed every treatment of nitric acid with alcohol, requires much caution, that we may eſcape accidents.

The reſidue of the diſtillation is acid, much changed in its properties, appearing more like vinegar, or even more reſembling the acid of ſorrel, or of ſugar. It burns to a coal, and produces, by a great heat in clofe veffels, carbonic acid, and empyreumatic oils, like all other vegetable fubftances.

The æther of all the proceffes here defcribed requires rectification, to clear it of acid and alcohol, which came over with it in the diftillations, or mix with it in the proceffes by digeftion. This rectification is performed by diftilling it from cauftic alkali. This reduces its quantity; for we muft not diftil more than two-thirds or one-half of the firft æther. To bring this to ftill greater purity, fome direct it to be mixed with one-fifth nitrous acid, and diftilled again,—taking two-thirds of the product fet apart, and rectified from cauftic alkali. The reft of what comes over is a lefs perfect æther,—the *mineral anodyne liquor of Hoffmann;* and the remainder in the retort is a *dulcified fpirit of nitre.*

Pure nitrous æther greatly refembles the vitriolic in lightnefs, inflammability, and flavour. This laft quality, indeed, is inferior to the vitriolic, being ftronger, and fomewhat pungent; the tafte is alfo more acrid; the colour inclines confiderably to yellow; it burns with a brighter flame than the vitriolic, and produces more fmoke, and leaves a ftain in the difh. When kept, it is apt from time to time to blow out the cork. This is attributed to æther not yet perfectly formed, and is faid never to happen, if the produce be carefully rectified from cauftic alkali, and if we do not take too much of what comes over.

The muriatic acid, in its ordinary ftate, exhibits no difpofition to act on alcohol, or any other inflammable fubftance. But the chemifts, curious after a knowledge of this new difcovered fluid, the æther, were eager to compofe one by this acid alfo, although their then received theories gave them little encouragement to expect it. Many attempts, however, were made, but long without fuccefs.

The fimple mixture and diftillation of the muriatic acid and alcohol has no effect. The mixture is indeed called the *dulcified muriatic acid,* but there feems no combination or change of properties. Nor

has better fuccefs followed the attempts to combine them in the inftant of the production of the acid, or by uniting their vapours.   At
laft, methods were difcovered, in which, by employing the acid in a
compound and peculiar ftate, a combination took place, and muriatic
æther was produced.   Of fome of thofe methods I fhall give you a
fhort account.

I. The Marquis de Courtanvaux mixed alcohol with a liquor called
the fmoking liquor of Libavius.   This, as you will learn afterwards,
confifts of muriatic acid furcharged with oxygen, and united with
tin.   Alcohol, being mixed with twice its weight of this liquor in a
retort produced heat and white fuffocating vapours, but accompanied
with an agreeable fmell.   When diftilled, we firft obtain a ftronger
alcohol, and then the æther appears, indicated by the fmell, and by
trickling down the fides of the receiver.   After fome time, the fmell
becomes fuffocating,—the receiver muft now be removed, otherwife
the æther will be tainted with many products, fuch as acid, oil, a
fubftance like butter, &c. which you will underftand afterwards.

The æther thus obtained muft be rectified from cauftic alkali, and
only half of what comes over muft be taken. ·

II. Similar to this is the procefs of the Baron de Born.   He ufes a
compound of muriatic acid and the flowers of zinc. *(Mem. des Scavans
Etrangers,* vi.*)* The acid obtained from twelve pounds of fea falt, being
faturated with the flowers of zinc completely diffolved, the folution is
evaporated till of the confiftence of greafe.   This is mixed very gradually with fix pounds of alcohol, and after digefting the mixture
eight days, and filtrating it, the clear liquor is diftilled, beginning with
a very gentle heat.   We obtain,

1. Water, amounting to almoft half of the fpirits.

2. An aromatic fpirit of wine,—the matter in the retort now
grows thick like melted wax.

3. Æther.—When this has all paffed, the matter in the retort is
dry, and the heat muft be increafed.

4. A fweet oil like effence of lemon. This will fwim on the æther, and is the laft volatile product.

Great care muft be taken that the heat be not too great before the matter in the retort becomes dry ; for it is apt to burft up in fudden clammy bubbles, which touching the colder parts of the retort, will fplit it.

The æther muft be rectified from the aromatic fpirit, by the gentleft heat of a lamp furnace ; and the fpirit which remains may be poured back on the refiduum in the firft retort, and more æther, &c. obtained, in the manner practifed for vitriolic æther, without end.

The author obtained by this procefs two pounds of æther, and four ounces of the fweet oil, both of which were remarkably fragrant.

III. Mr. Woulfe made muriatic æther by caufing the vapour of boiling alcohol to meet with the muriatic acid gas, as it was difengaged from fea falt, in a glafs veffel, from which a fyphon tube proceeded into another glafs containing alcohol. The tube reached almoft to the bottom of the alcohol, and the vapours which did not condenfe here efcaped, and went by another glafs tube into a fecond veffel containing alcohol, and what was not condenfed there went into a third, and fo on. By this management the union was effected, and æther obtained in each of thofe veffels, fwimming on the alcohol. Thefe portions were mixed and rectified by diftillation with a gentle heat, from cauftic alkali, being much contaminated with acid vapours. The whole procefs was extremely tedious and troublefome, requiring feveral cohobations, or returning the liquors back again upon the refiduum, before a tolerable quantity of æther was obtained. The veffels of alcohol were heated to a great degree by condenfing the vapours, and foon gave over condenfing. (See Phil. Tranf. 1767.)

IV. A Mr. Schroeter of Berlin prepared a muriatic æther by diftilling from a mixture of eight parts of fea falt, four of fulphuric acid, four of black manganefe, (a fubftance containing much oxygen), and three of alcohol. In this procefs, the muriatic acid was difengaged from the fea falt by the fulphuric, and, in its nafcent ftate, acted on the alcohol, being affifted by the manganefe (without which we know that it will not fucceed) in fuch a way as to produce abundance of æther. Mr. Pelletier of Paris, by a fimilar procefs, obtained four ounces of æther from eight of alcohol. The manganefe evidently appears to favour the production of æther exceedingly. If the ordinary muriatic acid be diftilled from manganefe, its properties are remarkably changed; and if employed in this ftate in Woulfe's manner, it produces æther with great facility. There is, therefore, fomething which it poffeffes, in this ftate, and in thofe of the muriats of tin and of zinc, that is fimilar, and which fits it for this preparation. This change on the muriatic acid will be minutely confidered in due time.

The æther obtained by any of thefe proceffes, when pure, is like the other two, immifcible with water except in a fmall degree, extremely light, highly odorous and penetrating, but much more offenfive to the lungs than they are. It is highly inflammable, and in burning has much fmoke, and emits a fmell as penetrating as fulphuric acid. It is lefs agreeable to the tafte, having the ftyptic tafte of alum.

Thus we fee that the three principal acids have in one refpect a fimilarity in their action upon vinous fpirits. Even the muriatic, which, in its ordinary ftate, fhews no action on inflammable fubftances in general, or this one in particular, can be put into fuch a ftate, competent to it as an acid, which enables it to contract an union with alcohol, or produce a change on it, fimilar even to that produced by the nitric, which acts the moft violently of all on them. This muft be confidered as a common property of the acids, which it behoves the chemical philofopher to inveftigate in his own way. A chemift is difpofed to afcribe it to fome ingredient common to the acids, and we are interefted to difcover what this may be.

This may be confidered as giving a theory of æthers, or an explanation how they are produced. They were difcovered at a time when

chemical fcience had made confiderable progrefs, and its cultivators
were eager to give a rational account of the many furprifing effects
or changes which it prefents to our view. Therefore attempts were
foon made to explain all thefe appearances.

One of the firft who attempted this was Mr. Macquer. He imagined that the production of thefe light oily fluids was nothing but
an abftraction of water from the ardent fpirits, and that this was
effected by the ftrong attraction for water, which is a diftinguifhing
property of all the acids. By this abftraction of water, he conceived
the alcohol to be reduced to its ftate of purity; and this he thought
was a compound of a fubtile oily principle and water ftrongly combined, and which, by the prefence and quantity of the water in its
compofition, becomes mifcible with water in any proportion. The
acids, by their ftrong attraction of water, rob the fpirit of a part of its
elementary conftitution, rendering it more inflammable and oily, and
no longer fo mifcible with water, but more volatile and inflammable.

But this theory is infufficient for explaining the phenomena. It
is incongruous with the general train of chemical facts, that depriving
ardent fpirit of water will diminifh its attraction for it. And it is
not fact that the acids which have the ftrongeft attraction for water
are the moft effectual for changing alcohol into æther. Acetous acid,
which has a weak attraction for water, produces more æther from a
quantity of alcohol than the foffil acids do; and the muriatic acid,
which attracts water with great force, produces æther with great difficulty, efpecially when employed in that ftate in which it attracts
water moft ftrongly. It fhould alfo be a confequence of Mr. Macquer's theory, that all æthers fhould be alike, which is very far from
being the cafe.

Mr. Berthollet, a chemift of the firft eminence, thinks that all the
æthers are formed merely by the addition of more oxygen to the compofition of the alcohol. This opinion feems chiefly founded on the
neceffity of employing acids, which abound in oxygen, and on the

phenomena and confequences of fome of thofe proceffes, which fhew the acid to be very much deoxygenated.

But I confefs that I view the formation of the æther in a different light.. When the proceffes are conducted in the beft manner, we have no appearance of deoxygenating the acid during the formation of the æther. The fulphurous acid does not appear till all the æther has paffed over. And we obtain the greateft quantity of nitrous æther, when we fucceed beft in preventing the explofion and the deoxygenation of the acid. When we neglect the precautions for preventing this, fuch as the keeping the mixture very cold, and making the additions very gradually, we have red fumes, indicating the deoxygenation of the acid, and we obtain lefs æther.

I am perfuaded that the æthers are compounds of the alcohol with a greater or lefs portion of the acid employed. The acids which are moft abundant in oxygen are the fitteft, for this reafon, that they have the ftrongeft action on alcohol and other inflammable fubftances. I believe that the acid is combined with the alcohol, fo as to be neutralized by it, while the attraction of the alcohol for water is diminifhed, for the fame reafon, and in the fame manner, as the attraction of both acids and alkalis for water and other fubftances is diminifhed when they are combined in forming a neutral falt. The alcohol, therefore, affumes that immifcible and oily nature which is obferved in it. That an oily appearance and confiftency may be produced and increafed in this way, is evident from the example of the *oleum vitrioli dulce*, which has more the appearance of an oil than the æther itfelf. Yet it is only æther fuper-faturated with acid. The æther is totally convertible into this oil by repeated diftillation with the acid; fo is alfo the nitrous æther. The prefence of an acid in the moft perfect æthers has been proved by Crell and Scheele. Both of thefe chemifts have fubftituted one acid for another in the fame æther, fo as to change one æther into another, in a certain order. Were pure oxygen all that is united with the alcohol in forming æther, all acids would be indifferent, and all æthers alike *.

Mr. Lavoifier has been the moft fuccefsful in explaining many phenomena in the action of the acids on alcohol and fimilar fubftances, and has made many judicious, accurate, and inftructive experiments with this view. His explanations are founded on a careful analyfis of the ultimate conftituent principles of alcohol, and of the vegetable fubftances from which it is produced.

Mr. Lavoifier's opinion was, that the vegetable fubftances in general, and thofe fufceptible of fermentation, fuch as fugar, mucilage, farina, and the like, are compofed of carbon, hydrogen, and fome oxygen, loofely joined, and in various proportions. By receiving more of the oxygen from the nitric acid, when it acts on them, or from the air, in the acetous fermentation, they are changed into vinegar, or fome other acid. When the nitric acid is made to act violently, it produces effervefcence and elaftic matter: 1. By the changes which the nitric acid undergoes into *acidum nitrofum*, or into nitrous air, according to the degree of deoxygenation; 2. By the change of part of the carbon into carbonic acid gas, in confequence of its clofe union with part of the oxygen of the nitric acid.

Alcohol is compofed of the fame principles which compofe thefe vegetable fermentable fubftances. But the principles are combined in the alcohol in different proportions from thofe which conftitute fugar, and the reft of them. In alcohol there is a lefs proportion of oxygen to the others, efpecially to the hydrogen. In confequence of this it is more inflammable and volatile than thofe other vegetable productions.

The carbonaceous matter in alcohol is clearly exhibited in the experiment of Prieftley and Lavoifier formerly mentioned, where the

---

* Do not the more brilliant flame, and the fmoke and foot, indicate a redundancy in the carbon rather than in the oxygen?

EDITOR.

vapour of alcohol was made to pafs through a red hot tube. It alfo prefents itfelf in the procefs for fulphuric æther, by imparting a black colour to the acid in the retort, in proportion as the æther forms, and oxygen is abftracted from part of the acid. Its place is fupplied by the carbon expelled from the alcohol by this elective attraction. At laft, the matter in the retort becomes coaly, and a confiderable quantity of very fine charcoal is elevated. It is probably thus partly retained and feparated from the other elements of the alcohol, by the ftrong attraction of the fulphuric acid, which, when the diftillation is too long continued, is imperfectly decompounded by it, and changed into fulphurous acid ; and the vapours of this volatile fulphurous acid, in conjunction with the carbonic acid gas, into which a part of the carbon is changed on this occafion, explain the violent and fudden ebullition of the matter in the retort, and burfting of the veffels, when the diftillation is pufhed far with too great a heat. And during the formation of nitrous æther, there is always a little effervefcence or a production of gas, formed from a part of the carbon changed into carbonic acid by fome of the oxygen of the nitric acid, which, being thus partly changed into azotic gas, contributes to make up the quantity of the elaftic gas. And along with thefe gafes, vapours of the nitrous æther, formed in confequence of its great volatility, alfo efcape. The matter of thefe different gafes, however, can be confined and made to remain combined with the nitrous æther for fome time ; of which we have an example in Mr. Navier's and in Mr. Beaume's proceffes for nitrous æther ; the ftrong preffure to which the materials are fubjected in the corked bottles repreffing their volatility,—but whenever the bottle is opened they are fure to efcape.

All the æthers therefore, as I faid before, appear to be compounds of the alcohol, and of a fmall portion of the acids made ufe of. A part only, and that very little, of the acid is decompounded, lefs or more, by the lofs of fome oxygen during the procefs ; and a part only of the carbon is taken from the alcohol. Æthers, therefore, contain a lefs proportion of this principle, and a larger one of the hydrogen than alcohol does. This idea appears probable, from the lightnefs and volatility of æther, and from the black colour and other appearances of the acid which remains in the retort in the diftillation to obtain fulphuric æther.

The experiments you have now feen with mixtures of alcohol and nitric acid confirm the character of this acid, which is eminent by a difpofition to act violently and powerfully on the inflammable fubftances, and others allied to them, which powerful action of it plainly depends on the great quantity of feparable and active oxygen which it contains.

Of this we have another eminent example in the action of fugar and nitric acid on one another.

Some of the qualities of fugar are fufficiently known to you, fuch as its folubility in water, and the cryftals called fugar-candy which it forms, when the folution of it, or fyrup, is properly cryftallized. Loaf fugar is only a mafs of very fmall cryftals cohering together, and therefore more readily foluble.

Sugar is entirely a vegetable production, and is found in the juices of many vegetables. It is commonly obtained from the juice of one plant, which abounds with it the moft of any, viz. the fugar-cane. But many other vegetables, or parts of vegetables, contain fome fugar. It is often cryftallized or concreted in fruits that are dried.

As fugar is always a vegetable production, not being found in any other part of nature, we may expect to find in it the fame elementary principles of which vegetables are compofed, or fome of thofe principles. Accordingly, when it is fubjected to the moft deftructive kind of analyfis by fire, it yields a large quantity of hydrogen gas, mixed with carbonic acid, and holding diffolved a fmall portion of the carbon not combined with oxygen : But a confiderable portion of carbon remains behind in the apparatus, in the form of charcoal. From the refult of this analyfis, it is plain that fugar is compofed of carbon, combined with hydrogen and with some oxygen at the same time. These two last ingredients are in their dense unelastic state. And all the three are combined with that weak attraction with which the elements of vegetables are known to cohere.

You can easily imagine that such a compound as sugar may be made to receive a larger quantity of oxygen than that which it naturally contains. And this is found to be true, when we apply to it the nitric acid; but no one can imagine, till he has seen it, the violence with which the action goes on. When a bit of loaf sugar is put into nitric acid, it is some time (about three or four minutes) before the combination becomes observable. Small bubbles begin to detach themselves, and immediately on their reaching the surface, acquire a deep blood red colour. The liquor becomes warm, and the emission of bubbles becomes more and more copious, till the whole is in violent ebullition, and the upper part of the vessel is filled with the blood red vapour, which becomes transparent; and the production of those red vapours goes on till the sugar is dissolved or consumed, and the liquor acquires the greenish colour of fresh made aquafortis.

If this mixture be made by employing common aquafortis, a gentle heat is required, and then the phenomena are nearly the same in kind, though in a much more moderate degree. What I chiefly mean to consider just now is the vapours which are produced. They are found to consist of oxygen and azote, the same which I have given as the constituent parts of the nitrous acid. But they are in a very different proportion, being nearly as 68 to 32, and require a great addition of oxygen to make them equivalent to nitric acid, or even to nitrous acid. When pure, they are not only transparent, but as colourless as common air, and are perfectly incondensible by pure water, or ordinary cold. The deep red colour acquired by the fumes arises entirely from the mixture with the vital air of the atmosphere. The gas combines with oxygen, and the two collapse into nitrous acid. If a glass jar, filled with vital air or oxygen, be inverted on water, and some of the gas obtained from fugar in the way now defcribed be let up into it through the water, we have an inftantaneous deep red cloud, and the water rifes to the top of the glafs, in confequence of the collapfing of the two gafes. It contracts no fuch union with azote or carbonic acid gas. From this account you fee that it is the fluid called nitrous air, or nitrous gas (as the French call it) which was firft difcovered by Dr. Prieftley, but produced by him by other proceffes than this. And by making a great number of experiments with it, he made many ufeful, interefting, and inftructive difcoveries. I fhall fhew the principal experiments with it foon,—when we fhall have an opportunity for preparing it in a more perfect and pure ftate than this. I fhall only remark on it at prefent, that it is the nitric acid changed to a much greater degree from its common ftate than it is in the acidum nitrofum, and this in confequence of having loft a greater proportion of its oxygenous principle: And therefore it has no perceptible acidity, and but little attraction for water. It retains, however, the difpofition to unite again readily with oxygen; and in uniting with it, forms firft the acidum nitrofum, and afterwards, with a larger quantity, the moft perfect nitric acid.

Thefe are the changes produced in the nitric acid while it and fugar act on one another.

The fugar too, as might be expected, undergoes a great change. It totally lofes its fweetnefs, and is converted into a perfect acid by its union with the oxygen of the nitric acid.

Mr. Bergmann, by diftilling aquafortis from one-fixth of its weight of fugar, obtained a prodigious quantity of the gas juft now defcribed, and reduced the fugar to a pure acid falt, called the ACID OF SUGAR. It cryftallizes in four-fided fpiculæ, terminated by a ridge.

The acid of fugar expofed to a heat gradually increafed to a red heat, firft efflorefces, lofing its water of cryftallization, and then becomes brown, and emits compounded vapours, which, by careful treatment, condenfe into the fame products that are obtained from all vegetable fubftances, or remain in the form of elaftic gafes; that is, empyreumatic acid and oil, foot, and a great quantity of carbonic acid and inflammable air. Part of the falt, however, fublimes, no way differing from its original form.

When examined in the way of mixture with the fubftances which we have already confidered, it exhibits feveral remarkable properties.

With the mineral alkali, it forms a falt of difficult folution, having an excefs of alkali in its cryftals.

With the vegetable alkali, it forms a deliquefcent falt, when perfectly neutral, but which cryftallizes when either ingredient exceeds in a certain proportion.

With volatile alkali, it forms a falt which cryftallizes in four-fided prifms. This falt is decompofed in a very fingular manner by heat, namely, by the deftruction of its acid, and we obtain very mild volatile alkali, formed by the carbonic acid which arifes from the deftruction of the acid. It is worthy of particular remark, that no deflagration appears in treating the acid of fugar, or its compounds, by great heats, nor is any azote obtained. This fhews that it does not owe its acidity to its containing nitrous acid, as was fuppofed when it was firft difcovered.

With lime, it forms a falt infoluble in water (if that name be competent to fuch a fubftance). It is very remarkable that the attraction of this acid for lime exceeds that of the three mineral acids. Lime alfo decompofes the three falts already mentioned. Acid of fugar, therefore, will detect lime in mineral waters, by taking it from every other folvent.

It forms a white powdery falt with magnefia, and with barytes, which laft earth decompofes the magnefian falt.

Acid of fugar diffolves in fulphuric acid, giving it a brown colour.

Nitric acid acts on it, and decompofes part of it, and we obtain, by diftillation, vinegar and carbonic acid.

If frefh nitric acid be diftilled from what remains, we have the fame products.

When this action of nitric acid and fugar on one another was firft difcovered by Profeffor Bergmann, it was fuppofed that the acid we obtain was a principle or production peculiar to fugar. It was therefore called the *acid of fugar*. But by fimilar experiments, which have fince been made on a variety of vegetable and animal fubftances, fubjected to the action of the nitric acid, we have learned that the acid acts on them all in a fimilar manner, that is to fay, with fimilar effervefcence and the production of fimilar elaftic fluids, and that many of thofe fubftances are thus changed alfo, at leaft in part, into an acid exactly fimilar to that got from fugar, although they are not at all fweet, nor appear to contain any fugar. This happens with the vegetable mucilages and glues, gum arabic, tragacanth, and others; alfo with ftarch, and the mucilaginous part of lemon juice. Its prefence in this laft was difcovered by Scheele; and he evinced its being the mucilaginous part of lemon juice which contained the faccharine acid, by cryftallizing the acid of lemons, and then examining this for the faccharine acid. It contained none. *Note,* That the citric or lemon acid is very eafily cryftallizable, by firft combining it with lime, and feparating them by means of the fulphuric acid. Even alcohol yields a little of this acid, when the action of the nitric acid on it is violent and long continued. Many animal fubftances yield it alfo, as Mr. Berthollet difcovered: And fome of them yield it in much greater quantity than fugar does. Sugar yields one-third of its weight. Wool more than one-half.

There is, therefore, no good reafon now for calling it the acid of fugar, efpecially fince it has been found ready formed in fome other vegetable fubftances, as in forrel. Since this acid has been well examined and characterifed, the falt of forrel has been found to contain it, and to derive its acidity from it; the falt of forrel itfelf being an alkali fuperfaturated with this acid. *Vide Bergmann on Elective Attractions.*

We may here further remark, that all the vegetable acids are more or lefs fimilar in their compofition to the acid of fugar, being compofed of hydrogen, carbon, and oxygen, in different proportions. All thefe falts are convertible into one another *in a certain order*, by the action of the nitric acid on them. The acid of tartar is changed into the oxalic or faccharine, the oxalic into the acetous, and this into the carbonic. But this department of chemical analyfis is yet in its infancy, and it will probably be a long while before any perfpicuous knowledge of it will be acquired. We run a great rifk of being led into important miftakes, by too confident application of imperfect and perhaps erroneous theory. We will correct with reluctance miftakes which refult from ingenious conceptions and laborious inveftigation. The prefent fubject in particular feems to encourage our refearches by great appearances of fuccefs; but the convertibility of acids, which at firft exhibited fuch uniform diftinctions, fhould make us extremely cautious in forming general conclufions. I have not room to enter into a difcuffion of fo many particulars. Nicholfon gives the current opinions on the prefent fubject with great candour and diftinctnefs.

Thus we have confidered the confequences of mixing fpirit of wine with alkalis and acids in different ways.

Of the compound falts, there are a few which can be diffolved in this fpirit; though the greater number cannot. The foluble are, the muriats of lime and magnefia, acetite of potafh, the acetite and nitrat of ammonia. The reft, in general, are not foluble, but, on the contrary, are precipitated by it from water. Hence it is that the fixed alkali which is diffolved in vinous fpirits, in the procefs for rectification, is of the utmoft purity. Mr. Woulfe employs this method for examining their purity.

Vitriolated tartar, after long digeftion with vinous fpirits, exhibits some remarkable appearances. We obtain æther, sulphurous acid, and volatile alkali.

None of the earthy substances have any remarkable action on vinous spirits.

Phosphorus unites with it imperfectly, but without any phenomenon very interesting to the chemist. Mr. Boyle may be consulted on the subject.

Sulphur does not unite, even by long digestion. Count Lauraguais combined them perfectly, when in the form of vapour, issuing from two retorts. They form a fœtid liquor.

Charcoal speedily clears spirits of wine from all distinctions of flavour, and makes all alike, and quite colourless.

VI.—OILS.

Under this division I comprehend, along with what are commonly called OILS, the solid fats of animals, and resins of vegetables, as being distinguished only by a slight difference in fusibility from the fluid inflammable substances of vegetables and animals, to which the term *oil* is in common language confined.

The bodies which belong to this division are far inferior to alcohol in simplicity and inflammability. Yet, when their inflammation is properly excited, they give more heat and light than alcohol does. They are, therefore, stronger fuels, or may be considered as more inflammable substances than alcohol is. And yet they are not (many of them at least) quite so easily set on fire, or brought into a state of inflammation. Most of them require to be heated more, and their inflammation in ordinary circumstances is not so complete. Their flame commonly throws out from the top of it a quantity of sooty matter, called lamp black, the inflammability of which is not exhausted. I call it sooty matter; but it is a matter considerably different from common soot, which is an article of the materia medica. This confifts partly of lamp black, or the half burnt oily matters of vegetables, and partly of many other fubftances, not inflammable, which are volatilized and expelled by the heat, along with the inflammable vapours. Soot generally contains an ammoniacal falt, formed of the volatile alkali, exifting or generated by the heat, in the vegetable, and an empyreumatic acid of the acetous kind. Lamp black is a lefs complex fubftance, being the fcorched or half burnt oily vapour, and is always produced when the flame is of fuch a large fize, or of fuch a form, that the air has not a fufficiently extenfive contact with it to act with full power on the whole of the vapour that compofes it. But when the flame is of a fmall fize, or when a ftream of air is made to rife up through the middle of it, as in Argand's lamps, or when a fmall ftream of air is forced through it with the blow-pipe, the formation of foot is effectually prevented. I obferve, however, that even a very fmall and clear flame from oil, without the leaft appearance of foot, carries up a minute portion of whitifh earth, which, during a long continued procefs with a lamp furnace, attaches itfelf to the bottom of the veffel that is heated by the flame. This is certainly part of the afhes, not volatilized, but merely blown out by the ftream of vapour.

The component parts of oils are beft difcovered when we gradually convert them into vapours, and caufe thofe vapours to pafs immediately through a red hot tube, without mixing with air, into a proper apparatus for collecting gafes. The refult of this operation is the production of an immenfe quantity of *hydrogen gas*, mixed with fome *carbonic acid gas*, and a fmall quantity of *watery vapour*. And in the tube and diftilling veffel, when they are allowed to cool, we find a portion of charcoal, which, when burned in the open air, yields a fmall quantity of afhes, and a ftill fmaller of faline matter, commonly fixed alkali.

This analyfis, therefore, fhews that the conftituent principles of oils are *hydrogen* and *carbon*, with a small proportion of oxygen, and a still smaller of earthy and of saline matter. The presence of the oxygen is evident by the appearance of the carbonic acid, which is formed by the union of that principle with a part of the carbon, and also by our getting a small quantity of watery vapour, which it forms by uniting with some of the hydrogen. The hydrogen gas obtained by this operation is necessarily of that kind which contains a small quantity of carbon dissolved in it, and which, on account of its being rendered denser by this admixture than a pure hydrogen gas, is therefore named *heavy inflammable air.*

When the oils are subjected to the treatment which was named CHEMICAL ANALYSIS, by the elder chemists, in which they were merely distilled in retorts and receivers, and the vapours of them thus condensed as fast as they were formed, and without being ever made red hot, the products obtained are considerably different; for this reason, that the principles of the oil are not so completely separated from one another. The greater part of the vapours are condensed into an oil, or oily fluid, which has very different properties from those of the original oil. And we get also some acid water, and some carbonic acid gas. And a small portion of charcoal remains in the retort.

Repetitions of the same process, with the same oil, diminish the quantity of the oil every time, rendering it more attenuated and volatile, and produce a little more water and a little more charcoal. And Mr. Lavoisier discovered that more water was produced every time from the same quantity of oil, when the distilling vessels were of a large size than when they were small. He therefore concludes that all this water did not pre-exist in the oil, but was formed in the distilling vessels from the hydrogen, which, uniting with the oxygen of the air contained in the vessels, formed the water, or a great part of it. And the acid found in this water, and which is of a peculiar kind, is undoubtedly formed of a part of the oxygen, loosely combined with a part of the hydrogen, and of the carbon, as it is in the vegetable acids. Dr. Crell of Helmſtadt made a very ſatisfactory ſet of experiments to inveſtigate its peculiar properties. They are publiſhed in the 70th volume of the Philoſophical Tranſactions. This acid is now. named the *ſebacic acid*, and the compounds it forms with alkaline ſubſtances are named *ſebats*.

The conſequences are different when we apply heat to the oils in the open air, ſo as to inflame them in the moſt perfect manner. The whole of the hydrogen and the carbon unite with the oxygen of the atmoſphere, and form water, or vapour of water, and carbonic acid in an elaſtic ſtate.

I have long been of opinion, that a great quantity of water ariſes from burning oils. A plate of very cold glaſs or metal being held for a moment above a very ſmall and clear flame, will immediately be covered with dew; and a bell-glaſs held over the flame, and kept cold by ſnow, will ſoon have drops of water hanging at its brim. But Mr. Lavoiſier has collected this water more accurately and completely, by means of a well contrived apparatus, and he finds that it exceeds the weight of the oil. Something, therefore, ſupplied by the atmoſphere muſt contribute to form it.

The only queſtion is, whether this water has been exiſting previouſly in the air, or has been formed in the manner aſſigned by Lavoiſier? The laſt is more probable; becauſe we cannot ſee how the heat produced by inflammation can make the air depoſit water which it is ſuſpended when colder *.

If the inflammation be improperly managed, the change and deſtruction of the oil is not ſo complete. The ſoot which is produced from all oils, when we attempt to burn them with too large a flame, is formed principally from the carbon.

* It is perhaps owing to this actual formation of dampneſs, or water, that Dr. Franklin found that all electrical experiments are almoſt ſtopped, while a burning candle is connected with the inſulated part of the apparatus. We have the means of deciding this queſtion. EDITOR.

To treat of the oils more particularly, they muft be diftinguifhed into two principal kinds, the AROMATIC and the UNCTUOUS ; to which it is ufual to fubjoin a third fection, named EMPYREUMATIC.

The *aromatic* oils are all formed by nature, and chiefly in vegetables. In thefe they are fecreted juices, lodged in particular parts or repofitories in the ftructure of the plant. All the aromatic oils make a ftrong impreffion on the organs of tafte and fmell, exciting on the tongue the feeling of heat and acrimony. And they have not that fmoothnefs and flipperinefs, when felt between the fingers, which is named *unctuofity*. They are alfo all capable of rifing in vapour, in the heat of boiling water, or at leaft along with the vapour of water. And they are very inflammable. The wicks of candles are often prepared for being quickly lighted up, by putting a fmall quantity of fome of thefe oils on the extremity of them.

The *unctuous* oils are alfo natural productions, and are found in vegetables and animals, in both of which they are alfo fecreted juices, or are formed and lodged in particular parts or repofitories.

When not corrupted, they have not any pungent tafte, or remarkable odour, but have unctuofity in a great degree, and for that reafon are much employed to diminifh friction in machines of all kinds. The heat of boiling water is not fufficient to convert them into vapour, and they are not fo readily and quickly inflamed as the aromatic oils.

The *empyreumatic* oils refemble the aromatic fo much by many of their chemical qualities, that they are not diftinguifhable from them by any general difference, except the circumftance of their origin. None of them are natural productions. They are all produced by art, and are either oils, changed from their natural ftate by the action of violent heats, or are entirely produced by heat from vegetable or animal fubftances which do not contain a formed oil before the heat is applied to them. We may alfo add, as an article in the character of the empyreumatic oils, that the odour of them is in general offenfive.

Aromatic Oils.

Let us now take a nearer view of the firſt of theſe ſections, the aromatic oils.

We find almoſt the whole of them in vegetables. There are a few examples of ſimilar oils got from ſome animal ſubſtances, but they are very few. It is by examining vegetables that we find a great number and variety of theſe oils. All the vegetable ſubſtances which affect our organs with any remarkable odour, or hot pungent taſte, produce theſe effects by the action of an oil, or oily principle of the aromatic kind, which they contain.

And, as in a great number of ſuch vegetables, theſe oils are the moſt remarkable, and the moſt uſeful and active matter which they contain, they are for this reaſon named the ESSENTIAL OILS, or ES-SENCES of ſuch vegetables.

The variety of them found in the different vegetables, or vegetable ſubſtances, is very great; but the greater number of them have not yet been found applicable to any uſeful purpoſe. And many others are contained in the vegetables in ſuch very ſmall quantity, that we cannot extract them except at a very great expence, far exceeding any value that can be ſet on them. Such, for example, is oil of roſes, which has the fragrancy of that favourite flower in the higheſt per-fection. Roſes contain more of it in the warmer climates, and yet even in theſe, ſo little, that it is valued at an extravagant price. Some of this oil is collected in India, partly from roſe-water.

The odour of theſe oils, which is in moſt of them ſtrong and fra-grant, is one of the moſt remarkable of their obvious qualities; and this odour is different and ſpecific in each particular oil. But they are diverſified by other properties beſides the variety of odours.

Some of them are extremely fluid, light, ſubtile, and volatile; others are more heavy and thick; and ſome are frozen in ordinary heats of the air. Many taste extremely hot, pungent, and acrid, when applied to the tongue. Some have a much milder taste. And it is difficult to give any general rule with regard to these qualities. It is very generally said, that the essential oils produced in colder latitudes are lighter than water, while those of the hot are so heavy as to sink in it. But this cannot be admitted as a general fact. The oil from parsley seeds sinks in water. There are too many exceptions to it, and the same oil does not always appear of the same density. Oil of cinnamon will float or sink, according as it has been drawn off by a gentler or a stronger heat. And I am inclined to believe that this inference, with respect to the weight of the aromatic oils of warmer latitudes, has been drawn chiefly from experiments made on the oils procured from the dry spices brought from the East and West Indies, in which the oil is grown thicker and heavier by age and evaporation, and by combination with oxygen, than it was in the recent and green vegetable.

All authors have observed, with regard to these oils, that it is difficult to preserve them long in perfection. The only way is to keep them in phials, with glass stoppers carefully ground to the phials, so as to shut them perfectly close. And further, it is proper to set them in a cool place. If these precautions be neglected, they are sure, after some time, to become less fragrant, less fluid, and generally to suffer a change of their colour. These changes happen to them more quickly when they are exposed to the air; and, in this case, a part of them commonly evaporates at the same time. These particulars, therefore, shew that the depravation they suffer, if kept with too little care, proceeds, in part at least, from exhalation of their more subtile and volatile part, upon the presence of which their odour and tenuity chiefly depend. This is confirmed by the nature of the operation which has been found most proper for restoring again to a state of perfection some part at least of the oils which have suffered this depravation. This operation is to subject them to distillation, along with some water, so that a part of the water and oil may distil over with gentle heat. These oils, when recent and in perfection, are, in gene al, very volatile. Oil of saffafras, for example, if dropped on a bit of paper, and held over a candle, at a great diftance, will quickly difappear, leaving no ftain. They evaporate flowly if expofed to air. And when the heat of boiling water is applied to them, they emit vifible fteams, and evaporate copioufly. This volatility they fhew moft remarkably when frefheft and in greateft perfection. When old and ill kept, we do not find them fo volatile ; at leaft it is not fo eafy to evaporate them entirely. But if fuch depraved oils be put into a retort with water, and part be diftilled off, the part diftilled is found to be much improved, and what remains in the retort to be grown worfe ; that is, thicker, heavier, darker, and lefs odorous. If the diftillation be not too long continued, the portion diftilled is equal in goodnefs to the oil in its recent ftate. By this operation, therefore, we recover, in a ftate of perfection, one part of our depraved oil. And if it be repeated feveral times, we increafe the tenuity and volatility of it, even beyond the natural or more ordinary ftate ; but, at the fame time, we ftill diminifh its quantity fo much the more.

Mere lofs of fome of the more fragrant part is not, however, all the change which the aromatic oils fuftain by expofition to the air. They actually combine with the oxygenous portion of atmofpherical air. This was obferved very early by Dr. Scheele and Dr. Prieftley. The thickening of oils and refins was one of the moft effectual means of phlogifticating the air in Prieftley's numerous and important experiments, and indeed one of the moft fpecious arguments for his opinions. The oil exhaled its fragrant ingredient, which was thought replete with phlogifton. The air was rendered unfit for the fupport of flame. Scheele, however, drew a more warrantable conclufion. He faw that the air was diminifhed in the fame way as by the abforption of hepar fulphuris. Therefore, he inferred that the *fire-air* was abforbed, and that the mephitic portion of atmofpherical air alone remained. He afcribed the change on the oil to this combination. This has at length been evinced by clear experiments in fome few cafes, in which the weight of the infpiffated oil was found greater than that of the frefh oil. When expofed in vital air, the change is much more rapid and remarkable. The fact is no longer doubted.

When we try how thefe oils are affected by mixture with other bodies, we find, in the firft place, that they can be combined in fmall quantity with water, by churning and maceration. By this operation, the water acquires a good deal of the flavour, and ftill more of the acrid pungency of the oil. But it does not contain any fenfible portion of it, nor does the oil lofe any weight. The mixture can fcarcely be called a chemical combination; for the water is foon covered with a fhining film, which thickens by expofure to the air (probably by abforbing oxygen); and the water in a fhort time lofes the greateft part of the pungent tafte it had acquired. If a fmall quantity of this water, while frefh made, and quite limpid, be agitated in a large veffel filled with vital air, it becomes milky immediately, and much air is abforbed. It would feem that it is this portion, fo mifcible with water, that acts the moft powerfully on oxygen, and combines with it.

Of the falts, the alkalis have a difpofition to unite with thefe oils. A compofition of this kind has long been defcribed by chemical authors under the title of *Starkey's foap*, formed of oil of turpentine, or of juniper, and the alkali of tartar. But they fpeak of much difficulty attending the procefs. This difficulty, however, proceeded from their taking the alkali of tartar in its ordinary ftate. Dr. Crell found that cauftic fixed alkali readily unites with the oil. This combination has not been much ftudied, except for medical ufes, and it does not feem to be of much importance in this refpect.

Combinations or mixtures of the volatile alkali alfo with thefe oils are formed for the purpofes of medicine, on account of their having fome medicinal powers fimilar to thofe of volatile alkali, viz. cordial, ftimulating, and antifpafmodic. We have examples of fuch combinations in Spiritus Volatilis Oleosus, and Spiritus Volatilis Foetidus, and Eau de Luce.

When the nitric acid, in a concentrated state, that is, having as little water as possible combined with it, is suddenly mixed with any inflammable body, its oxygen is so loosely combined, and retains so much of its latent heat, that it is immediately acted on, and its heat is extricated by combining with that body. If the substance is fluid, so as to allow an extensive surface of action, and not so volatile as to be dissipated in vapour at the first warming, the heat produced may increase to ignition. This is remarkably the case with the aromatic oils. Besides this, they shew a remarkable disposition to unite with oxygen. They are thickened by exposition to the atmosphere, as has been already observed, and this is attended by an absorption of vital air. They have also a low temperature of inflammation.

For all these reasons, when nitric acid is poured into an aromatic oil, such as oil of cloves, or oil of turpentine, the mixture generates heat, boils up with great violence, and bursts out into flame. When this phenomenon was first observed, it was seen that a strong acid was necessary, and as the process for the *spiritus nitri fumans Glauberi* is particularly directed to every circumstance that can ensure its concentration, this smoking acid was always employed: And it will always succeed, if the concentration be as perfect as we suppose. But you have seen that a very fuming spirit may be had that contains much water, merely by dissipating some of the active ingredient by a little vinous spirits, sugar, or such like. Strong, pale, or nitric acid will much more surely produce the required effect. The manipulation prescribed by Dr. Slare is also very proper, viz. to pour in about half of what we ultimately intend, and in four or five seconds after, to add the rest at once. When the heat produced by the first is at the height, and has expended part of the strength of the acid, we then add a parcel in its full force. Nor is the mixture of a little very strong sulphuric acid improper. This quickly generates a most inflammable vapour (fulphurous) which catches fire the more readily, as it meets with fo much loofely combined oxygen in the nitrous acid. Nay, the fulphuric acid alone, if clear and ftrong, will inflame the heavy aromatic oils, fuch as the oil of cloves,—and frequently too, in favourable circumftances, even oil of turpentine. The mixture, when it does not kindle, acquires by long digeftion qualities which greatly refemble the bituminous oils and folid bitumens. If the acid has been diluted, the mixture has a foapy appearance, mixing pretty intimately with water.

Even the muriatic acid may be combined with oxygen (by a procefs which will foon be explained to you) in fo abundant and loofe a manner that it will fire the aromatic oils with great readinefs.

The neutral falts have no action on eflential oils in mixture without fuch heat as to burn them.

Nor have any of the earths except quicklime. By long trituration it renders them mifcible in fome meafure with water, and is thought to difpofe them to a more ready yielding their moft odorous part by diftillation.

Eflential oils unite with fulphur very readily, and compofe balfams of fulphur, drugs of a very difagreeable fmell and tafte, which have remarkable effects on the nervous fyftem. The vapours are highly inflammable. They alfo diffolve phofphorus, and the compound becomes luminous by coming into contact with the air, and takes fire in very low temperatures. The warmth of the hand is in many cafes fufficient.

They are diffolved by alcohol, but are feparable by water, and in fome meafure alfo by diftillation. When we diftil the alcohol, however, from an aromatic oil or vegetable, the more fubtile and fragrant part of the oil generally rifes with the fpirit, if a very gentle heat be employed. On this account, the odorous fpirits diftilled from them may be fo prepared as to have a more fubtile and delicate flavour than the odorous waters derived from the fame oils, becaufe the heat of boiling water confiderably exceeds that of boiling alcohol. Thus are formed, from some aromatic oils or aromatic vegetables, the compound spirits and cordials of the apothecaries. Brandy is the spirit commonly used. Some of the finest perfumes are also prepared in this manner. *(See Beaumé.)*

Such is the general nature of the aromatic oils. There are a few, however, which differ so far from the rest as to require some notice of their particular qualities. These are distinguished by the appellation of CAMPHORS. But there is only one species commonly known, and which is always suggested by that term ; the others not being in use. The species I mean is that commonly used in medicine. It is got from a tree of the laurel kind, and has these qualities :

First, it is, like the other aromatic oils, highly inflammable, burning with a most brilliant flame, but producing a good deal of soot ; it leaves no ashes whatever ; it will burn on the surface of cold water ; it is dissolved in small quantity by hot water, and imparts to it its taste and odour ; it is readily soluble in alcohol, and separated by water. In all which respects it resembles the rest of the essential oils. It is particular by being always solid. When dissolved in alcohol in great quantity, it crystallizes by evaporation of the alcohol, as also when it is slowly separated by water, when the spiritous solution is set in a cold damp place. When heated, it does not melt, but evaporates, and sublimes in beautiful crystals. And this it does more readily and completely than the other aromatic oils, never leaving, when pure, the smallest matter behind. It is therefore one of those bodies which are more volatile than fusible under the pressure of the atmosphere. But by confining it much, it may be melted*.

---

* In the process for refining camphor, it is set in a sand heat, in very low flat matrasses, shaped like a flat turnip, and having a short neck about an inch and a half in diameter. This is shut by a bit of paper loosely twisted. In this situation I saw it boiling like water. The cake of sublimate formed very fast ; and though the vessel may be said to be open, and more than a hundred were on the furnace, there was only a very moderate smell of camphor in the laboratory.

EDITOR.

The relation of camphor to acids is more remarkable. It is diffolved by the fulphuric and nitric acids, but without violence; and with the laft it forms a fluid which appears like oil. Heat applied to this oil occafions the acid and camphor to act on one another; and the acid is imperfectly decompounded, and gives out red vapours. By repetition of this procefs, the camphor itfelf is changed into an acid which has peculiar properties, refembling thofe of the acid of forrel. Its properties have not been much examined, the preparation of it being very expenfive, and requiring eight diftillations with frefh nitric acid. But it quickly lofes this appearance by the application or contact of pure water to it, which immediately attracts the acid from the camphor again.

This habit of camphor in relation to the acids, efpecially the nitric, is its moft diftinguifhing quality. This acid acts with fuch violence on all inflammable fubftances, and particularly on the aromatic oils, and camphor being fo eminently inflammable, we fhould expect very different phenomena. The nitric acid fuffers none, or almoft none of the changes which refult from a feparation of oxygen. There is, therefore, fomething very peculiar in the conftitution of camphor,— but it is of difficult inveftigation. For, when a ftrong heat is applied to it, it flies off unchanged, and cryftallizes in the firft cooler place it comes to. Many chemifts think that camphor is the principle of aromatic oils and of refins, but on what grounds I know not.

There is only one fpecies of camphor ufed in medicine, or found in the fhops; but there are feveral kinds, which may be obtained from different vegetables, and even from fome aromatic oils already feparated from the plant, all of which have the properties I have now defcribed, and differ from one another only by odour. Newman gives experiments and examination of a camphor which cryftallized from the oil of common thyme; and he enumerates the roots of the cinnamon tree, zedoary, fchœnanthus, cardamomus, oriental mint, abrotanum, milfoil, daify, juniper, rofemary, falvia camphorata, la-

vender, hyffop, clary, maudlin, marjoram, &c. I am informed that
it has been obtained lately from the leaves of the pimento, or Jamaica pepper tree. If fo, we may foon expect it in great abundance,
that being a very common tree, with exuberant foliage. Camphor
is obtained from all thefe matrices much in the fame way as from
the laurus camphora, namely, by maceration in water, and then boiling the materials in an alembic, having the head occupied by loofe
ftraw. A good deal of the camphor fublimes, attaching itfelf to the
ftraws, and the reft goes over with the vapours into the receiver *.

As camphor is different from the reft of the aromatic oils by thefe
chemical properties, it is remarkably diftinguifhed from them alfo by
its medicinal qualities, being much lefs heating and ftimulating than
the effential oils, though at the fame time it has great powers as an antifpafmodic, an antifeptic, and a diaphoretic. By its being free of the
heating quality, it is fafe and ufeful in a great number of difeafes in
which the other aromatic oils are improper. But we muft not give it
in fuch large dofes as are faid by M. Fourcroy to be given in England.
Two fcruples at once are not fafe, except perhaps in mania. (*Vide
Dr. Alexander's Experiments*). Applied externally, in ointments or
other forms, it is very powerful in difcuffing or difplacing rheumatic
pains. But there may be cafes in which this ufe of it may be improper, as in external rheumatic pains of the thorax, unlefs other remedies are employed at the fame time.

I may add, that when burning, its light feems to be the fame, or to
confift of the fame proportion of coloured rays, with that of the fun.
All delicate colours, which appear different in candle light, appear of
the proper colour when illuminated by camphor.

And now I have faid enough of the aromatic oils. The manner

* From fome trials, I am difpofed to think that both the wood and the leaves of the
pimento, or Jamaica pepper, will yield a very good camphor: It is even deferving of
ferious trial, becaufe camphor would be very extenfively ufed in feveral manufactures if
cheaper. The pimento is fo abundant in the Weft Indies, that it would coft nothing.

EDITOR.

of extracting them from the vegetables which contain them is so fully described in every book on chemistry and pharmacy, and it is so commonly known and frequently practised by the apothecaries, that I need not take up much of your time with it here.

The most common operation by which they are extracted is distillation of the vegetable with water. A quantity of the aromatic vegetable is put into a common still, with as much water as floats it or covers it; and the distillation is begun immediately, or after a day or two. The hot water penetrates the vegetable matter, softens it, and dissolves more or less its aromatic parts, so as to disengage the oil in some measure. And while the distillation goes on, the oil is changed into vapour, along with a part of the water; which vapour of the oil is carried over with the vapour of the water into the refrigeratory. Thus the oil is distilled over faster than it could be with the same heat by itself, while, at the same time, the water prevents the vegetable, or the oil, from ever becoming hotter than 212° Fahrenheit, which it would certainly do, were it exposed to heat by itself. And experience has shewn that these oils are the more fragrant in proportion as they are obtained with less heat.

The oily substances called BALSAMS and RESINS belong to the same division with the aromatic oils, and resemble them very much by their principal qualities. They are found in a number of vegetables, and, like the aromatic oils, are secreted juices, deposited in particular spots, or particular vessels of the plants. They are in general more or less odorous substances, and also produce the sensation of taste, with more or less pungency and heat. They are all very inflammable, and burn with the same phenomena as aromatic oils, only they give more soot, and more fixed carbonaceous matter. Infused in water, most of them impart to it some taste and odour. They are affected by acids as aromatic oils are. They dissolve with ease in spirit of wine, and are separable from it again by water. So far you may imagine I am describing the aromatic oils.

The chief diftinction is the degree of fluidity and volatility. Balfams in general are not fo fluid or volatile as aromatic oils; and as there is great diverfity among them in the degree of thefe qualities, there are many that are commonly folid, and even confiderably hard. The terms of balfam and refin refer only to differences of this kind. The greateft part of what are called balfams have a fenfible degree of fluidity, and fome are almoft as fluid as fome of the thicker oils. Refins are folid and brittle in the ordinary temperature of the air; but if heat be applied to them, they melt into a vifcid oily fluid, which, fo long as it is melted by a gentle heat, is not diftinguifhable from what is called a balfam. But the balfams themfelves vary greatly in confiftence by age. By the evaporation of their more volatile and odorous parts, they always become more folid, and even hard.

Thefe fubftances in general are more difpofed than aromatic oils to unite with alkaline falts. And it appears that foap-boilers have found it their intereft, for fome time paft, to employ a proportion of common refin in the compofition of hard foap, by which they render it much more deterfive.

Befides thefe differences of balfams and refins from the aromatic oils, I juft now faid that they are lefs volatile. This is true, however, with refpect only to the greater part of their fubftance. When balfams and refinous fubftances are expofed to the heat of boiling water, or rather, are boiled with the water, they are always in part converted into vapour, and difperfe their odour around. Of many, a very confiderable part rifes along with the vapour of the water. This volatile part, when condenfed in diftillation, is a perfect aromatic oil; and the part which remains in the diftilling veffel is a refin, which becomes folid and brittle when cold, and is far lefs odorous and volatile than before. We have an example of all this in turpentine. If it be fubmitted to diftillation without addition of water, we firft obtain from it a quantity of an aromatic oil, fimilar to that which arifes when it is diftilled with water. But before it has all arifen, the refinous matter becomes too hot; for it is capable of being heated to a much higher degree than that of boiling water. The heat, therefore, accumulates in it, and foon begins to decompound and deftroy it; in confequence of which it is changed into a fpecies of empyreumatic oil, and a fmall portion of water, and of vegetable acid. And a charcoal remains in the retort, in greater quantity than that which is produced from the aromatic oils when they are treated in a fimilar manner.

It is worthy of remark, that when we diftil turpentine with water, the produce of the diftillation, together with the refin in the retort, weigh confiderably more than the original turpentine. The refin and the oil are immifcible with water. Water has, therefore, been fo combined as now to form part (nearly one-fixth) of the oil. Obferve alfo, that the oil is vaftly more odorous than the turpentine. This in fome meafure explains how effete aromatic oils are improved by diftillation with water.

There is a confiderable number of vegetables which contain balfams, or refins, all diftinct and diverfified from one another. And fome have been found ufeful in medicine, and in the arts. Their powers in medicine are, in general, fimilar to thofe of the aromatic oils, but they are not fo heating. In the arts, they are employed in varniſhes, paints, and perfumes, and other fuch compofitions: Alfo in natural hiftory, for preferving infects.

Thofe ordered by the colleges of London and Edinburgh to be kept in the fhops are,

TEREBINTHINA *Chia*, five *Cypria*. Lond.

——————— *Veneta*, (Ed.) From the larch of the Alps and Pyrenees. A kind from New England generally fupplies its place in the fhops.

——————— *Argentoratenſis*, (Lond.) Straſburg turpentine, prepared in different parts of Germany, from the firs which are native and moft common in England and Scotland.

BALSAMUM *Cauadenſe*, (E.) Another fir in America, the Virginian, or Canada fir-tree, yields a turpentine much fuperior, brought over under the name of *Balfamum Canadenfe.*

BALSAMUM *Commune,* (L.) The coarfeft, &c. from the *pinus fyl-veftris,* common in different parts of Europe.

———————— *Copaiba,* (L. and E.) From the Spanifh Weft Indies.

OPOBALSAMUM—*Balm of Gilead.* The beft kind exudes from the plant in Arabia, but is never feen in Europe. The inferior is feparated from the leaves and branches by light boiling in water, but is alfo extremely fcarce, fo as to be hardly procurable.

BALSAMUM *Peruvianum,* (L. E.) From an odoriferous fhrub in Peru, and the warmer parts of America, and extracted by coction. It does not unite with water, milk, unctuous oils, or wax. There is another balfam of Peru, of a white colour, and more fragrant, faid to be got by incifion.

———————— *Tolutanum,* (L. E.) From a tree of the pine kind in Tolu, Spanifh Weft Indies, and brought in little gourd fhells.

BENZOINUM, (L. E.) The juice of a large tree, in both Indies, and bearing our winters: But the benzoin is brought from the Eaft Indies only. It is ufed chiefly as a perfume. Water extracts very near as much of the flores by coction as is obtainable by heat.

GUMMI *Guaiacum,* (E. L.) Got by incifion from a tree in the warmer parts of the Spanifh Weft Indies.

———————— *Animé.* A refin, got by incifion from a large American tree. It diffolves totally, though not eafily, in fpirit of wine, and has a tranfparent amber colour.

———————— *Elemi,* (L.) A refin, brought from the Spanifh Weft Indies; and fometimes from the Eaft Indies. It diftils with water, gives a fragrant oil, and deferves more notice.

———————— *Hederæ,* From ivy.

GUMMI *Juniperi*, Exudes from the juniper, in warm climates.

SANGUIS *Draconis*, Brought from the East Indies.

SAGAPENUM, (L. E.) A concrete juice from Alexandria.

LABDANUM, (L.) Exudes upon the leaves of a small shrub in Candia, and other islands of the Archipelago, and is brushed off with a sort of rake, with leathern thongs for teeth. It sticks to the thongs, and is afterwards scraped off. It is mixed with much sand, and does not bear separation by extraction, without diminution of its fragrancy.

I observed before, that the balsams and resins are secreted juices in the plants which afford them. Many are collected by bleeding or extravasation, at natural or artificial wounds, as turpentine. Others are an exudation or excretion from the surface of the leaves or other parts of the plant, as in the moss rose, and labdanum. Others are obtained by boiling gently with water. And some resins are extracted from dried vegetable substances by the application of alcohol, and separated from the alcohol afterwards by gentle distillation, or by the addition of water.

And now I have given a description of the balsamic and resinous substances, which is general, and applies to all of them except a few species. In all parts of nature, we find her productions so greatly diversified, that it is impossible to give general characters and descriptions that will suit every particular. There are always some species, which must be considered as exceptions from the general nature of the rest. In this light must we view three or four resinous or balsamic substances, which I shall now mention. These are copal, benzoin or benzoe, and ambergrise.

1. COPAL is very transparent, and considerably hard, and is not dissolved, but only softened by alcohol. This distinguishes it from gum animé, which resembles it perfectly in external appearance. But it can be dissolved by some of the aromatic oils, and thus forms the most beautiful and durable varnish employed in the arts. It was invented

in France, and long known by the name of *vernis Martin*. In England it is called copal varnish, and is highly prized for its horny toughness. The art of preparing it is not commonly known.

When copal is treated with oil of turpentine in a close vessel, from which the vapours are not allowed to escape, they exert a great pressure, which prevents the boiling, and the mixture acquires a higher temperature. A very confiderable portion of copal is diffolved; and with the addition of a little poppy-oil, it forms an excellent elaftic varnish, inferior to the vernis Martin only in a tint of brownnefs, fcarcely perceptible.

Another good elaftic varnish is made of copal, by keeping it melted till an acid or four-fmelling aromatic vapour has ceafed, or become fcarcely fenfible. It muft then be mixed with an equal quantity of lintfeed oil, which has been deprived of all colour by long expofure to the fun's light. The varnished ware muft alfo be dried in the fun.

2. BENZOIN is obtained from a tree in India, and is a balfamic fubftance of the more folid kind. It differs remarkably from moft other balfams by the nature of its volatile part. It is not an aromatic oil, but a fubftance of a very peculiar kind.

Benzoin is commonly in the form of tears, like other refinous exudations. It is very fragrant when hard rubbed, or when touched by a hot iron, or if fprinkled on a hot plate; and is much ufed on the Continent for incenfing a room.

When merely heated, it fwells up and becomes very fragrant; but when thrown on hot coals, it burns violently, and the odour is too piercing. If flowly heated in a low pipkin, on which a cone of paper is fitted, a vapour arifes from it, which collects on the infide of this cone in fine white fpicular cryftals. Thefe are equally fragrant, but provoke coughing. They are very acid, with fome pungent bitternefs. This is *flowers of benzoin*, or the *benzoic acid*.

The fame cryftalline acid matter may be obtained by boiling powdered benzoin in a great deal of lime-water, evaporating to a fmall quantity, and then detaching the lime by muriatic acid. The benzoic acid then cryftallizes.

This is a fingular fubftance, holding a fort of middle rank between the aromatic oils and the falts. For it is highly fragrant, and inflammable, yet truly acid, uniting with all alkaline fubftances, and with the metals. It feems to maintain its character more firmly than any other vegetable acid. For it diffolves in the vitriolic and the nitric acid quietly, and without inducing that change which they fuffer from the lofs of a part of their oxygen. The nitric acid emits fome faint ruddy fumes indeed, but without effervefcence or commotion. A very fmall quantity of water added makes the compound of nitric and benzoic acids float atop like an oil; but a little more feparates the latter in filaments, and unchanged in its properties.

Benzoin, like other refins, diffolves in alcohol, and is feparated by water.

3. AMBERGRISE differs from the other balfamic or refinous fubftances chiefly by its origin. The greater part of what is brought to the market is found floating on the fea, fometimes in the northern parts of the Atlantic Ocean, but more frequently in the Indian Ocean, where it is fometimes found adhering to the rocks on the coaft. For a long time, very different opinions were formed of its origin. It was fuppofed by many to be a foffil fubftance, which had been wafhed out of its original place by the waves. More lately, it was alleged to be the production of a tree in America, from which the ambergrife iffued as the balfams do from other plants, and that it was carried into the fea by rivers. Had this been true, we fhould have heard of this tree long before now, and got the ambergrife from it. The lateft opinion is, that it is formed in the body of an animal. There are many reafons for believing that ambergrife is really formed in that fpecies of whale called the *phyfeter macrocephalus*, or *bottlenofe:* 1ft, Maffes of it have been found in the bowels of the animal; and though it has been alleged that the animal, which is very voracious, had in this cafe found it in

the fea, and had fwallowed it, this is a mere fuppofition. 2dly, There are often found mixed with it, little bones of fifh, and beaks, and feet of the *fepia*, or *cuttlefifh*, which is known to be eaten by that whale. 3dly, The fubftance of the ambergrife is of a brown colour, and has a confiftence like that of bees wax, but contains numerous white grains which are calcareous. This does not occur in any vegetable production. 4thly, A fmall mafs of ambergrife, which I faw in Apothecaries Hall at London, was like a gall-ftone. It had evidently a fort of nucleus, furrounded with thick concentric layers. I am therefore inclined to fufpect that it is a morbid concretion, formed in fome part of the alimentary canal of that animal, or in fome cavities which communicate with it, in the fame manner as the gall-ftones are formed in other animals.

We may further add here, that fubftances remarkable by a ftrong odour, are produced in a fimilar manner in feveral animals. Such are mufk, civet, and caftor. And there is in dogs a fimilar matter, which has an infupportable heavy fmell; and in infects of different kinds, as bugs, &c *.

That ambergrife, though an animal production, muft be confidered as a balfamic or refinous fubftance, appears from its properties. It has an aromatic odour, and it is volatile by heat, though not fo volatile as aromatic oils. It is alfo foluble in alcohol. Its general appearances, however, more refemble thofe of the bitumens.

Musk, Civet, and Castor, cannot properly be called either oils, balfams, or refinous fubftances. They are animal concreted juices, prepared by fecretion; but they contain an aromatic oily principle, which gives them their odour, and which rifes in diftillation with water. The

* There is a kind of whinftone rock on the coafts of Scotland, and alfo on the north fhore of the Frith of Forth, which has many fmall cavities, about the fize of fmall peafe, or pin-heads, many of which are filled with a fubftance having the colour, confiftence, odour, and chemical qualities of ambergrife. EDITOR.

quantity of it, however, is fmall, or the nature of it fuch that it cannot be collected by itfelf in form of an oil, but remains combined
with the diftilled water.

Before we difmifs the confideration of balfams and refins I muft
obferve, that the term *gum* is often, and very improperly, applied to
many of them in pharmacy; as to copal, gum-hederæ, guaiac, juniper,
animé, elemi, benzoin. When the term gum is ufed with propriety,
it is applied to fubftances totally different in quality from refin. A
gum is indeed a juice which exudes from plants, but it diffolves in
water, and not in fpirit of wine. On the contrary, it is feparated
from water by fpirit of wine. And it has no more inflammability
than any other vegetable fubftance. Gum-arabic is an example.

There is ftill another fet of products of vegetables to which the
term gum is improperly applied. They are likewife infpiffated juices.
Experiments fhew them to be compofed of a mixture of gum with
refinous, balfamic, or oily fubftances. Such mixtures are gum-ammo-
niacum, galbanum, fagapenum, afafœtida, opium, and feveral others.
The proper name for thefe is GUM RESINS. By applying alcohol, we
diffolve the refinous or oily part, and leave the gum. Water, on the
other hand, acts chiefly on the gum, and but very imperfectly on the
refinous or oily part. I fay no more at prefent on this fubject. We
are to treat more of thefe things when exprefsly confidering vegetable
fubftances, as gum is a merely vegetable fubftance, which does not
properly belong to any of the five claffes of the objects of chemiftry.

As among the refins, fo among thefe gum refins, there are fome
diftinguifhed by their peculiar nature or ufeful qualities. Such, for
example, is the fubftance called LAC, or GUM-LAC. It is produced
from the extravafated juice of fome trees in India, when they are
punctured by a fmall infect in their tender new branches or fhoots.
The infect is a *coccus*. The lac itfelf has all the chemical qualities of
a refin. It is of a deep red colour,—and makes the bafis of the finer
kinds of fealing wax, and of many varnifhes. The natural hiftory

of this fubftance, and of the infect which occafions its production, is curious. *(See Phil. Tranf. 1781.)*

The American concreted juice, called by the natives *caboutchouc*, is alfo a fingular fubftance, which I think belongs to the gum refins. It is the milk of a tree. It fhews that it contains a gummy or mucilaginous matter by the effect of warm water upon it, which foftens it and makes it fwell. And the refinous or oily principle in it appears by the action of the fulphuric æther, and by its inflammability. It is ufed for rubbing out pencil lines,—for injection bottles,—to form boots, portmanteaus, flexible perforated bougies, &c. But the manufacture of it is as yet extremely imperfect *. We get it in the form of little bottles. Thefe are formed by receiving the milk of the tree on clay moulds, by repeatedly fmearing them over with the exudation, and drying each coat by a wood fire, the fmoke of which gives it the dirty brown colour. When all is dry, the moulds are crufhed and wafhed out.

This fubftance melts, but without forming a perfect fluid ;—when cold again, its texture is quite changed, and its tenacity and elafticity are gone. It burns with violence and much fmoke, and leaves much coal.

No folvent has yet been found from which it can be feparated in an elaftic and uniform ftate. Nitrous æther diffolves it, and may be evaporated from it, but leaves it filamentous like paper or wafhed leather, permeable to water. Vitriolic æther faturated with water is faid to foften it fo that it can be eafily joined in any way. The infpiffated juice of the fruit of the briony has a confiderable refemblance to *caboutchouc ;* as alfo the juices of fome plants which grow in the Eaft Indies, fuch as the *ficus Indica*, and a plant defcribed in the *Afiatic Refearches* under the name of *urceola elaftica*.

* An ingenious chemift of Glafgow has difcovered a method of expanding it to any fize and thinnefs, as glafs is blown.   EDITOR.

Unctuous Oils.

These are very commonly called EXPRESSED OILS, from the operation by which the greatest number are extracted from the vegetable or animal substances which contain them.

Under this title I comprehend the solid fats of animals, which are but little different by their chemical qualities from the unctuous oils of vegetables : and there are vegetables which contain unctuous oils as solid as the animal fats.

These oils are distinguished from the aromatic by being mild, free of taste and smell, and feeling unctuous or greasy between the fingers ; and besides, they require a much stronger heat than that of boiling water to convert them into vapour ; and they do not take fire so readily as the aromatic.

They resemble one another upon the whole, more than the aromatic do. They are more nearly of the same gravity, being all lighter than water. And the greatest number of those which are commonly fluid are sluggish and thick, compared with many of the aromatic.

As the unctuous oils, in their perfect state, have no sensible odour, and are far less volatile than the aromatic, they do not soon suffer any remarkable loss or evaporation by exposure to the air, nor undergo the same change with the aromatic. But, in certain circumstances, they are liable to another sort of depravation, called *rancidity*. This appears when they have been too long kept, especially in a warm place, or in warm weather. They acquire a thicker consistency and offensive smell, and a great degree of acrimony, or at least a power to irritate the nerves of delicate stomachs with very great violence. The beginning of this sort of corruption is generally attended with a diminution of the colour of the oil. Thus, the fine oil of olives, when fresh imported, and perfectly found, is of a strong and bright yellow colour. When it begins to grow rancid, it becomes colourless, like water; and this is the cafe with many others. The nature of this corruption has not yet been examined or explained. But I believe it depends upon a beginning refolution and feparation of the principles of the oil from one another. I know that oils in a high ftate of rancidity generate flowly inflammable vapours. Bellows of an iron finery were often burft by the firing of an inflammabl e air produced within them from the oil with which the leather was anointed. This rancefcence feems chiefly owing to an extractive mucilage which unctuous oils contain; for one way of preventing or greatly retarding it, is to churn the oil with a great deal of warm water, repeating this operation till it comes off perfectly clear from the oil. This change in the mucilage feems a fermentation, occafioned by abforption of oxygen; for Scheele obferved, that oils becoming rancid abforb and fpoil air,—and it has been fince found that thefe oils become much fooner rancid in vital air. Oils and butter have been kept in water for fifty years perfectly found.

We find a great difference between thefe oils and the aromatic, in their difpofition to be affected by heat. I obferved before that they are not near fo inflammable; nor have they the leaft degree of volatility in boiling water, nor even in degrees confiderably above it. Moft of them contain a little humidity and mucilaginous matter, which, when the heat rifes above that of boiling water, produces a little crackling and boiling; but this is foon over, and then the oil is capable of being heated to 400 or 500 degrees. But, as the heat increafes, they begin to emit vapour and fmoke, acrid and offenfive to the throat and eyes. The remainder becomes thicker, and darker coloured, and capable of receiving more heat; and the vapour and fmoke become thicker, until at laft they break out into flame. But before this, the oil is fo hot, that tin and lead very eafily melt in it, and it is nearly as hot as mercury boiling.

In many employments of the fat oils in the arts, it is neceffary not only to clear them of the water they contain, but alfo to give them a boiling heat, which acts on some of their principles, so as to produce changes which fit them for particular purposes. As this heat is near to that in which they catch fire, the operation is frequently performed out of doors. If, in this condition, a shower of rain falls, there is great danger of the oil's being dashed out of the vessel. If it falls from a height, it penetrates to some depth, where it is blown up into bubbles of elastic vapour, and it explodes, dashing the oil about on all sides. But beside this, it is discovered that water is decomposed by oils in this temperature, and a vast quantity of *inflammable air* is generated. One drop of water will produce above an English pint. This is the chief cause of the explosion, and increases the danger by its inflammability. Great heat is necessary for this action.

When we simply distil these oils in a retort without addition, the greatest part distils over in the form of an oil, quite changed from its natural state. It has acquired a brown or black colour, and a penetrating offensive smell, mixed however with some fragrance, and an acrid taste. It is now become one of the *empyreumatic* oils, which are soon to be described. The quantity of it is somewhat less than that of the oil was before this operation, but the deficiency is made up by a portion of water obtained at the same time, and into which a part of the oil is changed, as I formerly observed. In this watery fluid is found the sebacic acid of Crell.

By trying to mix the unctuous oils with other substances, we learn, in the first place, that they do not communicate any taste or other sensible quality to water, as the aromatic do; nor can they be dissolved in alcohol; but they can be combined readily and intimately with the alkalis, and form with them perfect SOAPS, which can be dissolved, either in water, or in a mixture of water and alcohol. To produce a good soap, we must take the alkali in its pure or caustic state. Common alkali will not do. The usefulness of soap, and its importance as a manufacture, are very well known. It derives its solubility in water, with its detergent and penetrating qualities, from the caustic alkali which it contains. Nothing is fo detergent and penetrating as the cauftic alkali by itfelf. But the oil is neceffary in the compofition, in order to moderate the fharpnefs and activity of the alkali, and to give a flipperinefs to the clothes to which the foap is applied. Were the alkali ufed alone, the clothes could neither be handled with fafety in wafhing them, nor bear the hard rubbing and other mechanical violence neceffary to extricate the matter which renders them foul.

The general procefs for the preparation of foap is to boil the oils or animal fats with a ley made of pure or cauftic fixed alkali. The boiling is continued till almoft the whole of the water is diffipated. The vegetable alkali produces a fofter foap, and the foffil a hard one. Soft foap can be changed into hard by boiling it with a folution of common falt. A double exchange takes place, and we have a muriat of potafh and a hard foap. The procefs is a little delicate.

If we decompound the foap, and feparate the oil from the alkali, we find that the oil will now diffolve in alcohol by itfelf. It has undergone fome change in this refpect which we do not underftand. We have an eafy way of thus feparating the oil of foap for fuch experiments, by adding an acid to it. The cohefion of the oil and alkali in foap is eafily overcome by acids, which uniting with the alkali, immediately detach the oil.

From this effect of acids upon foap has been deduced the caufe of what is called *hardnefs* in waters. Waters are called hard, when they decompofe a little of the foap, and the oil comes to the top in a curdy form. We are told that fuch waters as do not diffolve foap are found to contain an acid, combined with fome fubftance, which does not adhere to it fo ftrongly as to hinder it from acting upon the alkali of the foap. Neutral falts do not render water hard. Macquer was led into the miftake that they did, by finding fea water generally in this ftate. But this is owing to Epfom falt contained in it. This falt, gypfum, alum, or metallic falts, always produce this effect. The addition of an alkali cures hardnefs in waters. This is too expenfive when water is to be ufed in large quantity, as in bleaching, and other manufactures. In fuch cafes, the choice of a good water is of immenfe confequence. Many chemical trials and tefts are delivered by authors; but foap itfelf is the beft of all.

The fame hard waters are unfit for fome operations in cookery, fuch as boiling vegetables foft. A little alkali may be of ufe here, and would improve the colour of the vegetables.

The effects of the mixture of acids with unctuous oils, are not fo violent in general as with the aromatic oils. When the experiment is made with the fulphuric and nitric acids, a part of the acids is de-oxygenated to a certain degree, but another part unites with the oil, and forms a compound which has fome folubility in water, and is named an *acid foap*. Strong nitric acid, however, mixes with the fat oils with violence and the extrication of great heat. Some of them, which are called drying oils, and are found to have a fuperior attraction for oxygen, are even fet on fire, when the operation is judicioufly managed, by pouring the fecond quantity of acid on the part that appears dry and charred.

Of the earths, the pure calcareous earth, or lime in its active ftate, fhews a confiderable difpofition to unite with thefe oils. We have an ufeful example of this in mixing any of them with lime-water, which will mix with olive oil very intimately by a little agitation in a phial, and the compound is found to be a moft excellent extemporaneous remedy for fcalds and burns.

Of the other inflammable fubftances, fulphur is fometimes combined with fome of thefe oils, for the purpofes of medicine, forming the BALSAMUM SULPHURIS CRASSUM. It has a heavy hepatic fmell. When the oil is very much faturated with fulphur by means of a confiderable heat, if the mixture be made to cool with extreme flownefs, it depofits a great part of the fulphur in very beautiful tranfparent cryftals of a dark ruddy brown colour. Sometimes the experiment appears not to have succeeded; and if we give the vessel a shake, it crystallizes in an instant, and so much heat is extricated that the mixture has sometimes taken fire.

So far as we have proceeded, the account I have given is applicable to the whole of these oils. But, in order to give more full information upon this subject, it is necessary to point out a few species, distinguished from the rest by peculiar qualities. These shew a little affinity with the aromatic, by containing a volatile principle, the presence of which in them is necessary to their fluidity, and which preserves them fluid in violent colds. It does not, however, give them an acrid and stimulating quality, for when they are quite fresh and uncorrupted, they are very mild; but it gives to some of them a perceptible odour. And, while it evaporates, in consequence of their being exposed to the air and to light, the oil becomes thick, and at last solid, for which reason, such oils are employed as paints or varnishes. This change is found to be accompanied with a copious absorption of oxygen from the atmosphere.

The oils that come under this description, are those of lint-seed hemp-seed, poppy-seed, and walnut-oil. On account of their drying up, and thus leaving a varnish on the surface of wood or other things to which they are applied, they are called DRYING OILS. Other unctuous oils never dry, but remain greasy, until they evaporate by decomposition, or are absorbed by dust, or other impurities. The drying oils also unite with acids a little more violently, and more like the aromatic oils than the other unctuous oils do.

Spermaceti and bees wax also belong to the section of unctuous oils, but differ by some particulars from the general description.

SPERMACETI is an animal fat, gotten from a particular species of whale, and differs from other fats, chiefly by having much more solidity and dry consistency in ordinary heats of the air. Hence it does not stain cloths, but rubs off; and in congealing, after being melted, it always displays a foliated texture. Dr. Crell made a very perfect soap with it.   It burns with a brighter flame than the greater part of animal fats.

Bees wax also differs from the unctuous oils, chiefly by having a greater degree of solidity in the ordinary heats of the air than most of them: by becoming ductile and plastic with a gentle heat before it melts; by giving a brighter flame ; and by its origin.   It is well known to be formed and used by the common honey bee in constructing its combs ; and has generally been supposed to be formed by the insect from the staminal dust of plants, which they are known to gather and carry to their hives.   But Mr. John Hunter lately found reason to assign a different origin to it.   He thinks that it is a sebaceous excretion from their bodies, and comes out from under the scaly rings which cover their hinder parts ; and that by the motion of those rings on one another, it is formed into the very thin plates with which they make up their cells, and incrust the inside of their hive, and any extraneous body in it which might give them offence, as a snail, &c.

Among the effects which these oils shew in mixtures, we must not omit to mention some consequences which attend mixing them with one another, as some of them in such mixtures produce effects which demand the greatest attention.   There are some of them which soon after they are mixed begin to act chemically on one another.   The mixed mass becomes first warm, then hot, and at last takes fire ; and, in some examples which happened in Russia, has set fire to ships of war, or to magazines of naval stores.

Four pounds of suet, rather greasy than firm, and half that quantity of lintseed, hempseed, or other drying oil, being mixed together, after a few hours become warm ; and if this generated heat be prevented from escaping, by wrapping up the mixture in flannels, it increases to inflammation.   The same effects follow when these oils are worked up into paints, with ochre and other colours; and when cloths, fresh smeared with them, are loosely rolled up in bundles, they grow hot, and scorch, and are thus consumed to ashes ; and they frequently burst into flame.

This propenfity to take fire feems owing to a ftronger attraction of the ingredients for oxygen when in their compounded ftate. Of this we have many inftances, fuch as fulphur, phofphorus, and mixtures of metals. The *black wad*, from Derbyfhire, was fold for a black pigment, till it was found that cloths painted with it, ground with oil, could fcarcely be laid up in the magazines without the rifk of taking fire fpontaneoufly.

The origin and ufes of thefe oils deferve our attention, as they are fubjects of an extenfive commerce. Though the aromatic oils are more precious, thefe are incomparably more valuable and ufeful ; but they are afforded by nature in far greater quantity, and therefore not fo highly prized.

In many of the warm climates, the olive affords an excellent and very perfect unctuous oil in great quantity. In Africa, the negroes extract from the kernel of a palm an oil of this defcription,—the *oleum palmæ*. An excellent oil, which keeps very well, may be got from the *arachis hypogaios Americanus* of Ray. The piftil and germen of this plant point downwards, and penetrate the ground, under which the germen ripens to a fort of nut, or hufky feed, called *ground peafe*. It is much ufed for fattening fowls and fwine. A bufhel of this feed cofts about eighteen pence, and will yield a gallon of oil of delicate tafte, and which keeps very well. *(Phil. Tranf. vol. 59.)* * Through the whole extent of Europe, and in many other

---

* The oil expreffed from the feeds of hemp, which has been carefully reared for this purpofe, and cleared from all bad ftalks, is, when frefh, almoft equal to the fineft butter in fweetnefs. The offenfive odour of hemp oil is partly owing to the hufk, and partly to a natural change which this oil and fome others undergo. The Ruffians, who are very nice, choofe for this purpofe the feeds before they are grown hard and dry, and do not allow either the plant or the feed to lie in a heap. The oil when expreffed is not perfectly fluid and tranfparent, but rather like honey beginning to grain. They put it into fmall bladders, and keep it hanging in running water. I have eaten it three months old, and thought it preferable to their beft butter.

EDITOR.

parts of the world. the fubjects from which we get thefe oils are the animals which are flaughtered for our food, and the feeds of fome plants,
rape-feed, lint-feed, hemp-feed, beech-maft, and walnuts,—and alfo
milk, the butter of which contains an oil of this kind as its principal
ingredient. But befides, expenfive voyages are made to the polar
regions, to kill whales of different kinds, for a plentiful fupply of the
oil they afford. The inhabitants of countries nearer the pole get
their oil entirely from different marine animals.

The ufes to which thefe oils are applied are numerous and important.

1*mo*, They make up a part of the food of mankind in every part
of the world in which they can be procured.

2*do*, They ferve by their inflammation to give us light. And the
Laplanders and Efquimaux ufe their fifh oil, not only as their beft
cordial and moft luxurious food, but depend. on it for both light
and heat in their fubterraneous dwellings during their long winter
nights.

3*tio*, Soaps, which are among the moft ufeful productions of the
chemical arts, cannot be made without thefe oils.

4*to*, They are alfo highly ufeful and much employed in all machinery, to diminifh friction.

5*to*, And laftly, fome of them are ufeful as paints or varnifhes.

Empyreumatic Oils.

As Newman, Lewis, and others, have treated of thefe oils feparately from the others, I have thought proper to follow the example.
This term is applied to all oils which have been forced to rife in vapour, and pafs over in common diftillation, with a heat greater than
that of boiling water, or which are produced by fuch a heat from
fubftances which were not oily before. When oily fubftances, ca-

pable of bearing a heat higher than that of boiling water, are expofed to it, and made to aſſume the form of vapour, and diſtilled, they always undergo a change from their natural ſtate. Oils, which naturally are quite bland, inſipid, and deſtitute of odour, become acrid and ſtimulating, and ſtrongly fœtid, and diſagreeable.

Empyreumatic oils may be diſtinguiſhed into four kinds :

1. Thoſe produced from balſams or reſins.

2. Thoſe produced from unctuous oils.

3. Thoſe produced from vegetable ſubſtances that are not of an oily nature.

4. Thoſe produced by heat from animal ſubſtances not of an oily nature.

The firſt kind have, with the empyreumatic fœtor, an odour of the aromatic oil, which the balſam or reſin contained. The ſecond have the ſmell of the ſmoke of a candle or lamp when blown out. The third ſmell of wood burning. The fourth of burning feathers, horn, hair, bones, and other ſuch animal ſubſtances.

When theſe oils are firſt diſtilled, they are dark coloured, and even dirty, by reaſon of charry matter which comes over with the oil ; but this oil being put into the retort, will riſe with a leſs heat, and comes over more volatile and attenuated, leaving a carbonaceous reſiduum. This operation, repeated ſeveral times, brings the oil to a greater degree of purity, and, which is important, of uniformity or ſameneſs,—in ſo much that it is highly probable that by repeated diſtillations all diſtinctions will be removed. Bees wax yields an oil which approaches the moſt rapidly to this ſtate,—and indeed, after two diſtillations, it can ſcarcely be diſtinguiſhed from an empyreumatical oil of olives that has been eight times diſtilled.

It may appear ſurpriſing to thoſe to whom it is new, that ſuch hard and dry ſubſtances as the hardeſt and drieſt wood, and the bones, horns, and other parts of animals, that have not the leaſt of an oily nature, fhould afford empyreumatic oils by diftillation with a proper heat. The fact, however, is eafily explained by the difcoveries and new principles of the modern chemiftry.

I lately remarked that all oils, in general, are principally compofed of hydrogen and carbon, with a fmall quantity of oxygen, and perhaps of azote, together with a very fmall portion of earth.

Now thefe principles, which have thus been difcovered to conftitute the oils in general, are contained in all the animal and vegetable fubftances, as will be explained to you more fully hereafter, not excepting the dryeft and hardeft of thefe. When fuch fubftances, therefore, are expofed to the moderate action of deftructive heat in clofe veffels, a part of the hydrogen, which is always a volatile fubftance, is volatilized, without being combined with that large quantity of caloric, which is neceffary for converting it into inflammable air. It carries up with it a part of the carbon, and being condenfed by cold in the receiver, it muft neceffarily be condenfed into an oil, which is black, in confequence of th carbon it contains, perhaps imperfectly combined with it, and has a penetrating difagreeable fmell, in confequence of the imperfect union of the hydrogen with the carbon and other principles which it contains. If there be azote in the compofition of the matter, which is the cafe with all the animal fubftances, and even fome of the vegetable, this principle, combining with fome of the hydrogen, forms fome volatile alkali; and if there be much of the oxygen, which is the cafe with all the vegetable fubftances, it contributes to form a quantity of carbonic acid gas, and very often an acid now called *pyro-lignic (pyro-xylic)*, which is neareft by its nature to the acetous acid.

The fame view of the nature of oils in general will alfo enable you to underftand why dark coloured acrid and fœtid empyreumatic oils are produced by fubjecting to heat for diftillation the mildeft and moft bland of the natural oils. The mildnefs and fweetnefs and unctuosity of these depend on the proper union and cohesion and proportion of their principles, all of which conditions are altered by the action of heat.

The empyreumatic oils agree with the aromatic in several particulars. A small quantity of them can be combined with water or dissolved by it. They are dissolvable by alcohol,—and they have not the lubricating quality of the unctuous oils.

All of them contain a very fluid and volatile oil, mixed with a grosser one. And the tenuity, volatility, and inflammability of the more volatile part of them, is in some cases very remarkable. Oil of wax is one of these.

One is valued in medicine, but not much used.—OLEUM ANIMALE, or OLEUM CORNU CERVI RECTIFICATUM. *(Pharm. Edin.)* Another, called *the oil of bricks*, because it is distilled from a red hot brick which has been thrown into olive oil and impregnated with it, is used by the seal cutters and lapidaries, for moistening the diamond powder with which they work, that it may adhere to their tools.

The principal example we have of the application of oils of this division to useful purposes is in the use of tar and pitch. TAR is the first production. It is procured from all the trees of the pine kind, by a rude kind of distillation. In Germany, Norway, and Sweden, where this timber is very abundant, it is piled up in billets in a sort of oven, which is covered with another oven, at a very small distance. The fuel is put into the interval between them. Thus the inner oven becomes a sort of retort, and the oily vapours having no other outlet, run from a gutter made in the floor of the oven. This exudation contains much soot intimately mixed, which makes the tar quite black. PITCH is made of tar, by separating the more volatile and fluid parts by evaporation or distillation. The tar is an empyreumatic oil extracted by heat, and contains a portion of the essential oil of turpentine, which is in all firs, and a quantity of the acid called now *pyro-lignic*. Hence tar-water, while it has that irritating acrimony

common to all the empyrumatic oils, is manifeftly acid, and affects
the vegetable purples.

VII.—BITUMENS.

The feventh fection of the inflammable fubftances contains BITU-
MENS.   Under this title I comprehend all the foffil inflammables, except fulphur, already defcribed.

Some of thefe are fluid and fome are folid.   The moft fubtile of
them is certainly that fpecies of inflammable air which often occurs
in coal-mines and other fubterraneous places, and which I had an opportunity to mention formerly, as liable to take fire from the flame of
candles, and to make dangerous explofions.   It muft be confidered as
an inflammable fubftance of fuch great volatility, that under no greater
preffure than that of the atmofphere, the ordinary temperature of heat
is much more than fufficient to preferve it conftantly in the form of
vapour.   The hydrogen gas is furely its moft abundant ingredient.   I
do not know that any perfon has made experiments with this kind of
air, to inveftigate the nature of it very particularly.   But I have no
doubt that it is heavy inflammable air, or that variety which contains
carbon diffolved in it *.

Next to this vapour, we may reckon the foffil oil called NAPHTHA,
and the varieties of it called PETROLEA.   The naphtha is defcribed as a
fluid of a very light yellow colour, or fometimes colourlefs ; and
which has a degree of fluidity and tenuity equal in appearance to
that of alcohol or æther ; but it is oily, and floats on water like æther.
It is alfo very volatile, and highly inflammable, catching fire at the
approach of flame rather more readily than alcohol.   The flame of it,

* What efcapes from the crevices of the Whitehaven coal-mines gives no indication
of any carbon.   The air in which it burns for a very long while does not affect lime-
water.

when not very fmall, is attended with fome fmoke or foot, as that of all other oils. The odour of it is ftrong and oppreffive to moft perfons. Such a fluid is faid to be gathered from the furface of the water of certain wells in Perfia, and in the duchy of Modena in Italy. But it is a rare production of nature.

The foffil oils called petrolea are not fo rare, but are inferior to naphtha in fluidity and volatility, though fome are ftill very fluid and volatile. The different varieties have different fhades of a yellow or brown colour, and a heavy, oppreffive, penetrating odour, and are highly inflammable. Such are found in feveral places in Italy, Sicily, Bavaria, France, iffuing from the crevices of rocks, or floating on the waters of fprings or wells, and readily take fire by the contact of flame, and burn on the furface of cold water.

The foffil inflammable fubftances which are next inferior to thefe in fluidity, are thofe to which the name of bitumen has been moft particularly applied. Such are a number of fpecies which have a confiftency refembling that of tar or pitch, or that of the vegetable balfams. They are diftinguifhed, according to their different degrees of fluidity or confiftency, by the names of *piffoleum, pix judaica*, and *piffafphaltum.* Some are little darker in their colour, or thicker in their confiftency than the coarfer petrolea, while the thickeft, or moft grofs, the piffafphaltum, is actually folid when cold, and requires heat to give it fluidity, in the fame manner as pitch. When cutting a level to a coal mine on the bank of the Severn, near the iron bridge, a fource of this kind was difcovered in a ftratum of free-ftone. It yielded this tar in confiderable abundance at firft; but this gradually abated,—and after two years feemed to be exhaufted. It had accumulated in the pores of the fand-ftone through a length of time, and had now drained off. A fimilar cafe occurred in Renfrewfhire a few years ago. St. Catharine's well at Liberton has yielded this fubftance for more than a century, and does not feem to abate. But the quantity is not great.

The very fluid bitumens are in general fuch rare productions of nature, especially the more subtile and volatile kinds of them, that they have been but little examined by the chemists. They have much of the appearance and properties of the empyreumatic oils.

The solid bitumens are AMBER and COAL. The common appearance of amber is too well known to need description. There are also varieties of it. The most valued pieces are of a light yellow colour, and either transparent, or agreeably clouded with white. But it is much more frequently found of a deeper yellow or brown, or even almost black and opaque. Amber is distinguished by an aromatic odour, which it emits when rubbed. This and some other particulars give it a great resemblance to some vegetable resins. But it differs from all that are known in some of its properties.

If it be heated gradually, and the vapours condensed in close vessels, it gives first a small quantity of water. This is followed by a salt, part of which is dissolved in the water, and part condensed into a solid form. Along with this salt, and after it, passes over a large quantity of empyreumatic oil, and some inflammable gas. Little charcoal remains. This salt is an acid, strongly tainted with the empyreumatic oil, more resembling the vegetable acids than any other. The oil has an exact resemblance to some petrolea, and if repeatedly distilled, becomes so like naphtha in every property, that they cannot be distinguished.

The natural history of amber, as found on the coasts of the Baltic, is given by Hoffmann and Newman; to which may now be added, that it has been found of late years abounding in Royal Prussia, near the shores of the Baltic, by digging pits of considerable depth, till they reach a stratum of forest trees, bedded in sand, or at least under a stratum of sand. The trees are all charred, and perfectly black. The amber is found among them in nests, and the working of these mines is found very profitable. The floating amber is, in all probability, torn up from the bottom, where the sea has penetrated to this stratum. (*Crell's Annals of Chem.*) It is found in general in small bits. Pieces that are four, or five, or six inches in diameter, are exceedingly rare; and if they happen at the same time to have beauty, are very highly valued. It has, therefore, been a difideratum among the chemifts to unite fmall bits of amber into larger maffes; and fome have been reputed to poffefs fuch a fecret. But it does not appear that the thing has ever been practifed, or that it is poffible.

PIT-COAL, the other folid bitumen, is of much greater value to the countries which poffefs it, in which it is well known to conftitute thick, numerous, and extenfive ftrata. And though inattentive obfervers confider it all as the fame, we find remarkable varieties of it in the different ftrata which it forms. I fhall here mention the moft diftinguifhed varieties only. There are many others intermediate between thefe; but we need not attend to them all.

1. *Cannel coal*, or *candle coal*; or, as it is named by our colliers, *parrot coal*.

2. Common *Scotch coal*, in which our colliers diftinguifh fome varieties; but we need not enumerate them.

3. *Fat*, or *caking*, or *blackfmith's coal,—fmithy coal*.

4. *Kilkenny coal*; or, as our miners call it, *blind coal*.

Thefe different kinds of coal are diftinguifhed by their manner of burning, and by the products they afford, or phenomena they exhibit, when they are fubjected to the operation called chemical analyfis.

The firft kind, or candle coal, kindles eafily, and gives in burning an extraordinary blaze of bright white flame.

The fecond, which is the moft common in this country, alfo gives in burning a good deal of white flame, though not fo much as the firft; and the cinder or charcoal that remains, when it ceafes to flame, is a better fuel, or contains more of the carbonaceous principle than the cinder or charcoal of the candle coal.

The third kind, the fat, or blackfmith's coal; gives lefs flame in burning than either of the two former; but in beginning to burn, or before it begins, it is in fome meafure melted by the heat; at leaft a bituminous matter that is foftened and melted by the heat, oozes out of it, and occafions the bits of it that are in contact to cohere together. So that, when a quantity of it, in fmall fragments, or perhaps partly in duft, is laid on a fire, and wetted with a little water, to make the fragments and duft enter into clofer contact, it very foon coheres together into one mafs, which being afterwards broken or divided a little, to let the air pafs through it, burns ftrongly and a long while, giving a great deal of heat before it be totally confumed. The cinder or charcoal of this kind of coal is very rich in the quantity of its carbonaceous principle.

This kind of coal is but rare in this country; but in fome of the principal coal countries of England, Newcaftle and Whitehaven, it is the moft abundant. It is the moft valuable of any, on account of its not fuffering wafte by being broken down fmall; for in the fmalleft fragments, it is equally good for houfehold purpofes as the large pieces; whereas the greater part of our coal, when broken down to duft and fmall fragments, becomes quite unfit for houfehold purpofes. By running like fand into all the interftices through which the air fhould pafs, it extinguifhes a fire inftead of mending it. In this ftate, therefore, it is not called coal, but *culm*, and is only employed for burning brick and lime, and for making falt from fea-water. The Englifh fat coal, though ever fo fmall, is never called culm, for this reafon, that by its caking quality, it makes excellent houfehold fires, and gives a great quantity of heat before it be totally confumed. It is alfo the moft thrifty, as it does not burn much when the fire is left at reft.

This kind of caking coal, when it is free from any admixture of fulphur or pyrites, is alfo highly valued by the blackfmiths, and is even neceffary in fome meafure to their operations. It enables them to make what they call a *hollow fire*. When they have occafion to heat a mafs of iron, or a thick bar, they put it into their fire, and cover it with a large round heap of this fort of coal, wetted on the outfide, and then by working their bellows for fome time, and other management, the coals that were in immediate contact with the iron are consumed, and a hollow is formed around it like an arch or little cavern, the sides of which are all composed of coal caked together. While the wind of the bellows is driven into this cavity, and circulates in it, a violent heat is produced all around the mass of the iron, and every part of its surface receives the necessary degree of it, which could not easily be obtained by using other coal that is unfit for forming the hollow fire. Such are the properties of the third kind of coal, the fat or caking coal, or smith's coal.

The fourth and last kind of coal I mentioned, the Kilkenny coal, or, as it is called by our colliers, the *blind coal,* differs greatly from the others by its manner of burning. It is more difficultly kindled, and when perfect of its kind, gives no flame whatever, but burns exactly like charcoal, and gives a great deal of heat before it is consumed. As it neither gives flame nor smoke, it is burnt in malt-kilns for drying the malt, the air which comes from it in burning being quite free from fuliginous vapours. It appears probable, therefore, that the volatile parts of this kind of coal have been separated or expelled from it by some operation of nature.

Such are the differences of these four kinds of coal in their manner of burning. They differ also, as I said, by the products they afford when they are subjected to distillation performed without addition, in retorts and receivers.

I shall first describe the effects of this operation when performed with the most common coal of this country, the common Scotch coal.

If the receiver be closely luted to the retort, and kept very cool, and provided with an air-pipe, we get, *first,* a small quantity of water, little different from pure water.

Secondly, While the distillation advances, and the heat is increased, more water comes over, attended with a brown oil.

Thirdly, The heat still increasing, less of the water comes over, but

much more of the oil, and at laft fcarce any thing but black oil is feen
to condenfe in the receiver.

The oil that comes at firft is thin or very fluid, and of a brown
colour. That which comes afterwards is thicker and darker coloured ;
and at laft it condenfes quite black, and as thick as tar. It continues
to come in the form of a thick fmoke, to the end of the operation,
that is, until the retort is heated to a degree of ignition, by which
the remaining matter is changed into charcoal, and no more volatile
matters can be expelled from it by an increafe of heat.

Fourthly, During all the time, when the oil is diftilling, efpecially
after the firft of it has been condenfed, an immenfe quantity of claftic
aëreal matter paffes out through the air-pipe, and may be collected in
inverted veffels. A part of it is carbonic acid gas, which precipitates
lime from lime-water. But the greater part is hydrogen gas, or inflammable air, not of the pureft and lighteft kind, but containing a
quantity of the carbon, or carbonaceous principle, intimately combined
with it. And therefore, when a mixture of it and oxygen gas or
vital air are fired together, the confequence is not the production of
water alone, but of water and of a quantity of carbonic acid.

The whole of the oil is of the empyreumatic kind, not unlike to
tar in colour and confiftence, only more fluid, and it has a very offenfive fmell, fimilar to that of the fmoke of coal fires when they are beginning to burn. Like other empyreumatic oils it is partly compofed
of a very volatile and fluid part, and a groffer one. They can be feparated by diftillation with a gentle heat, by which we get a portion
of the oil very fluid, and volatile, and tranfparent, and of a yellow colour, very much refembling fome of the petrolea. What remains in
the diftilling veffel becomes more like tar or pitch in its confiftence
and properties ; and is now manufactured, in order to be employed as tar,—the more volatile oil being at the fame time prepared for
making a fort of varnifh.

Along with thefe oils, I faid that a quantity of water is always extracted from the coal by the fame operation. And this water contains fome volatile alkali, which at firft is intimately combined with fome of the oil, but is feparated from it by rectification, &c. and afterwards can be employed in the manufacture of fal ammoniac.

Thefe are the products which we obtain by diftillation from the moft common coal of this country. But there is fome variation when we perform the fame operation with the other kinds of coal I enumerated.

The candle coal yields a great deal more of empyreumatic oil and other volatile products than the other kinds.

The common Scotch coal is the next to it with refpect to the quantity of empyreumatic oil, or volatile inflammable matter, which it affords.

The fat or caking coal affords ftill lefs oil, but a very rich charcoal.

And the Kilkenny, or blind coal, affords none at all, and fcarcely any volatile matter whatever. It may be confidered as a natural charcoal.

Such are the principal varieties of bituminous inflammable fubftances found in the earth. And when we ftudy their natural hiftory, and fome other particulars relating to them, we find great reafon to be perfuaded that all have derived their origin from vegetable matter.

In pit-coal many appearances occur which lead us to fuch a conclufion with regard to this fort of bitumen. It is common to obferve impreffions of leaves and other parts of vegetables in the ftrata, which lie above or below the coal; and very frequently in the coal itfelf we find a black matter, which has fo much of the ftructure of wood, and refembles fo exactly bits of wood which have been fcorched or burnt to a fort of charcoal, that there is no room to doubt of its origin. This matter frequently occurs in the coal which we burn here, and always forms a fort of thin ftrata intermixed with thofe coals, and compofing part of it, and occafioning it to fplit eafily, in the direction parallel to the two furfaces of the ftratum.

Sometimes little maffes of the coal are entirely made up of this fort of matter. I have been informed that fome of the coals of England contain a remarkable quantity of it, and are diftinguifhed by the name of *clod-coal*, and valued as the beft for melting iron from its ore. And the fame matter appears in a bit of coal which was brought me from the coaft of Greenland *.

From thefe and other facts and appearances, it has been concluded that pit-coal has been formed of vegetable matter, carried into the fea by rivers, and which having funk to the bottom in confequence of its being thoroughly foaked, has there undergone a degree of decompofition, and has been ftrongly compreffed by other materials depofited over it, and in fome cafes has been penetrated with mineral vapours and fubterranean heat, fo as to give it the various qualities and appearances we fee in coal.

The other facts, befides the appearances already mentioned, which have fuggefted fuch an opinion, are thefe:

1/t, Great rivers, in different parts of the globe, are well known to carry annually vegetable matter into the fea, efpecially thofe which have a long courfe through immenfe uncultivated tracks of the earth's furface that are overgrown with wood, as fome of the great rivers of North and South America, Africa, India, and the Ruffian territories. Great rivers neceffarily have a great part of their courfe through level countries, through which they make many ferpentine turns. And

---

* In 1759, coals were brought on board the Royal William, in Louifburg harbour, which had been bought for the captain's firing. When his fteward faw them, he refufed to take them, faying that they were only charred wood. The vegetable ftructure was very diftinguifhable in the greateft part of them. One piece in particular had unqueftionably been a large root of the common marfh iris. Its joints, and the commencement of the air veffels at the big end, could not be miftaken. Some pieces being firft made red hot under fand in a pipkin, fo as to diffipate a heavy ftinking fmoke, burnt afterwards with a fmell perfectly refembling that of wood charcoal. This coal had been brought in the boat from the mouth of the mine, in the face of a cliff, from which they had been let down by a pulley into the boat.

EDITOR.

they are conftantly undermining their banks in fome of thofe turns, and occafioning wood, leaves, mofs, and other vegetable matters, to fall into their ftream. Some of this matter floats for a long time, until it be fo thoroughly foaked as to fink to the bottom. But while it floats, it is carried down to the fea, and perhaps afterwards to a very great diftance, by tides and currents. Sometimes it runs aground in the fhallows that are at the mouths of fuch rivers, and gradually forms iflands in thofe fhallows, as at the mouth of the Miffiffipi; but the greater part is carried out to fea. Great quantities of timber are found floating in the northern feas, on the coaft of Iceland and Greenland, and the north coaft of Ruffia. All this, after floating fome time, muft fink to the bottom. In Iceland there is a large bay which is always full of floating wood, and fupplies the inhabitants with fuel.

2dly, The very circumftance of coals being formed into ftrata is ftrong in favour of this opinion, as we have the greateft reafon to be fatisfied that all ftrata have been formed of matter carried into the fea. But, befides, we find thefe ftrata of coal always intermixed with other ftrata, which have been manifeftly formed in the fea, as fandftone, limeftone, and clays of various kinds.

3dly, In fome parts of the world, among ftrata of the fame kind with thofe which commonly accompany coal, are found ftrata manifeftly compofed of wood, even trees compreffed and compacted together, fo as to form ftrata, bearing fome refemblance to thofe of coal, but in which the wood retains fo much of its original ftructure and fhape that it cannot be miftaken. There is an example in Devonfhire, called *Boveycoal*; and a ftratum of foffil wood in the north of Ireland.

All thefe reafons, therefore, leave little room to doubt of the origin of pit-coal in general, although, in many varieties of this bitumen, the firft contexture of the materials has been fo much abolifhed by immenfe compreffion, and the penetrating and diffolving powers of water and heat, and other caufes, that we hardly find any remains of it. It is probable too that many ftrata of coal have been formed of other vegetable matter, as mofs or peat,—carried into the fea during a long courfe of time, by rivers which have their courfe through extenfive tracks of the earth's furface, abounding with bogs and moors.

When we ftudy the varieties of the other folid bitumen,—amber, we can clearly trace its origin alfo from vegetable matter. It is not at all uncommon to find in amber, infects of various kinds, either fuch as creep on the trunks of trees, as ants, fpiders, caterpillars, or flying infects that frequently alight upon trees, or live upon them: Such are a variety of flies and moths. In the Britifh Mufeum there is a lizard in amber; a reptile which is well known to creep on the trunks of trees, or to dwell about their roots. Further, thefe things are immerfed in the amber, and every way furrounded with it; and, from other appearances about them, it is plain that the amber muft have been foft and vifcid, like the vegetable balfams, when they adhered to it, and when it involved them.

This, and the chemical qualities of amber, by which it refembles greatly the vegetable, balfamic, and refinous fubftances, gives fufficient proof that it was once one of thefe. It is well known that many foreft trees, in different parts of the world, at certain times of the year, fhed from ruptured parts of their bark, large quantities of balfams or turpentines, a part of which muft fall into rivers, or, if depofited in the foil, muft be wafhed out by them afterwards fome time or other, while they gradually change their channel by undermining their banks. This matter muft be rolled along by the river until it reaches the fea, or is left behind, in fome part of the bottom where the water has little motion, and by length of time, and the action of mineral vapours, affumes the qualities which we find in amber.

Such, therefore, appears to be the origin of the folid bitumens, coal and amber. That of the fluid may perhaps be thought more difficult to trace. But I think we have clear lights for this part of our fubject alfo. We muft, in the firft place, confider that the fluid bitumens bear a furprifing refemblance to the empyreumatic oils which may be extracted or produced by fire from the folid. Some fpecimens of the petroleum refemble exactly the empyreumatic oil of amber; others refemble the oils which may be extracted by fire from different kinds of pit-coal or foffil-wood. And all the fluid bitumens refemble thefe oils more or lefs. We muft alfo reflect on the evident figns of fubterranean fires which are frequent in different parts of the globe, particularly the hot fprings, many of which pour forth amazing quantities of hot water, and thereby fhew that the fires by which their heat is maintained have great power and permanency. This is further proved by the eruptions and fteams of volcanoes, which are very numerous over the face of the globe, the fires of which muft be at an immenfe depth below the furface, and muft have very extenfive communications, as is plain from the extenfive effects of earthquakes, which manifeftly depend on thefe fubterranean fires. If to this we add what we obferved in fome kinds of pit-coal, which fhew by their properties that the volatile parts have been expelled, it will be eafy to gather from all this fome highly probable conjectures concerning the origin of the fluid bitumens. We can hardly avoid being perfuaded that they have been produced by the long continued action of fubterranean heat on the materials of the folid bitumens, the volatilizable parts of which are gradually feparated from the reft by a fort of natural diftillation or chemical analyfis, and are driven upwards through the crevices or pores of the earth, until they come near to the furface, where they efcape from the further action of the fubterranean heat, and are within our reach. In confirmation of this fuppofition, it may further be remarked that fome of the countries in which thofe oils are found the moft frequent, are well known to be undermined by fubterranean fires, efpecially Italy.

And not only the fluid bitumens, to which this name is commonly applied, but that more fubtile fubftance, the inflammable air, or fire-damp, which occurs in mines, or breaks out at the furface of the earth in fome places, · can be imagined to have the fame origin. When we expofe the folid bitumens to the action of heat in clofe veffels, be-fides the empyreumatic oils which we obtain, and which refemble the fluid bitumens, we get alfo great quantities of inflammable air, or inflammable gas, which muft neceffarily be produced from folid bitumens or foffil wood, on many occafions by fubterranean fires.

Thus, therefore, the whole of the fubterranean inflammable matter that belongs to the title of bitumens may be imagined to derive its origin from vegetables. From whence the vegetables again derive theirs, will perhaps appear when we come to confider the matter of vegetables, and its production, as a particular fubject of chemical inquiry. *(See Note* 15. *at the end of the Volume.)*

CLASS IV.

METALLIC SUBSTANCES.

THE metallic fubftances are now to be confidered as the Fourth Clafs of the objects of Chemiftry.

They probably attracted the attention of mankind at firft by their fhining furface. But when their nature and properties became better known, they were valued as materials with which we can eafily execute many purpofes in arts which cannot be attained without them.

The qualities which render them ufeful and valuable are,

1*mo*, Their fufibility and malleability, by which we are enabled to melt different maffes of them into one, or eafily to divide large ones into fmall ones, and to mould or hammer, or otherwife work them into the forms which our purpofes require.

2*do*, The ftrong cohefion of their parts, in confequence of which they are the proper materials for all parts of machinery, or other works in which great ftrength or long duration is required.

3*tio*, The different degrees of hardnefs and elafticity which fome of them can be made to affume, render them highly ufeful to us on innumerable occafions, for the fabrication of tools, fprings, and many other implements neceffary in the arts.

4*to*, The denfity of their fubftance, which is impenetrable to water, and moft other kinds of matter, is one more of their ufeful qualities. And,

5*to*, and laftly, Their bearing fudden alterations of heat without being broken, is another.

Thefe feveral qualities render them ufeful for many different purpofes; and further, by their luftre and brightnefs, and the facility with which they are polifhed and wrought into ornamental forms, they are fit fubjects for many of the elegant arts.

The metallic fubftances are, for thefe and other reafons, much more the objects of attention and ftudy than the other claffes of natural bodies.

And yet we have gained lefs advantage from the refearches of many of the chemifts on this fubject than what might have been expected from the great labour which has been beftowed on it. The reafon is plain. The greateft number of experiments formerly made upon metals were prompted by the vifionary projects and notions of the alchemifts. Inftead of directing their inveftigations to the difcovery of new productions that might be ufeful in life, their fole aim was to convert cheaper metals into gold and filver. And it happened at the fame time, that the greateft number of thofe who were employed in this purfuit, were perfons of fo little education, and fo ignorant, not only of other fciences, but even of any rational principles of chemiftry itfelf, or of the nature of the materials which they employed, that they were for the moft part incapable of making experiments in a judicious manner, or of underftanding them properly when they had made them. Their extraordinary labours muft no doubt have led them to the knowledge of many curious facts; but every experiment which deferved attention was concealed by them with the greateft care, or if they mentioned, or pretended to defcribe any difcovery, they did it it fuch myfterious and ambiguous language, as muft render the ftudy of it infupportable to thofe that ever read any thing elfe; befides that they every where abound with abfolute falfehoods.

I fhall not, therefore, attempt the difagreeable and fruitlefs toil of inquiring into the meaning of their opinions and proceffes, but confine myfelf to fuch an account of the metals as may be deduced from clear and fimple experiments.

Upon this plan, I fhall firft mention the more general qualities of the metals, and afterwards defcribe the nature of each in a more particular manner.

1*ft*, The moft obvious general quality of the metals is their great weight or denfity, in which they exceed all other known matter. The heavieft ftones are not more than four times as heavy as water, or fcarcely fo much. The lighteft of the metals is feven times heavier.

2*dly*, They are remarkable alfo for a great degree of opacity, and a power of reflecting the light ftrongly from their furface; in confequence of which all mirrors are made of metallic fubftances, or are made to reflect light by their means. The thinneft leaves or films of metal have this power to a furprifing degree,—but we can in fome cafes reduce them to films of fuch exceeding thinnefs, that they tranfmit a very fmall part of the light which falls upon them.

3*dly*, Metals are diftinguifhed from all other folid bodies by fome qualities with regard to electricity, which give them the firft rank among *conductors* of electricity, and which fubject them on all occafions to be more affected by lightning than any other fort of matter which comes in its way. By conductors electricians mean bodies which are difpofed to receive readily and quickly the electrical fluid, and to tranfmit it in a moment through their whole fubftance and furface; fo that when we communicate a charge of electricity to any one part of them, it is communicated to every other part of them at the fame inftant of time ; or if we abftract electrical matter from any one part of them, every part of them will be found to be equally affected by the abftraction. Non-conductors are the oppofite to this : They receive difficultly, and tranfmit very flowly. For this reafon they are employed in electrical experiments for *infulating* bodies,—and alfo for collecting the electrical fluid when we defire to accumulate it or make it act on different bodies. The electrical fluid is collected in general by rubbing a smooth polished surface of these non-conducting bodies.

All other kinds of solid matter, provided they are made perfectly dry, have more or less of the *non-conducting* and *insulating* power; but the metals have none of it. The electrical fluid, when accumulated in any other bodies, has always a disposition to pass from these to the metals, and to be diffused from one part of them to another, with such astonishing velocity, that it can be conveyed by a wire to the greatest distance without requiring any perceptible time for its passage. Mr. Cavendish found by experiments, that iron conducts the electrical fluid 400,000,000 times better than water; that is, a current of this fluid will pass as readily through 400,000,000 inches of iron wire as through one inch of a cylinder of water equal in diameter to the wire.

4thly, Another quality of the metals in general is observed in their manner of assuming a fluid form by heat. They retain their opacity in their melted state, and form a fluid which has a bright and reflecting surface, and which is repelled by most other kinds of matter, or which does not adhere to them, and spread itself on them, as water and many other fluids do. Small drops of melted metal, therefore form themselves into little spheres, in consequence of the stronger attraction of their particles for one another than for the surrounding matter. This happens not only when the small drops of melted metals lie on the surface of various solid bodies, but also when they are immersed in fluids, as water, oils, melted salts, and melted earths, none of which will mix with the metals, so long as these retain their metallic form.

These several qualities of the metals,—their excessive weight,—the opacity and reflecting brightness of their surface,—their being such perfect conductors of electricity,—and their mercurial manner of fusion, as it is called, form the distinction between them and other bodies, and are the only qualities which are found in them all without exception.

But we may alſo take notice here of malleability, laminability, and ductility, as among the general qualities of the metals; for, though theſe qualities are not found in all of them, the greater number have them, and metals are the only bodies in nature in which they are found in an equal degree. Theſe qualities are diſtinct from one another, and do not always go together. Some metals have all the three in a high degree, e.g. copper, ſilver, gold, and platina, when pure. Others have two only, as tin and lead, which are malleable and laminable, but not ductile. Others, on the contrary, as iron, are ductile, but not malleable and laminable. Others, as zinc, are laminable, but not at all ductile, and ſcarcely malleable.

The moſt malleable metals, however, are liable to become rigid and hard by hammering, in conſequence of the expulſion of latent heat; and they muſt be ſoftened again by annealing, before we can proceed to hammer them further. I had occaſion formerly to explain annealing, and how it is performed. Tin and lead are annealed in ſome degree by the heat of boiling water, and by a ſomewhat ſtronger heat are perfectly ſoftened. Iron, copper, ſilver, and gold, muſt be made red hot. Platina is with great difficulty annealed.

Such, therefore, are the more obvious general qualities of the metals.

To take, in the next place, a more chemical view of them, let us attend to the effects of heat on the metals in general.

1ſt, Each of the metals is well known to require a particular degree of heat for its fuſion; and ſome require very violent degrees. Dr. Boerhaave's opinion or ſuſpicion, concerning the heat of ſome metals in their melted ſtate, is perfectly groundleſs.

2dly, After a metal is melted, if we increaſe its heat to a much higher degree, moſt of them may thus be changed into vapour; and in cloſe veſſels, ſome of them can actually be diſtilled or ſublimed; though in general the free acceſs of air to ſuch metals, when violently heated, makes them evaporate much more readily than they will do in cloſe veſſels.

But, in order to underſtand fully the action of the air on the metallic
ſubſtances, when they are ſtrongly heated, it is neceſſary to know that
they greatly reſemble the inflammable bodies, by having a remarkable
diſpoſition to attract oxygen, and to combine with it. And this combination, in ſome caſes, is attended with the ſame phenomena of heat and
light as in the caſe of bodies commonly called inflammable. For example, filings of zinc, or of iron, when thrown into a clear coal fire,
burn with a diſtinct flame. This inflammability appears from many
other facts and experiments; among others, from the effects of nitre, when
it is applied very hot to the metals: There is a deflagration, as it is called, in the ſame manner as when that ſalt is made to act on the inflammable ſubſtances, with this difference only, that, in the experiments
with metals, it is neceſſary to apply a much ſtronger heat to bring on
the action of the nitre and metal on one another. Filings of zinc,
when thrown into melted nitre, burn with a flame ſo brilliant and
dazzling, that the eye cannot bear it. If the nitre be red hot, filings
of iron will exhibit the ſame appearance, but it is not near ſo brilliant.
Filings of a mixture of lead and tin come next to theſe two in inflammability and brilliancy. The metals are thus ſuddenly changed into
an earthy-like matter, which is mixed with the remainder of the
nitre. I call it an earthy-like matter, this being its general appearance.
It is ſtill, however, very different, according as the experiment is made
with the different metals; but in none of them does it retain the appearance of any of the general qualities of metals which have been
already deſcribed.

A further proof of the affinity of the metals to the inflammable ſubſtances is that they are liable to ſuffer a change by the joint action of
heat and air, ſimilar to that which the inflammable ſubſtances undergo, a change analogous to inflammation. This change, in the caſe of the
metals, was called CALCINATION. A higher degree of heat is neceſſary
to bring on this change in moſt of the metallic ſubſtances than that
which is ſufficient for the commencement and continuation of the

inflammation in the bodies ufually called combuftible. And there is a certain latitude, or range of the fcale of heat, which is beft adapted to the calcination of the metals. In the lower degrees of this range, metals are calcined flowly and with difficulty. Near to the middle of it, the metals, in general, are the moft quickly or perfectly calcined. In the higher degrees, the calcination does not go on fo well. And when we increafe heat to an intenfity much above the range I now fpeak of, fome of the metals expofed to fuch violent heats will retain their metallic form, though they may eafily be calcined in inferior heats. The reafon apparently is, that in very violent heats the air acquires too much elafticity, and is too much rarefied, to act on them with fufficient power; and the action of the air is fully as neceffary to their calcination as the action of heat. If you would wifh to know in what part of the general fcale of heat this range fit for calcination is contained, I cannot be precife in pointing out its limits, as they are by no means diftinct. I can only fay that it appears to be comprehended within what are called the red heats, which, however, have many differences of intenfity. Below thefe red heats, and alfo in heats that are far above them, or what are called the white heats, calcination does not go on fo well, at leaft with moft of the metals.

The matter into which the metals are changed by *calcination* is alfo an earthy-like matter, fimilar to the matter into which they are changed by the action of nitre. The colour and other properties of it are different, according to the metal from which it is obtained, and partly too, according to the manner in which the calcination has been performed. This matter has been a long time called the CALX of the metal, and from it the operation by which it is obtained was called calcination. The term calx was chofen on account of a fuppofed analogy between the calcination of the metals and the burning of lime, though it is now well known that thefe two operations are of a totally different nature; and accordingly, the French chemifts propofe to fet afide thefe terms of calx and calcination, as very improper, and to fubftitute others which I fhall foon explain to you. In the mean time, we fhall proceed to make a few more remarks on the calcination of metals in general.

In the firft place, I wifh you to underftand, that metals, thus calcined, or changed into calces, are not always calcined to the utmoft degree of which they are capable. By a certain moderate action of heat and air, they can be changed into calces, which may be further changed by a continued or more effectual action of the fame powers. This gives occafion to a great diverfity in the calces of fome of the metals which are capable of thefe different degrees of calcination. A general account of which diverfity may be given in this propofition, That the farther the calcination is pufhed, the more does the calx refemble an earthy fubftance, or it is whiter, and the lefs difpofed to fufion by itfelf; and on the contrary, the lefs they are calcined they have the more colour, or they are lefs white, and retain more of their fufibility. Of this there are many examples, as in the calces of antimony, tin, and fome others.

But the nature of the calx produced, and the phenomena and quicknefs of calcination, are very different in the different cafes, partly in proportion as the heat is more or lefs ftrong, and more efpecially, as the operation is performed with the different metals. You will eafily perceive that there muft be great variety in the calcination of the different metals, if you attend to thefe particulars :

1ſt, The different metals are more or lefs calcinable one than another. Some with great difficulty, and they are but imperfectly calcinable. Others are more calcinable, but ftill are calcined flowly, and with fome difficulty. Others, again, are much more difpofed for calcination, and, in proper circumftances, are calcined eafily and quickly. While there are others which can be calcined to a certain degree only, but after this, refift the action of heat and air, without fuffering much further change.

2d, If you alfo reflect on the great difference of the metallic fubftances in their fufibility and volatility; fome being eafily melted with a very moderate heat, while others require the moft violent; and that fome can alfo be eafily converted into vapour, while others endure an intenfe heat a long time, without being volatilized by it;—you will perceive that the heats proper for their calcination muft bear different relations to the heats neceffary for their fufion, or converfion into vapour. Some, which require ftrong heats to melt them, are beft calcined in degrees of heat inferior to thofe neceffary to their fufion. Others, much more fufible, muft be melted and heated much above their melting point, before they can be calcined. And of the volatile metals, fome do not calcine faft, without applying to them fo much heat as converts them into vapour.

Further, the calcination, in fome cafes, goes on fo flowly, or to fuch a moderate degree only, that it is not attended with emiffion of perceptible heat and light. But, in other cafes, there is a plain appearance of inflammation, the calcining metal being feen to glow like a burning coal; and in others, a bright flame is produced, like that of fome volatile inflammable bodies.

What I have now faid will give you fome general knowledge of the phenomena which attend the calcination of metals in different cafes and varieties of this operation. But you may form a more diftinct notion of this variety by attending to fome examples which I fhall now relate:

1. Copper requires a ftrong heat to melt it, and therefore calcines beft in a heat inferior to its melting heat. It is not capable of being calcined faft, or to a great degree; and no light or heat is obferved to be produced by its calcination.

2. Tin is very fufible, and therefore melts before it calcines; but when a proper heat is applied to it, it calcines faft, and more perfectly than copper, and with appearances of combuftion. The calx, when well calcined, is very refractory, or hard to melt.

3. Lead, like tin, is very fufible, and therefore melts before it cal-

cines. When the calcination is performed with a ftrong heat, the metal fmokes all the while, or emits vapours; and it appears inflamed, or emits more heat and light than the furrounding matter,—and its calx, being a very fufible one, flows around it in a melted ftate.

4. Antimony is very volatile, and therefore emits vapours plentifully in a calcining heat, which vapours are at the fame time calcined, and the calcining metal emits light and heat.

5. Zinc is fimilar to the former, only more volatile, and in calcining, gives a dazzling light.

6. Lead and tin mixed. Mixing metals together increafes their difpofition to calcine. It increafes their fufibility alfo, and therefore diminifhes their cohefive attraction, which diminution of the cohefive attraction is probably a confequence of the chemical attraction of the two metals for one another, as chemical and cohefive attractions are generally antagonifts to each other. But by the diminution of cohefion the chemical attraction of the metals for oxygen will be increafed *.

As the matter into which fome of the inflammable fubftances are changed by inflammation can be reftored again to its former ftate, fo the calces of metals can alfo be reftored to their former ftate, and made to refume the metallic form. The operation by which this is effected is called REDUCTION, and is commonly performed by heating the metallic calx ftrongly, in contact with charcoal, or mixed with it. Thus the charcoal, or a part of it, is confumed, while the metallic matter recovers the metallic form. And to reduce a metallic calx in this manner, we can employ the charcoal of any inflammable fubftance from which a charcoal can be obtained, whether it be vegetable, animal, or foffil. There are even examples of metallic calces

---

* I may juft obferve here, that Dr. Black, in his earlieft lectures, afcribed much of the increafe of weight which is obferved in the calcination of metals, to the abforption of air, and its combination with the metallic fubftance. He obferved that this increafe is always fmall when the operation is performed in clofe veffels, and afferted that there would be none in veffels exhaufted of air. I have before me notes taken at his lectures in Glafgow, in 1762, containing all thefe remarks.   EDITOR.

being reduced, by inflammable fubftances, under other forms than that of charcoal; oils, and even alcohol, may be ufed. And nothing reduces the greater number of calcined metals better than hydrogen gas, when it is properly applied. This was firft taught us by fome of Dr. Prieftley's experiments. He placed different metallic calces under glafs receivers, or jars, filled with inflammable air, confined by water or mercury, and then applied to them the rays of the fun, collected into a focus, by a burning-glafs. The confequence was, the quick reduction of the metallic calx, while a quantity of the inflammable air difappeared.

Although any charcoal may be ufed in the reduction of a metallic calx, the moft proper and convenient are charcoals of vegetable fub- ftances; and the charcoal of tartar is reckoned the beft of all. It is, however, too coftly for common occafions, and is ufed only when the quantity of calx is fmall, and we defire to perform the operation neatly, and without lofs, as in affaying of ores, or in experiments with fmall quantities of metals;—and the operation is performed in a cru- cible. The reafon why charcoal of tartar is the beft is, that it contains an alkali in fuch quantity as renders it fufible: It therefore applies itfelf more clofely to the metallic matter, and alfo diffolves or melts any earthy matter which may happen to be prefent; in confequence of which fufion the particles of reduced metal more eafily fink to the bot- tom of the mixture, and unite there to form one mafs, called in this cafe a REGULUS. The charcoal of the tartar prepared for this purpofe is called BLACK FLUX.

But in the large way of working, when tons of metallic calces are to be reduced, they are fimply heated in contact with the fuel, and intermixed with it. In the great iron furnace, the ore, broken into fmall pieces, and mixed with fubftances which promote the fufion, is thrown into the furnace, and bafkets of charcoal, or coaks, in due pro- portion, are thrown in along with it. A part of the bottom of the furnace (which is alfo the narroweft) is filled with fuel only. This being kindled, the blaft of the great bellows is directed on it, and foon raifes the whole to a moft intenfe heat. This melts the ore immedi-ately above it; and the reduced metal drops down through the fuel, and collects at the very bottom;—The reft finks down, to fill up the void left by the confumed fuel and metal. Thus it comes in the way of the bellows; and here it is raifed to the fame intenfity of heat, and melts, and is reduced, in its turn. More ore and fuel are fupplied above; and the operation goes on, till the melted metal at the bottom, increafing in quantity, rifes almoft to the aperture for the blaft. It is now let out, by piercing a hole in another fide of the furnace, clofe to the bottom.

There is an opinion among the chemifts, founded, as they fay, upon experience, that the lefs calcined calces are moft eafily reduced, and that the calces of fome metals, if expofed to the action of heat and air for a long time, can hardly be reduced at all, or not without a con-fiderable lofs. Hence the notion of a *mercurial principle*, &c. in metals, which they imagined to be partly expelled and loft by violent calcina-tion. I fufpect, however, there is another more fimple reafon for the lofs or diminution of the metal, which is, that all the metallic fubftan-ces, whether calcined or not, have more or lefs of volatility, and are liable to fuffer a lofs by evaporation, when they are long expofed to the action of heat and air.

After thus defcribing the calcination of the metals in general, and the reduction of them again to the metallic ftate, I have been accuf-tomed to mention the opinions which have been formed of the nature of thefe operations, and the arguments and proofs on which thefe opinions are founded.

The moft diftinct and plaufible opinion, which prevailed among the chemifts for a confiderable period, was that of Dr. Stahl, fimilar to that which he entertained concerning the nature of the inflammable fubftances and of inflammation, viz. that the metals are compounded fub-ftances, confifting of that matter which was called the calx, and of the phlo-gifton; and that they had their metallic qualities from the principle of

inflammability ; and that during calcination, this principle was feparated from them ; and therefore the bafis or calx of the metal appeared in its feparated ftate, deprived of the metallic qualities ; but that
in the operation of reduction, the calx recovered again from the charcoal, or other inflammable matter, the phlogifton which it had loft,
and by this recovery, was reftored again to the metallic ftate.

This appeared fo far a plain intelligible account of the matter ; but
there was one material fact, which was a very great difficulty in the
way of this theory.   The fact I mean is, that the quantity of calx is
greater than that of the metal,—one hundred pounds of lead, for example, produces 110, or 112 pounds of calx.

Different attempts were made to get over this perplexing difficulty,—
fome of them very extraordinary, and almoft incomprehenfible, _e. g._
that the principle of inflammability was not only deftitute of weight,
but that it had the power of diminifhing the weight of bodies to
which it was added, &c *.

All thefe difficulties proceeded from their not having yet ftudied
the part which air performs during calcination.   This was not fufficiently done until of late, when the nature and powers of the different
airs, or elaftic fluids, became fo much the fubject of inquiry.

Among the experiments which this inquiry has occafioned, it was
foon difcovered by Dr. Prieftley, Scheele, and Beccaria, but efpecially
by M. Lavoifier, and other French chemifts, that when a metal is
calcined, it always abforbs and fixes a part of the air which contributes to its calcination and is neceffary to it ; and that the increafe
of weight in the calx is always equal to the quantity of the air abforbed.

* Nor could any valid objection have been made to this explanation, however unlike
our more familiar notions, had not Sir Ifaac Newton made experiments on pendulums
of all different kinds of matter, metals, and the calces of metals, and found that all vi-
brated alike, if of equal length.   Chemifts acquiefced, however, in this explanation by
Stahl, becaufe few, if any of them, were mathematical philofophers, and as few of the
mathematicians were experienced chemifts.   EDITOR.

Dr. Mayhow of Oxford obferved this in the cafes of antimony and lead, and fufpected that it was fo in all,—1674.

It has further appeared from fuch experiments, that it is always oxygen gas which the metals attract in this manner. They are calcined moft eafily and quickly in pure oxygen gas, and cannot be calcined in any other, except when it contains a mixture of the oxygen, as atmofpheric air does. When calcined in a limited quantity of atmofpherical air, therefore, they extract the vital part from it, after which the reft of the air has no more power to calcine them *.

It has alfo been proved in the cleareft manner that oxygen gas can be extracted from the calces of metals; from fome by heat alone,—the calx at the fame time recovering its metallic form. This happens to mercury, filver, gold, platinum. From others, the oxygen gas may be extracted by an elective attraction, as happens in lead, the calx of which, viz. minium, affords oxygen gas by the action of fulphuric acid affifted with a moderate heat. On the other hand, in the reduction of metallic calces by the action of charcoal, a great quantity of elaftic aëreal matter is extricated from the materials, which is carbonic acid, that is, a compound of oxygen and charcoal. All thefe particulars are now completely proved by many experiments; and upon thefe M. Lavoifier and his friends founded their new fyftem concerning calcination and reduction, totally oppofite to that of the older chemifts. The new opinion is, that the metal is not a compound, but a fimple body,—that the calx is compounded of the metal and oxygen extracted from the vital air,—that the heat and light are no proof of a principle of inflammability. They are extricated chiefly from the oxygen gas, which is fuppofed to have an extraordinary capacity for heat, and which, as being an elaftic fluid, contains a great deal befides in a latent ftate.

In reduction again, the new doctrine is that the oxygen is feparated

---

* A fimilar doctrine was maintained by Mr. Bayen, a French apothecary. *Journ. de Phyf.* III. 120,—IV. 487,—VII. 390.

from the metal, and nothing elfe happens or is neceffary to the recovery of the metallic ftate ; and therefore, in the reduction of mercury,
filver, gold, and platinum, which have but a moderate attraction for
oxygen, heat alone is fufficient to feparate it. Other metals cannot be
reduced by heat alone, on account of their having a ftrong attraction for
the oxygen, and retaining it too ftrongly to admit of its being forced off
by heat : But fuch are commonly reduced by the action of charcoal,
aided by heat ; and then the carbon attracts the oxygen from the metal,
and forms carbonic acid with it. The production of carbonic acid gas
in this manner has been afcertained by numerous and inconteftable experiments. And when inflammable air is employed, the bafis of this
air, the hydrogen, acts like the charcoal, by its ftrong attraction for
the oxygen that is in the calx. They unite together, and form water.
This alfo has been afcertained by many experiments.

A change of fome names has been propofed in confequence of
this theory, viz. *calx* and *calcination* are to be fet afide, and calcined
metals or calces are to be named OXYDS. *Des oxydes*, in French,
oxydum,-da, in Latin. But this term of oxyd is applied by the French
chemifts to every compound, whether metallic or not, that contains
oxygen in lefs quantity than that which gives acidity.

On the whole, this fyftem is much more directly and plainly fupported by facts and experiments than the ancient fyftem of the
chemifts.

The only fact which the French theory has not yet explained is,
the effect of light in reducing the oxyds of metals. There are feveral
examples of it, in confequence of which it is fuppofed, by the de-
fenders of the old doctrine, that the light unites itfelf to the metallic
matter, and thereby reftores it to the metallic form. But it muft be
acknowledged that when metallic oxyds are reduced by light, oxygen
gas is always detached ; and the French chemifts fay, that the light
affects the reduction, not by attaching itfelf to the metallic oxyd, but

by fome power which it has to occafion the feparation of the oxygen, perhaps by joining with it, and thus forming with it oxygen gas.

One fact more, which has been difcovered by the experiments lately made on the metallic oxyds, muft not be omitted, viz. That many of them, when once formed, are capable of attracting carbonic acid, and of uniting with it, and even of being .diffolved by it in water, like the alkaline earths.

I now proceed to confider the properties of metals refpecting their mixture with other bodies.

The firft clafs of the objects of chemiftry,—the falts, and of thefe the acids, are found to have the greateft activity with regard to metallic fubftances, and to produce the greateft variety of effects.

To give a general idea of thofe effects, I may begin by faying that there is a chemical attraction between metals and acids; that metals are capable of uniting with acids, much in the fame manner as alkalis and the alkaline earths are,—that they are difpofed to form with them faline compounds, many of which readily cryftallize; and in moft, the natural activity or corrofivenefs of the acid are very fenfibly abated by its adhefion to the metal, though, in general, not near fo much as in the falts compofed of alkalis or earths.

The chemical combinations of metals with acids are commonly named *folutions* or *diffolutions* of the metals, on account of their being commonly produced by putting the metal mechanically divided, or with an increafed furface, into the liquid acid, which acts on it as a folvent. The action of the acid is promoted by gentle heat, and alfo by annealing or foftening the metal before it is put into the acid. This laft practice is ufeful, probably, by diminifhing in the metal that fpecies of cohefive attraction called *hardnefs*, which is an antagonift to the power by which the diffolution is performed. At any rate, the fact is certain. If we take a few of the fmall caft iron tacks, (made of the white metal) which have not been annealed, and as many of the fame tacks long annealed, and put each parcel in equal quantities of acid, the laſt will be found much more diminiſhed in weight than the other in the fame time. Moreover, the unannealed nails feem to reſiſt moſt at the furface, and when half diffolved, preferve their form completely. In this particular metal, indeed, a fpongy nail remains coherent to the laſt.

In performing thefe diffolutions of the metals, we find alfo that in many cafes there is a mutual faturation, as with the acids and alkaline fubſtances; a certain quantity of acid being capable of diffolving, and converting into a faline compound, a certain quantity only of the metal; and the metal on the other hand, neutralifing or mitigating a certain quantity only of the acid. This does not happen however in all the compounds of metals with acids. In many of them there is no fixed point of faturation, the fame quantity of metal being capable of combining with different proportions of acid, and forming compounds which differ in appearance and other qualities, according to the pro-portion of acid which they contain. Even in this cafe, however, it is not fo properly faid that there is no fixed point of faturation. The faċt would be better expreffed by faying, that the faturation is not confined to one proportion. For, in thefe cafes, we find that when a certain proportion is attained, the addition of more acid, for example, dif-folves the falt, without changing its properties, till we reach another proportion, when a manifeſt change of chemical and other properties enfues. Sometimes we ſhall have a third proportion, in which a kind of faturation obtains. This refembles the combination of water and falt, in the ſtates of cryſtal and folution, and in the combinations of heat, inducing folidity, fluidity, or vapour. When the proportion of acid is large, the acid is lefs mitigated, and the compound is more foluble in water, and in fome cafes deliquefcent. When the quantity of acid is very fmall, it is much mitigated, and the compound has little or no folubility, the metal being only divided by the acid into a powder, and not diffolved. In this cafe, that is, when the acid divides the metal into a powder, but does not diffolve it or render it foluble in water, it is faid to *corrode* it, and the metal is faid to be *corroded*.

It muft be underftood further, that every metal is not equally dif-pofed to unite with each of the acids. Some of them can be combined with all the acids, or any of them ; but others with a certain number of them only. And there are fome which are not acted upon in the leaft, except by one or two, or by certain mixtures of the acids with one another.

Thofe metals which are capable of uniting with a number of the acids generally have the ftrongeft attraction for the muriatic, and next to the muriatic, for the fulphuric. The acid of fugar, or the oxalic, and the tartarous acid, and the phofphoric, are alfo ftrongly at-tracted by the metals in general. The nitric acid, although it diffolves a great number of the metals readily and eafily, is not fo ftrongly at-tracted and retained by them. It is more eafily feparated from them by heat or otherwife. The muriatic, when combined with them, is fo ftrongly retained, that it is not feparable by heat, but renders the metallic matter volatile along with itfelf.

Among feveral metals which can be diffolved by the fame acid, great differences are found in their attractions for it, and fome of them can be employed to precipitate others in a certain order. Zinc or iron, for example, can be employed to precipitate filver, or quick-filver, or copper, or lead, from the nitric acid. Lead precipitates cop-per, quickfilver, or filver. Copper precipitates thefe two laft ; and quickfilver precipitates filver. Zinc is therefore fuppofed to have the ftrongeft attraction for the acid ; next to zinc, iron,—next lead, cop-per, mercury, and filver.

There is a remarkable circumftance which diftinguifhes the folution of metals in the acids from the folutions of alkaline and other fub-ftances. While metals are diffolving in acids there is, in many cafes, a violent effervefcence,—an effervefcence fometimes as ftrong as that which attends the action of acids on common alkalis or calcareous earths. It is, however, quite different in its nature from the effervescence of alkalis or alkaline earths. It is not produced by the separation of fixed air or carbonic acid. The metals in their metallic state do not contain any. These effervescences are explained in a very satisfactory manner by a great number of the modern experiments, which shew that either a part of the acid or a part of the water is decomposed by the diffolving metal, which attracts to itself a part of their oxygen, and is thereby changed into an oxyd, (formerly called a *calx*) which oxyd, as a peculiar substance, is diffolved by the remaining acid, or is united to it. When the oxygen is taken up by the metal from a part of the water, hydrogen gas is produced in the effervescence, as you saw lately when iron was diffolved in diluted sulphuric acid. When the oxygen is taken up from a part of the acid, the acid is partly decomposed, and a different elastic fluid, or other matter is produced, peculiar to that acid. All this will be more fully explained as the cases occur. (*See Note* 16. *at the end of the Volume.*)

We find reason to be satisfied that this theory is well founded, when we examine the state to which the metals are reduced by being diffolved. A metal, while diffolving, is always calcined, or oxydated, more or less; in many cases, it is true, to a very small degree only, but in every case, some degree of this change does happen to it; for, if we precipitate the metal by the action of a pure alkali, we always get it in form of a calx; and we get it in nearly the same state if we expel the acid from it by the action of heat alone. And when, on the other hand, we change the metal into an oxyd by the action of heat and air, before we apply the acid to it, there never is an effervescence, the metallic matter generally diffolves quietly, in some cases more easily than the metal itself, when put into the same acid. But this happens only when the metal is calcined in the slightest degree possible. If much calcined, it diffolves more flowly and difficultly, and requires a greater quantity of acid to its diffolution than the metal itfelf would have required.

We may further remark here, that the different acids act on the metals with different degrees of quicknefs and violence. The action of the nitric is generally the moft quick and violent, and moft effectual in oxydating the metals. Next to this the fulphuric. The muriatic acid generally acts in a more flow and languid manner on the metals in their metallic ftate. This appears plainly to proceed from its having lefs of the oxygen to impart to the metal; for if the metal be firft flightly oxydated, no acid whatever unites with it fo readily and ftrongly as the muriatic acid. We have a proof of this, when we add muriatic acid to a metal already oxydated by another acid, as the nitric, or even the fulphuric acid. There are alfo particular ftates or conditions of the muriatic acid, in which it acts more readily and powerfully on the metals than it does in its ordinary watery ftate. We apply it, for example, in fome cafes, in the form of dry and hot vapour, as in cementation, defcribed in my general account of chemical operations. And there are other proceffes foon to be defcribed, by which we add oxygen to the muriatic acid. By this its power to penetrate and diffolve the metals is greatly increafed.

The compounds, or the folutions of metals with acids, are of different colours, when made with different metals, or with the fame metals in different ftates of oxydation.

Fourcroy (*Annales de Chymie, tom. x. on Mercurial Precipitates*) mentions an opinion of the French chemifts, that the colour of the folution or compound is fimilar to that of the oxyd contained in it, and that when, from fuch folutions or compounds, a precipitate is produced of a different colour, this proceeds from a change the oxyd has undergone in its degree of oxydation, while it was precipitated, or immediately after. Yet he fays, (page 309, *et alibi*) that Turbeth mineral, or fulphat of mercury, with excefs of oxyd or calx, is foluble in 2000 waters, and forms a colourlefs folution. Cauftic alkalis precipitate from this folution a grey oxyd. This colour indicates fome degree of deoxygenation. Fourcroy imputes this change to the cauftic alkalis: But this is no explanation.

The compounds of metals with acids may be decompounded in general without difficulty. The acid and oxyd are united together in the greater number of 'cafes, by an attraction that is not very ftrong, which is plain from its being overcome by many other attractions; by alkalis, the alkaline earths, and even by fome of the inflammable fubftances, which are difpofed to unite with acids. If frefh calcined litharge (which is an oxyd of lead) be ground to fine powder, and be put into a folution of lime in muriatic acid, we find the lime precipitated in a cauftic ftate, and the liquid now contains a compound of lead and the acid. Or if fea falt be digefted with litharge, we procure a cauftic mineral alkali. *(Scheele on Fire,* 174.) This explains the corrofion of the lead pipes which convey water. It very commonly holds both of thefe muriats in folution.

Heat alone is fufficient, in many cafes, to feparate acids from metals. All this proves that the attraction between them is not ftrong, in the greater number of cafes. It is probably in confequence of this moderate attraction, that in fome of the compounds of metals with acids, the acid is very little mitigated, the compound being nearly as corrofive, when applied to animal fubftances, as the pure acids themfelves. In this way we have obtained fome of the moft ufeful potential cauftics and efcharotics employed by the furgeons.

This being a very general property of the metallic falts, and being a property purely chemical, becaufe it is equally effective on the dead and on the living fubject, has attracted the attention of the chemifts, and they have attempted to difcover fome general principle on which it depends. Mr. Berthollet, unqueftionably one of the moft eminent, publifhed a differtation on this fubject in the *Journal de Medicine* in 1779. He afcribes it to the action of the acids on the phlogifton contained in animal and vegetable fubftances, and fupports this by a very judicious feries of experiments. The doctrine of phlogifton being exploded, this explanation, as far as it was valid before, is eafily accommodated to the new doctrines, and we muft afcribe the corrofive power of the metallic falts to the action of the oxygen which they contain. It is very true, that the action of oxygen produces very remarkable effects on animal and vegetable fubftances. But I do not think that they have a great analogy with the action of the highly corrofive metalline falts, fuch as that formed of mercury with the fulphuric, nitric, and muriatic acids, or that formed by filver and the nitric acid. Some of thofe falts are moft highly corrofive when they contain the fmalleft quantity of oxygen, a quantity fo fmall that they are fcarcely faline, and very ftrongly united. Others are more corrofive when they are truly faline, and contain not only oxygen, but alfo the diftinguifhing ingredient of the acid ; and others are fo far from being corrofive in either ftate, that they promote the healing of wounds and fores, feeming friendly to the functions of organifed fubftances. Perhaps the corrofivenefs is fufficiently accounted for, when we afcribe it to the acid itfelf. Nor is it incongruous with the general courfe of chemical facts to fay, that in fome of thefe compounds, new properties are induced, differing from thofe of either ingredient.

The other order of fimple falts, the alkalis, have not fo much diffolving power with regard to metals as the acids. They are more commonly employed to precipitate metals from acids, which they do by their great fuperiority of attraction.

The feparation of a metal from folution in an acid, prefents fuch a variety of phenomena, that I can fcarcely bring them under general rules, and fhall therefore notice them as they occur in treating the feveral metals. I may obferve, however, in this place, that in moft cafes, a fmaller quantity of alkali. is fufficient for feparating a metal from a given quantity of acid, than fuffices for faturating this quantity of pure acid. The precipitate is in the form of an oxyd, and in thefe cafes, has decompofed part of the acid. It is only the remainder that employs the alkali to feparate it from the metalline oxyd. How this is effected, even in thofe cafes where water is decompofed, and inflammable air is produced, muft be confidered as a difficulty ftill remaining in the new doctrines of the French chemifts.

The alkalis themfelves may alfo, in fome cafes, be made to act as folvents, and will diffolve moft metallic fubftances very perfectly. They act moft powerfully in their cauftic ftate, efpecially if applied with a melting heat. In the ftate of watery folutions, they act but weakly in general, though there are fome of the metals which they diffolve very perfectly even in this way. And if the metal be previoufly oxydated, efpecially by acids, there are few of them that will not yield to the action of alkalis.

Thus it happens, in many cafes, that when we precipitate a metal from the folution in an acid, by a cauftic alkali, if we put in too much alkali, we rediffolve the metallic oxyd. The volatile alkali does this in moft cafes. A certain preparation of alkali, which will be defcribed in due time, enfures the alkaline folution.

In a dry melting heat the alkalis are powerful folvents of the metalline calces, and form with them a variety of coloured glaffes.

Among the compound falts, borax is found ufeful in the melting of gold and filver, and in foldering thefe and other metals. Its ufefulnefs is remarkable, efpecially when filings or fmall pieces are to be united by fufion. It helps to make them melt more readily, and to join all together the more completely, and with the leaft lofs. The borax promotes the union of the melted particles or globules of the metals, by making their furfaces clean and bright, and preferving them in that ftate. The neceffity of this is plain in mercury, which the fmalleft foulnefs hinders from uniting. Its globules are then rather difpofed to feparate into fmaller globules when we attempt to bring them together. And the borax effects this, by covering and defending the metal from air, and thereby preventing the formation of any calcined cruft or pellicle, and likewife by diffolving any matter of this kind which is already formed; borax having much power to diffolve and melt the earthy fubftances and metallic oxyds. By fpreading itfelf alfo over the furface of the crucible, it gives a fmooth glazing, which allows the metal to be poured out.

The other neutral falts are not much difpofed to act on metallic fubftances, except when affifted by a ftrong heat, but, with its affiftance, thofe which contain the acids that have the greateft power over the metals are difpofed to act upon them, and diffolve or corrode them, more or lefs, the acids being loofened by the heat. Glauber's falt, vitriolated tartar, common falt, and digeftive falt, produce this effect. The effects of nitre and ammoniacal falts are more remarkable. Many of the metals, when they are thrown into melted nitre, are inflamed or burnt, as has been already feen, and the metal is reduced to the ftate of an oxyd. This oxydation concurs with the effects which you formerly faw in the mixtures of nitrous acid with alcohol, to fhew how eafily the nitric acid is decompofed. No part of it is to be found in the crucible, if enough of metal have been employed. We find only a cauftic fixed alkali, fometimes, indeed, tainted by the metallic oxyd which it has diffolved. Hence we muft conclude that the metal is oxydated by attracting the oxygen of the nitric acid, and thereby allowing the nitrous air or the azote to efcape. This efcape muft be confidered as the caufe of the deflagration.

The effect of common fal ammoniac upon the metal is eafily underftood. The attraction of the acid for the volatile alkali is fcarcely greater than for fome of the metals or their oxyds. Therefore, when one of them, in fine particles, is mixed intimately with fal ammoniac, and diftilled, we obtain the volatile alkali in a cauftic form; a proof, by the way, that metals do not yield fixed air in their effervefcence with acids. In this decompofition of the ammoniacal falt, the effect is brought about by the heat, which volatilizes the alkali. The vitriolic ammoniac exhibits the fame appearances.

The oxyds of fome metals, when put into a watery folution of fal ammoniac, act with great variety of appearances, which will be explained afterwards.

With refpect to the fecond clafs of the objects of chemiftry, the earths, the relations of thefe to metals in general has been, in fome meafure, pointed out already, and amounts to a very few propofitions only. I have already obferved, that fome of the alkaline earths have a ftronger attraction for acids than moft metals. I alfo had occafion to notice that melted earths and melted metals cannot be mixed, at leaft fo long as the metals retain their metallic form. But when they are in the form of oxyds or calces, they can eafily be mixed with melted earths or glaffes; and fome of them have even great power in this ftate to promote the fufion of fome of the earths, or to diffolve them in a melting heat. They are, therefore, added to fome of the compofitions for making glafs in fome cafes, in order to make it more perfect and tranfparent, and in others, to give it colours, in the imitations of the natural gems. When metallic oxyds happen to be mixed and combined in confiderable quantity with vitrified earths, they can be feparated from them by fufion with charcoal or black flux. Thus, a piece of fine Englifh flint glafs, or the falfe diamonds called *Paris paftes*, being reduced to powder, and urged with a ftrong heat, in contact with charcoal or black flux, will be decompounded, and we fhall find a button of lead at the bottom of the crucible.

Among the fubftances of the third clafs, the inflammable, fome ferve for the reduction of metallic oxyds. But, excepting this ufe of them, none but fulphur, phofphorus, and charcoal in fome cafes, are remarkable for a difpofition to unite with any of the metals.

Sulphur fhews a ftrong attraction for moft metallic fubftances, and may be united with moft of them, readily and intimately, in the way of fufion. And though the heat be not fufficient to melt the metal, but only the fulphur, if the metal be only divided into fmall pieces, the fulphur penetrates it, and entirely changes its appearance, examples of which fhall be given hereafter.

In making many of thefe compounds of fulphur and metal, a great quantity of heat is extricated from the materials, which make them become ignited, and glow as if they were inflamed. This was miftaken at firft for a real inflammation. The moft remarkable of thefe experiments were made by Van Trooftwick and Deiman. They mixed fulphur with different metals, in a clofed phial, or in an exhaufted tube hermetically fealed, or in a tube filled with hydrogenous or azotic gas, or carbonic acid. With an external heat, by no means very great, the fulphur melted, and, after fome time, the combination with the metal took place. In that moment, the whole broke out into a bright glow, in fome cafes brilliant, and almoft like a real deflagration. Three parts of copper to one of fulphur produced the brighteft light. Lead alfo, and tin, and efpecially zinc, produced a bright flame. *(See Note* 17. *at the end of the Volume.)*

In the new language of chemiftry thefe compounds are called *fulphureta*, for the Latin word, or, in the fingular, *fulphuretum*. And our chemifts, who have adopted this language, call them *fulphurets*.

Sulphur, in fuch experiments, not only fhews an attraction for moft metals, but like other bodies which have a power of this kind, it is found to attract different metals with different forces. And fome metals can be employed in many cafes to feparate fulphur from others, in a certain order. When this feparation is performed in the way of fufion, in a crucible, the fulphur, uniting with the added metal, which it attracts moft, forms a matter which flows uppermoft; the other metal feparates and falls to the bottom, or forms a REGULUS. But this method of feparating fulphur from metals is not the neceffary or only method in all cafes. There are many, in which the fulphur can be feparated by heat alone, and it is more ufual to expel it in that way.

Gold, platinum, and zinc, are exceptions to the general account I have given of the relation of fulphur to metals. Sulphur cannot be made to penetrate thefe three, or unite with them, if it be applied pure, and thefe metals are in their metallic ftate. But when fulphur is combined with an alkali, this compound acts much more power-ly as a folvent, in the way of fufion, upon metallic fubftances in general; and even the three I juft now mentioned are readily dif-folved by it. The different metals, when diffolved by this compound, unite with it with different degrees of force, in the fame order as fulphur does. Thus, we have it in our power to feparate the metals from it alfo by one another. The beft proportion for diffolving metals is equal parts of fulphur and alkali.

The management of phofphorus, in combining it with metals melted or made red hot, requires much caution. The beft way is to cut it into bits of four or five grains weight, and keep them under water, and, when they are ufed, to take out one at a time, wipe it dry with bibu-lous paper carefully, and introduce it into the crucible with a long pair of pincers, plunging it to the bottom. The greater part of it is always diffipated, but fome combines each time. Perhaps a compofi-tion, which will produce phofphorus, hard preffed into the bottom of a crucible, would anfwer better. The crucible being made haftily red hot, the melted metal may be poured into it, and the heat continued. The phofphorus would thus be prefented to the metal in its nafcent ftate, which is found to be favourable in moft cafes of difficult combi-nation. The fufibility of the metal is improved by this addition.

By Pelletier's experiments (*Mem. de l'Acad. des Sciences*) 100 grains of platinum are changed into an eafily fufible phofphoret, by the addition of eighteen grains of phofphorus,—and 100 grains of gold require only four grains of phofphorus to change them into phof-phoret. Thefe compounds, if expofed a long time to the action of heat, burn at their furface until all the phofphorus is confumed; and the above metals remain in their ordinary ftate.

In the experiments which have been made by mixing metals with one another, it appears that metals unite in general with one another, and this in every proportion. But there are fome exceptions, as iron and lead, iron and mercury, lead and cobalt, nickel and cobalt, cobalt and

METALLIC SUBSTANCES.

bifmuth. The mixtures are of different fpecific gravity from what correfponds with the ratio of the compofition. In general, the fpecific gravity and denfity are greatly increafed, fo that in many cafes the compound is denfer than the denfeft of the ingredients. Thus, the denfity of tin is 7363, and that of brafs is 8006. But the denfity of a mixture of two parts of brafs and one part of tin is 8916. The proportion of the compofition fhould have given 7793 *. They are alfo more fufible : Hence they are employed as folders,—for gold, we employ a mixture of gold and filver,—for filver, filver and mercury, —for copper, brafs,—and for brafs, another ftill more compounded, —for either lead or tin, a mixture of both.

Mixtures of metals, in general, can alfo be calcined more quickly than the fame metals in their feparate ftate; of which you have an example in the mixture of lead and tin, which burns like a bit of turf. If you would wifh to know the caufe, I fhall hazard a conjecture. I am inclined to think that this effect is produced by the chemical attraction of the metals for one another, which counteracts the cohefive attraction of each of them, and diminifhes its force, and thereby gives advantage to another chemical attraction,—their attraction for oxygen. That their cohefive attraction is diminifhed by their union with one another is evident, the mixture being always more fufible than the feparate metals †.

The method of feparating metals from one another again, is very various in the different cafes, and depends upon certain particular pro-

---

* It is alfo an almoft univerfal fact that the cohefion of the compound is greater than in the proportion of the compofition. In the ductile metals, whofe cohefions are not extremely different, the cohefion of the compound exceeds that of the firmeft of the ingredients. Thus the mixture of twelve parts of lead with one part of zinc is twice as coherent as the zinc. **EDITOR.**

† Yet the mixtures are generally more denfe and much more coherent than the ingredients feparately. **EDITOR.**

perties and differences of the metals which are to be feparated. Thus, in fome cafes, difference of fufibility is made the means of feparation : Thus lead is feparated from copper. In others, difference of volatility : Thus mercury is feparated from antimony. In many others, difference of folubility : Thus gold and filver are feparated by aquafortis, or aqua regia, or fometimes by fulphur ; or gold and iron by fulphur and lead.

Other feparating operations, which are frequently performed, depend on different degrees of their attraction for oxygen, and more or lefs difpofition to unite with it ; that is, on a difference in their tendency to calcination. Thus lead, and moft other metals along with it, are feparated from filver and from gold : And thus copper alfo is refined from admixture of the coarfer metals. In moft cafes of feparation performed in this manner, the procefs is called SCORIFICATION.

The mixture is expofed to a violent heat and to a current of frefh air, which caufes the furface to tarnifh, and then to gather a film or fcale of the more calcinable metal. This is ufually blown to one fide of the melted mafs, as it forms, by the bellows. It is fucceeded by more ; and this continues, till the lefs calcinable metal be as much purified as is poffible by this operation. What is thus blown or raked off, is called SCORIÆ, SLAG, DROSS. It is not always a flag or filth,—but a portion of the metal, of the fame purity, but no longer perfectly metallic, but a compound of the metal and the oxygen of the atmofphere.

The rufting of iron and other metals is a change of precifely the fame kind, and is produced by the fame caufe. It is very remarkable in iron, becaufe this metal alfo decompofes water very faft, and therefore rufts very faft in damp and warm air.

You are now informed of the general character and qualities of the metals. We muft now attend to their natural hiftory, or the different ftates in which they are found in nature.

The firft remark we have to make on this fubject is, that few of them are produced by nature in a ftate of purity. They are moft commonly found in the form of what are called ORES, which are compound minerals, in which the metal is intimately mixed with other substances, so as to have neither the malleability, nor the other qualities of the metals, except sometimes a degree of the shining metallic appearance.

Frequently the ore is an oxyd of the metal, and only requires the operation of reduction to be performed in order to give it the metallic properties.

The ores of metals are commonly found in the veins of the hardest mountains, and hardest stony strata. They are generally separated from the rock, being intermixed in the vein with a quantity of spar or quartz, or sometimes a softer matter. When a spar involves the ore, it is in some cafs a calcareous spar,—in others a fluor ; but more frequently a sulphat of barytes. When the ore is involved in quartz, it is sometimes a pure quartz, but oftener an intermixture of quartz with some or several of the spars ; and often also we find intermixtures of the spars without any quartz. These matters, thus accompanying the ore in the vein, are called in general the MATRIX of the ore, and by the English miners, the RIDER. The manner in which the ore and matrix are interspersed through the vein is altogether irregular. In some parts of the vein, the whole widenefs of it is filled with ore ; in other parts, with matrix alone ; in others the ore and matrix are found together, in all the variety of proportions and modes of intermixture that can be imagined. Such mixtures are called *brangled ores*.

The ore being separated and picked out from the matrix to be examined by itself, is commonly found to contain, not only the metal on account of which it is valued, but, along with it, some of the other metals, and a confiderable proportion of sulphur or arfenic, or both of these, and sometimes a small proportion of earth matter.

These ingredients are intimately united in the ore, that is, they are chemically combined with the metal ; and the operations by which the metal is extracted are procefses for feparating these matters from it.

But before we can well understand these processes, it is necessary to have some knowledge of arsenic. It will be found to be a metal; and it will be very convenient to make it the first subject of our study.

METALS.

GENUS I.—ARSENIC.

ARSENIC is a matter which resembles the saline substances so much by some of its properties, and the metallic by others, that the chemists were long in doubt to which class it should be referred. Dr. Boerhaave describes it among the sulphureous minerals, probably on account of its being often found in its natural state combined with sulphur. But when the arsenic is found pure, or is purified by art, it is widely different from sulphur.

The nature of arsenic was very little understood till Mr. Macquer published some papers in the Memoirs of the Royal Academy, containing a number of experiments he had made on this mineral; and more lately, Mr. Scheele, of Sweden, communicated his instructive experiments on it to the Academy at Upsal.

The ordinary appearance of arsenic is that of a compact heavy matter, of a white or yellowish colour, and having the glassy fracture, sometimes transparent, oftener of an opaque white. Exposed to heat under the ordinary pressure of the atmosphere, it becomes soft, or approaches to fusion, when very near the lowest degree of ignition. But in the same heat also, it begins to evaporate in white fumes, of a sickening heavy smell, thought to resemble garlic; and is thus totally converted into vapour by degrees, without becoming perfectly fluid. The fumes, if confined and condensed by cold, form a white powder, or chalky-like matter, which afterwards, if the vessel be of a proper shape, is softened and compacted by the heat into a whitish glassy-like substance. It is always by sublimation that it is brought to this form.

The qualities by which arsenic resembles the metals are, 1*st*, Its weight. 2*dly*, A capability to be metallized, or to assume some of the metallic qualities, such as the metallic opacity and bright reflecting surface, and a density or specific gravity similar to that of the metals. It also becomes a conductor of electricity, and fit for mixing intimately with the other metals, which it will not do in its ordinary state.

It may be so far metallized by several processes. Mr. Macquer describes one in which this effect is produced by means of oil and sublimation. I have succeeded equally well by subliming the arsenic repeatedly with charcoal dust.

But one of the best is Scheele's process, described by Bergmann. White arsenic is put into a crucible, with thrice its weight of black flux, and an inverted crucible is luted to it. The lower crucible is set on the fire, and slowly raised to a red heat. But the upper crucible must be defended from it, by means of an iron plate, having a hole exactly filled by the rim of the lower crucible. In this manner, the upper crucible will be covered within with a crust of regulus perfectly clean, and in a crystallized form. And it may be detached from it at once by a dexterous knock.

Arsenic, when thus made to assume the metallic appearance, is quite brittle, like several other metals, and its surface is liable to tarnish, so that it loses its lustre very soon, if exposed to the air. Or if we evaporate it in the open air, the vapour may be at once condensed into white arsenic; also if sublimed with fixed alkali. By these processes, therefore, it returns immediately to its former appearance.

These several particulars give us reason to consider arsenic as a substance of a metallic nature; and to view it, when in its common form, as in a calcined state, or as an oxyd of arsenic. But it differs

from the oxyds of the other metals, by having qualities decidedly
faline. We have a clear example of this in its folubility in water, and
in its action upon the alkaline falts, and upon nitre. The folubility
of white arfenic in water appears, if we beat it to powder, and boil
it in the water. We thus learn that it may be completely diffolved
in fifteen times its weight, and from this folution the arfenic may be
obtained in the form of cryftals. Arfenic unites alfo with watery
folutions of alkalis, efpecially cauftic alkalis, and with lime-water,
and its volatility is fomewhat repreffed by the union.

Its action upon nitre was thought the moft remarkable, as it fhews
that arfenic may be employed to decompofe nitre by expelling its
acid. Equal parts of nitre and arfenic being mixed in fine powder,
and expofed to heat in a retort, the acid of the nitre arifes very volatile and elaftic, and of a deep red colour, and does not condenfe unlefs
there be water in the receiver. It tinges water a fine blue. When
the whole acid is expelled, there remains melted a white mafs in the
retort, compofed of arfenic and fixed alkali, which diffolves in water,
and eafily cryftallizes into very regular cryftals. It melts in a crucible
and forms a tranfparent fluid, and is very fixed, emitting no arfenical
fumes; the arfenic appearing to be very ftrongly combined with the
alkali. Again, the qualities of this compound are quite different from
thofe of a compound formed by combining white arfenic with a
pure alkali. It requires for its production not only the acid of
arfenic, but fo much of it as to make it acidulous and cryftallizable.
The other compound is called by Macquer LIVER OF ARSENIC.
It might as well be called ARSENICATED ALKALI. It has a particular
weak but heavy fmell. If heated in a crucible, it is not fixed like
the other, but emits arfenical fumes in abundance. When diffolved
in water, it will not cryftallize, but when evaporated, becomes pafty
or gelatinous. Mr. Macquer was at a lofs to explain how this
happens, particularly why the cryftallizable falt formed by expelling
the nitric acid by arfenic fhould be different from common arfenicated

alkali ; and why arfenic does not decompofe common falt, though it decompofes nitre.

But the whole of this fubject has been cleared up by the experiments of Scheele, who made much more progrefs in difcovering the nature of arfenic, and has given us principles by which all the phenomena are explained. He learned, by a feries of inftructive experiments, that one reafon why the cryftallizable arfenical falt, and the common arfenicated alkali, have not the fame property, is, that the arfenic in the cryftallizable arfenical falt has undergone a change from the ftate of common white arfenic ; the acid of the nitre having acted upon it as it does upon fugar and fome other fubftances, fo as to change it into an acid. Of this he gave the moft fatisfactory demonftration, by applying the nitric acid to white arfenic by other different ways, by which he changed it into an acid, which he obtained feparate from any other matter ; and afterwards, combining this acid with the vegetable alkali in fufficient quantity, he formed a perfectly cryftallizable arfenical falt.

He contrived two proceffes by which he changed white arfenic into an active acid. The firft of thefe is entirely an imitation of the procefs by which fugar is changed into an acid, with this difference only, that fome muriatic acid is firft employed to diffolve the arfenic, that the nitric acid may act on it with more advantage.

Scheele's procefs is as follows : Into a tubulated retort, fitted with a receiver, put two parts of powdered white arfenic, and feven of muriatic acid, and diffolve by a gentle boiling heat. When all is diffolved, pour back what is in the receiver, and add three and one half parts aquafortis, and diftil. The nitric acid rifes in red fumes, and after fome time they ceafe. Now add one part arfenic, and one and a half aquafortis. Red vapours arife again. Diftil to drynefs, and make the retort red hot.

In the retort you have the arfenical acid, fixed in the fire, and deliquefcent in the air, and foluble in twice its weight of water.

Mr. Pelletier, however, fays that we fucceed equally well by ufing the nitric acid alone, in the proportion of fix parts of the acid to one of the white arfenic.   The acid comes off in red fumes of nitrous gas, and the white arfenic affumes the true characters of the arfenical acid (fo it is now called).   It muft be kept a good while in a ftrong heat, to expel all the redundant nitrous acid. *(Fourcroy* II. 507. *Ed.* 1786.)

This is an exact enough account of the phenomena, and perfectly inftructive in the nature of the operation ; but having practifed both methods, I agree with Dr. Scheele, that his previous folution in muriatic acid enables the nitric acid to act on a much greater quantity of the arfenical oxyd.

The manner in which this procefs produces its effect is fufficiently evident.   White arfenic muft be confidered as a metallic oxyd, containing a very moderate quantity of oxygen, and capable of a higher degree of oxydation.   I always viewed it in this light; and on this principle, I explained Mr. Macquer's experiments, and the effects which it produces with nitre ; a part of which is the change of the nitric acid into nitrous acid, in confequence of the abftraction from it of a part of its oxygen, attracted by the arfenic.   In Mr. Scheele's procefs, the nitric acid alone fupplies oxygen to the arfenic, and thus oxydates it to the greateft degree of which it is capable ; in which high ftate of oxydation, the abundance of oxygen which it contains gives it the qualities of an acid, and deprives it of attraction for other acids, but difpofes it to unite ftrongly with alkalis.   This is the proper explication of the procefs according to the principles of the new theory ; and there is an experiment defcribed by Mr. Scheele, which gives great fupport to the French explication of the phenomena, and eftablifhes it without a doubt.   This experiment is made with the acid of arfenic.   If fome of this acid be put into a retort by itfelf, and expofed to the action of heat alone, it endures a low red heat without change, or is only melted.   But if the heat be increafed, and continued, the greater part of the acid arifes gradually into the neck of the retort, in the form of a common white arfenic ; and while this happens, a very confiderable quantity of oxygen gas is produced or extricated from it. This is certainly a clear proof that a redundance of oxygen is contained in the acid of arfenic, confidered merely as an oxyd, fince we fee that when a part of it is feparated by the action of heat, the acid of arfenic returns to the ftate of white arfenic.

I may add here, the mention of one procefs more for obtaining arfenic in the ftate of an acid. It was difcovered and communicated by Mr. Pelletier. He performed Macquer's procefs, but ufed the nitrat of ammonia in place of common nitre. He thus expelled the nitrous acid, which paffed over in diftillation, and an arfeniat of ammonia remained in the retort. But by changing the receiver, and increafing the heat, he was able to make the ammonia or volatile alkali arife in a cauftic ftate from the arfenical acid, which remained in the retort.

This acid is now called the ARSENICAL ACID, and the compounds it forms with other fubftances are ARSENIATS, in the new language of chemiftry. Mr. Scheele made a great number of experiments with this acid, by combining it with other bodies, and inveftigating its properties. He informs us that it has not a ftrong tafte ; that it diffolves flowly in water, but may be diffolved in twice its weight of that fluid ; that in this fluid ftate it neutralizes the alkalis and alkaline earths, and diffolves many of the metals, or unites with them, with particular phenomena, which he defcribes. You may fee all this in his Effays, and an abftract of the whole in Bergmann's Treatife on Electric Attractions ; and in Nicholfon's Chemiftry, article *Arfenic.*

Scheele found, in attempting to diffolve fome of the metals in the watery folution of this acid, that it oxydated them, and was itfelf reftored to the ftate of white arfenic ; a proof that it does not retain the oxygen fo ftrongly as to prevent its communicating fome of it to other bodies that have a ftronger attraction for it. We have a more ftriking proof of this in the refult of feveral of the experiments Scheele made, by caufing the dry acid to act upon metals and other bodies. This acid is eafily reduced to a dry ftate, and can be melted into a tranfparent matter like glafs. If fome of this dry acid is beaten to powder, and then mixed with dry charcoal in powder, or with filings of different metals, and thefe mixtures are heated, there is, in many cafes, a ftrong deflagration, fuch as is produced by nitre with the fame fubftances; and the acid is in a moment changed, partly into white arfenic, and partly into pure or metallic arfenic, both of which are fublimed.

Mr. Scheele difcovered that Mr. Macquer's cryftallizable arfenical falt, prepared with nitre, is a compound not exactly neutral, but is a little acidulated by a fmall furplus of the acid of arfenic, and when it is made exactly neutral, it will cryftallize, but is deliquefcent. And he explains how it is formed in the procefs with nitre.

When arfenic is in its ordinary ftate of white arfenic, it is foluble, in fmall quantity, in a variety of fluids. Boiling water, I obferved before, diffolves one-fifteenth or one-twentieth. Spirit of wine alfo diffolves a little of it, and even oils. Aquafortis, or diluted nitric acid, alfo diffolves it with difficulty, but changes it more or lefs into acid of arfenic. The readieft folvent is muriatic acid, which diffolves oxyds of metals in general better than other acids do. It forms a compound, which can be diftilled or fublimed; and fometimes condenfes like oil. Fourcroy fays that calx of arfenic, that is, white arfenic, will not afford this oil, but that it is eafily got from one part of metallized arfenic and two parts of fublimate of mercury,—the mercury being revived.

Arfenic has a difpofition to mix in fmall quantity with earthy bodies in a vitrified ftate, or to act on them; and this is one of its ufeful qualities, a good deal of it being employed by the manufactures of glafs. White enamel is made with it, with which Delft ware is glazed; and alfo all the pretty ornaments, which were formerly twifted in very beautiful fcrolls in the ftalks of wine glaffes, &c.

Sulphur alfo can be combined with arfenic, as with the metals, and the compound is more fufible than arfenic alone. The fufibility is increafed by increafing the quantity of the fulphur, and the colour is yellow, orange, or deep red. It is called *yellow arfenic*, and *orpiment*, (the *fandarac* of the ancients) and *realgar*. The dangerous powers of the arfenic are confiderably abated in thefe compounds. Hence it is that the Chinefe, and other Orientals, form realgar into medical cups, and employ as a purgative, lemon juice which has ftood fome hours in them. The fulphur can be feparated again by fublimating the compound with potafh, and by other proceffes.

Arfenic can eafily be united with the metals in their metallic form, but only when it is itfelf metallized. The common way is to mix it with materials that will metallize it, and apply this mixture to the metal, with a proper heat, in the form of vapour. It whitens them, and makes them brittle. One of the diftinguifhing qualities of arfenic is that of uniting, when heated in any inflammable mixture, with fome of thofe fubftances, and flying off with it. Hence it is reckoned a purger or purifier of glafs; and is a powerful calciner, or fcorifier of the metals. But it is no lefs hurtful, on the other hand, in metallurgy, by carrying the metals off with it; to prevent which is one of the great operations upon the ores in metallurgy.

Origin, or Natural State of Arfenic.

Arfenic is fometimes found pure, or in the form of folid metallic arfenic, but oftener more loofely concreted, like a grey or black friable mater. But pure arfenic in any fhape is rare; though, in the ftate of combination, there is plenty of it in many of the ores of metals, efpecially thofe of cobalt, copper, filver, and iron. In the white pyrites, it is known by the garlic fmell when ftruck. It is moft plentiful in this mineral, and in the ore of cobalt. Exifting in fo many compounds, from which it may be expelled by heat, it abounds in volcanic countries. In Solfatara, it contaminates every volatile fubterraneous production, and is found in many of those forms into which we bring it in the operations of our laboratories.

There are also natural compounds of arsenic and sulphur, called *orpiment* and *zarnic*; but the greatest part of the orpiment in the shops is artificial. Arsenic for the use of the arts is prepared chiefly from cobalt ores and white pyrites, in Saxony, as a secondary business only, in the manufacture of zaffre and smalt. The arsenical fumes are collected in chambers, which act as subliming vessels, as we shall see presently. It is useful in the manufacture of glass, and in dying. Such, therefore, is the history of arsenic considered as an object of chemistry.

The knowledge of this mineral is necessary to the physician, both on account of its great efficacy in the cure of some diseases, when it is properly used, and also on account of its noxious powers, in consequence of which, it is sometimes given with the most criminal intentions. In such cases, the physician is called in to assist in forming a judgment whether arsenic has actually been given or not.

It has long been one of the secret remedies employed by some empirical practitioners,—externally, for the cure of cancers, and other obstinate ulcers; and internally, for the cure of intermittent and other fevers. And the ancient physicians, in some of their prescriptions, employed some of the natural compounds of arsenic and sulphur. In later times, the first example of its being publicly recommended as a remedy for the cure of fevers, is in the Memoirs of the Academy at Mentz, for the year 1757, by a Dr. Jacobi. But we are most indebted to Dr. Fowler for his late accurate trials of it. They were conducted in the most judicious manner, to secure exactness in the dose, and to ascertain the efficacy of the medicine; and they are related so fully and circumstantially, that they give complete information and satisfaction, with respect to every particular that is most interesting in the use of this powerful remedy. A better plan cannot

cannot be contrived for afcertaining the powers and ufes of the medicines we employ.

Phyficians and furgeons are fometimes called upon, in cafes of fuppofed murder by arfenic, to give their opinion; and the queftions commonly put to them are thefe:

1. Whether the appearances or fymptoms obferved in the dying and dead perfon give reafon to conclude that they were killed with arfenic?

2. Whether certain drugs or powders which were given to the dead perfon, or mixed with his food, and a part of which are committed to the phyfician to be examined, be arfenic, or contain arfenic?

It is neceffary to be cautious in giving our anfwer to the firft queftion, which feldom admits of a perfectly decifive anfwer, if the prefumption of poifon refts on the fymptoms alone; the fymptoms produced by arfenic being not unlike to thofe which appear in fome difeafes, fuch as the cholera. But thefe fymptoms may add to the proof which may arife from other evidence.

The fymptoms produced by a dangerous dofe of arfenic begin to appear in a quarter of an hour, or not much longer, after it is taken. Firft, ficknefs, and great diftrefs at the ftomach, foon followed by thirft, and burning heat in the bowels. Then come on violent vomiting, and fevere colic pains, and exceffive and painful purging. This brings on faintings, with cold fweats, and other figns of great debility. To this fucceed painful cramps, and contractions of the legs and thighs, and extreme weaknefs, and death.

After death, the inteftines are found inflamed and corroded; and fometimes inflammations and erofions of the anus happen before death.

In examining the dead body, we muft take care that we be not deceived by the diffolution of the ftomach by the gaftric liquor, and account it an indication of arfenic.

If we actually find arfenic in the ftomach or inteftines, or in the drugs or other fufpected matters which were given to the dead perfon, we can give a decifive anfwer to thefe two queftions. But we muft make ourfelves fure that what we judge to be arfenic be really fo.

We muft, therefore, take care to be well acquainted with the qualities of arfenic, by which it is diftinguifhable from all other fubftances. And its diftinctive properties are thefe:

1mo, It is a heavy fubftance, which may therefore be feparated by fkilful elutriation from animal or vegetable matter with which it may happen to be mixed in the bowels or in the drugs. Elutriation is commonly performed with water; but if the arfenic is mixed with oily or refinous drugs, it may be performed with alcohol. In examining the dead body, therefore, it may be proper to wafh out the whole contents of the ftomach and bowels into a bafon of water, and then, by careful elutriation, to try if any arfenic can be found in them. And in examining the drugs, if they are a mixture of different ingredients, we muft dilute or diffolve them, by grinding them a little with water or fpirits, and then elutriate.

2do, Arfenic, befides being a heavy fubftance, is volatile. When heated on a red hot iron, it evaporates totally before it be red hot, and goes off in white fmoke.

3tio, It is eafily metallized by mixing it with three times its weight of the black flux, and heating the mixture in a tube.

4to, In this metallized ftate, it eafily penetrates copper, when affifted by heat, and gives to the copper a whitifh colour like that of lead or tin. It muft be made of a dull red heat. This will completely diffipate corrofive fublimate, or other things which can whiten copper.

5to, In its metallized ftate, if it be fuddenly heated to a fufficient degree on a red hot iron, it takes fire, and burns with a flame, from which arifes a fmoke, which is white arfenic. Or, if the iron be not fufficiently hot to make it take fire, it fimply evaporates, and gives vapours which have an odour like that of garlic. The same odour is perceived, if we mix white arsenic with an equal weight of charcoal dust, and throw a little of the mixture on a burning coal, or on iron strongly heated, so as to set the charcoal dust on fire. This experiment has been often misunderstood.

Having had occasion some time ago to exercise myself in these experiments, and to try with how small a quantity of arsenic they might be made, I found I was able, by means of a small tube, to get metallized arsenic from one grain weight of white arsenic; and with this metallized arsenic I made the other experiments.

Remedies to save, if possible, the life of a person who has taken Arsenic.

The first symptoms which the arsenic produces shew plainly that for some time after it is taken, it acts on the stomach and intestines, as an highly irritating, inflammatory, corrosive substance. But if the patient survives the first violent effects, the poison being evacuated out of the bowels, the symptoms which appear afterwards are those of excessive debility, and a great irritability of the intestinal canal, and of the whole system. The degree of debility is particularly remarkable. It not only is evident from the languor, distress, and feebleness of the patient, but also from the state of the pulse. I never felt a more feeble pulse than that of a person in this situation. All this is attended with a sort of paralytic affection of the limbs, and a degree of marasmus.

The methods commonly recommended to save the life of the person in the first of these states is, to give plenty of milk and oil, as obtunding remedies, and which help to wash and carry off the arsenic out of the intestines, while vomiting and purging continue.

A better practice, however, might be substituted for this. Arsenic being a heavy substance, is not easily washed out by milk, and it may probably coagulate the milk by its acidity. Oil will not mix with it after it is wet. I should prefer mucilage, taken in large quantities, and if it do not pafs off quickly, I would promote its paffage by means of a purgative, fuch as Glauber's falt, or fal catharticus amarus. A friend of mine once gave whites of eggs with fuccefs.

In the fecond ftage of the diforders produced by arfenic, which is commonly of long duration, a mild diet of milk is proper. The frequent ufe of opiates, to relieve from conftant diftrefs, and after fome time electricity, are very ferviceable. De Haen found electricity one of the beft remedies for the cure of the diforders occafioned by lead. Mineral waters have been recommended, efpecially the fulphurous waters; and to imitate thefe, hepar fulphuris diffolved in water may be employed. But this practice is founded upon project and fpeculation, not upon experience.

Metallurgy.

The nature of arfenic being now explained, you are prepared to underftand the operations by which metals are extracted from their ores.

The firft of thefe operations is to feparate the ore as much as poffible from the fpar, or other ftony matter that accompanies it.

When the ore is found in the vein in large maffes, or fills up the whole, or a confiderable part of it, many feet wide, which is often the cafe, it may be cut out tolerably clean, or free from the adhering matrix; or, after it is taken out of the vein, any part of the matrix that adheres to it may be ftruck off. But, in many mineral veins, the ore is interfperfed through the matrix in very fmall maffes, and fo entangled in it that they cannot be feparated in this way.

The expedient to which the metallurgifts have recourfe in this cafe is, to take advantage of the great difference of fpecific gravity between the ore and the ftony or earthy matter in which it is involved. They break or bruife the whole into fmall fragments, like gravel or fand,

and then expose it, upon a board or shallow trough, to the action of a small stream of water running over it with moderate velocity, while the broken ore is gently stirred at the same time. The moving water carries along, and washes away the earthy and stony particles before it can move the heavier particles of the ore. This is called *washing*, or *dressing* the ore; in Latin, *elutriatio*. There are many ingenious and simple ways of performing this process, of which you will find a distinct account in *Agricola de Re Metallica*. This process is easy, when the ore is found in sand or earth, which is often the case with regard to some metals; but others are, in general, found in a solid matrix, not divisible by water alone, or diffusible in it. It must therefore be divided, or broken down mechanically. This is done, in some cases, by flat iron mallets, or by means of stamping mills. It often happens too, that the matrix is much harder than the ore, and then mechanical pulverization alone will not answer. This difficulty can often be obviated, by burning or calcining the mineral with a brisk fire of short continuance, and throwing water on it while red hot. The effect of this upon the stone or matrix is great, if it be calcareous or sparry, but less, yet very considerable, even upon quartz. Thus, therefore, this sort of ore, in a reduced state, can be pulverised and dressed with success. But, although the skill and address with which this art is practised enables the miners, in many cases, to separate the ore from very great quantities of the stony matter, some ores cannot be powdered and washed without loss of some of the metal; and, where circumstances are unfavourable, a great quantity is lost, in consequence of its being broken so small that the water carries it away. Therefore, in such cases, it is found more expedient to procure the separation by bringing the whole into fusion. This is much practised in Germany, in the management of some ores, but very little, if at all, in this country. They perform the fusion in the ordinary furnace, and mix the ore and stones of different kinds. When the whole is thus hastily melted, the ore, as it consists of metallic substances and sulphur or arfenic in an uncalcined ftate, feparates from the melted earthy matter, and is collected at the bottom. It alfo undergoes fome change by this hafty fufion; fome of the fulphur and arfenic is diffipated; and fome of the more calcinable metals are calcined and mixed with the fcoriæ. But it ftill retains fo much of its impurities, that it may be confidered only as a fort of ore. This operation is called *crude melting*, or *crude fufion*. It is not for all forts of ore, but only for thofe which are entangled with the matrix, fo as not to be conveniently feparable by wafhing, but capable of perfect fufion by proper additions.

When the ore is freed from all the earthy matter, which can by this or the former operation be feparated from it, the next fet of operations which it undergoes are intended for feparating the fulphur, or arfenic, or both, if it contains them. It is neceffary to feparate thefe with milder heat, and the action of air, before it be melted, on account of their ftrong adhefion, and their occafioning much lofs of the metal, by carrying fome of it off in vapours, and alfo occafioning more to calcine and mix with the fcoriæ. The ore is, therefore, firft expofed to the action of a gentler red heat, long continued, which is called *roafting* of it *(uftulatio)*, moft commonly in heaps, with fuel intermixed, in the fame manner as bricks are burned. Sometimes this is done in a furnace of a particular ftructure, contrived with great ingenuity, for collecting and preferving the arfenic, and even fome volatile and very calcinable metals, contained in the ores of that metal for which the whole procefs is chiefly carried on.

Some ores require feveral repetitions of this procefs of uftulation before they are fufficiently freed from volatile matters.

When thefe operations are finifhed, the metal remains more or lefs calcined, and therefore requires, in the next place, that the operation of *reduction* be performed. This is done in particular furnaces, of different ftructures, according to the degree of heat neceffary, and the mode in different places. A confiderable variety are defcribed by Agricola, from which those at present in use do not differ essentially, except in the very large size, to which some are now carried. In these, the ore is melted in contact with the fuel, that the whole of the oxygen and sulphur, still combined with the metal, may be absorbed by the burning fuel, and carried off in the form of carbonic acid and sulphurous acid. Proper fluxes are also added for any earthy matters that still remain; and also earths, and even metals, are sometimes added, which still more completely absorb the sulphur. The desired metal, thus freed from extraneous matters, collects at the bottom, and the earthy and sulphurated matter floats above as a scum or flag. This is let off, in some cases, by a tap-hole, at some distance from the bottom. In other cases, it is drawn off with rakes, or driven off by the bellows. The melted metal is then let out by a lower tap-hole, in the great furnaces, or lifted out of the small furnaces with ladles. The flag is, in many cases, rich in some other metal, which has been scorified in this process for the principal metal, and is worked off for it in other furnaces.

The next operation generally is *refining* the metals.

Many ores either do not require, or do not admit of all these operations, but agree best with the omission of some, and variation of others, as I shall observe more particularly hereafter.

In treating this subject, it is usual to mention the art of ASSAYING ores and metals, which is an assistant to the art of separating metals from their ores, and of refining them. The art of assaying is the art of performing all the operations I have been describing, upon small quantities of the subjects, and in a very short time, in order to judge of the nature of an ore, the metals it contains, the best manner of working it, and the profits it will yield. It is also necessary to the art of coining, and to the regulation of the fineness of plate.

The practice of this art requires the greatest degree of attention and accuracy in the artist, to avoid the smallest loss of any part of the subject he is working upon, by any mismanagement; as a small loss of this kind would occafion very great errors in the calculations of the profitablenefs of working the ore, or of the value of the fubject of the affay.

To defcribe the art of affaying would require much more time than can be fpared in an elementary courfe of chemiftry. I muft, therefore, refer you to thofe authors who treat it in the moft judicious manner. I need fcarcely name another, after mentioning Cramer's *Ars Docimaftica.* He was a moft excellent chemift; and his work is complete with refpect to the *docimafia viâ ficcâ.* Of late, however, another method of affaying, *viâ humidâ,* has been much cultivated. The beft writers on this affay are Bergmann, Fordyce, and Woulfe. Mr. Fordyce was, I believe, the firft who thought of this as a complete mode of affay. It certainly gives a more complete analyfis than the dry way; but is too tedious for the neceffary difpatch of bufinefs.

The dexterous ufe of the blowpipe is of unfpeakable fervice in the examination of ores and minerals. By a proper application of the flame, we can either burn off all inflammable matter, or we can affift the fubftance on which it is directed to unite with, or act on the inflammable matter of the charcoal on which it lies. When we envelope the body in the blue point of the interior flame, the exterior flame keeps off the air, and allows the combination with the charcoal to take place, and to continue. But by directing the very point of the flame on the body, we confume all inflammable matter by the application of unfaturated air to it. Therefore, when we operate on metallic ores, we can, in an inftant, either calcine or reduce the metallic part.

The operations of metallurgy, in the great way for commerce, are too various, and too complicated, to be defcribed and explained in an elementary courfe of chemiftry, and at any rate, cannot be introduced at prefent, becaufe, fince the propriety of them depends on the nature of the various metals contained in the ores, which is yet un-

known to you, you could not underftand the moft perfect account
of them. On this fubject, which comprehends almoft the whole of
our fcience, you muft confult the *Ars Metallica* of Agricola, and the
Metallurgy of Schlutter. Of this laft work, which is a very excellent
performance, Macquer has given very good extracts in the articles of
his chemical dictionary.

The very fhort account which I have now given you of the art
of metallurgy was a neceffary preamble to the philofophical ftudy of
the different metals, confidered as the fubjects of chemical fcience. It
was, I confefs, fomewhat irregular to include arfenic, one of their
number, in this preamble ; but fo many of the metallurgic proceffes
are performed in order to get rid of this troublefome ingredient, which
is not fought for on its own account, nor was even thought metallic,
till of late, that we could not proceed intelligibly without a certain
knowledge of its peculiar qualities.

I fhall now enter profeffedly on the inveftigation of the peculiar
properties of the different metals, following, in each, the method of
treatment already familiar to you, and I fhall conclude each article
with a fhort account of its natural ftate, and the metallurgic proceffes
for obtaining it in its pureft metallic form ;—proceffes which, although
fomewhat alike in all, muft yet be modified, in compliance with the
peculiarities in the chemical properties of each.

In profecuting this important part of the courfe, which, with pharmacy, conftituted almoft the whole of ancient chemiftry, I fhall take
the metals in that order in which I think their properties moft eafily
explained. This is almoft the fame with that of their difpofition to
be calcined or burned, which are equivalent changes of their appearance and properties. Of all the metals, gold, or platinum, is the moft
tenacious of the metallic form ; and, on the other hand, the moft calcinable or combuftible of them all is the mineral called *manganefe*,
now found to be a metal, having fingular properties, which are moft

curious in themselves, and are of great affistance to us in explaining and eftablifhing fome of the leading doctrines of chemiftry.

For thefe reafons, manganefe fhall take the lead; and I fhall confider the metals in the following order. Magnefium, iron, mercury, antimony, zinc, bifmuth, cobalt, niccolum, lead, tin, copper, filver, gold, and platinum. Having already confidered arfenic I now proceed to

GENUS II.—MAGNESIUM.

This is by no means a rare production; and it has been long in our hands, and much employed, although its nature is but lately underftood. It often occurs in our mines: Few are altogether without it; and fome contain very great quantities. Its common appearance is a hard fubftance, of a grey colour, and frequently of a chocolate colour, which may almoft be called black. It is eafily known by its foiling the hands exceffively, like a dark brown greafy pigment. The Germans, for this reafon, call it *braun ftein*. It may be feen at every potter's kiln, where it is ufed for giving the black, or dark purple glazing to the very coarfeft of our earthen ware. It is very abundant in Derbyfhire, in a foffil called *black wad;* and was, for a while, fold for a coarfe pigment, till the knowledge of fome of its remarkable properties put a ftop to its employment in this way. For its other varieties, I refer you to Kirwan's or Cronftedt's *Mineralogy*. It is employed for tinging glafs purple, and to compofe the enamels that are ufed in decorating the glazing of fome earthen ware, fuch as the tiles with which our chamber fire-places were formerly lined,—and indeed for the figures on much of the common kinds of Delft ware. It is alfo ufed for a purpofe which feems the oppofite to thefe, namely, for clearing glafs of all colour, efpecially that green colour with which it is generally tainted.

The chemifts knew nothing farther about manganefe. They fupposed it to be an ore of iron, but of a bad quality, because little or no iron is obtained from it.   Some, however, have long suspected that it was a metallic calx, of a peculiar kind ; and by such curious naturalists, it has been a good deal tortured.   It has been described under various names,—*black wad* in England ; *braun stein* among the German miners ; in Latin it has been called *manganesium, magnesium, magnesia nigra,* &c.   Chemists have now agreed to call it MAGNESIUM ; and under this name I shall speak of it.

It is needless to mention the many vague and random experiments which were made on it during the nonage of chemical science.   We never had any clear notion of its properties, till Mr. Scheele took it in charge ; and he has given an account of his examination in an admirable paper, read to the Royal Academy at Upsal, and printed in the Transactions, in the Swedish language, from which it was soon translated into German, and thence into French, and here into English.   Whatever is the subject of Mr. Scheele's examination, he shews in his investigation such an uncommon degree of industry, of ingenuity, and of chemical skill, that you cannot have a better pattern for the proper conduct of chemical researches.   I have often quoted this excellent author, and shall yet often refer you to his ingenious dissertations.   To him, as I have said, we owe our first clear knowledge of this mineral.   But it was Dr. Gahn, another chemist of the same academy, who first demonstrated magnesium to be a metal, by reducing it completely to the state of a hard and brilliant metal, by treating it properly with charcoal in a most violent and long continued heat.

Magnesium has no volatility in its natural state.   But, if exposed for a long time to a red heat, it suffers a considerable loss of weight. If this be done in a proper apparatus, we obtain from it a prodigious quantity of aëreal matter.   This, when examined, is found to be vital air, of the purest kind,—when the magnesium itself is pure, which, however, is rarely the case.   It is frequently tainted with *iron,* and with arfenic, and with earthy matters; carbonic acid, therefore, often taints the aëreal product. Many chemifts affirm alfo, that the pureft manganefe always emits azotic gas in the beginning of the experiment, and that a ftronger heat is neceffary for detaching the vital air. This, however, is its moft abundant product; and it is at prefent the fubftance from which it is obtained for moft of our experiments, and alfo for moft important purpofes in the arts.

This effect of heat on magnefium induces us to confider it as a metallic oxyd, fimilar to many others very familiar to the chemift; but it differs from them in a remarkable circumftance. When this great quantity of vital air has been obtained from it, the remainder is of a much brighter colour than before; and in fome cafes, in which we have expelled the greateft quantity poffible, it becomes perfectly white. The oxyds of other metals are always of the deepeft colour when the greateft quantity of oxygen has been taken from them.

When magnefium has been thus deprived of fome oxygen, it fpeedily recovers it again, merely by expofing it to the common air. The rapidity with which it attracts the oxygen of the atmofphere, is indeed very remarkable; a few hours will fometimes fuffice for reftoring it to its former ftate of repletion with oxygen; and in thus recovering its natural fhare, it alfo recovers its natural colour, and becomes black as before.

When oxygen, already in a gafeous form, is made to combine with other matter, there is always an extrication of heat and light. The prefent cafe is no exception; when circumftances are favourable. When a confiderable quantity of magnefium, which has been deprived of its abundant oxygen, is expofed to the air in an extenfive furface, it recovers it with heat, and even incandefcence. It cannot flame, becaufe it is not volatile.

This furprifing property of magnefium occurred to Dr. Scheele in the courfe of his examination. Having expofed a quantity of black magnefium to a red heat for a quarter of an hour in a phial, the mouth

of which was ftopped with a piece of chalk, he turned it out, ftil
white on a piece of paper. It fo:: deflagrated, and it became black.
If it was kept fhut up till quite cold, and then turned out on a hot
plate, it immediately became red hot. It does fo in an inftant, if loofely poured into a jar containing vital air, and may be feen white, as it
enters this air, and black before it reach the bottom, when the jar is
tall.

Thefe and other analogous phenomena made Dr. Scheele fuppofe
that black manganefe was a fubftance which had attracted phlogifton;
and this opinion, natural to a chemical philofopher of that day, was
confirmed by a feries of the moft curious experiments, made by
treating magnefium with the more active chemical fubftances. Of
thefe the acids afford the moft inftructive phenomena, which I fhall
take pretty much in the order in which they occurred to Dr. Scheele,
led by the train of his fpeculations and conjectures, prefuming that
the procedure of fuch a mind will not be unpleafant to you.

Magnefium, in its natural black ftate, diffolves in the acids but flowly, and with difficulty, and in fome it does not diffolve at all. I lately
had oceafion to remark that the metals, when very much calcined,
that is, when combined with a great quantity of oxygen, have their
attraction for acids greatly diminifhed; and there are feveral examples of their being thereby rendered altogether infoluble. Manganefe
in its natural ftate is fimilar to thofe oxyds; but if fome vital air be
expelled from it by heat, or if it be taken in its white ftate, it will be
diffolved in fuch as did not act on it in its black ftate.

The fulphuric acid is altogether inactive on it in its natural ftate;
but if it be treated with charcoal, and made by heat to give out a
prodigious quantity of carbonic acid,—and if it be then put into fulphuric acid, it diffolves very readily. It is pretty much the fame with
refpect to the other acids. If ftrong fulphuric acid be boiled for fome
time on black manganefe, a fmall portion is diffolved, the acid having
expelled fome oxygen by the affiftance of the great heat.

If precipitated by a cauftic alkali from any of thofe folutions, it is always white, or at leaft, is far from its ordinary dark colour,—and in this ftate it diffolves readily in all the acids.

Thefe facts make it clear that the folubility of magnefium in the perfect mineral acids depends greatly on its degree of oxydation. It is infoluble when the oxygen is abundant, but foluble when this is diminifhed to a certain degree : What proportion of the abundant oxygen muft be abftracted has not yet been afcertained with precifion. It appears alfo, that charcoal has a greater attraction than magnefium for oxygen,—at leaft when its attraction is aided by the volatility of oxygen augmented by heat.   But there are fome other circumftance in the relation of magnefium to the acids, which are not a little myfterious and puzzling.   Although the fulphuric acid does not diffolve manganefe, the fulphurous acid diffolves a part of it very readily,— yet the falt is not of the kind we expect from fuch ingredients,—it is a permanent fulphat of magnefium,—and if we precipitate the magnefium by an alkali, it is in its deoxydated ftate.   Therefore the acid in it is not the fulphurous but the fulphuric, having its full fhare of oxygen.

Perhaps this phenomenon explains the folubility.   The black manganefe may be fuppofed to have a portion of its oxygen redundant, while the fulphurous acid is deficient in oxygen.   The attractions may be fuppofed nearly equal, fo that when the acid has taken as much of the redundant oxygen as completes it to the ftate of fulphuric acid, this acid diffolves what part of the manganefe has been fo much deoxydated as to become foluble.   This explanation is rendered very probable by the fubfequent experiments of Dr. Scheele.

An ounce of aquafortis and twenty grains of manganefe being mixed, no folution follows.   But upon adding about a drachm of vinous fpirits, and making the mixture lukewarm, there is an effervefcence and eruption of nitrous gas,—and the manganefe now diffolves completely ; and after this, four additions, of twenty grains each, will be diffolved. The mixture grows very hot. I imagine that the effervefcence is chiefly produced by the carbon of the alcohol, and partly perhaps by the oxygen gas or vital air.

Mr. Scheele effected the folution of manganefe in nitric acid in different ways,—by means of fugar, or effential oil, &c. But a much more curious means occurred to him. He, and indeed other chemifts before this, had noticed the remarkable effect of the fun's light, in blackening the vitriolic acid, and in caufing the pale nitric acid to become ruddy and fuming; and they attributed this to the introduction of phlogifton, becaufe fulphur was produced, and becaufe fugar produced the ruddy fumes. Obferving that fugar enabled the nitric acid to diffolve manganefe, he tried whether the fame effect might not be alfo produced by the fun's light. Pouring half an ounce of nitric acid on twenty grains of manganefe, and agitating the mixture, it continued quite inactive. He then fet the matrafs in the ftrong light of the fun, on midfummer day, and in an hour's time, he had a perfect folution. He added twenty grains more, which were alfo diffolved,— and continued adding till no more was taken up. He precipitated the manganefe by means of a diluted folution of potafh,—it was perfectly white. He confidered it as manganefe united with phlogifton and fixed air. He mixed it with one-fourth part of nitre, and gave it a red heat in a retort, and got a quantity of nitrous gas. The nitre in the retort was alkalized.

Reflecting on thefe experiments, fhewing the folubility of manganefe in pale nitric acid by means of inflammable bodies or light, we have reafon to conclude, that thefe bodies act firft on the nitric acid, producing their ufual effects, viz. abftracting or expelling a part of its oxygen, and changing it into a more volatile acid. It then becomes a folvent for the manganefe.

But in whatever way we diffolve it, we always find it in a deoxygenated or white ftate, when we precipitate it with an alkali; and in this ftate it will diffolve without difficulty, if done foon,—but if allow-ed to remain fome days, it becomes black and infoluble.

I have called a portion of the oxygen combined with manganefe *redundant*. It is unqueftionably provided with it, to the very point of faturation; and what has been laft accumulated feems to be weakly united: For it is eafily abftracted from it by other bodies,—particu-larly by fome inflammable fubftances. Of this we have.a remark-able example, by mixing a quantity of black wad in powder, with as much boiled lintfeed oil as will juft make it ftick together in clots. If this mixture be made up into a heap, and fet in a warm place, as on the hearth, before a glowing fire, fo as to become lukewarm, it will take fire in about half an hour; and it deferves remark, that this combuf-tion is accompanied by a fmell by no means offenfive, though very ftrong. It is fomewhat aromatic, and altogether unlike the fmell of lintfeed oil burning with a fmothered heat. I imagine that the redun-dant oxygen is combined with the oil, which we know to have a con-fiderable attraction for it, and to abforb a great deal from the at-mofphere. This combination is accompanied by the extrication of heat, in the fame manner as in the mixture of nitric acid with fuch oils. Only, the oxygen of the magnefium is probably neither fo co-pious nor fo loofe as in the nitric acid. During this combuftion, the carbon and hydrogen of the oil fly off with the oxygen of the manga-nefe, forming carbonic acid and water; and the manganefe foon recovers its abundant fhare of oxygen again, by expofition to the atmofphere.

The mixture of manganefe with muriatic acid exhibits phenomena ftill more remarkable, and indeed is one of the moft curious facts in chemiftry. When the common muriatic acid is applied to black manga-nefe, it diffolves it, flowly indeed, but eafily, and in confiderable quan-tity, and needs no addition of inflammable matter. But it adheres fo feebly, that even water fuffices for precipitating the greateft part of it, in its ordinary form. This eafy folution made Dr. Scheele fufpect that muriatic acid, in its ordinary ftate, contained fome phlogifton, naturally combined with it, which reduced the manganese to a less calcined state, and thus disposed it to dissolve so easily in the acid. This conjecture was confirmed by the phenomena which the mixture presented when a gentle heat was applied to it. He learned that a part of the acid is greatly changed from its ordinary state, and is gradually converted into a penetrating vapour, of a yellow colour, and most insupportable suffocating smell.

This gas is one of the most remarkable objects in chemistry. It is with difficulty obtained in any other form than that of an elastic aëreal matter, having scarcely any attraction for water. The ordinary muriatic acid gas has a strong attraction for water, and is, by this means, obtained in the ordinary form of a watery acid: But when distilled from black manganese, it will scarcely unite with water, and unites only in very small quantity, and is easily separated again, or rather is with great difficulty preserved in a combined state. A freezing cold, indeed, will condense it into a sort of soft or solid matter, but when the cold abates, it immediately reassumes the gaseous form. It may, however, be condensed, in consequence of its attraction for other substances, such as the alkaline salts, and especially the inflammable substances, such as oils, animal and vegetable substances, some bitumens, and metals in their metallic state. It acts so strongly on many of those substances as to inflame them. Thus phosphorus, plunged into this gas, takes fire immediately. Sulphur also, if hot enough to begin to melt, instantly takes fire. Even charcoal, if in exceeding fine powder (such as may be got by washing the nitre out of gunpowder) when made warm, and thrown loosely into this gas, instantly kindles. I have met with some assertions that the vapour of alcohol also takes fire in it, but I have not found it so, either with the vapours of alcohol, or that of the vitriolic æther. Nor have I found that it kindles some of those compound hydrogenous gases which are so ready to take fire. The abominably fœtid gas, containing phosphorus and hydrogen, fires pretty readily in this muriatic gas,

but requires particular management, and a particular ftate of the ingredients, which you w^{...} underftand as we proceed.

Moft of thefe curious facts occurred to Dr. Scheele, although in a
more complex manner, becaufe he was not thoroughly inftructed in
the nature of this vapour. But he faw enough to make him conclude
that the muriatic acid, in the ftate of a yellow vapour, is deprived of
a part at leaft of the phlogifton which he fuppofed it to contain when
in its ordinary ftate. He alfo found reafon to conclude that it has a
ftrong difpofition to reunite itfelf to that principle ; and in confequence
of this attraction, it acts on bodies which contain it,—and this with
great vivacity, when they hold it in abundance, and but loofely combined. He accounts in the fame way for its action on the metals, by
which it is eminently diftinguifhed from the common muriatic acid,
which, as I have told you, acts on them in a languid manner, if in
their metallic form, though it diffolves thefe calces very readily,—more
readily indeed than the other acids.

Induced by all thefe facts, Mr. Scheele called this vapour or gas the
dephlogifticated muriatic acid. We need not wonder that he employed
this language, and viewed the phenomena in this light ; for, when
he made and publifhed this examination of manganefe, the phlogifton
was every where admitted as a principle in chemiftry, and Dr. Scheele
was the very firft that expreffed any diffatisfaction with the fimple
form in which this doctrine was delivered by Dr. Stahl. I have already had occafion to mention the ingenious and fanciful modifica-
tions of the original doctrine of Stahl, which this excellent chemift
attempted to eftablifh, but which he would have been the firft to
abandon, had he pufhed fome of his own experiments one ftep farther.
I hold it, therefore, to be unpardonable arrogance in the French chemifts to fay that no man can entertain the belief of the exiftence of
phlogifton who has a grain of common fenfe. Scheele's differtations,
of every kind, will ever ftand in the very firft rank of chemical writings. By the natural progrefs of all knowledge that is founded on

experiment, we have come to interpret the many difcoveries of Dr. Scheele in a different way; but the difcoveries remain the fame, and they are his, and refulted from deep and ingenious meditation. He confidered all the phenomena which we derive from the privation or abfence of oxygen as proceeding from the addition or prefence of phlogifton; and he afcribed to the abftraction of phlogifton what we know to be owing to the acquifition of oxygen.

Since the time of Dr. Scheele, all thefe phenomena of the muriatic acid and manganefe have been maturely confidered, and carefully inveftigated, and clearly explained, principally by the chemifts of France. We now hold that the change of appearance and properties which the muriatic acid fuffers depends on the addition to it of a great quantity of oxygen, which it acquires from the manganefe. For this reafon, the muriatic vapour which I am now confidering has got the name of OXYGENATED MURIATIC ACID. Mr. Kirwan calls it the OXY-MURIATIC ACID. You certainly recollect that I have feveral times had occafion to mention a particular ftate of this acid, in which it was furcharged with oxygen; and in confequence of this redundancy, had fome fingular properties. It was thus that I explained the procefs for preparing muriatic æther by means of the fmoking liquor of Libavius, and feveral experiments of Berthollet, eftablifhing the conftitution of volatile alkali, with other things of fimilar nature. It was this preparation which I then had in my thoughts.

We have proof that the acid, in Dr. Scheele's experiment of diftilling it from manganefe, has abftracted oxygen from that mineral. Filtrate what remains in the retort, and add a very pure fixed alkali,— the manganefe is precipitated in the form of a white powder or mud; a fure fign of its having loft fome of its oxygen. If the precipitate be haftily wafhed and dried, and then urged by a ftrong heat, in a proper apparatus, we may perhaps ftill obtain fome oxygen, by the extreme force of heat, but the quantity will be exceedingly fmall.

I may now repeat, in a more comprehenfible and inftructive manner, what I formerly mentioned when I gave you an example of the manner in which the French chemists extended the doctrine of La voisier to almost every chemical phenomenon in this world, namely, their account of the constitution and formation of the volatile alkali. When the pure or caustic alkaline gas is made to mix with the oxy-muriatic gas, there is an immediate decomposition of both. The re-dundant oxygen of the latter seizes on the hydrogen of the volatile alkali, and forms water; and the azotic gas is set at liberty; and the oxy-muriatic gas is changed into common muriatic acid, by the de-parture of the redundant oxygen. If we have made use of a watery solution of the caustic volatile alkali, and make the oxy-muriatic gas pass through it, we have an effervescence, occasioned by the extri-cation of the azotic gas. All this was discovered by Dr. Scheele, and is related in his Essays.

Some of the French chemists, and particularly Mr. Berthollet, the most eminent of them, have followed Mr. Scheele in the investigation of the properties of this remarkable gas, and have made some very curious and important discoveries relating to it.

I shall briefly mention some of its leading properties, referring you for farther information to Mr. Berthollet's most excellent dissertation *Observations sur quelques combinaisons de l'Acide Marin dephlogistiqué, ou de l'acide muriatique oxygéné, par Mr. Berthollet.* It is one of the best pieces of experimental philosophy that has appeared in any lan-guage.

1. The oxy-muriatic gas retains its affinity for alkalis, but wonder-fully diminished, and modified by very particular circumstances. It unites, he says, with mild alkalis without effervescence. I have not yet examined this with sufficient care, but am disposed to doubt it. It is scarcely possible to procure an alkali perfectly saturated with car-bonic acid. I never saw potash that was so but once. This gas con-tains so little saline matter, and its attraction is so weak, that it must of necessity attach itself in preference to such of the alkali as is in a

cauftic ftate ; and I apprehend that this may have been enough to abforb the whole. The vaft quantity of water alfo, that is neceffary for
condenfing the gas, may abforb a great part of the carbonic acid that
is really detached from the alkali ; and the fact is, that heat applied
to this mixture will detach much carbonic acid from it.

2. When the oxy-muriatic gas is condenfed by a very diluted folution of cauftic potafh, the alkali became turbid, and depofited fome
earth, and fome faline cryftals. The mixture being evaporated by a
moft gentle fand heat, afforded two falts,—namely, the ordinary falt
of Sylvius, and another, eafily feparable from that falt, becaufe it diffolves much more copioufly in hot than in cold water. When the
procefs is well performed, we obtain about four parts of falt of Sylvius
and one of this new falt.

3. This falt has many fingular properties. Its cryftals are hexa-
edral prifms, or more frequently confufed laminæ. It has an unpleafant
maukifh tafte, and raifes a feeling of coolnefs on the tongue, as nitre
does. It contains the acid in its highly oxygenated ftate. An hundred
grains of the dry falt yield 75 cubic inches of vital air by means of
heat, and yield it more eafily than nitre does. Mr. Berthollet, obferving the vital air fo copious in it, and fo loofely combined, mixed
it with charcoal, and tried whether this would detach it by the affiftance of heat. It deflagrated with prodigious violence, and the acid
was not deftroyed, but only reduced to its ordinary form ; for the refiduum was the ordinary falt of Sylvius. Therefore no oxygen had
operated in this detonation, except the portion obtained from the manganefe. Mr. Lavoifier found that 100 grains of the falt contains 37
of oxygen, which requires 14 of carbon to feparate it, and produces
51 cubic inches of carbonic acid.

4. Mr. Berthollet, from thefe and other analogous experiments, in-
fers that all the oxygen which produced that change in the marine acid
is concentrated in this falt, and that it contains no other, nor any acid
in its ordinary ftate. He therefore calls it the OXYGENATED MURIAT

of potafh or of foda. One part of this falt contains all the redundant oxygen that is furnifhed by fix of oxy-muriatic acid. I may remark that the nitrofe acid, in its union with alkali, exhibits phenomena pretty fimilar. We obtain a true nitre, and a nitrous gas.

The union of the ingredients of this oxy-muriat feems very flight. Expofition to the air feems to decompofe the acid : For in a few days, or even hours, the falt changes to the ordinary falt of Sylvius. This decompofition happens more fpeedily in a watery folution of the falt ; and we fee a continual fimmering on its furface, by the efcape of minute bubbles of elaftic matter. This is increafed by expofure to the rays of the fun fo as to be like an effervefcence. This is vital air, of the pureft kind. It would feem that the cauftic alkali acts too powerfully on the bafis of the ordinary muriatic acid, and thus diminifhes its attraction for the oxygen ; or the oxygen exifts in it, perhaps, in a femi-elaftic ftate. Mr. Berthollet alfo thinks that the vaft abundance of water neceffary for abforbing this gas prevents a clofer union of the acid and alkali. He found that the combination could not be effected unlefs the folution of the alkaline falt be extremely diluted. If this liquor be evaporated, preferving it at the fame time from the action of light, or the naked fire, it undergoes, at a particular period of the evaporation, a fudden change, by which the above falt is formed ; and that after this, neither the liquor, nor any of the falts which it affords by evaporation, are poffeffed of the peculiar powers of the oxy-muriats.

Mr. Scheele difcovered in this oxygenated acid another moft remarkable property, namely, a power to deftroy all vegetable and animal colours, and even thofe which are moft permanent, prepared for the purpofes of dying. It whitened or bleached vegetable fubftances in a furprifing manner. Mr. Berthollet firft thought of applying it to this ufe, and found that bees wax, brown linen yarn, and cloth, are bleached by it in a few hours or minutes, as effectually, and with as much fafety to the ftaple of the goods, as if they had been expofed to the fun and air, with that intention, for as many weeks.

In confequence of this great difcovery, trials have been made, with a view to the employing it in the art of bleaching fine linen, or cotton cloths, threads, and light manufactured goods. Mr. Berthollet has publifhed the procefs, as it has been practifed in fome manufactories in France; and his account of it is tranflated into our language *.

The firft trial of it in the great way, however, was (according to the beft of my information) made in Scotland, with the affiftance, and under the direction of Mr. Watt, who had been at Paris, and had converfed with Mr. Berthollet, and immediately formed the defign to try if the powers of this acid could be employed in practice; for Mr. Berthollet had only confidered it as a project in fpeculation. Mr. Watt had an opportunity foon after to make his experiments in the bleachfield of a friend at Glafgow; and fince that time, the procefs has been applied to this purpofe in many other bleachfields. It is beft adapted, however, to the bleaching of thin goods, fuch as lawns, and fine muflins, and fine thread, and ftockings. Thefe require lefs of the acid than coarfer goods, the thick and harfh threads and fibres of which are penetrated with difficulty, and alfo contain much colouring matter,

---

* Therefore I need only mention here, that Mr. Berthollet produces the oxygenated vapour of muriatic acid, or oxygenates the acid, during its formation. He mixes, in dry powder, fix ounces of black manganefe with 16 of common falt. A tubulated diftilling veffel is prepared, whofe pipe is inferted into a receiver with two necks, and into the oppofite neck is inferted the pipe which conveys the gas into the veffel containing the liquor which is to abforb it, or the matters which are to be bleached by it. This powder is put into the retort, and then there is poured on it 12 ounces of ftrong fulphuric acid, diluted with nearly as much water; and the ftopper is put in. Gas is immediately produced in vaft abundance, and paffes into the reft of the apparatus. The two-necked receiver condenfes the ordinary muriatic acid; but the oxygenated gas paffes on to the abforbing veffel. No heat is applied till the emiffion of vapour becomes very gentle. The heat is cautioufly and flowly raifed to boiling, and continued till the two-necked receiver grows hot. This finifhes the diftillation. The abforption is promoted by churning vanes, which are worked by fome machinery. Thus the abforbing liquor is more fpeedily and thoroughly impregnated. (Annales de Chymie, t. ii.)

which foon exhaufts the ftrength of the gas. Attempts have alfo been made to bleach rags for paper. The methods employed are frequently to wet the matter to be bleached, by dipping it into water, while it is at the fame time expofed to the vapour of the acid; or water which has been made to abforb fome of the vapour is applied to it; or it is fteeped in a folution of fixed alkali, or in lime-water, which has abforbed fome of the vapour. Such folutions are found to have the bleaching power, although the acid be faturated with alkali.

Different methods are employed by different manufacturers, according to their notions of their refpective efficacy; and each manufacturer has his noftrum, which is a fecret. At prefent, none of them, I believe, ufe the fimple oxygenated acid, which was formerly prepared for them, as a fteep. The fmell which it occafions is abominable, and cannot be cleared from the hands for many days. The workmen therefore will not fubmit to it. They either ufe the vapour procured by Berthollet's procefs, with alternate dippings into water,—the whole of which operation is performed in a clofe chamber by the intervention of machinery; or, more commonly, they employ folutions of alkali impregnated with the vapour. A chemift in Glafgow has made a great improvement upon the whole procefs, by employing lime inftead of alkali. He thereby prepares the drug in a very concentrated, and even a folid ftate.

Mr. Berthollet's trials to combine this volatile acid with fixed alkali have alfo produced fome other furprifing difcoveries. Obferving how much the oxyd itfelf was difpofed to give out its oxygen to inflammable fubftances, and how loofely it is combined in the acid overcharged with it, while this acid ftill retains its relation to alkalis, he tried the effect of the neutral falts produced by it. When the oxygenated muriat of potafh is ground in a mortar with fulphur, fmall explofions happen under the peftle, which affect the hand of the operator like an electric fhock; and if haftily ground, with ftrong preffure,

the whole will explode. Nay, if the falt alone be thrown into fulphurous acid it will explode.

Mr. Berthollet alfo examined, in the fame way, the fuper-oxygenated muriat of foda. This deliquefcent neutral falt gives out its
abundant oxygen by mere expofition to the air, and changes to common fea falt. When treated in clofe veffels, 100 grains yielded 75
cubic inches of vital air, of the greateft purity, with much lefs heat
than nitre yields it. It was now a common muriat of foda.

He compofed a gunpowder with the muriat of potafh ; and it was
faid to be vaftly ftronger than the nitrous gunpowder. Trials were
made of it at Woolwich ; and it was found that it was really ftronger
in fmall charges,—but there was no fenfible difference in great quantities, fo that it was not thought worth while to profecute the difcovery further *.

Such gunpowder has qualities which make it inferior to the common. The very loofe combination of this oxygen muft caufe the
powder to become effete. The refiduum of the detonation will be
digeftive falt, inftead of the hepar fulphuris left by the ordinary powder. This will be more hurtful to the fire arms than fixed alkali. It
would alfo appear, from the accident which happened in the preparation, that it is more difpofed to explode in the operation of grinding.

From the whole of what I have faid on this fubject, it appears that
the nature of the muriatic acid is very fingular, when we compare it
with the other two foffil acids.—Thefe, when they have not the fufficient quantity of oxygen, are more volatile than ordinary, and their
attraction for water is diminifhed. The muriatic acid, on the contrary, has its volatility increafed, and its attraction for water diminifhed, by receiving enough of oxygen.

Mr. Scheele's experiments and reafonings have alfo fhewn that

---

* The oxy-muriats of lime, barytes, and ftrontites, deferve a trial for a fulminating
compofition.   EDITOR.

there is a fingularity in manganefe, when we confider it as a metallic'
calx. It is moft coloured, or darkeft, when moft calcined, and becomes
white by abftraction of oxygen, contrary to other metallic calces.
And it gives colour to its folutions, and to glafs, when it is highly
oxydated ; but when this high degree of oxydation is abated, the colour difappears. The oxyds of other metals in general give moft colour when they are leaft oxydated. This fingularity in manganefe
is finely feen by treating a fmall portion of it with the blowpipe. A
fmall bit of the fal microcofmicus being melted on charcoal by the
blowpipe, if a minute portion of the black manganefe be added, the
globule acquires a red colour, fo much the fuller, as we have added
more of this oxyd. If we keep this globule in the middle or blue part
of the flame, where it is defended from the action of the air, the red
colour vanifhes, by the reduction of the calx by the charcoal. If we
now direct the point or exterior part of the flame to the under fide of
the globule, the effect of the charcoal is prevented, and the reduction
already operated is deftroyed, and the globule becomes red. We may
thus change the colour as often as we pleafe. The fmalleft particle of
nitre being put to the colourlefs globule makes it red in an inftant, by
calcining the manganefe.

A globule of manganefe and borax preferves its dark red or black
colour under this treatment, probably by keeping it from touching the
charcoal.

We have an example of the colour it gives in its highly oxydated
ftate, in a watery folution of a compound of manganefe and fixed alkali, analogous to a *liquor filicum.* This combination is beft formed by
mixing the manganefe with nitre, and giving the mixture a mild melting heat. The nitric acid is expelled by the heat, leaving the manganefe well ftored with oxygen. The alkali remains combined with the
manganefe, and forms with it a dark green or blackifh mafs, which is
foluble in water,—and gives it a green colour. We fhould perhaps account it a blue, becaufe in a day or two a yellow powder falls down,

And the folution is blue. The combination is but loofe. Water feparates it, firft of a violet colour,—then red, which grows brown,—and laftly black. If fulphuric acid be added to feparate the alkali, the folution becomes colourlefs, the acid diflolving the deoxygenated manganefe. Such a variety of colours have procured to this folution of nitre alkalized by manganefe the name of *chamelion minerale.*

If we add a few drops of a folution of hepar fulphuris to the folution of manganefe in the fixed alkali, it no longer exhibits the changes of colour. It produces this effect by its ftrong attraction for oxygen, by which it deprives the manganefe of that precife quantity which enabled it to impart thofe colours.

It would appear that manganefe clears glaffes of the yellow and green colours which tinge them, by yielding a quantity of oxygen fufficient for calcining the colouring matters. When too much manganefe is employed, after having difcharged the colours produced by other metallic contaminations, it communicates its own colour,—a purple *.

Liquid phofphoric acid, as produced by the fpontaneous decompofition of phofphorus without heat, diffolves manganefe, giving a fluid of a rich red colour. If this be kept in a phial clofely fhut up, it lofes its colour,—but refumes it by expofing the whole to the air in the procefs of filtration. This may be repeated as often as we pleafe, and no matter is feparated by the filtration. This is undoubtedly pro-

* This property feems to me to have been longer known than is commonly fuppofed. Pliny fays, " Mox, ut eft aftuta et ingeniofa folertia, non fuit contenta nitrum mifcuiffe. " Cœptus addi et *magnes lapis :* quoniam in fe *liquorem* vitri quoque ut Ferrum trahere " creditus." I think it probable that *magnes lapis* is manganefe. The fentence has no truth, if the loadftone be meant, for this deftroys all beauty in glafs. Foffils were then little known, and Pliny was not a fkilful foffilift,—*liquorem* may be a corruption for *colorem* or *livorem.* He is fpeaking of the perfection to which the manufacture had been brought, and fays that the moft valued was what is clear and colourlefs as cryftal, and then he relates the above practice.

duced by abſtraction of oxygen from the manganeſe, and reſtoration of it by the atmoſphere.

Acetous acid diſſolves the black oxyd, and, in this ſtate, very readily produces æther, when treated with alcohol. It alſo gives moſt beautiful cryſtals, when employed to diſſolve copper.

Mr. Milner of Oxford publiſhed a paper in the 79th volume of the Philoſophical Tranſactions, giving the general reſult of a number of experiments he had made with manganeſe, by making it red hot, in tubes of iron, or of earthen ware, and forcing the ſteam of water and of other things to paſs over it, or through it; and among theſe, he tried the vapours of the pure or cauſtic volatile alkali. It always happened, that when the vapour of this alkali paſſed through the red hot manganeſe, it was partly converted into vapour of nitrous acid. This effect was conſtant; but he did not attempt to examine and aſcertain how much of the nitrous acid might be produced from a limited and known quantity of the volatile alkali.

This remarkable and intereſting experiment is eaſily explained by Mr. Cavendiſh's diſcoveries, and the new theories in chemiſtry to which they have given riſe. He diſcovered that water is produced from inflammable air and vital air; and that nitrous acid can be produced from azotic gas (phlogiſticated air) and vital air. Therefore, to underſtand Mr. Milner's experiments, ſince we know that the volatile alkali is a compound of hydrogen and azote, we need only to ſuppoſe that part of it is totally decompoſed and deſtroyed by the action of the oxygen contained in the manganeſe. Part of it, uniting with the hydrogen, forms water or watery vapour; and part, uniting with the azote, forms vapours of nitrous acid.

He obſerved alſo that the ſteam of water, paſſing through red hot manganeſe, promoted very much the extrication of thoſe gaſes which it affords by heat *.

* *January* 1796. There is a rumour that the French have manufactured ſaltpetre, during a part of the war, by obtaining nitrous acid from the vapours of volatile alkali, forced to paſs through red hot manganeſe. AUTHOR.

Having mentioned the moſt important properties of this remarkable ſubſtance, we ſhould next conſider the methods of reducing it to its metallic ſtate. This has been but little ſtudied; and its metallic properties are very imperfectly known. The reduction of manganeſe to a regulus is a very difficult proceſs. This was firſt accompliſhed by Mr. Gahn of Sweden.

A crucible muſt be lined with wetted charcoal, (rendered very denſe, and compacted by beating), leaving a ſmall hollow in the centre, to receive a ball of the oxyd, made up with oil into a compact paſte. Charcoal is put in above it, and well compacted by beating, and another crucible is luted on, alſo filled with charcoal. The moſt violent heat muſt be given for an hour and an half. A button is thus procured, of a dull iron colour, commonly rough on the ſurface. This ariſes from minute globules of manganeſe leſs perfectly reduced. It requires even more heat than iron for its fuſion ; and no chemiſt has been able to join pieces in this way. All ſaline fluxes are found hurtful to the operation.

The metal quickly calcines by expoſure to the air, and is ſoon reſolved into a black oxyd. To preſerve ſpecimens in a metallic ſtate, it is neceſſary to varniſh them ; for even the air of a phial ſuffices for ruſting it in a few hours. I have ſome ſuſpicion that it abſorbs azote as well as oxygen, for I obſerve that it leaves a ſmaller portion of that gas unabſorbed than any other proceſs I have tried. 200 grains of regulus increaſed in weight 76 grains in the open air. Another 200, from the ſame maſs, expoſed in vital air, increaſed only 65 *. The ſpecific gravity of magneſium is about 6,35. It is quite brittle ; and uſually contains iron, which is not ſeparated without extreme trouble.

* This conjecture of Dr. Black's is corroborated by the obſervation of Mr. Seguin, that black manganeſe yields azotic gas in low heats.   EDITOR.

GENUS III.—IRON.

AFTER magnanefe, we are now to confider iron.

I choofe to defcribe iron next, and among the firft of the metals, for many reafons. It is the moft abundantly produced by nature. It is alfo by far the moft ufeful among the metals and of the greateft importance, by having a combination of properties, which make it an excellent fubject for the ingenuity and induftry of man to work on. Thefe properties are, a ftrong cohefion of its parts, by which it excels all the other metals; and which, in fome ftates of the iron, is attended with great toughnefs and ductility; in others, with exceffive hardnefs; and in others, with great elafticity *. In thefe different ftates, or intermediate ones, it is excellently adapted to different ufes. We can alfo foften it by heat, and thus work it eafily, under the hammer, or by the compreffion of rollers, into different forms; and when ftrongly heated, different pieces of it can be joined firmly together. We can alfo have it in a fufible ftate, and caft it in moulds.

Iron is fingular among the metals, and indeed all other bodies in nature, by being the only one affected by the magnet; and as the magnet can be made to communicate to iron its own polarity, and other magnetic qualities, this metal becomes thereby ftill the more ufeful. Artificial magnets can be made of it, far exceeding the natural ones in ftrength. A very fmall magnet, turning freely on a point, is the needle of the mariners compafs, and pilots our fhips through the tracklefs ocean.

While iron is cold, though it be tough and flexible, it is too hard and ftrong to be workable under the hammer, as fome metals are; and, befides, it foon lofes its toughnefs when we hammer it ftrongly,

---

* An iron wire, of one-tenth of an inch in diameter, will carry 450 pounds. A copper wire not quite 300. AUTHOR.

its latent heat being thus eafily driven out of it. It retains its latent heat and toughnefs much better while it is drawn into wire, and it can therefore be drawn into very fine wire. Its cohefion is greatly increaf- ed by this operation. When it is red hot, it is very malleable, and al- moft plaftic like clay, and can be beaten by hammers, or compreffed by rollers into very thin plates.

It is one of the moft refractory metals, when in its pureft ftate. The moft intenfe fires can hardly melt it, when it is expofed fimply to the action of heat in clofe crucibles. In its unrefined ftate, it can be melted eafily, and perfectly, by a violent heat, in the tempera- ture 17,977° Fahrenheit, or 130° Wedgewood; and the contact of the fuel increafes its fufibility. Mr. Lewis fays that gypfum has the fame effect.

When pure or tough iron is heated to a white heat, 90° of Wedge- wood, different pieces of it can be made to cohere and unite perfectly into one piece by hammering them into clofe contact. This is a com- mon operation of the blackfmiths: They call it *welding*. Some fand, or pounded fand-ftone, is thrown into the fire in performing this ope- ration. The fand melts on the furface of the heated iron, by uniting with a little of the metal that is calcined, and forms a liquid glazing, which defends the metal from farther calcination, and keeps the fur- face of the different pieces in a condition for uniting together, when they are properly applied to one another. Sand is of the fame ufe here as borax is in foldering. This liquid glazing readily flies away from between the pieces, when they are hammered into contact.

The only defects of this metal are thofe of being too eafily calcin- able, and too liable to the action of different folvents. It is capable of being calcined fo fuddenly and violently, as to give all the appear- ances of inflammation: As,

1ft, When heated to a white and dazzling heat, and fuddenly ex- pofed to a ftream or blaft of air, brilliant fparks are darted from it, which the metallurgifts call *brandifhing*.   Thefe are figns of the iron being inflamed, and calcining rapidly.

2*dly*, The fiery fparks produced by the collifion of flint and fteel are another example of it.   They are fmall parts or rags of the fteel, torn off by the flint.   By the violence with which they are feparated from the mafs, and the extrication of latent heat, they are heated red hot ; and in falling through the air, are blown up into an inflamed ftate.   That it is not a fimple ignition or incandefcence, occafioned by the great heat generated by friction, is evident from this, that if the fame experiment be made in fixed air, they are then feen juft like fo many ignited particles of fand, and the light is feen only in the very fpot where the collifion is made ; whereas the fame fparks in free air are brighteft when they have got to fome diftance, and feem to brighten by degrees as they recede from the ftroke.   Sometimes a brilliant fpark is not feen at all, till at a confiderable diftance ; and it branches and fplits into two or three fparks.   Alfo, when the two fets of particles are examined with a microfcope, thofe that were driven off into fixed air are plainly thin rags of metal, which have been fcraped off by the flint, and rolled up into a fort of fpiral ; whereas thofe ftruck off into good air are fcorified and bliftered, and can be crufhed to powder between the fingers.   Nay, if a fine fteel wire be twifted round a thick one, and if the whole be thruft red hot into a veffel containing vital air, the fine wire will take fire, and burn away like a fquib, crackling and brandifhing with great brilliancy.

3*dly*, Similar fparks are produced by throwing fine iron filings through the flame of a candle.

4*thly*, Larger maffes of iron are eafily made to burn in the fame manner, by heating them in vital air, or introducing them red hot into it.   It is this violent inflammation that makes the fpark of iron ftruck off by flint kindle gunpowder with fuch abfolute certainty. Were it merely red hot when ftruck off, it would be fo much cooled in its paffage, that it would produce no fuch effect.   Other metals, equally or perhaps more inflammable, will not do, being too foft for producing the firft heat by the ftroke.

Iron expofed to a mild heat gradually increafed, is calcined at the furface, without producing any heat or light. The firft phenomenon of this calcination of a bit of polifhed iron, is the appearance of rainbow colours on its furface, before the iron is red hot, but juft approaching to that heat. The fucceffion of colours is faint yellow, golden colour, purple, violet, and deep blue, weakened infenfibly to a water colour, which is the laft fhade diftinguifhable before it is red hot. After this, a dufky cruft and friable fcales appear. When long calcined, the colour is deep red. All the calces of iron, if much calcined, are either dufky yellow, red, or purplifh. Hence they are called *croci*.

This fucceffion of colours admits of a very fatisfactory explanation, by the difcoveries of Sir Ifaac Newton concerning the colours produced by tranfparent thin plates. The colours fucceed each other in that order in which they fhould follow, were they produced by a tranfparent plate, gradually increafing in thicknefs. Now we know that this fucceffion of colours terminates in the production of a fcale of calcined or oxydated iron. We alfo know that the calces of metals are frequently tranfparent coloured glaffes, and in all cafes, compofe fuch glaffes by admixture with earths.

When iron is thus calcined, the power of the magnet over it diminifhes; and when the calcination is carried to a certain length, becomes imperceptible. The fcales are ftill attracted, but much more weakly than iron; and if further calcined, there is no more fenfible attraction. Nor is the ruft, formed by long expofition to the air, or any of the natural calces or ochres of a yellowifh or reddifh colour, attracted fenfibly, in the common way of making the experiment. They are eafily brought back again, however, to a lefs calcined, or lefs oxydated ftate, fo as to recover the black or dufky colour of the fcales, and then they are attracted by the magnet. We need only to mix them with a little charcoal duft, or any animal or vegetable matter, and make them red hot.

Iron is not only diffolved by feveral acids, but it is acted upon by all the faline fubftances without exception. They either corrode the metal, or diffolve the ruft or calx. Water alone, or moift air, has confiderable effect, and is perhaps the chief agent in all thefe corrofions. It has, therefore, long been a defideratum to preferve the furface of polifhed iron from corrofion by the air and weather, i. e. from rufting. The only method is to cut off the communication by fome varnifh. Oily or greafy matters are the fitteft, efpecially fuch as harden in the air. Pure lintfeed oil, or hempfeed oil, are therefore preferable, when fpread over the iron, by rubbing it with a clean flannel cloth dipped in the oil, and rubbed on fo as to leave no fenfible quantity. The very operation of polifhing has a confiderable effect, for it is always by the means of cutting powders, rubbed on with greafe of fome kind or other. The blue colour produced by heating the iron is alfo thought to defend it.

All the acids act on iron, and diffolve it. Even the carbonic acid, efpecially if affifted by water, diffolves it. For this reafon, it is ufual, in the manufactories of polifhed goods, to throw them into limewater, or water having lime in it, to preferve them from rufting during the time that they are in the hands of the workmen. The fulphuric acid diffolves the iron the moft readily and completely. But it muft not be applied in its ftrongeft ftate. It is neceffary that a good deal of water be added at the fame time. I had an opportunity not long ago to fhew you how the iron is diffolved by fulphuric acid thus diluted, and the great quantity of inflammable air which is produced during the diffolution. The production of this inflammable air is undoubtedly beft explained by the new theory of the French chemifts. It proceeds from the water, which the iron decompofes.

The proofs of this are, 1ft, When we force the ftrong fulphuric acid to act on iron filings, without adding water, but fimply by heat, we do not get inflammable air,—we obtain fulphurous acid. And the acid, deprived of part of its oxygen, becomes fulphurous acid, which does not act readily on iron. This action requires a very great heat; and when the mixture is diftilled to drynefs, we obtain a fublimate of fulphur, deprived of all the oxygen with which it formed the acid. The folution of the iron, therefore, required more oxygen than was fupplied by that part of the acid which really diffolved it, and combined with it. Even this ruddy compound of acid and iron is not completely neutralized, it being a very deliquefcent falt, whereas the green coloured falt, produced from a diluted acid, even falls to powder by evaporation.

2dly, Water alone can be made to act on iron, fo as to produce inflammable air in great abundance. Even in the ordinary heat of the atmofphere, it produces fome inflammable air, and changes the iron into a ruft or calx. Iron filings, digefted in pure water, are converted into calx, and a quantity of inflammable air is produced. Red hot iron, plunged into water, caufes a thick black fcale to form inftantly on the iron, and much inflammable air is difengaged. This fcale is fimilar to that which forms on iron fimply heated in the air. This, I believe, fuggefted to Dr. Prieftley a much more effectual way of applying the water, namely, in the form of fteam, when the iron itfelf was red hot. It was fent through a porcelain or copper tube, in which iron wire is coiled up, and it converted the whole into the fame calx; and inflammable air was difengaged. Water confifts of oxygen and hydrogen, the bafis of inflammable air. It appears, therefore, that the calx is produced by the decompofition of the water, which oxydates the iron, while the inflammable air efcapes.

When the diluted acid acts on iron, as the folution goes on, a double operation feems to be going forward. There is a copious formation of a black mud, which is of the fame nature (at leaft a great part of it) with the oxyd formed by iron and water. This feems to be a different procefs from the faturation of the acid : For, if the acid has been but little diluted, the folution ftops, but is renewed by adding more water. The water of the acid is employed in forming the martial falt ; and now the acid has not enough to continue this formation, till frefh water be added. But the mud is an oxyd deriving its oxygen from the water only.

When the fulphuric acid is faturated with iron, and the effervefcence ceafes, the folution is muddy for fome time, but at laft becomes clear, and of a light green colour. The matter which made it muddy is obferved partly at the bottom, and partly at the top. It is of a blackifh colour, and refembles the duft of what is called *black lead*, or *plumbago*. Profeffor Bergmann analyfed this blackifh fediment, and found that it has really the qualities of plumbago, which had been analyfed before by Mr. Scheele. It is compofed of carbon, united to a fmall quantity of iron. If we take either black lead of the fineft kind, or the black matter now fpoken of, and expofe it to the moft intenfe heat in clofe veffels, it fuffers little or no change. But in open air, it is all confumed, except an ochry calx, variable in quantity.—But much of it efcapes, by being volatilized without change. It detonates with nitre, requiring a very great proportion of nitre to finifh the detonation entirely. We obtain carbonic acid by this procefs ; and the nitre, befides being alkalized, contains fome ochre, and is greatly diminifhed in quantity. It would feem that thefe things happen, becaufe this combination of carbon requires a heat for its combuftion fo much greater than charcoal does, that the nitre muft be raifed to fo high a temperature that much of it is diffipated in vapour. Plumbago may therefore be confidered as a fpecies of charcoal, which has a fmall quantity of iron for its bafis, inftead of the fmall portion of earth and falt contained in common charcoal.

To free the folution of iron from this black matter, it muft be filtrated, and then evaporated. It gives green cryftals, called *green vitriol*, or *fal martis*. Great quantities of it are ufed in fome of the arts by the name of *copperas*; but it is made by a much cheaper procefs. Thefe cryftals contain much water. They undergo the watery fufion and fpontaneous calcination. The matter which remains, after the water is evaporated from them, is called *calcined vitriol*; and, according to the degree of heat employed, vitriol is calcined to whitenefs,—to light red,—to rednefs,—and laftly, we have *colcothar* of vitriol. The heat applied enables the iron to act powerfully on the acid, and to attract from it a part of its oxygen. Hence the metal becomes highly oxydated, and in that highly oxydated ftate it has a deep red colour. From this compound the vitriolic acid was formerly obtained; whence its name. And it is ftill ufed in fome proceffes, in which the action of the vitriolic acid is neceffary, as in the old procefs for aquafortis.

I prefume that you now fee the propriety of confidering vitriol, and other metallic falts, formed by an union with acids, as falts of a fecondary order of compofition; one of their ingredients, namely, the acid, being avowedly a compound of oxygen and another bafe. Hence arifes the difference of the falts formed of the fame metal and different acids. We muft therefore confider vitriol and colcothar as very different from the black *æthiops martialis*, from which the carbonic acid has been expelled. This is a compound of iron and oxygen, but colcothar is a compound of iron, and oxygen, and fulphur *.

The nitric acid, the next in order, is feldom applied to the metals in its ftrongeft ftate. In that ftate, it contains too little water to diffolve the compounds which it forms with the metals. In diffolv-

* There is here fome difference of opinion among the chemifts, or at leaft in the way of conceiving this circumftance. By proper treatment, all the acid may be expelled from the green vitriol by heat, and the iron is left combined with about onehalf its weight of pure oxygen. The name of *colcothar* is given alfo to this oxyd by feveral eminent chemifts; while others confider colcothar as this very oxyd, with this dofe of oxygen, but farther combined with a fmall portion of fulphuric acid. They

ing metals with it, we commonly employ it diluted with an equal quantity of water, and in this diluted ftate it is called *aquafortis*.

If we attempt to diffolve iron filings in aquafortis, in the fame manner as when we diffolve them in diluted fulphuric or muriatic acid, the action of the materials is rapid, the mixture becomes hot in a moment, and there is an outrageous effervefcence, or a fort of explofion. The acid is changed into elaftic red vapours and incondenfible gafes. The iron attracts the oxygen of the acid with too much violence, and in too great quantity. It decompofes too much of the acid, and is itfelf oxydated to fo high a degree, that it requires a very large quantity of the acid for its diffolution. The acid employed is, therefore, not fufficient for diffolving it, and the iron is only corroded.

To diffolve iron well in this acid, we muft moderate the action of the metal and acid on one another, by preferving them cool, and adding the iron very gradually, and with much lefs furface for the acid to act on. In this way, an ounce of aquafortis diffolves only 26 grains, fomewhat lefs than one-eighteenth.

There never is any inflammable air produced during this diffolution of the iron in the acid of nitre, nor in diffolving any other metal in that acid. The reafon is plain. The iron, while diffolving, is difpofed to attract a quantity of oxygen, but it eafily obtains it from the acid. The nitric acid abounds with this principle; and contains it in a more loofe, and feparable, and active ftate, than the fulphuric or muriatic acids do in their ordinary form. The iron, therefore, gets it

confider green vitriol as another oxyd, having a much fmaller dofe of oxygen (about $\frac{1}{17}$) combined with a much larger dofe of fulphuric acid. They think that iron has only thefe two ftates of oxydation, and that the apparently different ftates in which we fee it, are only mixtures of thefe oxyds. This doctrine or view of the matter is to be found very diftinctly expreffed in a differtation of Mr. Prouft on Pruffian blue. This variety of opinions is very embarraffing, when we read the explanations given of other phenomena; and a theorift may often fhift his ground in defcribing the internal procedure. **EDITOR.**

readily from the acid, and no part of the water feems to be decompofed, nor is any inflammable air produced.

When the iron is diffolved in this manner, it firft tinges the aquafortis of a fine green colour, and fome thin red vapours of the nitrous acid make their appearance. And by afterwards gradually adding a little more iron, the acid becomes faturated, and will not diffolve any more. Then the red vapours gradually difappear by flying away; and the colour of the folution changes to a pale brownifh yellow.

The green colour which appears at firft is therefore produced by the prefence of the nitrofe acid in the folution, fince the colour difappears as foon as this acid has time to evaporate entirely. You no doubt remember, that the nitrous acid always gives a green or blue tincture to water with which it is diluted, and that when this volatile acid is allowed to evaporate, the colour difappears.

You will now better underftand the reafon of the violent outrageous effervefcence which happens when, in place of taking this method, we throw a quantity of fine iron filings into the aquafortis. The acid is allowed, in this cafe, to act on a very extenfive furface of the iron, and to come into contact with a great quantity of it at the fame time. It therefore diffolves a fmall part of the metal, and corrodes the reft with the greateft rapidity, or in a moment of time. But this rapid diffolution and corrofion is attended with two accidents: 1mo, The fudden change of the nitric acid into the nitrous acid, by abftraction of its oxygen; and 2do, With the extrication of a great quantity of latent heat from the iron. The mixture becomes fo hot in a moment, that the glafs is too hot to be touched. This heat gives great elafticity, and a deep red colour, to the nitrous acid vapour; and thus is produced that fort of explofion.

We cannot obtain cryftals from this folution by evaporating it with that intention, but I have got remarkable cryftals from it by accident, of an amethyftine colour.

Chemifts have often obferved a fmell of volatile alkali in the de-compounding metallic falts by fixed alkalis and by lime. It paffed unheeded, being attributed to impurities in the alkali or lime, which are known to be generally contaminated by a minute portion of vola-tile alkali.

But, in the decompofition of a frefh-made nitrat of iron, this fmell is very remarkable, if the arbitrary circumftances of the experiment chance to be favourable. Thus, take two drachms of aquafortis, and dilute it with four of water, fo that the action on the metal may be very moderate. Diffolve two drachms of clean iron filings in this diluted acid, keeping all as cold as poffible, and when the folution feems ended, add three drachms of flaked lime. There will be little efferveſcence during folution ; and the lime will immediately occafion a very ſtrong fmell of volatile alkali. The formation of this gas is fo copious that we can even make its efcape vifible. Moiften the infide of a tall glafs (like a beer glafs, or ſmall cylinder) with oxygenated muriatic acid, and hold this over the mouth of the folution glafs, and it will inftantly be filled with a white cloud, and the fal ammoniac will fall down like fnow.

This is one of the moft interefting experiments in chemiftry,—and eminently inftructive. There was not an atom of volatile alkali in the materials. It muft, therefore, be produced by a feparation, and a new combination of fimpler fubftances contained in them. We can now guefs at the hidden procefs, by what we know of thofe materials. The filings of iron decompofe water, by attracting its oxygen, and fet the hydrogen at liberty. We know alfo, that, by the fame ftrong at-traction for oxygen, it decompofes the nitric acid, and fets its azote at liberty ; or, at leaft, in a ftate of redundance, as an ingredient of ni-trous gas. We do not know of any other fubftances being difenga-ged from their former combinations in this experiment. It would not, therefore, be unreafonable, though perhaps a little hafty, to conclude that thefe two fubftances have united, and by their union, compofe

volatile alkali. Dr. Prieftley, by paffing the electric fpark through volatile alkali a great many times, either extricated or produced a vaft
quantity of hydrogenous gas. It is very true that Dr. Auftin
(*Phil. Tranf.* 1788.) could not combine thofe gafes by heat, nor by
the electric fpark, fo as to produce ammonia: But he could not
expect it, fince the electric fpark feparated them. Scheele found that
when metallic oxyds are reduced by ammoniacal gas, a great quantity of azotic gas is *extricated.* His former experiments did not permit him to fay that it was *produced*, becaufe they had convinced him
that it is a primitive fubftance. Bergmann, therefore, was difpofed to
think that volatile alkali confifted of the two fubftances, azote and hydrogen.

Dr. Auftin could not combine the gafes by heat. This is not an objection. Heat combines oxygen and azote in the fame manner as it
promotes all combuftions. But here is no combuftion. There are
many other examples where the bafes of the elaftic gafes combine,
although the gafes themfelves will not, being hindered by their combination with caloric. I take this to be one. The two fubftances
meet in the very inftant of their extrication from their former combination. Dr. Auftin fucceeded, however, when only one of the fubftances was in the gafeous ftate. Iron filings, moiftened with water,
were put into a fmall quantity of azotic gas, confined by mercury. The
volatile alkali was produced.

The operation of the quicklime in the experiment which led me into
this difcuffion is very evident. It prevents the inftant union of the volatile alkali with fome of the acid; or it decompofes the nitrous ammonia as faft as it is formed. The want of this prevents the appearance
of the volatile alkali in every folution of iron in nitric acid.

On the whole, I confider this as a very diftinct fact in confirmation
of Bergmann's fagacious conjecture, and a corroboration of the proofs
which Mr. Berthollet has given of the fame opinion. I have already
mentioned feveral of them, in my account of hydrogen, drawn from

a variety of fources. I believe I forgot to mention a very diftinct one, which, I think, admits of no other explanation. When the vapour of oxygenated muriatic acid is paffed through the pure vapour of volatile alkali, both difappear, and we get a watery folution of common muriatic acid, and a quantity of azotic gas; and heat is produced. The redundant oxygen of the muriatic acid attracts the hydrogen of volatile alkali, and forms water, fetting the azote of the alkali at liberty. There was nothing elfe in the materials that could yield azote, nor any other way in which it could be detached.

The muriatic acid diffolves iron with effervefcence, occafioned by the production of inflammable air, which is the fame in quantity and quality, or very nearly the fame, with that produced by the fulphuric acid and iron. When the iron is diffolved and the acid is faturated, we have a folution which has a green colour, like the folution in fulphuric acid, and in which there is alfo fome *plumbago*, which makes it muddy, and muft be feparated by filtration. To have thefe phenomena in the beft manner, ufe the following proportions: Iron, one ounce; muriatic acid, fix ounces (equal in bulk to $5\frac{1}{4}$ of water); and add $2\frac{1}{2}$ ounces of water. Thus we have it of a clear green. It readily affords cryftals of the fame colour.

Thefe cryftals are not affected by heat in the fame manner as the vitriol or fulphat of iron. We can gradually expel the fulphuric acid from iron, by the force of heat, and the iron remains in the ftate of an oxyd. But when we attempt to expel the muriatic acid in the fame manner, a great part of the iron is volatilized by it. Thus, if we expofe a compound of iron and muriatic acid to heat in a retort, and gradually increafe the heat to the greateft degree which the retort can bear, no part of the acid arifes pure; but what rifes firft, however, is lefs charged with iron, and condenfes into a yellow liquor. What follows this is combined with more iron, and condenfes into fhining deep yellow or black fcaly cryftals, in the neck of the retort. In the bottom of the retort remains a mafs, concreted into fcaly cryftals *. It confifts of a greater part of the iron, retaining a portion of the acid, but not enough to give it volatility. All this agrees with other phenomena, in fhewing that metals have a ftronger attraction for the muriatic acid than for either of the other two foffil acids.

The muriatic acid diffolves iron readily alfo, when this acid is combined with water in its oxygenated ftate ; but then there is no effervefcence of inflammable air. The only effects produced are, that the iron is calcined and diffolved, and the acid returns to the ftate of common muriatic acid. The folution is, therefore, not different from one produced by diffolving iron in common muriatic acid, excepting that no inflammable air is produced. The reafon is evident. The oxygenated muriatic acid contains plenty of oxygen to fatisfy the attraction of the iron for this principle. No part of the water therefore is decompofed, nor any inflammable air produced from it. In this refpect, the oxygenated muriatic acid is fimilar to the nitric acid,—in the action of which upon metals inflammable air is never produced.

Thus, I have defcribed the action of the three foffil acids on iron, when applied to this metal in its ordinary metallic ftate. If the iron be previoufly calcined before thefe acids are applied to it, it can be diffolved by them in this ftate alfo, but with more difficulty, and in lefs quantity ; much more acid being required to diffolve an ounce of calx than an ounce of iron. And fuch diffolutions are performed without effervefcence, except when the calx happens to be combined with carbonic acid.

The muriatic acid diffolves the calcined iron better than other acids. But even of this acid a much greater quantity is required to diffolve calcined iron than metallic iron, and always a greater quantity in pro-

---

* This fublimate is deliquefcent. When mixed with thrice its weight of fulphuric æther, and digefted, the æther takes all the colour to itfelf. Seven parts of alcohol added to this, forms Beftufcheff's drops, a medicine of high character on the Continent. It becomes colourlefs by a few minutes expofure to the fun's light.

EDITOR.

portion as the iron is more calcined. The colour of the folution alfo is different from that of the folution of metallic iron. Inftead of being green, it is yellow. There is no inflammable air produced on this occafion : nor is it to be expected, for the iron being ufed in the ftate of a calx, or already combined with oxygen, has no power to attract oxygen from the water, or to decompofe any part of it.

This yellow colour, which is very like that of the folution of gold, has often mifled vifionary and ignorant chemifts, and made them imagine they had difcovered gold in earths and ftones, and other minerals, of which there are a great number containing iron in the ftate of a calx, and therefore difpofed to diffolve in muriatic acid, and to tinge it of the fame yellow colour.

The vegetable acids act alfo on iron, and by uniting with it, give it a foluble faline form. But they diffolve it much more flowly than the foffil acids, and require to be affifted by digeftion. They produce folutions of a green colour at firft, but changing, by time and the action of the air, to a ruffet, or rufty colour. This, or fome fimilar change of colour, happens to all the folutions of iron, when they are kept for fome time, and efpecially if they are expofed to the air. And if they were faturated with the iron at firft, they are fure to depofit a part of it, in the form of ochre, or of a rufty coloured calx. Fourcroy fays that the calx of iron, diffolved in the acetous acid, gives us a folution of a fine red colour. The feparation of part of the iron in this manner is much promoted by boiling fuch folutions in an open veffel, or by evaporating and diluting them repeatedly. By adding more acid, we can in moft cafes rediffolve the iron or calx, which is feparated, and make the folution clear again.

I have been long accuftomed to explain this feparation of part of the iron from its folutions and change of colour which thefe undergo, by fhewing that it proceeds from a farther oxydation which happens to this metal after it is diffolved. During the effervefcence which appears while the iron is diffolving in the fulphuric or muriatic or acetous acids, this metal is oxydated to a very moderate degree only ; but after it is diſſolved, it undergoes a farther oxydation. The diſſolved iron continues to act ſtill, though much more ſlowly than at firſt, upon the water, and to attract oxygen from a ſmall part of it, or perhaps from the air contained in the water. And the iron, by becoming more oxydated, has its attraction for acids diminiſhed. It requires a greater quantity of the acid to keep it diſſolved. Hence the quantity of acid which was ſufficient at firſt is inſufficient afterwards ; and ſome of the moſt oxydated part of the iron ſeparates in form of ochre or calx.

That this is a juſt account of the phenomena is proved by ſeveral facts. 1ſt, Recent ſolutions of metallic iron emit an odour of inflammable air, and continue to emit it, though not ſo ſtrongly, for a long time after they are made. 2dly, Changes of colour, which happen in them by keeping, are always ſuch as to end in the colour produced by diſſolving a calx in the ſame acid. 3dly, An old ſolution, which has ſuffered this change of colour, will reſume its primitive colour, if ſome freſh iron be diſſolved in it by means of a little additional acid. The freſh iron, while diſſolving, reduces the other to a leſs calcined ſtate, by taking from it a part of its oxygen. 4thly, The calces of iron, otherwiſe produced, are well known to be leſs diſſoluble than iron. They require a greater quantity of acid to diſſolve them, in proportion as they are more calcined. And they have much leſs attraction for acids than iron has which is leſs calcined.

From this change which happens to the ſolutions of iron, you will eaſily underſtand the origin of the great quantity of ochre, or yellow calx of iron, which mineral ſprings, containing vitriol of iron, depoſit in their courſe. In the ſame manner is produced the durable yellow ſtain which all the ſolutions of iron leave upon white cloths, on account of which ſome are uſed in painting linens. The metal is introduced in a diſſolved ſtate into the very ſubſtance of the fibres, and there, by the action of air, which calcines it to a great degree, it is depofited in the form of ochre, which never can come out until it is again diffolved. This alfo points out the proper method of removing thofe ftains, viz. by applying fuch acids as can fafely be ufed to re-diffolve the iron. The muriatic acid is the moft powerful diffolver of the calces of iron; and when applied weak, effectually takes out the ftain, without injuring the cloth in the leaft. Some vegetable acids alfo, fuch as the acid of tartar, are very effectual; and even the fulphuric acid may be ufed for this purpofe, if applied with caution, and in a proper ftate of dilution. The nitric acid has little action upon the calces of iron : It is rather difpofed to render them more difficult of folution; and is actually employed for this purpofe in fome proceffes.

You will alfo underftand the propriety of the procefs given in the laft edition of the Edinburgh Pharmacopœia for *Tinctura Martis* *.

Alkalis in their ordinary ftate fhew little power to act upon iron. We commonly evaporate watery folutions of the alkalis in iron veffels, when we wifh to have the alkalis in a dry ftate; and they neither corrode the iron nor are fenfibly tainted by it. If we add alkalis to folutions of iron in acids, they precipitate the iron. If the fulphat of iron is recently made, the precipitate is bluifh green. It is in that ftate of very moderate oxydation to which the iron was reduced during its diffolution. But if it ftand a while expofed to the air, or when the water in which the precipitation is performed contains plenty of good or refpirable air, the precipitated metal attracts more oxygen, and changes its colour, and becomes ochry or rufty coloured. Scheele found that the rapid exhibition of this colour was the fureft teft of the prefence of vital air in water. (*On Fire*, &c. No. 94.) It is evident that the yellow colour which the oxyd affumes is in confequence of its becoming more oxydated, or abforbing more vital air. It is alfo very much difpofed to abforb carbonic acid, when in this ftate.

In fome cafes of the precipitation of iron by an alkali, the precipitate can be re-diffolved by adding more alkali. The moft noted ex-

* Vide Difpenfatory.

ample of this is in the precipitation and rediffolution of the nitrat of iron. The blood red folution thus obtained is called the *tinctura martis alkalina* of Dr. Stahl.

But among the moft remarkable effects produced by alkalis on the folutions of iron, are thofe of the PRUSSIAN ALKALI, or LIXIVIUM SANGUINIS.

Lixivium Sanguinis.

This is a preparation of alkali difcovered by a Pruffian chemift named Diefbach. He evaporated blood to drynefs, and mixed it with half its weight of fixed alkali; and roafted or charred the mixture with a ftrong heat, till little or no more fumes arofe from it. He diffolved this charry matter in water, and filtrated the folution, which had a deep brown colour, and a ftrong difagreeable fmell. When the lixivium is thus cleared of impurities, and brought to a proper degree of concentration by boiling, it is then called LIXIVIUM SANGUINIS, PREPARED ALKALI, and PRUSSIAN ALKALI.

If this lixivium be dropped into a frefh made folution of iron in fulphuric acid, to which a few drops of muriatic acid have been previoufly added, we fhall fee each drop of the lixivium produce a fæcula of the moft rich and beautiful blue, which will flowly defcend to the bottom, and there collect. It may be feparated by decantation, and repeated wafhings with water. Without the muriatic acid, it is generally tainted with a brown or yellow ochry matter. This fæcula is manufactured for a pigment, by the name of PRUSSIAN BLUE.

The moft minute quantity of iron in a mixture becomes difcernable by this addition, which is therefore a trial of the prefence of iron in mineral waters.

The colour of this precipitate is not liable to fade in the air; neither is it altered by diluted acids. It manifeftly depends on a matter which is joined to the iron, and precipitates with it; the iron having received it from the alkali by a double elective attraction; the alkali combining at the fame time with the acid which had diffolved the iron.   Thefe are the manifeft facts.

This is further proved by an experiment defcribed by Mr. Lewis, in his *Materia Medica*, and alfo by Macquer.   In this experiment, this blue precipitate is decompounded by digefting it with a diffolved alkali.   The iron is fuddenly deprived of its colour, and the alkali fhews that it has attracted a part of the tinging matter, for it will now precipitate iron blue.   This fhews that a particular matter is combined with the iron in the blue precipitate, which matter can be taken from it and reftored again.

A careful examination of the phenomena in this experiment fhews that, on the one hand, the alkali poured on the Pruffian blue not only abftracts the colouring matter, but alfo diffolves fome of the oxyd of iron ; for an acid added to it will make it depofit fome Pruffian blue. This feems to happen, becaufe the added acid firft diffolves this oxyd, and the folution is immediately acted on by the Pruffian alkali, and thereby precipitated blue.

On the other hand, there is the fame evidence that the ochry matter left by the alkali is not an ordinary calx ; for when an acid is poured on it, we have a blue precipitate.   Mr. Berthollet fays, that after repeatedly wafhing the calx left by the alkali, with water, which comes off clear, we ftill at laft obtain a blue precipitate.   How this is produced will be explained afterwards, but in the mean time, we fee that fome of the tinging matter remains combined with the iron.

Therefore the alternate treatment with acids and alkalis, and the concomitant production and removal of the blue precipitate, muft have a period ; we fhall fee prefently that the colouring matter is ultimately decompofed by this treatment.

But we have alfo learned by experiments that there is not enough of this matter in the lixivium fanguinis, as it is ufually prepared, to convert the whole of the precipitated iron into blue precipitate, and that a good deal is in the ftate of an ordinary oxyd.   Hence we perceive the ufe of the muriatic acid in preparing the blue precipitate. It is neceffary for re-diffolving that part of the iron which is precipitated in the form of ordinary calx, and which fpoils the colour. There is no danger that the acid will diffolve any part of the blue precipitate: It refifts the action of acids in their diluted ftate.

The nature of this matter which is joined to the iron in the blue precipitate is fingular and furprifing. It appears to be a fort of ammoniacal compound. It contains a volatile alkali intimately combined with a very peculiar volatile matter, which is inflammable; the vapour of it readily taking fire, while at the fame time, it partakes alfo of an acid nature, or has an attraction for alkalis, alkaline earths, and metals.

But when this acid fort of matter is combined with alkalis alone, it adheres to them with a force which is very weak. It diminifhes very little their alkaline qualities, and is eafily feparated from them by the weakeft acids, even by fixed air. There is only one way by which it can be made to adhere ftrongly to the alkaline falts, that is, by adding to the compound fome of the metallic fubftances, particularly iron very flightly oxydated. A fmall quantity of the iron is diffolved by the compound of alkali and tinging matter; and thus is formed a compound of three ingredients, all of which cohere together much more ftrongly than the volatile tinging matter and the alkali did before by themfelves. Mr. Macquer firft gave us fomewhat of a diftinct notion of it, by demonftrating the double exchange which happens in the formation of the blue precipitate. Mr. Scheele difcovered the particular fubftance in which the colouring power refides, and inveftigated feveral of its chemical relations. His experiments are to be feen in his chemical effays, tranflated by Dr. Beddoes, 1786. Mr. Berthollet profecuted the difcovery of Scheele, and has greatly added to our knowledge of this remarkable fubftance.

The triple compound I juft now mentioned is formed at once, either by adding a calx of iron to the lixivium fanguinis, or by mixing a folution of pure fixed alkali with the blue precipitate, as in that experiment I mentioned a little ago ; and thus the alkali can be completely faturated and neutralized. The folution of it in water has a yellow colour, from the iron which it contains. It can be feparated from water by evaporation, and in fome meafure cryftallized into flakes or fcaly cryftals ; or it may be made to feparate from the water, and to cryftallize fuddenly by means of alcohol. If into a faturated folution of Pruffian alkali, fomewhat infpiffated and filtrated, we pour an equal meafure of alcohol, the mixture becomes immediately thick by fhaking, and the precipitated matter appears a foft granulated fubftance, like foaked rice. By long keeping, fome fpirit feparates to the top, and the granulated matter becomes fcaly.

I faid juft now that the common procefs for preparing the Pruffian alkali is to mix the alkali with dried ox-blood, and burn the mixture to a charcoal, from which the alkali is extracted with water : But feveral other ways to prepare this alkali have been difcovered.

All the animal fubftances, from which a volatile alkali can be formed or extricated by heat, will ferve as well as dried blood to be mixed and charred with the alkali. Horn fhavings, hair, wool, pounded bones, &c. are extremely fit for this purpofe. And Mr. Scheele prepared a very good Pruffian alkali, by making fixed alkali and charcoal duft red hot, and then thrufting down a little mafs of fal ammoniac to the bottom of the crucible, and continuing the mixture for a fhort while in a low red heat. This fhews that a volatile alkali is neceffary to the production of this tinging matter *.

This effect of the lixivium fanguinis on folutions of iron is the foundation of the art of preparing the colour called Pruffian blue, for painters, and other artifts. The chemift who firft difcovered it lived at Berlin, and kept it fecret for fome time, that he might fell the blue at an high price ; and hence it is called Berlin, or Pruffian blue. Many chemifts have made experiments on it, and endeavoured to inveftigate

---

* A mixture of chalk with the fal ammoniac greatly improves the procefs, and is the moft certain and expeditious method for obtaining Pruffian alkali. EDITOR.

the nature of it, and to improve the art of preparing it, of which you may see many examples in the Memoirs of the Academy of Sciences; but the greatest progress has been made by Mr. Macquer, by Scheele, and by Mr. Berthollet. (See *Dictionary of Chemistry, Scheele's Essays,* and the *Annales de Chymie,* also *Note* 18. *at the end of the Volume.*)

I must not, however, omit informing you that the Prussian blue of the shops has other ingredients besides the vitriol and Prussian alkali. This would produce a pigment almost black, so intense is its colour. A great quantity of alum is used in the preparation. Its acid adheres very loosely, and saturates the alkali which is not already combined with the Prussic acid. Of this there is always a great portion, even after the most careful preparation. The alum being thus decompounded, its earth, which is a snowy white, dilutes the intense blue, without discolouring it. It is better liked in this state : But surely the painters will find it much less powerful in mixture with other colours.

Mr. Macquer contrived a very ingenious method of depositing this most beautiful colour on cloth, paper, &c. The stuffs are first soaked in a solution of vitriol and alum. When thoroughly impregnated, they are dipped in the lixivium of Prussian alkali. Any ochry matter in the precipitate may be removed by dipping into a vitriolic or muriatic acid, greatly diluted.

Such are the different ways of precipitating iron from acids by alkaline salts. We may next subjoin to these precipitations an account of the effect of astringent vegetables upon the solutions of iron.

By astringent vegetables are meant such whose juices have the power to contract animal fibres, and which, when tasted, occasion a sensation in the mouth as if they produced that effect there. They seem to draw the mouth together. More or less of this quality is found in a great number of vegetables. It is remarkable in sloes, and many other unripe fruits, and in all parts of the oak,—the leaves, the bark, the juices, and the galls, which are a tumour produced by extravasation of juices. Galls are the most powerful astringents.

If a decoction of any of those vegetables be thrown into any solution of iron in an acid, the mixture immediately becomes a dark blue, or black, and its transparency is greatly diminished. This may be called a precipitation, although a great portion of the black fæcula will remain suspended for any length of time in close vessels ; for the greatest part of it may be separated by filtration.

I introduced the mention of this effect of astringents among the precipitations of iron, because, although it is not commonly a precipitation, but generally a mere change of colour, especially at first, there are however some astringents, logwood, for example, that actually separate the iron from the black sediment, especially when much diluted.   Besides, a little acid destroys the colour, and a little alkali restores it : Indicating, I think, that the astringent matter was imperfectly separated from the acid,—seeing that more acid destroys the colour, by bringing the iron into more perfect solution.   The added alkali restores things to their former state, merely by saturating the acid.

These experiments shew some sort of power in astringents to loosen the cohesion of iron with acids.   There are other experiments which shew that they have a disposition to unite with the iron themselves, and to dissolve a small quantity of it.   Several chemists, therefore, have chosen to consider the matter which thus affects the iron as an acid ; and they have named it the GALLIC ACID, because most abundant in nut galls ; and they think that the iron has a stronger affinity with this than with the sulphuric acid, and precipitates with it from vitriol.

A surprisingly small quantity of dissolved iron, and astringent vegetable matter, have a perceptible effect on one another in this manner. If a little infusion of green tea, for example, be spilled on a knife, it soon becomes tinctured with an inky blackness.   The astringent matter of the tea acts on the knife, and dissolves a minute portion of the iron, with which it produces that appearance ; or if a knife, or any thing made of iron, is laid on a wet table, the table, especially if oaken, is stained black where it is touched by the iron. Hence this experiment too, as well as that with the Pruffian alkali, is ufed as a teft of the prefence of iron in mineral waters.

The black colour alfo given to filk, wool, leather, &c. by the dyer, depends on this effect of aftringents and iron. The principal materials employed in thofe arts are always fome of the compounds of iron with acids, (generally vitriol or fulphat of iron) and fome of the aftringent vegetable fubftances. Common tanned leather requires only vitriol, becaufe it is already impregnated with the juice of oak bark, one of the moft powerful aftringents.

We alfo derive from thefe difcoveries the art of making the ink now ufed for writing, which is an object well worthy the attention of the chemift, as upon the perfection and durability of it may often depend the prefervation of important records, or valuable manufcripts. The art of making good ink has accordingly been ftudied by feveral chemifts, (*See Caperarius de Atramentis*) and by none fo thoroughly as by Mr. Lewis, who publifhed his experiments in his *Philofophical Commerce of Arts*, to which I refer you upon this article. He found that galls, when found, are the beft of all the aftringent vegetable fubftances, efpecially for durability. Logwood concentrated the colour, by combining more iron in the compofition, by nearly one-fifth. One part of recently made vitriol to three of galls feemed the beft proportion. More vitriol made the writing blacker at firft, but greatly impaired its power to refift the action of air and light. Gum-arabic both hinders the ink from fpreading, and proves a varnifh, which defends the compofition from the air. The ancients wrote with levigated charcoal, the moft indeftructible fubftance we know; and accordingly, the writings found in Herculaneum are ftill a perfect black. They were pigments. Their defect is, that being fuperficial, they may eafily be erafed, whereas the ink of modern times penetrates into the paper. Dr. Lewis combined the two inks very fatisfactorily, by mixing fome fine lamp black with his ink, in fuch quantity only as to make it very fenfibly blacker at firſt. This ink withſtood all action of the weather. It ran from the pen with abundant eaſe. I recommend this paper of Dr. Lewis as being alſo a good example of the chemical inveſtigation of any complicated ſubject, and for many judicious and uſeful incidental obſervations through the whole. I ſhall only add to what he recommends, the uſe of a ſmall quantity of cloves, and to keep the ink in a cool and dark place, and not too long. The reaſon of this laſt caution is, the corruption of the vegetable aſtringent. The effect of the cloves is to retard this, and to prevent the formation of mould upon the ſurface *.

* As vinegar has not the power to diſſolve and diſcolour this black compound, Dr. Lewis recommends vinegar or wine as the beſt ſolvent of the materials. It undoubtedly becomes more durable, but it is found to ſoften the nib of the pen ſo much that it is quickly worn away by the paper. The great art in ink-making is to have a ſuperabundance of aſtringent matter, to counteract the diſpoſition of the iron to a farther calcination, which renders the ink brown. It would be a great improvement in the manufacture of writing paper, if ſome aſtringent matter could be introduced. A little ardent ſpirits effectually prevents the ſpoiling of ink by keeping, but makes it ſink and ſpread. Corroſive ſublimate prevents mouldineſs *completely*.

A good proportion for writing ink.

Raſped logwood	-	-	-	- ·	1 ounce.
Beſt gall nuts in coarſe powder	-	-	-	3 ounces.
Gum-arabic in powder	-	-	-	2 ounces.
Green vitriol	-	-	-	-	1 ounce.
Rain water	-	-	-	-	2 quarts.
Cloves in coarſe powder	-	-	-	1 drachm.

Boil the water with the logwood and gum to one half; ſtrain the hot decoction into a glaſed veſſel; add the galls and cloves; mix and cover it up. When nearly cold, add the green vitriol, and ſtir it repeatedly. After ſome days, decant or ſtrain the ink into a bottle, to be kept cloſe corked, in a dark place.

A ruddy friable gum reſin, not unlike gum-kino, from New Zealand, was given me by a friend. Finding its taſte remarkably aſtringent, and the whole very mucilaginous, I mixed ſome of it with ink powders, and found that it improved the ink in a very remarkable manner, both deepening the colour, and giving it body. Writings with it, and with ink made by Dr. Lewis's receipt, on a card, being expoſed to the weather on a ſouth wall, this addition was found to have encreaſed its durability ſurpriſingly.

EDITOR.

Of the compound falts, I obferved formerly that none are fo remarkable for their effects on the metals as nitre and fal ammoniac.

Nitre deflagrates violently with iron in a ftrong heat, and is alkalized, or its acid is decompounded and diffipated, as in the other cafes of deflagration with this falt. The alkali remains mixed with the calx of the metal, and is cauftic, in confequence of which it acts on the calx, and unites with a part of it,—rendering it foluble in water to a certain degree along with itfelf; but this folution is not permanent.

Sal ammoniac and this metal act remarkably on one another. I pointed out the reafon of this in treating of metals in general, namely, the volatilizing power of the muriatic acid. Common fal ammoniac contains the acid which has the ftrongeft attraction for the metals; and this acid is combined with the weakeft of the alkaline falts,—the alkali which has the weakeft attraction for acids in general. Accordingly, if we mix together equal weights of fal ammoniac and iron filings, and grind thefe materials well in a mortar, they foon after begin to act;—the matters grow warm,—the more readily, if a little humidity be added. The fal ammoniac begins to be decompounded; the acid joins itfelf to the iron; and the volatile alkali is detached, and fhews itfelf in a feparated ftate by its pungent odour. There is alfo feparated at the fame time a quantity of inflammable air, the odour of which mixes itfelf with that of the volatile alkali. This inflammable air is produced by the decompofition of fome of the water. The action of thefe materials, and the decompofition of the fal ammoniac, are not complete, however, until we apply heat to the mixture in glafs veffels. Then the fal ammoniac is totally decompounded, and a cauftic volatile alkali rifes, which it is almoft impoffible to condenfe completely, on account of its being accompanied with inflammable air. The whole of the acid remains combined with the iron in the retort.

All thefe effects are alfo produced on fal ammoniac by the calces of iron, with this difference only, that it requires a larger quantity of the calx to decompound the whole of the fal ammoniac, and that, during the action of the materials, there is no heat or inflammable air produced.

But when fal ammoniac and iron are made to act on one another, it is not ufual to take fuch a large proportion of iron, or of calx of iron, as will decompound the whole of the fal ammoniac. There is a preparation obtained from fal ammoniac and iron, ordered in both the London and Edinburgh Pharmacopœias, under the title of FLORES MARTIALES. This is produced by mixing with the fal ammoniac fuch a quantity of iron filings, or of a calx of iron, as is fufficient for decompounding only the half of the fal ammoniac; and this mixture is expofed to heat in proper veffels. Firft, we get a cauftic volatile alkali, detached from that part of the fal ammoniac which is decompounded. After this has come over, the heat is increafed, to make the entire part of the fal ammoniac arife, which requires a much ftronger heat. The vapours of it condenfe into a folid matter, or what is called FLOWERS; but while it arifes in this manner, it carries with it a part of the compound of acid and iron that was formed in the beginning of the operation. This compound is in fome degree volatile, when there is enough of acid combined with the iron. A part of it, therefore, rifes, and is condenfed with the fal ammoniac, and forms the reddifh and yellow flowers to which we give the name of *flores martiales.*

It is found that an oxyd, inftead of the metal in its pureft ftate, enfures the defirable appearance of this preparation,—the red and yellow colour of the flowers. Therefore, when filings of iron are employed, it is of confiderable ufe to expofe them for fome time to the damp air of a cellar. An oxyd ftill more replete with vital air, fuch as colcothar is, more certainly produces red flowers. *(See Note* 19. *at the end of the Volume.)*

Iron in its calcined ftate can be mixed in fufion with vitrified earths;

and the glafs compound is tinged of various colours, according to the quantity of the iron, and the degree to which it is calcined. The colour which this metal moft readily communicates is a green, like that of its folutions in acids. Such a green is feen in the common glafs of which bottles are made, and in crown or window glafs; and the green phials of the apothecaries are tinged with iron. But when the calx of iron is very highly calcined, it will fometimes give a dull yellow colour, and a red; and in all cafes, when much of the iron is added to the glafs, the colour is increafed or darkened, fo much as fometimes to produce almoft a blacknefs.

It is owing to the prefence of this calx, in fome degree or other, in almoft every foffil body, that our clays are in general fo much difpofed to vitrify, when urged by great heats. A very minute portion of it gives an incipient vitrification to the beft of them, and produces the compactnefs that is fo defirable in all, with very moderate heats. It is this which gives the red colour to our common brick and tile.

Such bricks are unfit for the conftruction of furnaces, becaufe a moderate heat makes them run into flag. The greenifh yellow colour of the fmall Dutch bricks is not owing to a want of iron, for they contain a great deal, but to the prefence of magnefia. The greyftock bricks of London have fupplanted the red bricks, which are now to be feen only in the old houfes, but are inferior to them in their power to refift the weather.

Experiments made with the different earths have fhewn that the flinty is the moft difpofed of any to unite with calx of iron in the fire, and be melted by it. Hence the ufe of it in welding iron.

The effects of fulphur and iron on one another deferve particular notice.

Sulphur unites very eafily with this metal in the dry way, or by fufion, as with moft others. And it penetrates iron more quickly than moft other metals, fhewing thereby a ftronger attraction for it. A bar of iron, thruſt red hot into melting ſulphur will preſently melt and drop off like wax held in hot water, with a bright dazzling glow.

Alſo, if one part of flowers of ſulphur be ground with five parts (by weight) of fine iron filings, and the mixture be put into a thin glaſs veſſel, ſuch as a matraſs, or cucurbit, and if this be held over clear burning coals, we ſhall obſerve the ſulphur begin to ſhew ſoftneſs very ſoon, ſticking to the ſide of the glaſs, long before the heat has got to that degree that would otherwiſe produce this effect on ſulphur unmixed. It does not riſe much in its temperature for a good while. Then, it ſuddenly becomes fluid all over, and the filings almoſt undiſtinguiſhable; and preſently, the heat riſes rapidly, the maſs fixes, and becomes red hot. This is one of the beſt examples of the emiſſion of latent heat. It takes place when the chemical combination is accompliſhed; and this required previous fluidity. But the temperature is too low for the compound being fluid; it therefore congeals, and latent heat emerges. If there be too much ſulphur, this takes fire. This compound of iron and ſulphur is a hard brittle ſubſtance, of a metallic appearance.

In conſequence of this diſpoſition to unite with ſulphur, iron is often employed to ſeparate ſulphur from other metals, in the way of fuſion. The iron, uniting with the ſulphur, forms a very fuſible compound, which flows uppermoſt: The other metal ſubſides to the bottom.

But beſides the way by fuſion, there is another way of applying ſulphur and iron to one another, ſo as to make them act, It may be called the humid way. The iron is taken in filings, and the ſulphur in powder, and they are mixed together with as much water as makes the mixture into a paſte.

If this mixture be ſhut up cloſely, to prevent its communication with the atmoſphere, the ingredients begin, after ſome hours, to act upon one another,—the iron is corroded into a black mud or paſte, in which the ſulphur and it are ſomehow combined, for the yellow colour of the fulphur entirely difappears.   This pafte has a furprifingly ftrong attraction for vital air, and can be employed to abforb it when we wifh to know what quantity of vital air is contained in atmofphe- rical air, or other fuch mixtures of different airs.   Mr. Scheele ufed it as a eudiometer, in a fet of experiments which he continued for a whole twelvemonth, to learn what proportion of vital air was contain- ed in the atmofphere of Stockholm in all the different feafons and months of the year.

A drachm of flores fulphuris, mixed with two drachms of filings of iron, and as much water as will make it a thick gruel, muft be put into a little difh, connected with a piece of cork, fo as to float on wa- ter.   If we invert upon this a jar capable of holding 66 ounces of water, (nearly two quarts), all the oxygen contained in the atmofpheric air, which fills this jar, will be abforbed by the mixture of iron and fulphur,—the water of the tub rifing up into the jar, and occupying its place.   When the filings are extremely fine and clean, and are mixed with the dry fulphur by grinding, and then a moderate quantity of water quickly mixed with thefe materials by kneading brifkly, and this preparation put into a hot floating difh, it will abforb the vital air fo rapidly that it will take fire.   This may be enfured, by fmearing with this magma fome fmall twigs.   This expofes a great furface, and foon heats.

If fome pounds of this mixture are expofed to the atmofpherical air, they will attract the vital air fo faft from it as to grow hot and to take fire.   They grow hot alfo, although they be not expofed to the open air, if the quantity of them be fo great as 50 pounds or more; for, in this cafe, there is fo much air entangled in the pafte in making it up, that the abforption of this air is enough to produce a confider- able heat, and even to fire it.

After this abforption and heat is over, if we examine the remaining matter, we find the fulphur changed more or lefs into fulphuric acid, which remains combined with the iron, and forms with it a vitriolic

compound, more or lefs perfect. It is evident that the vitriolic compound is formed in confequence of the ftrong attraction, both of the
fulphur and of the iron for the oxygenous principle, which they draw
partly from the water, but principally from the air. By this principle
the iron is flightly calcined, and the fulphur is changed, at leaft in
part, into fulphuric acid. The decompounded water, when there is
enough of it, gives inflammable air, and the air from which the oxygenous principle is attracted, gives out a great part of its latent heat
or caloric, in the form of fenfible heat, which heats the materials
fometimes to that degree as to make them take fire. When the materials are in large quantity, and covered up, or buried, fo as to have
little communication with the air, there is more water decompounded,
and more inflammable air produced. When, on the contrary, the
quantity of them is fmall, and they are confined with a quantity of
air, they draw the oxygenous principle chiefly from that air, and decompound the whole air which the veffel contains *.

A fimilar change of fulphur happens in fome kinds of pyrites,
which are natural compounds of iron and fulphur. And the difcovery
of this has given origin to the art by which common vitriol, or the
fulphat of iron, is prepared in large quantities, and at a cheap rate.
I formerly obferved, that the manner of preparing it, by diffolving
iron in fulphuric acid, is only practifed as a procefs in pharmacy, and
the vitriol, or *fal martis*, fo prepared, applied only to purpofes of
medicine, but that a much cheaper way of obtaining this compound
has long been practifed, to fupply the confumption made of it in feveral arts. By this cheaper method, it is prepared chiefly from fome
particular kinds of pyrites, which, when expofed to the action of the

---

* Confult on this fubject Mr. Lavoifier's Effays, publifhed in 1777, where the whole
procedure is analyfed with great precifion and diftinctnefs. Le Sage, *Elem. de Chymie,*
i. 42.—Werner on the *Origin of Volcanoes.*—Hopfner's *Magazine of Natural Hiftory.*—
Lemery, *Mem. Acad. Paris* 1700.

air, and of humidity at the same time, undergo a change, by which the sulphur becomes sulphuric acid, and we get a vitriol, or sulphat of iron, in place of the pyrites, or compound of iron and sulphur. Humidity alone is not sufficient. The pyrites has not the power of decompounding water. Cramer says he observed immense quantities of pyrites all along the English shores, especially about Harwich, which naturally lie in a bed at the depth of some feet below the surface of the sand. As long as they lie there, though they are constantly wet with sea water, they continue entire and quite insipid, but when thrown up to the air by the waves, they are crumbled down, and converted almost entirely into a heap of little crystals, in a fortnight. This is indeed the case every where, for the pyrites is always exposed to abundance of humidity in the bowels of the earth, but nevertheless retains its form until it is dug up and exposed.

This change, which the pyrites of iron is liable to undergo, is called *vitriolization*,—it is said to *vitriolize*.

It is proper to remark, however, that only particular kinds of pyrites are liable to it, not every kind. Some kinds, suchas the cubical pyrites, common in slate, and some others, withstand the action of air and humidity a long time without suffering a change, or if they are changed, it is very slowly, and only into a hard rust at their surface, but without affording vitriol.

The cause of this difference among the varieties of pyrites has not been discovered. I suspect that it depends upon the proportion of the sulphur to the iron, and that those which contain most sulphur are most liable to vitriolization.

Process for Vitriol from Iron Pyrites.

The account which I have given you of the chemical procedure will naturally direct you to the proper methods of accelerating the operation. I have only to remark, that some kinds of pyrites require a preparation, by roasting in a heap with a small quantity of fuel. This is found to dispose the harder pyrites to a more rapid decomposition;

probably opening their texture by the expulfion of fome of their volatile ingredients.

The pyrites being thus in a proper ftate for vitriolization, the reft of the procefs is the fame for all kinds. A piece of ground is chofen, having a very gentle declivity, juft enough to make the water run in one direction. This is laid out in narrow quarters, like a kitchen garden,—the furrows which feparate them having the direction of the flope. The ground between is covered with firm clay, well beaten, fo as to be quite impervious to water. The furface of each quarter has a flope on each fide toward the furrows. Another furrow is led acrofs the lower extremities of thofe which feparate the ridges; and this furrow terminates in a great ciftern.

The pyrites, broken into fmall pieces, are fpread on the quarters or ridges, and made up into beds about two feet thick. They efflorefce; and the rains wafh off the faline cryftals. The folution trickles off into the gutters, which conduct it down the flope into the crofs drain, and by it into the ciftern. In very dry weather it may even be neceffary to fprinkle the pyrites on the ridges. As the folution is very weak, efpecially in the rainy feafons, it is frequently thrown again on the vitriol beds by a forcing pump, and it comes off them ftill more impregnated with the falt, till at length it is fit for boiling.

For this purpofe, it is pumped up into the boiling houfe, which is built over the ciftern, and is reduced to a ftrength proper for cryftallization, by boiling off the fuperfluous water. From the boilers the liquor goes into the vats or cifterns, where it is allowed to cool and to cryftallize.

There are fome pyrites which are fo much difpofed to vitriolize, that heaps of them are liable to grow hot and take fire; and this has happened on different occafions to fome kinds of pit-coal abounding with pyrites, and which had been collected together in large quantity. At Ayr, for example, an immenfe body of coal, built up for exportation, took fire; and in Dublin it happened to a large quantity of Englifh coal, which had been provided by the corporation to prevent scarcity; and this happened in both cases after a heavy shower. Vessels loaded with such coal have unfortunately been set on fire sometimes in that way.

In many coal pits the pyritical rubbish, which is mixed with the coal, and would spoil its sale, is picked out and left below. When these pyrites have been accidentally in too large heaps, they have frequently taken fire, and have set fire to the coal works. George Agricola, who wrote in the 15th century, speaks of the fire in the coalwork at Dysart in Scotland as a very old thing. It is not yet extinguished. At Kilkerran, in Ayrshire, a very extensive coal-work has been burning now for fifty years. At Johnstown, near Paisley, a stratum near seventy feet thick was set on fire by the rubbish, and continued burning with incredible fury, having a great current of fresh air. In one place there was a face of coal red hot for near 100 yards, and the flame ascended through one of the pits to the height of above 100 feet. It was at last extinguished by drowning it; but whenever the air is admitted, it takes fire in a few days.

From such examples some have endeavoured to account for hot springs, subterraneous fires, volcanoes, and earthquakes. They have even attempted to imitate some of these convulsions of the earth. Lemery was, I believe, the first who attempted this imitation. (*See Mem. Acad. Par.* 1700.) It has often been done since. The general process has been, to mix as large a quantity as can be conveniently had of clean iron filings, with somewhat more of sulphur, and as much water as will make a firm paste, and to bury this in the earth, (first wrapping it up in a cloth) and ram the earth firmly above it. In a few hours it grows warm, and swells so as to raise the ground. Sulphurous steams make their way through the crevices, and sometimes flames appear. Rarely is there any explosion; but when this happens, the fire is vivid and brandishing,—and the heat and fire continue for sometime, if the quantity of materials has been great.

We cannot expect more,—we fee the production of internal heat, fufficient for more violent operations,—but other circumftances muft concur. If water get in by a crevice into a cavern containing red hot materials, there feems no bound to the effects which it may produce.

The phofphorus of urine can alfo be combined in fmall quantity with iron, by melting the phofphoric acid with iron and charcoal duft. The compound is more fufible than iron, and is not decompounded but with great difficulty, a thing we fhould not expect of a fubftance fo volatile and inflammable. It makes iron cold-fhort, that is, it will not bear hammering out when cold, but cracks all over.

Relation of Iron to other Metals.

Iron can be mixed with all the other metals, excepting lead and mercury. It fhews a difpofition to unite the moft readily and ftrongly with arfenic, and with the regulus of antimony, and with gold. None of the mixtures of it with metals have been found ufeful, except perhaps manganefe, which is fuppofed to improve iron, but its effect is not fufficiently afcertained by experiments.

Iron is, in general, much impaired in its moft valuable qualities, cohefion, malleability, and toughnefs, by any metallic mixture. Arfenic in particular unites moft readily, and adheres moft obftinately, and makes it red-fhort, that is, apt to crack under the hammer while forging, and even to fall in pieces.

Ores of Iron.

Iron is feldom found pure and metallic, but examples of this however have occurred. A mafs of 1000 pounds weight was found in Siberia: Its interftices were all filled up with glafs, of a beautiful bright green, and fo friable as not to cut the fingers. This circumftance fhews that it has been the effect of fire, and that the glafs has been quenched in water while yet red hot. Another mafs, much larger, has been found in Paraguay. Such maffes are faid to be not uncommon in Senegal.

(See Note 20. at the end of the Volume.) Mr. Margraaf has found malleable iron in ftrings at Eibenftock in Saxony. Thefe are the only
known inftances; but the ores of this ufeful metal are plentiful in all
parts of the world. And befides, iron is found mixed with many of
the other productions of nature, fuch as the ores of other metals, coloured clays, and boles. Indeed few earths are free from it. The
garnet, the emerald, the ruby, topaz, fapphire, and amethyft, appear to
derive their colours from it. It is found even in the afhes of vegetable and animal fubftances,—in the blood, milk, flefh, and hair. It
has been found in plants which have been raifed from a feed in diftilled water. Some foffils containing iron are of great ufe, although not
as ores of iron, fuch as emery for the lapidary; hæmatites for the bur-
nifhers; the red, yellow, and blue ochres for the painters; and the
loadftone for the navigators.

As the ores of iron are very common, only the richeft or moft profitable are wrought. In this country, the minerals melted to extract
iron from them are diftinguifhed into three kinds.

1ft, Iron ores found in veins.

2d, Iron-ftones, interfperfed through ftrata, or forming thin ftrata
among others.

3d, Bog ores, intimately blended with clay, and with the remains of
animal and vegetable fubftances; not mixed, but in many cafes chemically combined.

All thefe are in general oxyds of iron, in different ftates of oxydation; and fome are combined with fixed air, and with fmall quantities
of fome of the earths. When they are much oxydated, and nearly
pure, or with very little of the earthy fubftances combined with them,
they have a dull and deep red colour, which is the natural colour of
the oxyd of iron highly oxydated. Or if fome fixed air be united with
them at the fame time, they are yellowifh: Thefe are the fpathofe
ores of iron, or carbonats of iron. If, on the contrary, they contain
the iron but little oxydated, they are grey or black, or approach by

their colour to iron itſelf. They are alſo variouſly cryſtallized; of theſe the bloodſtone, or hæmatites, is the moſt beautiful. Their appearance is alſo very much affected by the mode of their concretion, and by the admixture of earthy matter. White ores and micaceous ores often contain manganeſe, and are thought to yield excellent iron.

The hard pyrites, though it contain much iron, is not treated as ſuch: It is conſidered only as an ore of ſulphur.

To extract the iron, the minerals are firſt roaſted. This operation appears to be neceſſary or uſeful, in extracting iron from many of the minerals from which it is got. It is probably uſeful by expelling ſulphur or ſulphuric acid.

The next operation is to reduce the roaſted ore to a metallic ſtate, or metallize it completely.

This is done in a high furnace, ſhaped internally ſomewhat like two round crucibles joined by their mouths, the uppermoſt without a bottom. The bottom of the furnace is hollow, having on one ſide a hole through which the blaſt of a bellows can be directed a few inches above the bottom. On the other ſide is a larger hole, through which the melted metal can run entirely off. A heap of bruſhwood is thrown into the furnace. It is kindled, and immediately covered with charcoal or coak; and then baſkets of ore and fuel are added till the furnace is filled,—the bellows being worked all the while. The tap-hole of the furnace is ſhut up with clay. Every thing being looſe and open, the flame is forced by the blaſt of the bellows through the whole furnace, and a violent heat is raiſed all over it. As the fuel conſumes, the matters ſubſide; and baſkets of ore and fuel are repeatedly thrown in to fill up the furnace. The earthy matters and ore are melted by the intenſe heat into a glaſs; and ſince the burning fuel is every where in contact with the ore, and the coal has a much greater attraction for the oxygen than the iron has, the metal is reduced, and ſinks to the bottom of the furnace under the melted glaſs produced by the fluxing of the earths. The metal is now defended from the action of the bellows. When the workman thinks that there is enough collected, he makes a hole in the clay with the tap-iron, and the metal runs out, and is followed by the glaſs or ſlag.

A conſiderable quantity of limeſtone is mixed with the ore in this operation ; ſeldom leſs than one-ſixth of the ore. Its uſe is partly to promote the fuſion of the earthy matters mixed with the metalline oxyd, and partly to combine with the ſulphuric and ſulphurous acid ; by which means it is carried off in a vaporous form, but takes with it ſome iron. It is thought that a better quality of iron is obtained by the large uſe of lime, even when not required as a mere flux.

It is alſo found, in all caſes, of great advantage to the quality of the iron to uſe a very great quantity of fuel. This, unqueſtionably, produces a more complete abſorption of the oxygen and reduction of the metal. A great allowance of fuel brings more metal from the ſame ore, and makes it more fuſible after repeated meltings. It is very grey, ſhewing ſhining black facets, and talky-like matter, and is ſofter and eaſier cut than white caſt-iron which has been made with leſs fuel. But the room occupied by the fuel cauſes the furnace to produce leſs iron per day ; while the blaſt or quantity of air conſumed remains the ſame. Therefore, this iron is much more coſtly. In the Scotch furnaces, the iron-ſtone requires between three and four times its weight of ſoft pit-coal, and each ton of iron produced requires the conſumption of 672,000 cubic feet of common air. Some fuels, particularly blind-coal, require much leſs. Much more is required in ſummer than in winter.

The quality of the iron is affected alſo by the nature of the fuel. Wood charcoal is beſt for producing iron to be refined and made into bars. Charred pit-coal is beſt for iron to be employed in caſt work.

When firſt melted down from the ore in the great furnaces, the metal is in the ſtate called pig-iron, or pot-metal, and is fit for caſt work ; in which ſtate it is perfectly fuſible, but is either quite inflexible and brittle, or has very little flexibility and toughneſs. It is uſeful, however, for many purpoſes. In conſequence of its fuſibility, it can be caſt in moulds, to form large maſſy pieces, and inſtruments of a very great ſize, which are thus formed at a far leſs expence than if it were neceſſary to forge them; beſides that ſome, on account of their large ſize, could not be forged. But fuſible iron is unfit for anchors, and other ſuch inſtruments, which muſt be exceedingly ſtrong and tough.

I ſhould obſerve, in this place, that the caſt iron obtained by this proceſs appears in three ſtates, which are very diſtinguiſhable, both in appearance and qualities, according to the manner in which it has been manufactured.

1ſt, WHITE CAST-IRON, extremely hard and brittle. It can not be filed, or bored, or repaired in any way, nor bend. It cannot bear very ſudden changes of temperature, without cracking. Its ſtructure is cryſtallized, with very ſmall brilliant facets.

2d, GREY CAST-IRON, of a granulated texture, and often plated, but of an unequal dull colour. It is much more coheſive than the former; and is therefore employed for artillery. It is alſo ſofter, and may be cut and bored, and even turned in a lathe.

3d, BLACK CAST-IRON, the moſt unequal in its texture, the moſt fuſible, and therefore often mixed with the white to make it ſtand a repeated melting. It is leſs tenacious than either of the former.

The firſt ſeems to contain much unreduced oxyd, and even ſome earthy matters, which vitrify along with it. Theſe are entangled among the truly metallic parts, and thus defend them from the action of the air. The ſecond kind has little unreduced oxyd, or earthy matter, having been prepared with more fuel and greater heat, but there is more of the inflammable matter of the fuel combined with it. The laſt alſo contains much of this, and derives other varieties from the nature of the ores. It makes the beſt refined or malleable iron.

Iron is changed from this fuſible and rigid ſtate into tough and malleable iron, in the forge, or finery as it is called, and by an operation which is conſidered as a refinement of it.

Until lately, the only manner of performing this operation was by heating two or three hundred weights of it at once, in a hollow forge, in which the iron was covered with a fire of charcoal, and lay in a bed of charcoal duft and afhes until it juft melted, with no more heat than what was barely fufficient; and it was kept in this heat for fome time, —the wind of the bellows being directed through the charcoal, fo as to have fome effect on the furface of the metal; and being ftirred now and then, it was taken out a toughifh cavernous mafs, and was compacted by hammering.

But within thefe few years another procefs was contrived by a Mr. Cort, which admits of an immediate infpection and examination of all the phenomena; the metal not being covered with the fuel, but expofed to the view of the operator and obferver.

Here the iron is melted in a reverberatory furnace, and a ftrong flame is kept blowing over it, fo as to keep it in a moft intenfe heat. It is all the while ftirred about with a rake, fo as to bring every part of it to the furface in fucceffion. By this treatment certain impurities of the iron are diffipated in a way that we do not yet underftand. The fact is, that it gets into a particular ftate, which is indicated without any danger of miftake, by the iron becoming thick like gruel, fwelling up a little, and the furface of it all at once becoming of a dazzling brightnefs, by a low flame which gleams upon it. It is now faid to be *brought into nature*.

When the iron is thus *brought into nature*, that is, has loft its fufibility, and is become malleable, it is collected into two maffes in the *laboratorium* of the furnace, by raking it together, and compacting it with the ftrokes of an iron club or mace. Thefe maffes are called *loops*. They are very porous and fpongy, and contain a great deal of fcoriæ, or femi-vitrified matter, mixed with the iron. The heat to which they are expofed is then increafed for a fhort time, until thefe fcoriæ are completely melted, at which time the loop is taken out, and immediately expofed to the ftrokes of a large and heavy hammer, wrought by a water wheel or fteam engine. This comprefses the mafs, unites the parts of it more clofely, and forces out the greater part of the vitrified matter, or *finery cinder*, as it is called, out of its pores. The mafs is turned under the hammer, until it is formed into a fhort and thick prifm or cylinder called a *bloom*, the parts of which are not yet fo clofely united as they ought ; but by heating it again, and fubjecting it again to the great hammer, or pafling it between rollers contrived for this purpofe, the parts of it are made to enter into clofe cohefion, the liquid drofs completely fqueezed out of its pores, while it is at the fame time forged into bars of different kinds.

The iron made by this procefs, from very ordinary iron, fuch as was known to yield bad bar iron, proved equal to the beft Swedifh iron. Very fevere trials were made by Government, as it was recommended for the ufe of the navy, for anchors, bolts, ftraps of blocks, and other ufes, where toughnefs was an indifpenfable requifite. But, to make bad iron fo eminently good, required much labour and expence. Caft iron, which yielded better bar iron by the ufual procefses, was amenable to the fame degree of finenefs with much lefs trouble. I mention this circumftance, to fhew that the procefs is general, and depends on principle.

By this procefs, the expence of wood charcoal, which is a very dear fuel in this country, is faved ; the whole operation being performed by the flame and heat of pit-coal.

By whatever procefs fufible iron is refined to the ftate of tough iron, a confiderable part of it (one-third commonly) is fcorified or calcined, forming a fufible calx or oxyd, called FINERY CINDER. It is exactly fimilar in its appearance to the fufible oxyd formed from iron when it is burnt in vital air. The production of this finery cinder is abfolutely neceffary ; and the fkilful workmen fay, that for the right performance of this procefs, it is neceffary that the refining iron be well foaked, and mixed to a certain degree with this finery cinder; for which reafon, although they let fome of this cinder run out at a tap-hole of the furnace or forge, when it accumulates too much, they never let it all run out, but retain a proper quantity in the furnace or forge all the time they are working.

But it is alfo certain that the iron lofes a quantity of inflammable matter, which is contained in it while in its hard and brittle and fufible ftate.

When we try to diffolve fufible iron in acids, we find it more difficult to diffolve it, and there is an incomparably greater quantity of plumbago left from it than from tough iron. You know that plumbago is principally compofed of carbon, or carbonaceous matter; but there is, befides, often fome fulphur or fome phofphorus. The change of fufible into malleable iron appears, therefore, to depend chiefly on the deftruction of the plumbago, or of the greater part of it. And it is eafy to fhew in what manner this happens in the refinement of iron. Dr. Beddoes has endeavoured to explain how it happens in Mr. Cort's procefs, and he has explained it in fome meafure, but affumes a principle which I cannot admit.

The principle affumed by Dr. Beddoes and the French chemifts is, that the unrefined fufible iron, befides containing a quantity of carbon, alfo contains iron in the ftate of an oxyd, and that the oxygen of this ill reduced iron acts on the carbon, and produces the increafe of heat, and the intumefcence, or fermentation, and the eruption of elaftic fluid and of flame.

But it always appears to me inconceivable, that there fhould be prefent in fufible iron, at the fame time while it flows white hot from the great furnace, both an abundance of carbon and an oxyd of iron, without acting on one another originally, efpecially in the grey iron, which moft abounds in carbon.

There is no occafion for having recourfe to fo unwarrantable a fuppofition. It is evident that the effect of a great part of Mr. Cort's procefs is to calcine a part of the iron in the firft place. While the workman is bufily employed in ftirring it over from one fide of its bed to the other, feparating it into fmall parts, and mixing the whole carefully together, a quantity of the iron (to which the air is admitted all the while) muft be calcined or oxydated ; he then increafes the heat a little, to make the oxygen of this calcined iron act on the carbon of the uncalcined. This increafes the heat in the iron, and produces the fermentation, or formation of carbonic acid gas, which, while it rifes, carries fome carbon with it,—and hence the blue flame. The nicety of this operation depends on knowing how far to carry the calcination of part of the iron, that there may be enough of oxygen to confume the whole of the carbon. If more iron than enough be calcined, it is fo much loft.

Such a perfuafion fuggefted to Mr. Reaumur a project to foften or give toughnefs to caft iron by cementation. He conceived iron to differ from other metals by combining with much more inflammable fubftances than was neceffary for its ductile form, and expected to abforb or confume this by proper fubftances. He therefore cemented caft iron with earthy powders ; and after the cementation is carried a certain length, the iron yields to the file, if flowly cooled, but if tempered, is quite hard. It has fome malleability when red hot, but none when cold. If it be ftrongly heated, it flies to pieces under the hammer. Thus it plainly refembles fteel. Continuing the cementation longer, it becomes now fo foft as to be eafily cut, filed, bent, or forged, and does not become much harder by temper. In proportion as it becomes foft, it lofes its fufibility, and the internal parts can be melted out ; fo that in every refpect it is fimilar to forged iron, only it is not fo compact or folid. All thefe changes feem accountable by a gradual exhalation of the carburet included in the crude iron. This may be affifted in Reaumur's proceffes by the cementing matters, either by their abforbing this matter as it exhales or comes to the furface, or perhaps diminifhing the attraction, by prefenting to the

iron or the carbon fome vapour for which they have fome attraction. But we do not know any matters that are much preferable
to others which are not obvioufly improper. Though Mr. Reaumur
did not fucceed in eftablifhing fuch an art in France, it is now practifed in Britain.

The very beft kinds of refined, or bar iron, are very ftrong and tough
when cold, and bear to be bent backwards and forwards very much
before they break, and they are alfo very malleable when hot. Sweden
is remarkable for producing iron, from fome of its ores, of this very
beft quality. But the greater number of iron ores do not produce iron
fo good, though it be good enough for common ufes.

The blackfmiths diftinguifh the common kinds of iron into two
varieties, of which however there are many degrees. To thefe two
varieties they give the names of red-fhort iron and cold-fhort iron.
The red-fhort is tough and flexible when cold, and works very well
when ftrongly heated, but if under-heated, it is brittle under the hammer ; and as there are many varieties, the fmiths make trials at firft
to learn how the iron is beft wrought. The cold-fhort iron, if cooled
flowly after being heated, proves quite brittle while cold, efpecially
thick bars of it, and acquires flexibility by being plunged red hot into
cold water, but is always deficient in it. In its brittle ftate it breaks
fo as to fhew facets or plates remarkably large in the furfaces of the
fracture.

Profeffor Bergmann made a great number of experiments to learn
the caufes of thefe different qualities of iron, and thought he had difcovered the caufe of cold-fhortnefs, afcribing it to the prefence of a
metal which he called *fiderum*: But this has been found to be a miftake. Mayer fhewed that it is a phofphat of iron. Mr. Cort's procefs removes or prevents this bad quality.

When a piece of tough hammered iron is broken by bending, it is
found to be fibrous ; but if the fame piece be kept red hot for fome
time, and again broken, the fibrous appearance is almoft gone. It

would feem that it gets it by the action of the hammer, and that iron confifts of malleable infufible ftuff, mixed with a remainder of fufible iron. Such a mafs, when heated red hot, muft work under the hammer, like a pafte in kneading, the fluid part being fqueezed along between the ductile parts like fo much greafe. Perhaps this is the fource of that fhining or glowing fkin which it gets in a welding heat, which enables two pieces to be ftruck together by the force of blows.

I have now defcribed the progrefs of iron from the ore to that ftate of this metal which appears to be the pureft and moft perfect ftate of it, and in which it is called tough iron, or bar iron. I muft, in the next place, give fome account of STEEL, which is iron in a different ftate, and poffeffed of properties by which it is much better adapted to ferve fome of our purpofes than tough iron is.

Steel is refined iron, intimately combined with a fmall quantity of pure carbonaceous matter, or the carbon of the French chemifts.

Slender bars of iron are put into a crucible, and covered with charcoal duft rammed clofe round them. The crucible is luted, and fet in a furnace, where it is expofed to a ftrong red heat for fome hours. When the iron is taken out, its furface is frequently found rough and bliftered : This I believe is owing to the pufhing the operation a little too far. It is now converted into fteel, and confiderably increafed in weight.

At Newcaftle, and other places in England, they have furnaces for this purpofe, in which large quantities are made. Thefe have the form of a large oven or arch, terminating in a vent at the top. The floor of this oven is flat and level. Immediately under the middle of this floor there is a long arched fire-place with grates, which runs quite acrofs from the one fide to the other, fo as to have two doors for putting in fuel from the outfide of the building. A number of vents or flues pafs from the fire-place to different parts of the floor of the oven, and throw up their flame into it, fo as to heat all parts of it equally. In the oven itfelf there are two large and long cafes, or boxes, built of a good fire-ftone, and in thefe boxes the bars of iron are regularly ftratified with charcoal duft, ten or twelve tons of iron at once, and all is covered with bed-fand. The heat is continued five or fix days and nights without intermiffion. And it requires as long a time for the furnace to cool again, before the fteel can be taken out, and the boxes filled with frefh bars of iron.

This procefs is a cementation, and the carbonaceous matter muft be introduced into the iron in the form of heavy inflammable air. You muft recolleƈt the experiment of Mr. Morveau, in which he expofed a diamond to intenfe heat, fhut up in a fmall cavity in tough iron. The diamond vanifhed, and the iron around it was converted into fteel. Thefe faƈts fhew very plainly what happens. The charcoal, or carbon in fubftance, combines with the metal.

Accordingly, a chemical analyfis of fteel fhews this completely, by giving us the carbon again in fome other combination, as will appear in a variety of ways. It appears by the refults of its expofure to great heats. It is gradually deprived of its qualities of fteel by long continued red heat, and it emits carbonaceous matter, if not carbonic acid. A fine fteel wire burns with amazing brilliancy in vital air, and produces both fixed air and inflammable air. This carbonaceous matter is difcoverable in fteel by the aƈtion of acids, particularly aquafortis. When fteel is diffolved in an acid we obtain much more plumbago from it than from the worft kinds of tough iron, although not fo much as from the grey caft iron. If this experiment be performed with fulphuric or muriatic acids, the difference is not fo remarkable, but not lefs real. In thefe acids, the folution proceeds chiefly, or primarily, from the decompofition of the water, and inflammable or pure hydrogenous gas is produced. This combines with part of the carbon in its nafcent ftate, and compofes the heavy inflammable air, which is really obtained in the folution of fteel in fulphuric acid, whereas pure tough iron gives us hydrogenous gas.

With aquafortis, the produce of plumbago is much more confiderable, when the procefs is carefully managed. The nature of it is alfo more diftinctly feen ; and I am perfuaded that iron conftitutes an ingredient of it, and that the French chemifts are in the right when they call it a *carburet of iron*, though others fay that the iron exifts in it only accidentally. In the trials I have made, its quantity or proportion is too conftant to be accidental.

Steel, therefore, appears to be pure iron combined with inflammable matter. By pure iron, I mean the metal completely reduced, that is, cleared of all oxygen, and of all earthy and combuftible matter ; in which refpect it differs from caft iron, which undoubtedly contains much earthy matter, united with an oxyd of iron in the form of glafs, which is entangled in the pores, and cannot be fqueezed out without a vaft deal of hammering and labour. The caft iron refembles the fteel in this, that, like it, it contains much inflammable matter, which it got from the fuel, and which remains defended from the action of the air by the very fluidity which it occafions, and which is expelled only by ftirring, and prefenting a great furface for a long time to the action of the air.

This notion is confirmed by the fact that a flender rod of iron, plunged into white or grey caft iron in fufion, is converted into fteel *.

The qualities of fteel, by which it differs from iron, are,

1. It is more fufible. Tough iron cannot be melted in the moft violent fires of common furnaces. It only becomes very foft, but never fluid. But fteel can be melted perfectly in crucibles, if expofed to the moft violent heats of common furnaces.

2. In its folid ftate, fteel is more rigid and hard than iron, nor can it be fo much foftened by heat, without lofing its tenacity and flying in pieces under the hammer. It requires therefore more care and attention to forge it well than to forge iron.

* Should not finery cinder, and all pure oxyds of iron, when reduced by charcoal and proper fluxes, in the nice experiments of the laboratory, be fteel, and not iron ? Are they always fo ? I think not. EDITOR.

3. Steel is much more readily broken by bending it than tough iron. It does not bear to be so much bent backwards and forwards. When a bar of it is broken, the surface of the fracture is quite different from that of iron. A bar of tough iron shews by its fracture that it is composed of fibres, the surface of the fracture being very rough, with the ragged ends of them. But steel, when broken, shews that it is composed of very small grains, of a plated structure; and presents a whitish grey surface, much more plain than that of the broken iron.

The most useful qualities by which steel excels iron are the strong cohesion of its parts*, and the extraordinary hardness it acquires when made red hot, and very suddenly cooled. It is thus made so hard as not only to cut iron with ease, but steel itself in its softer state.

This excessive hardness is attended with perfect rigidity and inflexibility, which makes such hard steel in some measure brittle. Files, which are hardened to this degree, can be broken by a fall. But the artists temper this hardness more or less, to fit the steel for different purposes.

This is done by heating the hard steel again. If its surface be made clean by grinding or polishing, then, when it is heated again, it will acquire a straw colour, which will gradually proceed to a full gold colour, with ruddy purple streaks, which afterwards become full purple, violet, and deep blue. These colours direct the artist in what state he shall arrest the temper, by dipping the steel into water or grease. The first appearance of yellow fits it for the edges of chizels and punches, which are to be employed upon iron itself; the full gold colour, or the beginning of purple, fits it for chizels which are to be employed on the softer metals; a little more of the purple fits it for common edge tools; and the violet or blue fits it for watch springs. When clouds of a dingy yellow are appearing among this blue, it is becoming too soft.

---

* A steel wire of one-tenth of an inch in diameter, will just break when loaded with 900 pounds, if properly tempered, and more than 700, if soft.   EDITOR.

It would be a very defirable thing to combine this extreme hard-nefs of fteel with the toughnefs and tenacity of iron. The only way we can do this is by welding them together. It is thus that our edge tools are made. A bit of fteel is welded to the iron, on that fide of the plate or bar which is to be worked into an edge.

There is another way, which is peculiarly ferviceable on particular occafions. We can convert the furface of any piece of iron into fteel by cementation, which we can ftop before it penetrate fo far as to make the whole become fteel, and brittle. This is called CASE-HARDENING. The piece of work, when very nearly finifhed, is covered with a pafte made of charcoal or other combuftible matters. Long expe-rience has produced an univerfal preference of animal charcoals, or rather the crude fubftances. Horn or hoofs, chopped hair, bone fhavings, and fome other fanciful ingredients are made into a pafte. The iron is covered with it, and the whole wrapped up in clay. This is firft dried and hardened before the fire, and then put into the forge, and kept of a low red heat for an hour or two. When taken out, it is fuperficially fteeled. In this manner are almoft all the parts of gun locks treated. Befides the fuperficial hardnefs acquired by it, it is well known that they have incomparably lefs friction than when only iron; and they are much lefs liable to rufting.

The extreme hardnefs appears to me to depend on the extrication of latent heat. And the abatement of this hardnefs by the temper feems to be produced by the reftoration of a part of that heat.

With refpect to the qualities of fteel in mixture, they are precifely the fame as thofe of iron, except the fmall differences proceeding from the excefs of inflammable matter which fteel contains. It is more fu-fible, more inflammable, not quite fo readily rufted or diffolved; and it can be reduced to iron again, by abftracting, or otherwife feparating this inflammable matter.

Befides the fteel produced from bar iron by cementation, we fome-times have it directly from the ore. The fpathofe ores in Carinthia afford it at the first. The caft metal from them is melted in a pot. By throwing a little water on it, a plate is congealed on the furface, and lifted off; and this is repeated till all is expended. Thefe plates are then melted again, and kept in a melting heat for a long time, which requires a continual increafe of heat; and they are found to be fteel. An ore in the ifland of Elba, and fome ores in Barbary, have the fame qualities with the German ores now mentioned. We do not well know the various procefles, many of which feem very fanciful.

The fineft and beft kind of fteel is called *caft fteel;* and the procefs for preparing it was invented in England. All fine inftruments and razors are now made of it. *(See Note 21. at the end of the Volume.)*

I conclude this article by referring you to Fourcroy, who has treated this important fubject with the greateft diftinctnefs, and a very fair and copious narration of facts, fo that I think a judicious chemift will find himfelf quite independent of all his theories, which feem but ill fupported.

Medicinal Virtues of Iron.

This metal has always been efteemed as a valuable article of the materia medica; and furnifhes, in its preparations, medicines of great efficacy. There feems to be no foundation, however, for the opinion which chemifts entertained with regard to it, that according as it is differently prepared, it acquires powers over the body of a different, and almoft oppofite nature.

Iron itfelf, and moft of its preparations, taken into the body, has often a very manifeft conftringing effect, which is ftill more remarkable in feveral of its preparations, when applied externally, in cafes of hemorrhage from wounds, &c. They act with fome degree of acrimony and pain, in contracting the veffels from which the hemorrhage proceeds. They have often had this effect, when given in cafes where it was not practicable, or not eligible, to apply a remedy directly to the part itfelf from which the hemorrhage proceeded; as in dyfentery, hemorrhoids, too profufe flow of the menfes, lochiæ, &c. Thefe obfervations firft eftablifhed the character of iron as an aftringent medicine. But its effects upon other occafions feemed to be of a different kind. Both the metal itfelf in fubftance, and all its preparations, were long noted as fome of the moft powerful remedies by which the menftrual evacuation may be reftored, after it is fuppreffed; and for promoting the motion of fluids through the whole body. The faline preparations of iron, if exhibited along with a confiderable quantity of water, increafe very fenfibly the evacuations in the form of urine, and of perfpiration; and, if given in a greater dofe, have a quick effect in exciting difcharge from the guts. On account of thefe effects, the metal was alfo fuppofed to be endued with an aperient or deobftruent virtue, attributed to one of the ingredients in its compofition, viz. the phlogifton; while the aftringent virtue was referred to another,—the earth. And the chemifts endeavoured, by a variety of proceffes, to feparate them, or obtain fome medicines poffeffed only of the aperient quality, and others only aftringent. And fome even imagined that they performed this in fome degree; fo that we ftill find, in fome difpenfatories, preparations diftinguifhed by the titles of aperient or aftringent. It is, indeed, certain that fome of the preparations contain a larger proportion of the inflammable metal than others; but it appears from the confent of practical phyficians on the fubject, that this only renders the medicine more or lefs efficacious,—thofe containing the greateft quantity of this fubtile principle producing their effects more fenfibly and certainly, but not differing in the manner of their operation from others.

The powers by which iron produces its effects are, in the firft place, its aftringency, or conftricting virtue, a power of bracing the fibres over the whole body, diminifhing in fome degree the cavities of the veffels, and increafing the force wherewith they comprefs the fluids. Secondly, a powerful ftimulus, friendly to life. By this it excites or promotes circulation, and all thofe motions in the body by which the alimentary parts are converted into good fluids, and the fluids preferv-

ed in a found healthy ftate; and by which all fecretions, both of recrementitious and ufeful fluids, and all abforptions, are performed in different parts of the body. By this laft property, iron differs from lead,
which, together with aftringency, fhews a power rather of diminifhing
and deadening motion, and therefore is a much more powerful, though
dangerous aftringent. The aftringent quality of the preparations of
iron fhews itfelf, by the tafte of acerbity, which draws the mouth together, and by Dr. Hale's experiments with Chalybeate water.
Their ftimulating effects become fenfible, if a large quantity of them
be taken: They prove purgative, or even emetic, with a good deal of
heat and difturbance. Thefe effects fhew, that when applied in confiderable quantity to the nerves of the ftomach and guts, they not
only ftimulate thefe, fo as to produce a confiderable increafe of the
fecretions performed in thofe parts, but alfo excite a motion by which
they are expelled. Their ftimulating effects are likewife extremely
evident over the whole fyftem, when ufed in fmaller quantity, fo as
not to produce difturbance in the ftomach and guts. This ftimulus is
fhewn by the warmth it diffuffes over the whole body of the perfon who needs it, by giving a frefher colour to the fkin, by the vi-
gour and alacrity with which it infpires, in confequence of exciting
a more brifk circulation. Phyficians have doubted whether this action
of iron proceeded from its entering the lacteals, and being mixed
with the blood; becaufe they obferved that perfons put under a courfe
of fuch medicines very evidently void a confiderable quantity of
ochre, or calx, in the fæces. And, indeed, it is certain that the ftimulus and conftriction with which it affects the ftomach muft have
immediate effect on the whole fyftem, on account of the remarkable
fympathy daily obferved between the nerves of the ftomach and
thofe of every other part almoft of the body. But it is very probable
too, that fome part enters the blood, becaufe, although falts obtained
from iron readily depofit a confiderable quantity of ochre, it is not
found that they can be entirely refolved in this manner. But in what-

ever way it act, it is certain that a very quick change is wrought in the blood and circulation. The belly is, indeed, often bound in some measure, and the coftivenefs is produced, probably by quick abforption of the thinner parts of the food; but it is eafily obviated by a fmall dofe of rhubarb.

Upon account of thefe two qualities, iron and its preparations are among the moft powerful remedies in weaknefs and laxity of the folids, and in watery difpofition and poverty of the fluids, together with all difeafes depending upon fuch a ftate of th. body —fuch as difpofition to dropfy, and anafarca, or watery fwelling, fluor albus, gleet, hyfterical and hypochondrical difeafes, indigeftion, diabetes, diarrhoea, and the too copious flow of the menftrual flux, when thefe proceed fimply from a weaknefs and relaxation of the folids. In the chlorofis, it not only removes the languor, but reftores the proper flow of the menfes, by the ftoppage of which this difeafe is either produced, or very much increafed.

In autumnal intermittent fevers, when obftinate, and the body much wafted, it prepares for bark.

The preparations from iron are not advifeable, if the humours are confiderably acrid, or when there are appearances of a putrid tendency. In fuch cafes the nervous fyftem is generally too irritable.

When the veffels are very full, iron medicines are hazardous, for the fame reafons; and indeed it is always a proper precaution to take off fome blood before a courfe of them.

In perfons of ftrong and rigid fibres, and thofe fubject to violent cramps, thefe medicines have often left fenfible effects, in a difpofition to thefe complaints.

In ulcerated lungs, and old and obftinate obftructions of the vifcera, they are ferviceable, and it is always proper to give them in fmall dofes, and continue them fome time.

The moft certain preparations are; limatura ferri,—ferrum faccharatum,—crocus martis aperie s,—crocusmartis aftrin gens,—tincturæ martis,—fal martis,— flores martiales.

GENUS IV.—MERCURY.

MERCURY is diftinguifhed by being always fluid, and being by far the moft volatile of all metallic fubftances. And thefe two qualities have laid fome chemifts under difficulties and doubts with refpect to its proper rank among the metals. They made it a queftion whether it ought to be called a metal, or femi-metal; and many have thought it could not properly come under either one or other denomination. But there is no reafon to doubt that it is a metal. All the metals are fufible, or capable of fluidity, and volatile too, as well as it, and it differs from them, only in being more fufible and volatile than any of them. We know now that it becomes folid, and proves very malleable, when cooled to a fufficient degree, viz. —40° of Fahrenheit's thermometer.

Chemifts have given much of their attention to mercury, partly with alchemical views, and partly with a view to medicine. The alchemifts probably expected fuccefs in their attempts upon it, on account of its great weight, by which it approaches more nearly to gold than the other metals then known,—its fpecific gravity being 13568 *. And it became very much the object of pharmaceutical chemiftry, foon after it was difcovered to be a fpecific for the venereal difeafe, and before the beft manner of applying it in the cure of that diftemper was well underftood. Its action on the human body, as it was employed at firft, was attended with great inconveniences, and permanent injury, and even fatal confequences. It was thrown in haftily, and in great quantities, until it brought on a plentiful falivation, which was fo violent, that it kept the patient in mi-

---

* A cubic foot weighs 13,568 ounces avoirdupois, or 848 pounds.

EDITOR.

fery fo long as it lafted. And moreover, the mercury, when employed in this manner, is liable to ftimulate fome other fecretary organs befide the falivary glands, with a degree of violence that is not without danger.

The chemifts imagined that thefe troublefome effects depended upon fome noxious corrofive principle in the compofition of mercury, and therefore tried innumerable proceffes with it, and changed it into a thoufand forms, to feparate, if poffible, the noxious corrofive parts of it, and increafe the medicinal powers of the reft. But the opinion upon which they proceeded appears to have been erroneous; and their labour has been in a great meafure loft. Many of their preparations, it is true, are ufeful, and applicable to particular purpofes for which crude mercury is not adapted; and in feveral, they reduced it to the moft efficacious forms; but in none of them have they taken away its power to produce noxious effects. Thefe are only to be avoided by attention and fkill in the management of it. When employed in the venereal difeafe, it never fhould be thrown in fo haftily as to produce a great falivation, by which it will run out of the body again as faft as it is thrown in. We generally give it in fuch quantities as to bring on a flight fpitting and affection of the gums; for unlefs this fign of its action can be perceived, we cannot be fure that its powers are employed; but it never fhould be made to act with greater violence. And we muft be careful that it do not excite other evacuations, and run off by them. We muft be alfo careful to remove or obviate any difpofition to inflammation, which may appear by inflammation actually prefent, or by a fullnefs and hardnefs of the pulfe. The firft impreffion of the mercury always increafes the inflammatory difpofition, and unfortunate confequences often follow. The fortunate ufe and effects of mercury, therefore, depend more on the judicious management of it, than on its having been prepared by elaborate proceffes: At leaft, if we confult experienced practitioners, who know more of difeafes than of chemiftry, the moft will give it as

their opinion, that the plain pill or ointment are equal, for the cure of the
venereal difeafe, to all other preparations. It is neceffary, however, to
make it undergo at leaft the procefs of *trituration*, and that with the
greateft care, and in the moft perfect manner, that the parts of it may
be divided and attenuated, and that very fubtilely, to make it act certainly upon the human body. When thrown in undivided, it runs
through the inteftines, and produces no effect, or very little, and very
uncertain. It was the fafhion about fifty years ago to take mercury
in this manner, from an opinion that it had extraordinary good effects, by virtue of a fpirit, or effluvium, that was fuppofed to be extracted from it by the inteftines, though no part of the mercury itfelf,
or none of its groffer parts, entered the blood. It happened, however, in fome cafes, that even the crude mercury, thus given, brought
on a falivation; but from its having produced this effect in a few cafes
only, there is reafon to believe that it had lodged in the inteftines,
and was divided by their motion.

When we defire to give it with certainty of its producing an effect,
and of the effect being proportional to the quantity thrown in,
we muft take care that it be prepared at leaft by the moft careful divifion or feparation of its particles from one another. One way of
doing this is, by rubbing it long and patiently with vifcid and other
fubftances, which, by interpofing themfelves between the divided parts,
hinder them from readily joining again. The fubftances fuitable to this
intention, are axunge, turpentine, and other balfams; mucilages, that
is, mixtures of gummy fubftances with water; vifcid fweets, as honey;
even earthy powders fometimes, and water. Boerhaave firft produced
a black powder from mercury by agitating it in water; and Dr.
Prieftley has made many experiments upon this black powder and its
production, and has difcovered a number of curious particulars relating
to it. You will fee them defcribed in the volumes which he has
publifhed, to which I refer you.

In whatever way mercury is thus divided, it acquires a fmall de-

gree of folubility in the animal humours, and affects the tongue with
a fenfation of a particular naufeous tafte, which fhews that fome of it
is diffolved by the faliva. This power in our fluids to diffolve mercury
probably depends upon the ammoniacal falt they contain, for common
fal ammoniac plainly acts upon mercury when rubbed with it.

But it is alfo believed by many, that water, unaffifled with any
other matter, has fome power to diffolve it, or to receive from it fome
medicinal efficacy. Gaubius, the French phyficians, and many others,
aver, that when it is boiled in water, it gives to the water the power
of killing worms in the inteftines of animals. Such water has been
recommended by many as an effectual and fafe anthelmintic; and this
power it can only derive from fome fmall part of the mercury diffolved in it, though, if there is any, it muft be exceeding fmall, as the
mercury does not fuffer a perceptible lofs. The water certainly acquires what is called the metallic tafte, not fenfible at firft, but fufficiently fo in a few hours, and for a long while after.

When mercury is expofed to the action of heat, it foon begins to
evaporate and boil like water. This happens when the mercury is
heated to about 600° of Fahrenheit's thermometer *.

The vapour of it, when mixed with air, does not immediately calcine, as the vapour of fome of the other volatile metals does. If immediately condenfed, it returns again to the form of mercury. And
this metal may be diftilled in a retort, without fuffering any material
change. We can therefore purify it by this procefs from admixture
with other metals, moft of which it is capable of diffolving or mix-

---

* It evaporates in temperatures far below this, when the preffure of the atmofphere is
removed. If a well filled barometer be kept near a window in a warm room, the part
of the tube above the column of mercury becomes dim within, on the fide next the
window. When this is viewed through a magnifying glafs, it will be feen owing to
fmall fpherical drops of mercury, which the coolnefs of this fide of the tube has condenfed.
If the barometer be inclined, fo as to make the mercury reach the top of the tube, it will
lick them all up, and the tube becomes quite clear.

EDITOR.

ing with.   And it happens to be mixed with them fometimes, either by accident or by defign, in different operations in which it is employed.   When thus mixed, it is called foul or impure mercury.   The figns of this foulnefs are,

1ft, A flat furface in the phial, not convex like that of pure quick-filver, the convexity of which fhews that it has more cohefive attraction than the impure.

2dly, The formation of a pellicle on its furface, in confequence of which a fort of wrinkles are formed on it in moving it gently.   This pellicle, adhering to the furface of the veffel, leaves a train behind the mercury, when it is made to run from one part of the glafs to another.

3dly, The formation of a black powder upon the furface of mercury, when it is fhaken in a phial with good or refpirable air, and ftill more quickly with vital air.

It fometimes has the appearance of a pellicle on its furface from duft or greafinefs, or dampnefs adhering to it.   In this cafe, ftraining it through thick linen or leather, or paffing it through a paper cone, will make it clean.   But if it becomes foul again, there is reafon to conclude, that other metals are the caufe of its foulnefs, and it muft be made clean.   There are various methods for doing this.   Simple agitation, in contact with atmofpheric air, will produce this effect. It is a general fact, that all mixtures of metals attract oxygen more ftrongly than the metals when feparate.   This is remarkably the cafe with mercury.   Agitation with air quickly produces a black powder, which, when feparated by filtration and expreffion, is found to be chiefly a calx of the other metals ; and we can obtain the mercury perfectly pure by this procefs.   But fome of the mercury is entangled in the powder, and fome is really oxydated : It is not, therefore, the beft procefs for purifying mercury ; and it is but a lately difcovered one. The ufual procefs has been by diftillation.   Iron filings are added, and with good effect.   Their ufefulnefs is commonly imputed to their attraction for the metallic impurities. But lead is the moſt common impurity; and iron has no attraction for it. Others ſuppoſe that they ſerve only to prevent any drops of the boiling mercury from ſtarting over into the neck of the retort. But though ſand anſwers this purpoſe, it has not the effect of iron, which gives the mercury a remarkable degree of brightneſs and mobility.

Iron calcines moſt eaſily and rapidly, and muſt therefore be a powerful opponent to the calcination or oxydation of the mercury, quickly abſorbing any vital air that may be in the apparatus. The greaſy matter too, which is in all filings of iron, contributes in the ſame manner to prevent the oxydation of the mercury. I have very frequently perceived a ſtrong ſmell of volatile alkali in the veſſels in which mercury has been diſtilled with iron filings. I apprehend that this is formed by the union of atmoſpherical azote with hydrogen diſengaged by the iron from moiſture contained in the veſſel and materials.

Though this metal does not calcine very perceptibly, when only changed into vapour to be immediately condenſed again, it is not incapable of being calcined.

There is a proceſs deſcribed in chemical books, by which it is changed into calx or oxyd by the action of air and heat. The heat applied muſt be ſuch as will keep the mercury conſtantly circulating; and it muſt be applied without interruption for ſeveral months, to produce a moderate quantity of calx. This is performed in a matraſs of a low flat ſhape, with a long neck, and a capillary opening at the extremity. Some chemiſts alſo recommend a ſmall opening below. The mercury is kept continually, though gently boiling; and after a long time, a red calx or oxyd appears on its ſurface, which continues to increaſe: It is called *mercurius calcinatus, vel precipitatus, per ſe.* The combination with vital in the formation of this precipitate, may be very plainly ſeen by tying an inflated bladder to the top of the matraſs. It will gradually collapſe, and in a ſhort time be quite flaccid. Chaptal, a great practical chemiſt, ſays that it abſorbs nearly eight *per cent.* It is faid that a fmall portion of gold greatly expedites the calcination of the mercury.

The calx is much lefs volatile than mercury in its metallic form, and even endures the firft beginning of ignition, and melts into a beautiful glafs; but if the heat be increafed above this degree, the calx is gradually converted into vapour, and at the fame time returns again to the form of quickfilver; for thefe vapours, when condenfed, afford quickfilver inftead of calx. While this reduction of it is performed, a great quantity of very pure vital air is obtained from it, each ounce of oxyd yielding about fifty cubic inches. This method for producing pure vital air was firft difcovered by Dr. Prieftley and Dr. Scheele. But it is to Mr. Lavoifier that we are chiefly indebted for the diftinct account of this procefs, and for the ingenious and moft important inferences which may be made from it. I may venture to ftate this as the moft interefting fact in modern chemiftry. Mr. Lavoifier faw it in all its importance, made a multitude of experiments on it, which are models of judicious procedure, and evince the foundeft judgment, as well as the greateft ingenuity.

According to him, this calx confifts of mercury and oxygen, in the proportion of twenty-eight to one nearly. This oxygen was imbibed from the air in the veffel. It appears from Dr. Prieftley's experiments with mercury, agitated in a phial, that the production of the black powder (which is alfo a lefs perfect calx) is always accompanied by the abforption of the vital air of the atmofphere. For Dr. Prieftley found that the air of the phial was phlogifticated, that is, by what we know now of the matter, the azote was left alone. And when he afterwards agitated the mercury with vital air, he found it not phlogifticated, but confumed or abforbed. Lavoifier obferves that this abforption goes on during the tedious procefs juft now defcribed for the calcination of mercury; for the calcination foon ftops if the capillary tube be fhut up. In this cafe, if we pufh the heat a little farther, part of the calx is reduced again; and if we go too far,

we burst .the veffel.   When thefe facts are compared with the extrication of vital air in the reduction of the *mercurius calcinatus per fe,*
the inevitable conclufion feems to be, that the union between mercury
and oxygen is fo flight, that a very low temperature is fufficient for
feparating them.   Mr. Lavoifier varied this experiment in many ways
that lead to the fame conclufion.

No fubftance has been fo much tortured by the chemifts as this
metal.   It has been united or mixed with almoft every fubftance in
nature, and it appears to form fome combination with them all ; for
by fimple trituration with any fubftance whatever, even dry fand, the
mercury lofes its metallic luftre and its fluidity, and becomes black ;
at the fame time it becomes very fenfibly fapid, and affects the nerves
of the ftomach and the fecretory organs.   But this is owing to its di-
vifibility, and its difpofition to oxydate when prefented to the air in
fuch an extended furface.

All the more powerful chemical agents combine with it very readily ;
and the combination affords a great many remarkable and inftructive
appearances.   I fhall firft confider its relation to the acids.—

1ft, The vitriolic acid has no action in the cold, not even though in
its ftrongeft ftate.   If we dilute it with water, it cannot act on mercury, even though affifted by heat.   In order to combine the acid with
this metal, we muft take the acid in its ftrongeft ftate, and employ
the affiftance of heat to make it act.   It will then diffolve half or two
thirds of its weight.

When the acid becomes near boiling hot, it begins to diffolve the
mercury with effervefcence.   This effervefcence is occafioned by the
change of a part of the acid into fulphurous acid, which change always happens to a part of this acid, when made to act on metals in
its ftrong ftate, without addition of water.   The diffolving metal, not
having water to act upon, and to fupply it with oxygen, attracts a part
of the oxygen of the acid.   This, in confequence, is decompofed, and
returns more nearly to the ftate of fulphur from which it was originally

produced. And there are even many examples of the change of a part of this acid into perfect inflammable fulphur, when it is made to act in its ftrong ftate on fome of the metals. On account of the change which a part of the acid undergoes in this manner, we muft ufe more of it than what actually remains afterwards combined with the oxyd of the mercury. In proportion as the *oxydated metal* and acid unite, they form a faline compound of difficult folubility, which is accumulated around the mercury in form of a white matter; and if we have ufed a proper quantity of fulphuric acid, the whole is converted into a white faline mafs, whofe regular cryftals are four-fided columns, terminated at each end by pointed pyramids. Seldom, however, does it attain this complete form.

This is a fingular falt, and prefents an appearance of much importance to be attended to in the metallic falts. It feems to be a mixture of two very different fubftances. For if boiling water be applied, we feparate from this mafs a portion having a yellow colour, and almoft perfectly infoluble in water, and not containing a particle of fulphuric acid. Its colour occafioned its being called *turpethum minerale ;* and it was conceived to be a perfect calx of the mercury, like that obtained by the action of heat and air. It is indeed found to be a perfect oxyd, united only with the oxygen which was feparated from that part of the acid which efcaped in the form of fulphurous acid, that is, deficient in oxygen. This turbith, when urged by ftrong heat, is decompounded, affording mercury in its metallic form, and vital air.

The other portion feparated by the boiling water confifts of the oxyd now defcribed, combined with the acid, and is a real *fulphat of mercury.* It cryftallizes in flender deliquefcent needles. Still this fulphat exhibits nearly the fame properties with the mafs from which it was feparated. For long boiling feparates the acid from part of it, and leaves it in the form of a turbith; and I believe the fame effect will be produced on what has now been feparated, and that the whole may in this way be changed into turbith. When the turbith and water firſt poured on it for waſhing it have boiled together for ſome time, an efferveſcence takes place, and ſome elaſtic matter eſcapes, which I have not examined. In theſe proceſſes, it is plain to obſervation that the firſt boiling water ſeparates the greateſt part of the acid; but if the mixture be allowed to cool, the calx reſumes a conſiderable portion of it, and the paleneſs of colour ſhews that the turbith is not nearly perfect. The water containing the acidulous ſulphat ſhould therefore be poured off immediately.

With reſpect to the other part of the ſaline maſs, which has been cleared of the mercurial oxyd, and is a true ſulphat of mercury, I muſt obſerve, that it is a very acrid mercurial compound, and ſpeedily deſtroys all organized ſubſtances *. It is much leſs volatile than quickſilver; and when urged by a ſtrong heat, it ſublimes without decompoſition. The truth is, that the ingredients differ ſo little in volatility that they ſhould riſe together, and we ſhould not expect any decompoſition. But if the vapours come in contact with the atmoſphere, they ſeem to be completely decompounded, and this in a variety of ways, yielding mercury, cinnabar, ſulphur, and ſulphurous acid, according to circumſtances.

The ſulphat of mercury is decompounded by alkalis. Fixed alkali precipitates the mercury of an orange brown colour, when mild, but more approaching to yellow, when cauſtic. Volatile alkali, in its cauſtic ſtate, will ſcarcely decompoſe the ſulphat; but by the help of

* The various appearances which this ſalt exhibits ariſe from a different proportion of ſulphuric acid which it may contain. In its yellow ſtate, it contains leaſt of all; when white, it contains more. Waſhing the white maſs with boiling water makes it yellow, the water taking off the more ſoluble part. When this is evaporated and ready to cryſtallize, or would cryſtallize by cooling, the addition of a very minute quantity of acid prevents the precipitation till farther evaporated,—ſhewing plainly that the greater ſolubility ariſes from an exceſs of acid. Conſult Fourcroy for a very diſtinct account of all theſe ſtates.

EDITOR.

a double elective attraction, it does it very readily, when united with excefs either to carbonic or muriatic acid. Lime-water alfo precipitates the mercury, of a brown or yellow colour, according as little or much lime-water has been employed.

I obferve that fome chemifts affert, that by repeated diftillation of fulphuric acid from mercury, in this procefs, the mafs becomes more and more fixed, and melts into a blood red fluid before it evaporates. *(See Cornette Mem. Acad.* 1799.) I have not obferved this : Nor has Mr. Bayen, who has made a great many very judicious and interefting experiments on this fubject.

The nitric acid diffolves mercury more quickly and readily than any other acid does. The moft convenient ftrength of the acid is when diluted with about an equal weight of water ; in which diluted ftate it is called aquafortis. It is properly diluted when an ounce of it can diffolve an ounce of mercury.

This folution exhibits no remarkable appearance at firft. Generally the metal lies quiet for fome little time at the bottom ; and all that we obferve is a green colour in that part only of the folution, where it is in contact with the mercury. This colour gradually fpreads farther up, and fmall bubbles appear rifing from the mercury, but are abforbed as they afcend, till the green colour produced by them at laft reaches the furface. They then fly off in the ufual form of red fumes. If water be carefully added to the mixture, before the green colour reach the furface, the bubbles will rife a little way in the water before they are abforbed, and they tinge it green, and promote the mixture of the acid liquor with the water above it. By very careful management of the procefs in this way, Bergmann has fometimes obtained a complete folution, without any efcape of red fumes. This, however, is a very rare cafe, and the folution has rare properties. Had the water not been added, and had the fumes broken out, and a complete folution been obtained, the addition of fo much water to the folution would have caufed a great portion of the mer-

curial nitrate to fall down; but the folution obtained without effervefcence will bear any dilution with water.

The more common phenomenon of the procefs is a fpeedy formation of the ruddy effervefcence. The rife of the bubbles from the
mercury is accompanied by a confiderable extrication or formation of
heat; and when the folution is carefully obferved, the bubbles are
plainly feen to form in all parts of the green liquor. In fhort, the
acid is changing from nitric to nitrous, and an elaftic matter breaks
out that is deficient in oxygen. The folution grows very warm, and
the fumes become very copious, and blood red. Hence we muft conclude, in conformity to the numerous facts already mentioned on different occafions, that the metal is now acting on the acid, and decompounding either the whole, or only a part of it, by abftracting from
it a portion of its oxygen.

When thofe fumes, inftead of being allowed to efcape into the atmofphere, and mix intimately with it, are made to pafs through a
narrow glafs veffel, the ruddy colour foon difappears in the tube, and
the expelled gas becomes quite tranfparent there, and almoft colourlefs.
It may be collected in the fame way that we have employed for other
gafes, by making it pafs into inverted jars filled with water, and
ftanding in a ciftern of that fluid. Taking this method, we find that
all the acid qualities are gone. The gas has fcarcely any tafte. It
has very little attraction for water. Diftilled water diffolves about
one-eighth of its bulk, but retains it fo weakly that a fhort expofure
to the air diffipates it completely.

This colourlefs, elaftic, aëreal matter, was firft defcribed by Dr.
Prieftley, who called it *nitrous air*. The French call it *nitrous gas*.
The moft common name of it in England, however, is ftill *nitrous
air*. It is now well known to be a portion of the acid of nitre, dif-
guifed or changed by the lofs of a great quantity of its oxygen, which
the diffolving quickfilver has abftracted from it. One proof of this is,
that if we feparate the quickfilver from the acid, either by precipitat-

ing it with a pure alkali, or by the action of heat alone, we get, by
either way, an oxyd of the mercury, which is not unlike to the oxyd
obtained by the action of air and heat.   And like that oxyd, it yields
a large quantity of pure vital air, when heated red hot in a retort, the
mercury at the same time refuming its metallic form.   We have another proof, by mixing this elaflic fluid with atmofpherical air, or
vital air, which has much more effect.   Either of them, by fupplying
oxygen, changes the nitrous air into nitrous acid, which fhews itfelf
in red vapours, and is immediately attracted and abforbed by the
water.   Much lefs of the vital than of the atmofpherical air is fufficient to produce this change on the nitrous air; and it is evident,
from other experiments, that the atmofpherical air produces its effect
only by virtue of the vital air which it contains.

When thefe red vapours are formed, a confiderable heat is produced in the glafs.   This heat is fuppofed by the French chemifts to be
a part of the heat, or matter of heat, or calorique, which was contained
in a latent flate in the compofition of the two elaftic fluids.   (See
Note 22. at the end of the Volume.)

By means of this nitrous air, Dr. Prieftley very ingenioufly contrived to diftinguifh wholefome or refpirable air from the different
kinds of tainted or noxious air, and endeavoured to meafure the different degrees of wholefomenefs or purity in common atmofpheric air.
His method of doing this is defcribed in the volumes he has publifhed; and his apparatus is fuch as enables him to fubject a fmall
quantity of air to this fort of trial or teft, as he calls it.   The principle
of the method is briefly this.   Nitrous air is the vapour of nitric acid
deprived of part of its oxygen.   If, therefore, a cubic inch of it be
mixed with as much vital air as will make up the deficiency, they
will unite and form nitric or nitrous acid, mifcible with water.   If,
therefore, a cubic inch of this gas be thrown up into a jar inverted
over water, and containing an air confifting of as much vital air as
is required by this nitrous air, and three cubic inches of any other

gas immifcible with water, whatever may be the bulk of this compound air, it will be reduced to the bulk of three inches. If, therefore, nitrous air be very uniform in its compofition, a fingle experiment will inform us what bulk of vital air it will unite, and difappear with, by abforption in the water. Therefore, throwing up nitrous air, till there ceafes to be a diminution, will inform us how much vital air is in the mixture. Nay, as this requires time, and certain complicated precautions, a feries of proper experiments will teach us the proportion of vital air, by means of the contraction occafioned by mixing a given meafure of nitrous air with any quantity of compound air; and we may proceed on this principle, that 100 parts of nitrous air abforb 48 of pure vital air, fo that both difappear.

Dr. Prieftley's apparatus confifts of a tube, in which the bulk of the mixed airs is meafured. It is about three feet long, and one-third of an inch wide. In examining atmofpherical air, or air like it, he mixes two ounce meafures of fuch air with one of nitrous air in a fmall jar; and foon after throws up the mixture into the long tube, the cavity of which is divided into ounce meafures, and each of thefe into a hundred parts by a fcale, and he expreffes the refult of the trial, by telling the number of thefe parts which the mixture fills after the abforption is over. The beft air, therefore, gives the fmalleft number.

In trying air better than atmofpherical, he adds to one meafure of fuch air two meafures of nitrous air, which quantity of the nitrous air is enough for the pureft vital air.

Other authors have endeavoured to make improvements in the manner of employing nitrous air for this purpofe. The Abbe Fontana has ftudied this point with particular care, and has defcribed his *apparatus*, or the *eudiometer*, as it was firft called by Landriani, in which there are many neat contrivances. Mr. Cavendifh alfo has given fome accurate and fkilful remarks in the 73d volume of the Philofophical Tranfactions. And Dr. Ingenhoufz has made fome

improvements in the manner of ufing the apparatus. But the contrivance which appears to me the moft fimple and convenient, as well
as the moft exact, is that of Mr. Sauffure.

He ufes a phial, which he calls the *receiver*, containing about five].
and a half ounces of water. Its diameter and height are nearly equal,
and it has a ground ftopper well fitted. He has another fmall phial,
called the *meafure*, which contains fomewhat lefs than one-third of
the former. He has alfo a fmall funnel, and a fmall exact balance.
(*Sauffure Voyage dans les Alpes, p. 514.*)

Having prepared the requifite quantity of nitrous air on the fpot,
he proceeds as follows : 1*ft*, Having filled the recipient with water,
and holding it under water in the bucket, he introduces into it two
meafures of atmofpheric air, and one meafure of nitrous air. He
then fhuts the recipient with its ftopper, and fhakes it under water,
opening it now and then to let in the water. This fhaking and opening is performed juft three times in every experiment, in order
that all may be perfectly fimilar. 2*d*, The recipient is now ftopped
clofe, and its outfide is wiped quite dry. It is then weighed, and its
weight compared with what it would be if three meafures of common
air had been admitted. The difference in his apparatus amounted
commonly to one ounce fix drachms and forty grains. N. B. His
meafure contained one ounce fix drachms twelve grains of water.
Suppofing, therefore, that two parts of nitrous air combine with, or
are decompounded by one of oxygenous gas, the quantity of this gas
in common air is, by the above refult, juft one-fifth of the whole common air, or 340 in 1704: More exactly, it is 22 *per cent*. By fuch
experiments he learned that the air of the valleys among the Alps, and
at Geneva, is better than that on the tops of high mountains by about
nine grain meafures of oxygenous gas in the above two meafures.
This is probably owing to the action of plants during fun-fhine.

I am perfuaded that this apparatus is more ufeful than the expenfive
and fragile eudiometers confifting of tubes and ftop-cocks. But when

I reflect on the unavoidable differences in the proportions of the ingredients of nitrous air extemporaneoufly prepared, and on the different propenfities of ordinary water to abforb or emit elaftic fluids, I
cannot think that thefe eudiometrical experiments are a proper foundation for any accurate judgment of the falubrity or unwholefomenefs
of airs. And I fhould be forry to fee much dependance had on them.
I have always confidered them as too delicate for the hand of any
perfon but a judicious chemift, perfectly at leifure. The odds of 10
or 20 grains in 1740, is an error from which it would be difficult to
fecure ourfelves. Yet even this is a very great part of the greateft differences that have been obferved. It is alfo very inaccurate to confider this experiment as a teft of the wholefomenefs of air, and to
call the inftrument a eudiometer. Chemically fpeaking, it only meafures the quantity of oxygenous gas contained in every air. We know
very well, that the commixture of fome exhalations, particularly of
flowers of the lily kind, in a quantity too fmall to be perceived by
fuch a teft, gives the air a power of affecting fome of our organs in a
way which, though not immediately deadly, is yet extremely preju-
dicial to good health.

Accordingly, the experiments made to examine the goodnefs of air
by employing nitrous air, do not always agree exactly together, even
though made with the fame airs and materials, and the fame apparatus. And when we wifh to be exact, it is neceffary to repeat them
feveral times, and to take a medium of the refults. And when we
choofe to compare two portions or fpecimens of atmofpheric air with
one another, the experiments for this purpofe fhould always be made
at the fame time, and in the fame place, and with the fame nitrous air
recently prepared; experience having fhewn that nitrous air is fenfibly different in its quality, as it is prepared at different times, and in
different places. This is now underftood to proceed from the more or
lefs violent change which the acid fuffers when we are preparing the
nitrous air. A part of the acid always undergoes the changes you

have feen, but a fmall portion of it is completely decompounded, the whole oxygen being taken from it, and then what remains of this portion is azotic gas, which cannot be brought back to the ftate of nitric acid by fimple mixture with refpirable or vital air. There is only one way by which we can bring it back to the ftate of nitric acid, or convert it into that acid, that is, by mixing three meafures of it with feven meafures of vital air, and then promoting the union of the two airs, or their action on one another, by a ftrong heat, or by repeated flafhes of electrical fire, in the manner practifed by Mr. Cavendifh, in the courfe of thofe curious and important experiments which I have frequently referred to as the great fupport of the new chemical doctrines and difcoveries. Now, when metals are diffolved in nitric acid, fome fmall portion of the acid is, as I faid juft now, fo totally deprived of oxygen, that it is changed into azotic gas, and this happens more or lefs according to the violence, rapidity, and heat, with which the diffolution is performed; and therefore the nitrous air which we get, turns out different on different occafions, by its containing different quantities of azotic gas, and being more or lefs fit for the examination of the wholefomenefs of refpirable air.

By this trial air is judged to be wholefome, or fit for refpiration, in proportion to the quantity of nitrous air which it can decompound; and this is known by the diminution of bulk which happens in the mixture of the two airs as confined by water. If there is no diminution of the two airs, the one which was mixed with the nitrous air is totally unfit for the fupport of animal life, even for a moment; and if there is but little diminution, an animal will die in it in a very fhort time.

When common refpirable air is examined in this manner, by mixing it with an equal meafure of nitrous air, the diminution of the mixture, when fhaken with water, has been obferved to be 90 or 100, or 110, or even 120 parts in 200 of the mixture *, with frequent variation however of the refult, though the trial be made with the very fame materials, and in the fame place.

For all thefe reafons, I think that it may be worth while to try whether Mr. Scheele's way of examining the goodnefs of atmofpherical air, or the quantity of vital air it contains at different times, and in different places, may or may not be preferable to this. Scheele's way was, as you may remember, by abforbing the vital part by humid fulphuret of iron filings.

To finifh thefe remarks on the changes which a portion of the nitric acid undergoes while it diffolves the metals, we may notice here, that part of it, on fome occafions, is reduced to an intermediate ftate, between the ftate of nitrous air and that of pure or perfect azotic gas. In this intermediate ftate, it is a colourlefs elaftic fluid or gas, like nitrous air, but has different properties.

1*mo*, It is much more abforbable by water, and can be feparated again from the water by heat, unchanged.

2*do*, It is not converted into nitrous acid by the admixture of vital or of atmofpherical, or of nitrous air; or to exprefs it more accurately, it does not unite with the oxygenous gas in thefe airs by fimple mixture.

3*tio*, Though it be totally unfit for fupporting animal life by refpiration, it brightens and enlarges the flame of a candle immerfed in it;—and inflammable air mixed with it and fired, explodes much in the fame manner as when it is mixed and fired with atmofpherical air, but with a green flame. But it has very little power to maintain the combuftion of charcoal, or of fome other combuftible fubftances.

4*to*, It may be produced or obtained from nitrous air, by expofing.

* In fome of Dolomieu's experiments, the lofs was from 118 to 85, but the heat of the air was very different when fo great a difference was obferved. In the firft cafe the wind was north-weft, and the thermometer 55½. In the fecond it was a firocco, or fouth-eaft wind, and the thermometer 82.

such nitrous air to the action of substances or mixtures which have a strong attraction for oxygen, such as the sulphuret of potash, the humid sulphuret of iron, the muriat of tin, &c. From several phenomena which have occurred to me in experiments, I apprehend that the most expeditious way of obtaining it pure, would be to employ the neutral salts formed by the alkalis and the sulphurous acid. I imagine that it is by such intermedium that even the sulphurets effect its production. Such matters first change the nitrous air into this intermediate kind, in two or three days, and afterwards, by a continuance of their action, change it completely into azotic gas, the bulk of it being at the same time very greatly diminished. *(See Note 23. at the end of the Volume.)*

It may be also produced, and in a purer state, by some particular modes of managing the action of nitric acid upon the metals, which we shall notice soon.

It was first discovered by Dr. Priestley, who gave it the name of *dephlogisticated nitrous air*, on account of its promoting the burning of a candle, and having the power to brighten and enlarge the flame of it. He discovered several different ways of producing it, and made many experiments with it. But its nature has been best illustrated and explained by the society of philosophical chemists at Amsterdam, in consequence of an ingenious inquiry, and a great number of experiments they made with this view. (See a very good account of these experiments in the *Analytical Review*, vol. 17. p. 356.)

They are of opinion, that, like nitrous air, or nitrous gas, it is a compound of azote and oxygen, combined by chemical attraction, but that it contains less oxygen than nitrous air does, though a greater quantity than that contained in atmospherical air. But besides, in the air of the atmosphere, the oxygen is closely combined with the latent heat only, or is in the form of a gas, simply mixed with azotic gas, whereas, in the sort of air which we are at present describing, the oxygen is combined, as I said just now, with the azote, by chemical attraction.

The chief reason for thinking the union of azote and oxygen different in thefe two compounds is, that the peculiar qualities of this laft take place at a pretty precife proportion of the ingredients, fomewhat like other chemical combinations, when a mutual faturation takes place. I cannot fay that this is a very ftrong argument, although it is not without fome weight. But it is more difficult to account for its increafing the flame of a burning body, while nitrous air, which contains more oxygen fimilarly combined, fo far from having fuch effect, extinguifhes flame in an inftant.

The proportions of oxygen in thefe different gafes, or airs, are thus ftated by thefe gentlemen, from fome accurate experiments :

1mo, The atmofpherical air is known to conta'n 27 or 28 parts of oxygen gas, or vital air, in 100 parts of the air.

2do, The air or gas we are now defcribing contains 37 or 38 parts of oxygen in the 100.

3tio, And nitrous air contains 68 parts.

Thefe gentlemen have given a new name to the gas we have juft now defcribed : They name it *gafeous oxyd of azote*. But this name is juft as proper for nitrous gas, which is, like the other, *azote, in the ftate of an oxyd*, or combined with oxygen, but with lefs of it than what is neceffary to form an acid. It would have been better, therefore, to have named the gas we have been confidering, *the lefs oxydated nitrous gas*, or, *the lefs oxydated oxyd of azote* *.

They have alfo made an ingenious attempt to explain its peculiar properties, efpecially the power it has to promote the burning of a candle, and to brighten and enlarge its flame, although breathing animals cannot live in it a moment. Thefe apparently inconfiftent qualities they impute to the peculiar mode of combination of the elements of this gas, which are fo ftrongly united, that the oxygen,

---

* Surely the name *nitrous oxyd*, given it by Chaptal and others, is more agreeable to the general plan of the French nomenclature. This compound feems to be in the fmalleft poffible degree of oxydation. EDITOR.

though it has power to act on hydrogen with the affiftance of heat, has very little power to act on carbon ; and accordingly, the gas has very little power to promote or fuftain the inflammation of charcoal. But in the refpiration of animals, it is neceffary that a carbonaceous principle be difcharged from the blood, which carbonaceous principle is diffolved and carried away from the lungs, by refpirable air in the form of carbonic acid gas ; whereas the gas we have now defcribed cannot ferve this purpofe. But it can ferve for fuftaining the flame of a candle, for this reafon, that tallow, and all other oils, contain a much larger proportion of the hydrogenous principle than of any other in their compofition : And the oxygen of this gas is in that ftate which difpofes it to unite readily with hydrogen, while the hydrogen is feparating itfelf from other matter, and before it has time to attract latent heat, and to expand with it into a gas. This ftate of hydrogen was named by Dr. Prieftley the nafcent ftate of inflammable air. We have already had feveral examples of a greater facility of union in this particular condition of things *.

The quickfilver, while it produces thefe changes in a portion of the acid, is itfelf changed into an oxyd, by its union with the oxygen which it has attracted ; and this oxyd of the quickfilver is then combined with the remaining acid, or is abforbed by it †.

---

* Perhaps this explanation may not be thought to agree very well with the proportion of oxygen that appears, by the experiments of Lavoifier, La Place, and Monge, to be required for the inflammation of equal weights of hydrogenous gas and charcoal ; the laft requiring fo much lefs than the former. Nor does it agree with the fuperiority that is affigned to carbon, in affinity for oxygen.   EDITOR.

† This experiment, the folution of quickfilver in the nitrous acid, is, perhaps, the moft perfpicuous and inftructive example of the general nature of metalline falts, and for this reafon has met with particular attention from Mr. Lavoifier and his coadjutors. The character of the compounds, in their different ftates of oxygenation, are more diftinct than in any other example. There feems to be feveral degrees of this, and each

The compound, or mercurial nitrate, being fufficiently foluble, part of it will, in a day or two, cryftallize at the bottom of the folution. The reft continues diffolved by the water of the acid, and forms a heavy, tranfparent, and nearly colourlefs fluid, very acrid and corrofive with refpect to animal and vegetable fubftances. A drop of it makes an indelible black mark on the fkin. The falt which it contains is fometimes more, fometimes lefs foluble in water, according to the quantity of acid with which the metal is combined, and in proportion as it is new, or has been long kept; for the fame thing happens to it in fome degree as to the folutions of iron. The metal, which, while it is diffolved, is evidently oxydated to a certain degree, by decompounding a part of the acid, becomes fomewhat more oxydated afterwards in the folution; it therefore becomes lefs foluble, or requires a greater quantity of acid to keep it diffolved, and with that acid it forms a compound not quite fo foluble in water as the fimilar compounds which contain the mercury in a lefs oxydated ftate.

I may farther obferve, that the folubility of the nitrate of mercury varies exceedingly with the temperature of the water. Water boiling hot takes up a great deal more, which is depofited on cooling, in cryftals. Nay, for the fame reafon, the acid will diffolve more of the mercury, by the affiftance of heat, and this is fpeedily let go in cooling, and is of a different kind, as we have feen already, from what is obtained from the metal when it is diffolved in a cold temperature, being more oxydated. Alfo, if we add a great quantity of diftilled water to a ftrong fluid folution, we have a precipitation of an oxyd, yellowifh, if hot, and white, if cold: A fmall redundancy of acid prevents this precipitation; fo various are the appearances of this *Proteus metallorum.* But thefe differences are fimilar, as I have faid, to what we fee in iron; and they are

has a uniform effect, even on the ftructure of the cryftals. They are alfo very diftinguifhable, by the differences in their affinity for water, being fo fingularly and uniformly affected by its being hot or cold.   EDITOR.

produced in the same way, or by the same causes, only more diversified.

The nitrate salt of mercury has attracted much notice. It was tortured by the alchemists into a thousand forms, being the produce of their favourite mercury, on which they professed to build their greatest hopes, and of the wonderful acid, which manifested the most remarkable and important properties. This precipitate is white or yellow in its first preparation, in the circumstances I have considered : But, if long continued in a low heat, it becomes of a full red, perfectly similar in appearance, and also similar in many of its properties, to the *mercurius precipitatus per se.* It is called RED PRECIPITATE.

This compound is much used in medicine and surgery, as an escharotic, or cautery, for destroying fungous flesh, &c. It has also become of great value to the philosophical chemist, as affording the readiest method for obtaining oxygenous gas in its purest state. It is, therefore, of importance to prepare it in the most perfect manner, and particularly, to prepare it so as to be uniform in the proportion of oxygen it contains.

When the mercurial nitrate is simply evaporated to dryness, especially when the precipitation has been partly effected by cold water, it is white, or yellowish, for which reason it has been called NITROUS TURBITH. If it be exposed to a very low red heat in this state, a considerable portion of acid is driven from it, and becomes of the fine full red colour, fit for the market. When this is done in a sand bath, it frequently happens that before the central parts and surface have acquired the proper colour, the part at the bottom has been deprived of its oxygen, and the mercury revived. Dr. Higgins, therefore, performs this process in a deep porcelain dish, under a muffle, applying the heat all around, and covering the dish with a glass plate. A small portion of mercurial muriat usually exhales in this process. It arises from impurity in almost all the nitric acid that we can get ; even a little of the nitrate itself sometimes goes off. Chaptal, to have it in perfection, with a rich colour, pours fresh nitric acid on it, and diftils it to dryneſs, three times.

I ſaid that, in the ſtate of a nitrate, this compound exhibited certain uniform differences in the form and ſtructure of its cryftals. When the ſolution has been made without the affiftance of heat, and the mixture kept cloſe during the proceſs, and if it be allowed to evaporate ſpontaneouſly, the cryftals are thin ſquare plates, whoſe four ſides are formed into an edge, like that of a ſtone-cutter's chizel, and having the four angles cut off; that is, the cryftals are formed of two tetrahedral pyramids, joined at their baſe, and having the angles cut off. *(See Romé de l'Iſle, No. 38. VI. 11.)*

But if the evaporation has been promoted by confiderable heat, the cryftals are flat-pointed ſpiculæ, having the flat ſide elegantly ſtriated in an oblique direction, like a ſword blade which has been ground by holding it very obliquely on a rough ſtone. If the ſolution has been effected by heat, the cryftals are alſo flat ſpiculæ, like lancets; but the ſtriæ are now parallel to their lengths. This pointed form of all the cryftals has been much infiſted on by the mechanical chemifts, in their attempts to account for the corroding powers of the mercurial nitrate, but, unluckily, the moſt corrofive are the cryftals in flat plates.

Mr. Fourcroy has treated this ſubject with the ſame diftinctneſs and perfpicuity as the ſolution in ſulphuric acid; and his analyſis of the different ſtates of the nitrate ſeems to me very judicious and inſtructive.

The muriatic acid, in its ordinary ſtate, has little power to act on mercury in its metallic form. This is a confequence of the weak power, or rather *want of power*, in this acid to oxydate the metals; and they muſt be oxydated in ſome degree, in order to form compounds with acids. But the muriatic acid oxygenated by manganeſe, readily unites with metals in general, and forms a ſaline compound, the oxygen being in this caſe ſupplied to the metal by the acid. We can therefore eafily obtain a compound of mercury and the muriatic acid, by employing this acid in its oxygenated ftate. Or, if the quickfilver be firft oxydated by air and heat, or by other acids than the muriatic, that acid will then readily unite with it, and even feparate the other acids from it.

Whether it has in this manner feparated the mercury from another acid, or diffolved the mercury calcined by the action of heat and air, no difference can be perceived in the muriat. If for example, *mercurius calcinatus*, and red precipitate (accurately prepared) are diffolved in feparate quantities of muriatic acid, no difference whatever can be perceived.   This muft be received as an abundant proof, that red precipitate contains no nitric acid, or any thing that it may not acquire by fimple expofition to heat and air.

The compound of the muriatic acid and mercury is foluble in water, but not fo eafily foluble as the compound with the nitric acid.

The ufual method, therefore, for combining quickfilver with the muriatic acid is, firft to oxydate or diffolve the metal with another acid, the nitric for example, and then apply to it the muriatic, which can be applied effectually by feveral different ways.   We can apply it pure, or combined with an alkali, as in common falt.

When applied pure to the mercurial nitrate in folution, a copious precipitate, or rather coagulum, is immediately formed. This is ufually called the white precipitate of mercury.   If we apply common falt, or any earthy muriat, we have fimilar precipitate of mercury, and a neutral or earthy falt in folution.

If, however, we ufe the oxygenated muriatic acid, we do not obtain the fame precipitate, becaufe the mercurial muriat in this cafe is very foluble in water, whereas the common muriat is far lefs fo, indeed fcarcely foluble, requiring almoft 2000 times its weight of water, according to Beaumé and Lemery.   The muriat, and the oxymuriat (as the French call it) of mercury, differ alfo in many other important properties.

But when the muriat of mercury is formed by thefe methods, the feparation of it afterwards from the other acids, or falts, which are in the mixture, is very difficult. Thefe methods, therefore, are not convenient for practice, and a different one is commonly followed, by which this muriat is formed, and at the fame time *feparated by fublimation* from the other materials employed in the procefs; this muriat being totally volatile in a moderate heat.

The proper materials for this purpofe are,

1*ft*, The quickfilver, reduced either to an oxyd, or to a faline compound, by the action of the nitric or fulphuric acid.

2*dly*, Materials which, when heated, will fupply to the quickfilver a proper quantity of muriatic acid in the form of dry vapour. You will find in Neuman's Chemiftry a copious lift of all thefe mixtures for procuring a combination of mercury with the muriatic acid.

The procefs which was formerly moft generally recommended in chemical books, was to mix a dry nitrate of mercury in powder, with an equal quantity of common falt, and as much exficcated fulphat of iron, and to fublime this mixture. This is a fure procefs for obtaining a perfect corrofive muriat of mercury.

Neuman fays, and I believe he fays truly, that thofe procefles in which the nitrous acid has acted, produce a more complete union of mercury with the acid. The mercury is more completely oxydated by this previous treatment, and will therefore unite with a greater quantity of the acid, and make a more active or corrofive compound. The truth is, that the metal attaches to itfelf a greater quantity of unmixed oxygen than it can obtain in any other way, and it is on the abundance of oxygen that the corrofive nature of the muriat entirely depends. But the procefs now practifed is fimpler.

The quickfilver is firft combined with the fulphuric acid, and the fulphat mixed with dry common falt in fine powder, and this mixture is fublimed. This is Neuman's feventh procefs. The fulphuric acid quits the quickfilver to join itfelf to the alkali of the common falt, forming with it Glauber's falt, or fulphat of foda; while at the fame time, the muriatic acid unites with the oxyd of quickfilver, and they fublime together in the form of a corrofive muriat of quickfilver. The fulphat of foda not being volatile, remains at the bottom of the fubliming veffel.

This compound, when well made, diffolves totally in water, but requires about nineteen times its weight of water.

When it is lefs foluble than ufual, we find that it will effervefce with nitric acid, being in a ftate which refembles in fome degree another preparation, to be mentioned prefently, in which the mercury is redundant, corroded, but not perfectly oxydated; and therefore it decompofes the nitric acid. It diffolves more eafily, if fome muriatic acid be added, or fal ammoniac, or common falt; the muriatic acid in thefe falts acting in fome degree, in confequence of the force with which it attracts the quickfilver. It is pretty extraordinary that two volatile fubftances, often fublimed together, fhould become more fixed: Yet this happens with mercurial fublimate, and fal ammoniac.

It muft be confidered as a triple falt. It is an ancient preparation, and was called *fal alembroth, i. e.* falt of the wife, or of the artifts, by the alchemifts. It cryftallizes readily, by cooling the folution made in hot water.

This preparation was called the *corrofive fublimate of mercury*, both on account of its being in fact very acrid and corrofive, in confequence of the quantity of muriatic acid it contains, and alfo to diftinguifh it from another fublimate of mercury which refembles it a little in external appearance, but is quite free from any corrofive quality. It is now called the *hydrargyrus muriatus*. This *mild fublimate,—mercurius fublimatus dulcis*, or *calomelas*, or *hydrargyrus muriatus*, as it is now called, is produced by a procefs which was contrived at random, in order to correct the corrofivenefs of the former preparation.

Corrofive fublimate is carefully mixed, by trituration, with an equal quantity of mercury, or with as much as can be ground with it, fo as completely to difappear; and this mixture is then fublimed. It rifes very imperfectly fublimed in the firft fublimation, and therefore of a dirty colour. Repeating the whole procefs makes it whiter, and by a third fublimation, we have it in its beft ftate,—a white ftriated mafs. The procefs is exceedingly hazardous to the workman who mixes the materials, for the dry powder is apt to be received into the lungs, and a very fmall quantity may be fatal. This is remedied in a great meafure by adding a little water.

The compound is a *corroded mercury*, quite infoluble and infipid, having no more tafte than the mildeft preparations. In it the mercury is much lefs oxydated than in corrofive fublimate.

I called this a random procefs, becaufe at the time of the firft preparation of this drug, the chemifts had no knowledge of any properties of mercury which could lead them to expect an intimate combination. Nay, to this hour it is not eafy to fay what is the exact nature of it. The mercury is here not merely oxydated, but combined with the acid, in a ftate intermediate between that of an oxyd and a muriat. Beaumé, a very judicious and experienced pharmaceutift, affirms, that by reafon of the great difference in the volatility of corrofive fublimate and fweet mercury, the two cannot be uniformly mixed by any number of fublimations, and recommends, as the only way of freeing the drug from corrofive quality, to boil it with water, and a little fal ammoniac, by which all the corrofive part is diffolved.

This compound is not very different from what may be obtained by diffolving mercury in the nitric acid, for it may be faturated with metallic mercury without allowing the formation of nitrous gas; and then pouring muriatic acid on the folution, a falt immediately falls down, fcarcely diftinguifhable from mercurius dulcis.

The extremely corrofive nature of the mercurius fublimatus is very evidently owing to the action on the inflammable matter in the fubftances corroded by it. Mr. Berthollet, while he held the Stahlean doctrine, attributed the corrofive quality of all metallic falts to their action on the phlogifton. His chief arguments were drawn from his experiments with this compound. When digefted with vinous fpirits, a confiderable portion became fweet mercury, and fome was revived. When diftilled along with oils, the mercury fublimed in a metalline form, and the oil was rapidly deftroyed, and left a prodigious quantity of charcoal. He now explains the fame phenomenon by the powerful action of the oxygen. No doubt there is a rapid combination of oxygen with the inflammable matters. But we do not fee how the combination of it with the metal tends to promote its combination with other fubftances; and I am by no means certain that the corrofive falts are more fo than the acids themfelves. It were, perhaps, as juft to fay, that the natural corrofivenefs of the acids is lefs mitigated by a combination with metals than with alkaline fubftances. Nor is Mr. Berthollet's opinion confiftent with the inoffenfive, nay fanative quality of the martial and faturnine compounds.

It is found that there is no intermediate ftate of oxydation between this of corrofive fublimate and that of mercurius dulcis. When too little mercury has been triturated with the corrofive fublimate, the mild fublimate is ftill formed perfectly mild; but, under it, next to the veffel, we find the undecompounded corrofive fublimate.

Thus we have feen how mercury may be combined with each of the three foffil acids, and with fome in different proportions, and in different ftates.

Mr. Margraaf difcovered that it may alfo be combined with the vegetable acids and with the fixed alkalis. And others fince, by following his method, have found that it can be joined with a number of other acids.

Mr. Margraaf's method is to precipitate the mercury from a nitrate of mercury, by a pure fixed alkali. The precipitate diffolves with effervefcence in acetous acid, in vinegar, in the acid of lemons and of forrel, and even in Rhenifh wine. It requires digeftion, but has no difficulty, and a great deal is diffolved. See alfo *Fourcroy*, IV.

277. alſo the publication of the Dijon academy. The acetate of mercury is the baſis of Keyſer's pill.

With reſpect to the order in which the acids attract mercury, I ſhall only obſerve, that of the foſſil acids, the ſulphuric and the muriatic attract the metalline oxyd more ſtrongly than the nitric, as appears by the proceſs of the corroſive ſublimate. The ſulphuric acid finding the mercury properly oxydated, unites with it very readily, without the aſſiſtance of heat. Theſe two acids can alſo be combined with the mercury, by double elective attraction, or double exchange, when neutral ſalts which contain them are added to the ſolution of mercury in nitric acid, e. g. vitriolated tartar, or Glauber's ſalt. or vitriolic ammoniac ;—and in the ſame manner, any of the compound ſalts which contain the muriatic acid.

Mr. Chaptal gives another proceſs extremely ſimple, and recommends it as producing a very perfect mercurius dulcis. Decompoſe mercurial water (ſolution of nitrate of mercury) by a ſolution of common ſalt. This throws down a white precipitate, which gives by ſublimation a mercurius dulcis. Scheele practiſed the ſame proceſs.

And by a ſimilar double exchange, this metal may be eaſily combined with the acid of tartar, or with that of ſalt of ſorrel, with which acids it forms a pearly precipitate, that is probably the preparation of mercury uſed in China.

It can alſo be combined in the ſame manner with the acid of phoſphorus.

Let us next conſider how it may be ſeparated again from the foſſil acids.—I ſaid already, that neither the ſulphat nor the muriat of quickſilver can be decompounded by heat ;—the acid and metal in theſe compounds riſe in vapour together.

It is not ſo with the nitrat. The quickſilver remains in the form of a red oxyd, which, if the acid be expelled from it as completely as poſſible, by the continued action of heat, becomes very ſimilar to the calx or oxyd of quickſilver prepared by the action of air and heat.

By the deep colour of the vapours of the nitrous acid thus expelled from the quickſilver, as well as the appearances during its diſſolution, it is evident that the metal retains a great quantity of the oxygen of the acid. This oxyd is much leſs volatile than mercury, but if made red hot, it is totally converted into vapours or elaſtic fluids.

In collecting theſe vapours we get by a proper apparatus,

1mo, A quantity of nitrous acid in red fumes, but too ſmall to be condenſed.

2do, A ſmall portion of the quickſilver, which is combined with a part of the acid vapour, and forms with it a yellow ſublimate in very ſmall quantity.

3tio, The greater part of the quickſilver gradually diſtils over, reſtored completely to its fluid metallic form,—and

4to, Along with this a large quantity of vital air is obtained. This vital air is much purer than that which is got from nitre. Still, however, it is not perfectly pure. It is always tainted with a minute quantity of azote; and, I believe, with a ſmall portion of the nitrous acid, and even of the quickſilver itſelf.

This manner of preparing vital air was firſt diſcovered by Dr. Prieſtley.

He obtained it from *mercurius calcinatus*, by means of a burning glaſs, on the 1ſt of Auguſt 1774, and ſoon after, from red lead moiſtened with nitrous acid, by diſtillation in cloſe veſſels. Next month he went to Paris, and communicated his diſcovery of this wonderful fluid to Mr. Lavoiſier, and other members of the Academy of Sciences, then aſſembled at his houſe, who were ſurpriſed and delighted with the diſcovery. Returning to Britain, he engaged in ſo many reſearches that this had but a ſhare of his thoughts. Mr. Lavoiſier, with perhaps as much curioſity, had more of a philoſophical and ſcientific mind, and was a much better reaſoner than Dr. Prieſtley. He took a juſter view of this experiment, and ſaw at once its vaſt importance in philoſophical chemiſtry, and he repeated Dr. Prieſtley's experiment with an attention and exactneſs in all his meaſures, that was·

altogether his own. He was the firft who made it certain that mercury and other metals have the power to decompound the acid of nitre, and that we are thus enabled to obtain from it two different fubftances in their feparate ftate, namely *nitrous air* and *vital air*. We obtain by itfelf nitrous air or nitrous gas, while the mercury is diffolving; and we obtain the vital air afterwards, when we expofe the diffolved or corroded mercury to a ftrong heat. And if we mix thefe two elaftic fluids together, they immediately unite, or their fundamental matters or bafes unite, and form again nitrous acid.

It is plain that I here fpeak of pure azote and pure oxygen. But it is very difficult to obtain thefe in a gafeous form, and at the fame time perfectly pure. The proceffes for obtaining them always leave a taint of the fubftances from which they are obtained. It is not difficult to procure the azote very pure. Lime will free it from carbonic acid, and hepar fulphuris will free it from oxygen. It is more difficult to procure vital air perfectly pure, and, efpecially, free from azote. The red nitrate of mercury affords the beft procefs I know for it. But I have generally found it tainted with nitrous air, and with azote. I call your attention to this circumftance, becaufe many of Dr. Prieftley's experiments, by which he ftill thinks that the theory of Stahl is fupported, have had refults which were certainly owing to fuch impurities. I have particularly in my eye at prefent thofe which he publifhed 1792.

Thus, therefore, we can by heat alone recover mercury, which has been combined with the nitric acid.

We cannot in the fame manner feparate the other two foffil acids from mercury alone. I obferved already, that they adhered to it too ftrongly; but we can feparate them by other fubftances which have a ftronger attraction for them than mercury has, and fome of thefe bodies ftill more readily feparate the nitric acid.

The fubftances which may be thus employed to attract thefe acids more or lefs perfectly from the quickfilver, are very numerous.

1st, Water applied in large quantity, and especially if heated, will in some cases produce a partial separation. This is remarkable with respect to the sulphat of quicksilver.

In washing the *mercurius vitriolatus* to make turbith, it is plain that the first boiling water takes from the mercurial calx the greater part of the acid. If, indeed, the mixture be allowed to cool, the calx resumes a great part of the acid again, and becomes soluble. I have examined again and again the quantity of acid it contains in these different states with care, and the differences are very sensible. It is difficult to say whether the heat increases or diminishes chemical action in this experiment. It is owing, in all probability, to the different proportions that are necessary, in different temperatures, for saturation, when we mix the compound of oxyd and acid with the compound of water and acid. If this proportion vary very much with respect to the last of these compounds, and very little with respect to the first, the phenomena will be what we observe. The nitrate is not so liable to be decompounded by water; but if largely diluted, it suffers also a little decomposition. The muriat cannot be decomposed by pure water alone.

2do, The precipitates by the fixed alkalis are of a reddish brown colour, approaching thereby to that of a pure oxyd of mercury. The reason is, that they are nearly pure oxyds, but not perfectly pure; they retain a small portion of the acid. The fixed alkali has taken from them nearly the whole, but left a minute quantity, which adheres strongly to the oxyd.

But the precipitates by the volatile alkali are remarkbly different from these, and from one another, the nitrate giving a gray, and the muriate a white precipitate. The cause of this difference is, that the volatile alkali has much more power to separate nitric acid from quicksilver, than to separate muriatic acid from it. The volatile alkali has a stronger attraction for the nitric acid than for the muriatic acid; and, in addition to this, the mercurial oxyd does not retain nitric acid so strongly, by far, as it retains the muriatic acid. In the white, or muriatic precipitate, therefore, the mercury retains a confiderable part of the acid, but not enough to render it foluble in water.   In the dark-gray precipitate from the nitric acid, not only the acid is almoft totally feparated, but a great part of the oxygen and the quickfilver is reftored, in part, to its metallic ftate, by the fuperfluous volatile alkali employed in forming the precipitation.   The volatile alkali contains hydrogen to attract the oxygen.   There is, however, combined with the mercury, a very fmall remainder of the acid, and fome of the volatile alkali.   And it alfo retains a fmall portion of the oxygen.

There arofe great varieties and uncertainties on the feparation of mercury from corrofive fublimate, by means of volatile alkali, depending much on the management.   The reafon is, that the acid feems to attract the mercurial oxyd more powerfully than it does the volatile alkali : For if mercury be diffolved with fal ammoniac, we inftantly difengage the alkali in a cauftic ftate.   When this alkali is applied to corrofive fublimate, it takes up fuch acid only as is not neceffary for faturating the mercury, and with this forms fal ammoniac, leaving the reft in the form of *mercurius dulcis.*   Lemery obferved this, by collecting and fubliming the precipitate.

3tio, Quicklime, too, has power to attract the acid from quickfilver, and to precipitate it even from thofe acids with which it is moft ftrongly combined.   The change of colour, when much lime-water is added, feems to proceed from fome action of the lime on the mercury which has not been examined.   Half a drachm of corrofive fublimate, thrown into a pound of lime-water, forms a yellow precipitate.   This water is ufed, before fubfidence, for cleaning the fkin from pimples, tetters, &c. by the name of Aqua Phagedenica.

Mr. Bayen difcovered that the precipitate of mercury from the acids by means of alkalis, particularly the mild volatile alkali, or by lime-water, when ground with a fmall portion of fulphur, form an exploding compound, fimilar to *pulvis fulminans.*   If expofed to heat, flowly increafed, till it melt, it explodes with great violence, leaving a residue, which fublimation fhews to contain mercury and fulphur, by forming a good cinnabar. Something approaching to this had been obferved by Brugnatelli, and others, and alfo very remarkable detonations by concuffion. (*See Note* 24. *at the end of the Volume.*)

4*to*, Precipitation of mercury with milk may be mentioned here. A faturated folution of mercury with nitric acid, mixed with hot milk, gives a rofe-coloured precipitate of mercury; but when the milk is but juft warm, it is of a purple colour. The intenfenefs of the colour, and the firmnefs of the coagulum depended on the heat of the milk. They are foluble in all the acids, and are mild tafted. Other metallic folutions form a coloured coagulum with milk, in the fame manner.

Many of the other metals have a ftronger attraction for acids than mercury, but thefe, in general, precipitate it in its metallic form.

If a piece of clean copper be moiftened with the nitrate of mercury in folution, and, after remaining a few minutes, be dipped in water, the furface will be found whitened; and if it be then rubbed hard with a clean and foft linen cloth, it will be as bright and white as filver. Putting this in a low heat, approaching to incandefcence, the whitening film evaporates, and leaving the furface corroded by the nitrous acid. Had the copper been allowed to remain long enough in the folution, it would have been diffolved entirely, and running mercury would have been found at the bottom of the veffel. (*See Note* 25. *at the end of the Volume.*)

Such is the variety of appearances this metal can be made to affume by the action of falts,—red, yellow, white, black powders, white tranfparent foluble falts. In fome of thefe ftates it is much more fixed than before.

But however it is thus changed, it can be eafily reftored again to its natural form. It returns readily to its ufual form, when expofed to heat, along with fubftances which can attract the acid from it, provided they feparate the whole acid, or nearly the whole, and reduce the mercury to the ſtate of a pure oxyd. This oxyd, as I obſerved before, when heated to a certain degree in cloſe veſſels, recovers its metallic ſtate. The ſubſtances added are iron filings, or fixed alkali, or quicklime, &c. The reſtoration of the mercury is not called reduction,—but REVIVAL, and the mercury is ſaid to be REVIVED. This refers to its name ARGENTUM VIVUM.

When an oxyd of quickſilver is revived by the ſimple action of heat, and without any addition it happens often that beſides the vital air, which it affords in great quantity, there is mixed with this a ſmall quantity of carbonic acid. This is ſuppoſed to have been attracted from the atmoſphere when the oxyd was too much expoſed to it, &c.

Neither mercury nor its oxyds can be made to combine with earthy ſubſtances, being too volatile.

Of the inflammables, ſulphur alone is diſpoſed to unite with mercury. Other inflammables are only employed to divide it. Sulphur may be combined with mercury, either by long and patient triture, or, more readily, by the aſſiſtance of heat, viz. by melting the ſulphur, and pouring in the mercury warm, and ſtirring well. The compound in either caſe is of an intenſely black colour;—hence, probably, called *æthiops.* When prepared by triture alone, the mercury is gradually divided and diſappears, the mixture becoming grey and blackiſh; but although the mercury appears to be perfectly incorporated with the ſulphur, it does not acquire the full black colour until it has been kept for ſome time, eſpecially if kept warm. Indeed, I find that the union by trituration is rendered at leaſt three times more expeditious, and the combination more perfect, by a heat juſt enough to make the ſulphur emit a ſenſible ſmell;—perhaps a little more heat would ſtill improve the proceſs. But ſtill, after the longeſt keeping, the mercury and ſulphur are more eaſily ſeparated than when mixed by fuſion. When the mixture is made by heat, they are more perfectly combined. But if they be made too hot, heat and flame are produced, ſeemingly by a fudden extrication of heat in the very act of union. This again arifes from two caufes: 1ft, The compound may have a lefs capacity for heat than the fum of the ingredients. 2dly, In the act of combination the mixture becomes ftiff, and fometimes hard. The heat therefore which makes an ingredient in their fluid forms, as latent heat, now emerges in fufficient quantity to kindle the fulphur. One part of fulphur is capable of forming a perfectly uniform æthiops with five or fix, or even feven parts of mercury. They are united together in this proportion for obtaining another preparation of mercury, viz. CINNABAR or VERMILLION, by a brifk fublimation, in oval or oblong veffels of coated glafs, or of earth.

This preparation forms a very confiderable manufacture, chiefly carried on in Holland, to fupply the market with the great quantity ufed in painting, and for medicinal preparations; and the defcriptions given by authors of approved proceffes are numerous, and have confiderable varieties. The moft valuable property of the compound is the richnefs of its colour. When this is very deep or full, a moderate quantity of this expenfive article bears dilution with a great deal of other colour. The fulnefs of colour correfponding alfo to its purity, becomes a general teft of its goodnefs. The attention of the artifts is, therefore, chiefly directed to this circumftance. It is found beft to ufe an æthiops prepared by fufion. That obtained by trituration and digeftion is apt to feparate, in part, in the fublimation. Stahl's procefs, or that in the London Pharmacopœia 1788, feems to me the cafieft and beft.

The mercury is made hot, and poured into the melted fulphur, and the mixture carefully ftirred. The flame fhould be extinguifhed by covering the mouth of the pot. The compound is then powdered and fifted, and rammed into an oval fhaped veffel, fo as to fill one-third of it. This is fet in fand up to the neck, and has its mouth open, in order to allow the efcape of a fuliginous vapour of uncombined materials.

When thefe fumes are condenfed, they form a mafs as black as pitch. It is compofed of the fuperfluous fulphur, and this is blackened by a fmall quantity of the metal loofely combined with it. The fublimation is carried on with a brifk heat, and it produces a very good cinnabar. Beaumé fays that two or three fubfequent fublimations greatly improve its colour.

The fublimate is a folid cake, of a blood red colour, which grinds to a red, fo much the brighter as it is more levigated. It may be made fo fine as to become too pale. Chemifts have wondered at this red colour, imputing it to fome kind of oxydation; and indeed an experiment of Lemery's feems to confirm this. Vitriolic acid renders it white, and fit to afford turbith; and the operation is accompanied with effervefcence. Now we know that the acid does not attack it till it be oxydated.

Hoffmann mentions a procefs by which he produced cinnabar without fublimation. He mixed running mercury with the volatile tincture of fulphur, that is, the compound of fulphur and volatile alkali. The mixture was long agitated violently. It foon became black, and at length grew a beautiful red, and a perfect cinnabar. Beaumé defcribes the fame phenomenon as a difcovery of his own.

Mercury is alfo faid to be affected by the fat oils. In particular, it is faid to acquire fome ductility, if poured boiling hot into lintfeed oil. I have not obferved any thing of this fort, nor can I indeed have any notion of the ductility it can acquire. A folid that is brittle or friable may become ductile: A fluid cannot; it may be rendered fluggifh, clammy, and perhaps vifcid or tough, like melted glafs, or fealing wax; but I have never obferved this change produced on mercury by the treatment now mentioned.

All thefe combinations of mercury with fulphur are decompofed, and the mercury revived again, by diftillation with quicklime, fixed alkali, iron filings,—all of which retain the fulphur. The mercury obtained from cinnabar is thought to be the pureft that can be had; and that obtained by means of iron filings is thought to excel the reft.   I am fenfible of its having a more perfect colourlefs brilliancy.   Its denfity is alfo the greateft.

Relation to other Metals.

Mercury unites with all the metals' except iron, arfenic, and platina.   With the regulus of antimony, indeed, it unites very imperfectly, as we fhall fee in a little ; and its more intimate union with the other metals will alfo be confidered in their places.

Although arfenic does not unite with mercury, it affects the mercurial compounds, and is itfelf affected in a particular manner.   If reguline arfenic be heated with twice its weight of the mercurial muriat (corrofive fublimate) in the way of diftillation, the vapours condenfe into an oil-like liquor, part of which grows gelatinous, and almoft like very foft butter which has been melted, being grumous or a little granulated, which I take to be an incipient cryftallization. When this has finifhed, the vapours condenfe into metalline quickfilver.

This is a curious enough procefs.   Neither the mercurial muriat, nor the regulus of arfenic, are fo volatile as this compound ; and the regulus is not fo volatile alone as mercury is.   The fluid and gelatinous matter are a compound of arfenic and muriatic acid.   Water decompofes it in the fame way as it does other metalline falts, that is, it feparates it into two parts, one of which is nearly in the ftate of an oxyd, and the other contains this oxyd diffolved in the acid.   This oxyd is an intermediate ftate between that of the arfenical acid and an arfenical oxyd.

I may remark on this occafion, that mercury clears the acid of arfenic (white arfenic) from fulphur and other inflammable matters which frequently make it foul.   With the fulphur it forms an æthiops, which yields cinnabar in the ufual way.   I do not fee fo well how mercury can feparate the other impurities ; and it is the more remarkable, becaufe all other metals (filver and gold excepted) render arfenic foul.

I may alfo remark, that all the metals, when they decompofe the muriat of mercury, form the fame kind of butter-like compound; and it is moft violently corrofive, deftroying all organized bodies in a minute. Leaks in the diftilling apparatus are, therefore, productive of the moft fatal confequences to the operators.

Mercury is fufceptible of mixture, in fome manner, with all the other metals. And fuch mixtures, called *amalgamas*, are of different confiftency, according to the proportion of the ingredients. If there is little of the other metal, they are fluid; if more, they are foft, like pafte or butter. If there is lefs mercury than an equal weight, they form a fort of folid, more or lefs brittle and friable, in proportion as there is more or lefs mercury. The colour of thefe mixtures is always white. In amalgamas that are very foft or fluid, the mercury is in two different ftates. Part of it is fluid, and nearly pure; the other part is united to the metal in a folid form, in the fame manner as water is united to falt in cryftals of falts. This folid matter, compofed of the metal and part of the mercury, is felt like gritty grains in the amalgam: Or fometimes it actually forms oblong cryftallizations in it. Thus, if we diffolve gold in boiling mercury, fo as to form a foft amalgam, and then allow it to cool, the gold cryftallizes or concretes with part of the mercury into fpicular concretions, which are felt by the fingers. Hence amalgams, if fluid, can have part of the mercury imperfectly feparated by expreffion through leather; but long triture, or long digeftion, feems to produce more intimate union. It can be feparated, however, by diftillation.

Upon the difcovery of thefe preparations of mercury, in mixture with the different metals, have been founded the arts of feparating gold and filver, but efpecially gold, from their matrices, (of which hereafter); alfo the art of foiling glafs for mirrors; and that of gilding filver, copper, or brafs.

The procefs for foiling glafs mirrors is performed in the following manner. The glafs is laid flat on a very fmooth and firm table. A fmall border is made all round the edges, to hold in the mercury. The glafs is then wiped very clean; and fometimes rubbed over with a piece of foft leather, on which a little amalgam of mercury and tin has been fpread. It is then covered all over with tinfoil; and mercury is poured on it, till the border is full. The mercury penetrates the tinfoil, and makes it lie clofe to the glafs, like a wet cloth. Indeed mercury penetrates the tinfoil juft as water penetrates and foftens a cloth. When this has remained fome time, the mirror is flowly raifed on one edge, that the fuperfluous mercury may run off; and it is allowed to ftand in that pofition fome days. Thus all the mercury not combined with the tin gradually fubfides to the lower edge, leaving only what is neceffary for keeping the tinfoil adhering to the glafs plate.

Glafs globes are coated on the infide, by warming them very much, and pouring in fome fluid amalgam, which adheres to the glafs like water. By turning the globe round and round, the whole infide may be thus fmeared with the amalgam, fo as to look like a metal globe highly polifhed; but when held between the eye and the window, we difcover that it is covered only as with a cobweb of the metal.

Silver and brafs are gilded by fmearing them with an amalgam of mercury and gold. This makes them as covered with filver. The piece is then put into an oven, and the heat evaporates the mercury, leaving the gold adhering to the piece, in the form of a dirty brown powder. This is now gone over with a fine burnifher, made of fteel or of hæmatites, highly polifhed. This flattens down all the particles of the gold, fpreads them out, and caufes them to adhere, while it alfo polifhes them. Copper and brafs acquire a fkin that is undiftinguifhable from gold. But this gilding on filver has always a brown colour, very diftinguifhable. Moiftening the furface of the piece with aquafortis caufes the gilding to be more uniform and adhefive: For whenever the mercury touches the wet furface, the metalline nitrate adhering to it is decompofed by the copper, and fome mercury is left on it in a metalline ftate. The buttery amalgam unites with this very readily, and when the mercury is expelled by heat, the gold remains. Without this precaution, parts of the furface refufe to let the amalgam attach itfelf to them. In confequence of the fame properties, it may alfo be applied to fome ufeful purpofes in medicine and furgery,—to diffolve a lead ball, a piece of leaden tube, or probe, in the bladder; alfo to take off a ring from a fwelled finger.

All the metals exhibit a peculiar appearance, when employed to decompofe the muriat of mercury. They unite with the acid in a foft buttery form. Another general circumftance in this decompofition is, that this butter is more volatile than the muriat, and even than the mercury, although the metal employed may be much more fixed. Thefe compounds will be mentioned in courfe; but I thought it proper to make you notice thefe general charaters.

Ores, or Origin.

Mercury is fcarcer than gold. The only mines are at Almaden, in Spain; Idria, in Hungary; in the Duchy of Deux Ponts; in Friuli, in Italy; and in the Eaft and Spanifh Weft Indies. Small quantities are fometimes found in France and Britain. Virgin, or fluid mercury, is often interfperfed through other minerals; but the moft common ore is cinnabar. The fluid mercury is ofteneft found where cinnabar is moft abundant. I think that the rarity of this metal, and even of fmall portions of it mixed with other ores, is an abundant proof that there is no fuch thing as a *mercurial principle*, the bafis of metallization. Were it fo, we fhould find imperfect metals very frequent, and in different ftages of the progrefs, to the ftate of perfect metal.

The account given of the chemical properties of mercury makes it eafy to perceive how we fhall obtain it from its ores. All the proceffes, performed at all the mines, are diftillations, managed with more or lefs art, and varied according to the nature of the ores. Even where found in a fluid form, it is generally in globules, too fmall to be feparated by the hand, or by pounding and wafhing. It is much eafier to do it by diftillation. The cinnabar ores are generally mixed with fo much calcareous matter that no addition is neceffary for feparating the fulphur. When this is not the cafe, (as at Deux Ponts), flaked lime is mixed with the ore in the retorts. Thefe are placed in a row, in a long furnace like a gallery. Their necks project through holes in the fides. To thefe are fitted receivers containing water, into which the mercury falls. At Almaden in Spain, a long row of earthen aludels go from each retort, down a gently floping terrace, and then up a fimilar terrace, and terminate in a chamber, having a concave floor, and a tall chimney. The diftillation is forced by confiderable heat, and the vapours condenfe in the aludels, both in the defcent and fubfequent afcent. What are not condenfed there are condenfed in the chamber. The floping pofition of the aludels caufes the mercury to run out of the higher into the lower, on both fides of the valley formed by the two terraces, and it all collects in the aludel at the bottom, This one has a fhort pipe, or neck, in its under fide, which is plunged into a hole cut in the ftone which forms the gutter between the two terraces. The gutter is filled with water. From this conftruction, it is plain the aludels have no communication with the air, except by the chamber. The diftilled mercury, therefore, runs out of the lower aludel of each row, fills the hole in the gutter, and then runs over, with water above it, into the gutter. This has a very fmall declivity towards one end. The mercury, therefore, runs along this gutter into a gutter at one end of it. Thus the diftillation is carried on with rapidity; and there is no rifk of burfting the veffels, or of the vapour efcaping by any leaks.

Medicinal Preparations of Mercury.

The very great activity of mercury on the human body, and its effects when exhibited as a medicine, are fo remarkable in the treatment of many difeafes, that phyficians have turned it into almoft every poffible fhape; and the mercurial preparations are almoft innumerable. In this refpect it has but one rival in the foffil kingdom, viz. antimony, the preparations of which are no lefs various. To pafs them by would be highly improper; but to defcribe them all would require a winter's feffion, and would be no lefs improper. I muft content myfelf with confidering them in two points of view. I fhall defcribe fuch of them as indicate any fingular chemical property of mercury and antimony, by which your chemical fcience is improved. I fhall alfo take particular notice of any connection or dependence that I can difcover between the medicinal effects of the preparations, and the general chemical refemblances that may obtain among the methods of preparing them. This will direct the medical chemift to fuch treatment of the metals as are the moft likely to produce the defired medicinal effect. Thus will the philofopher be inftructed, and the artift directed. *(See Note 26. at the end of the Volume.)*

GENUS V.—ANTIMONY.

AFTER mercury, I fhall next confider the brittle and volatile metals, which, on account of their wanting ductility, are called imperfect, or femi-metals; and antimony *(ftibium olem)* being one of thefe, we fhall take it firft. It is the one to which the chemifts and phyficians have paid the greateft attention, torturing it into an endlefs variety of forms.

ANTIMONY is a dull filver-coloured metal, of the fpecific gravity 6,7. Its furface exhibits facets, and a ftar-like figure. This appearance proceeds from a cryftallization, or arrangement of its parts in cooling. It is totally deftitute of malleability, and can eafily be beaten to powder. It is poffeffed of fufibility and volatility. The vapours of it calcine to a white fmoke, and condenfe on the furface of any cold body expofed to them. Or if they are collected, and fomewhat confined, in the cavity of a veffel applied to receive them, they condenfe

into a matter compofed of minute fpicular cryftals, very like fnow, or
flores benzoini, called *nix antimonii*,—fnow of antimony ; or by fome
chemifts, filvery flowers,—*fleurs argentines*. You may fee all thefe
phenomena by heating a little bit of antimony with the blow-pipe on
charcoal. There is here an evident inflammation. The white fmoke
is partly compofed of an oxyd, formed by the action of the air, and it
adheres to any cold body in form of a white powder. It is pretty
highly oxydated. But when the heat of the mafs abates, and the vapours arife more flowly, and are lefs calcined, a part of them condenfes on the furface of the metal itfelf, appearing like a covering of
fnow.

We can alfo calcine this metal with an inferior heat, fo as not to
convert it into vapour. It muft be beaten to powder, and fpread out
in a broad and fhallow veffel, and expofed to a very gradual heat.
The furface of the metallic particles foon becomes tarnifhed and
dufty, and they are by degrees converted into a duft, or earthy-like
powder ; firft dark gray, or afh-coloured, but by continuance of the
calcination, yellowifh, and at laft white.

Thefe calces, or oxyds, are remarkable by proving all volatile, if
expofed to the action of heat and air at the fame time, and when the
vapour of them is condenfed by cold and a proper apparatus, they form
the filvery flowers.

Some chemifts have thought that there was a refemblance between
antimony and arfenic ; and they certainly refemble one another in this
refpect, that the calces of both, when in a ftate of moderate calcination, can be converted into vapour very eafily. But the calx of antimony differs from white arfenic in this point, that whereas white
arfenic will evaporate in clofed veffels, or by the action of heat alone,
without requiring the affiftance of frefh air, the calces of antimony
will not evaporate, except when expofed to the action of both heat
and air. If the accefs of frefh air is excluded, they endure a very violent heat : And the effect of the heat in this cafe is to melt them in-

to a glafs. They melt the more eafily the lefs they have been calcined. And the glafs thus produced has always a deep ruddy brown
yellow colour, which is fo much the deeper and darker, in proportion
as the calx was lefs calcined.

All fuch calces or glaffes are reducible to a metallic ftate: The lefs
calcined, however, the more eafily and completely. The reduction is
performed by fufion with charcoal duft, or black flux, or fixed alkali
and foap. So far of the effects of heat.

The greater number of the acids can be made to act on antimony in one way or another. Some attack it readily in its metallic
ftate; others better when it is flightly or moderately calcined,—but
when it is greatly calcined few of them are difpofed to unite with it.
Dr. Pearfon in his differtation on James's powder (*Phil. Tranf.*
1791.) gives an account of a feries of experiments made with the
view of comparing their different folubility; and it appears that the
degree of calcination has a very fteady connection with the folubility
of the calces.

The action of the fulphuric acid on antimony very much refembles its action on mercury. It muft be ftrong and hot. The action
is attended with effervefcence and the eruption of fumes of fulphurous acid; and towards the end of the procefs, actual fulphur. The
effervefcence being frothy, it is very troublefome, rifing in the veffel
and running over. After fome time, this ceafes, and the liquor may be
faid to boil rather than to effervefce,—it is ftill however the fame elaftic
gas which it emits. At length, the metal is completely taken up, or
changed, and we have a white precipitate, and a folution of a fulphat
of antimony. The precipitate is nearly a perfect oxyd. The folution will not cryftallize by evaporation; but when reduced to a dry
mafs, it immediately deliquiates again. The folution may be decompounded by alkalis, and gives an oxyd extremely difficult of reduction.

Nitric acid can fcarcely be called a folvent of antimony, for its form never difappears. After remaining fome days in the acid, the metal is only found divided, fwelled up, and covered with bumps like colli-flowers, juft as in its calcination ; the laminæ are feparated, and the interftices filled with a white oxyd. All this is accompanied by a great change in the acid,—red fumes are emitted in abundance, and the liquor acquires a green colour. In fhort the acid is decompofed. When the materials are examined, we find the metal completely changed. There is a matter lying at the bottom, which, on examination, is an oxyd of the metal, with fome excefs of oxygen. The liquor contains a nitrate of antimony. Water caufes fome precipitation ; and the re-mainder is the folution of a very deliquefcent falt, which can be de-compounded by an alkali. This precipitates an oxyd of antimony, which it is very difficult to reduce to the metallic ftate.

Muriatic acid acts very flowly and languidly on antimony ; but it acts on it, and diffolves it with a flight effervefcence. If, however, it be digefted on antimony, it diffolves it, and retains it, when nitric acid, either ftrong or weak, is added to the folution ; thus, making it evident that it attracts the metal. With care, the muriat may be made to cryftallize in fpiculæ, but it is extremely deliquefcent.

The action of the marine acid on antimony is much promoted by adding to it a fmall quantity of nitric acid. When this mixture is poured on the metal, we have at once a brifk folution, and the co-pious eruption of nitrous fumes, even although the quantity of nitric acid fhould not amount to one-tenth of the muriatic. It is better however to add one-fifth. This mixture diffolves a confiderable quan-tity ; and the muriat yields a very deliquefent falt, which is alfo very fufible and volatile,—and is parted by water, like the nitrate, yielding an oxyd, and retaining the true muriat ; and this may be decom-pounded by an alkali.

Thefe phenomena, exhibited by antimony in the nitric and muriatic acids, are eafily underftood. The nitric acid contains oxygen very weakly united. Part of it unites with the antimony, and the oxyd, being very little foluble in water, lies at the bottom of the liquid. The fumes are nitrous gas. The muriatic acid has its ingredients united too firmly, and does not therefore act on the metal without long digeftion, and even then acts feebly. But a little nitric acid fupplies it with as much oxygen as fuffices for the oxydation of the metal; and the fumes which break out are nitrous. Indeed the fimple addition of colourlefs nitric acid to the muriatic immediately emits the fumes. When we have got the metal into the ftate of an oxyd, the folution goes on apace. In confirmation of this explanation, I muft obferve, that the oxygenated muriatic acid diffolves the antimony with great facility, and produces the very fame compound with that juft now defcribed.

This, however, is not the ufual or moft complete way of effecting a combination of antimony with the muriatic acid. The proceffes in common ufe are different, but they depend on the fame principles.

In one of thefe proceffes we mix fixteen parts of corrofive muriat of quickfilver with fix parts of the metal of antimony, both in powder, and apply to this mixture a fubliming heat in a retort and receiver. The metal is in this cafe both oxydated and at the fame time combined with the dry and ftrong muriatic acid of the corrofive muriat of mercury. The oxydation is effected by the mercury, which being itfelf in a highly oxydated ftate, or combined with a good ftore of oxygen as well as with acid, transfers both the oxygen and acid at the fame time to the antimony, which has a ftronger attraction for them both than the quickfilver has. The antimony therefore is both oxydated fufficiently, and combined with the acid,—the quickfilver refuming its metallic form. In this, as well as in the fimilar procefs with arfenic and mercurial muriat, the mercury requires a greater heat to raife it in vapour than the antimonial muriat; and this laft is even more volatile than the mercurial, though antimony is more fixed than mercury.

The new compound differs from the corrofive muriat of mercury by being more eafily melted than volatilized, whereas the corrofive mercury is more volatile than fufible. The antimonial cauftic or corrofive therefore rifes in vapours, which condenfe firft into a fluid matter, on fome of the cool parts of the retort, but when they reach parts ftill colder, the fluid matter quickly congeals into a folid icy-like fubftance, which can be eafily melted again, and congealed at pleafure, melting with a gentle heat almoft like tallow or butter. Hence the ancient chemifts gave it the name of *butter of antimony.* Its colour is dark at firft, but, by rectification, it becomes white or colourlefs like ice; a little fuperfluous metallic matter and fome impurities being thus feparated from it.

When we make further experiments with this compound, we learn that it has a ftrong attraction for water. A very fmall quantity of water diffolves the greater part of it, or reduces it to a liquid form; and it readily attracts fo much water from the air, if it be expofed in open veffels. As it is more convenient to keep and ufe it in a liquid form, it is commonly fold liquid, by the name of *antimonial cauftic.* In the laft editions of the London and Edinburgh Pharmacopœias, this compound is called *antimonium muriatum.* It is very acrid or corrofive with refpect to animal and vegetable fubftances. A little of it applied to the fkin very foon burns or deftroys the part. Whence antimonial cauftic is employed to deftroy fungous excrefcences from ulcers or other parts.

Water, added in large quantity to this compound, affects it in the fame manner as it does the fulphat of mercury. The water attracts and diffolves the acid, leaving the metal in the form of a white oxyd, which, although wafhed with repeated additions of hot water, retains ftill a fmall part of the acid. It is called *powder of Algarotti,* from the name of an Italian phyfician, the inventor; alfo *mercurius vitæ,* though it contains no mercury. Nor does it contain any acid, if prepared with proper care. It fhould be a pure oxyd, pretty copioufly oxydated.

Antimony unites alfo with the vegetable acids. The acetous and tartarous acids, and the acid of vinous liquors, diffolve it, and become violent emetics. But thefe acids diffolve it much better when it is flightly oxydated. They are medicines of great activity, and fhould therefore be very accurately prepared. Unfortunately, it is very difficult, if poffible, to reduce them to the fame ftrength when prepared by different proceffes; and the phyfician prefcribing his dofe by the effects of a preparation to which he has been accuftomed, may produce very unexpected effects, when the fame dofe is given of the medicine otherwife prepared. I think that the precipitate from the antimonial muriat has a greater probability of being uniformly the fame than any other oxyd of the metal. It is, therefore, a good bafis for thefe vegetable additions.

Of the neutral falts, only nitre and fal ammoniac act upon antimony. The firft deflagrates, and the heat of deflagration volatilizes fome of the metal. We thus get an oxyd whiter and lefs fufible in proportion to the quantity of nitre. Nitre two parts, to antimony one part, gives an oxyd very white and unfufible. The alkali of the nitre, mixed with this oxyd, is called *reguline cauftic.*

Sal ammoniac prefents nothing particular, or different from its action on metals in general.

Of the inflammable fubftances, fulphur readily unites with this femi-metal in the fire, and forms a fufible compound, which bears a great refemblance to the femi-metal by itfelf; but the fulphuric compound has lefs brightnefs, and a darker colour, and more flender cryftallizations.

The fulphur can be feparated, and the metal recovered in its pure ftate by different proceffes; as,

1/t, By roafting the fulphurated antimony, fo as to evaporate the fulphur, we get the metallic matter in the form of an oxyd, which retains very little of the fulphur, and may be reduced to the metallic ftate by melting it with black flux. The firft part of this procefs requires patience and attention, on account of the volatility of the metallic matter. I fhall take further notice of it afterwards.

2dly, Another method, which is fhorter, is to melt the fulphurated antimony with an equal, or half its weight of black flux. The alkali joins the fulphur, and a part of the metal in its pure ftate feparates to the bottom of the veffel. But by this method we obtain only a part of the metal. A confiderable part is diffolved by the alkaline fulphuret ; and if a pure alkali is ufed in place of black flux, it diffolves the whole.

3dly, We can feparate the fulphur by other metals, many of which have a ftronger attraction for it than this femi-metal has. Iron is commonly ufed, either in filings, or fmall nails, or the fcraps from the tinplate-workers. A ftrong heat is neceffary. Thus the antimony fettles to the bottom, the fulphurated iron flows uppermoft, and is feparated more eafily, if fome faline matter has been added to promote the fufion.

When this metal happens to be combined with fulphuret of alkali, it can hardly be feparated again, except in this manner, by melting the compound with other metals which have a ftronger attraction for the fulphuret. Other metals thus employed to feparate the antimony from fulphur, or alkaline fulphuret, are liable however to mingle with it a little, and to make it impure. For it is difpofed to mix with all other metals, if properly applied to them with a melting heat. Iron appears to have the ftrongeft attraction for it. With all the tough metals, it produces compounds more or lefs brittle. And befides this effect on iron, it deprives the iron of its magnetic qualities and difpofitions to be attracted by the magnet. No other metallic fubftance affects iron in this manner *.

The method of feparating it from the different metals is various, according to the nature of the metal with which it is joined.

* I find that lefs than three parts of antimony to one of iron will not deftroy the magnetifm of the iron.   EDITOR.

From gold or platinum it is feparated by evaporation, or the action of heat and air.   From other metallic fubftances it is beft feparated by the addition of fulphur, or, in fome cafes, by fcorification, as from lead.

Such is the chemical nature of this femi-metal.   It is one of thofe which are produced by nature in large quantity in fome places ; and the ore of it is every where nearly of the fame kind.   Sometimes, though rarely, it has occurred in a ftate of purity ; but fuch fpecimens are very rare, and the quantity trifling.   The moft common ftate in which it is found is combined with fulphur ; and I believe there is fome variety in the proportion of the fulphur to the metal in the different ores.   Moft of them contain a little arfenic :   Its prefence is eafily known by the white and filvery appearance of the ore, and by the orpiment which it yields by fublimation.   The appearance of thefe ores is very like that of the artificial compound of the metal and fulphur.   And the only operation which the metallurgifts perform with thefe ores before they fend them to the market, is to feparate from them the earthy and ftony matters, or matrix, which happens to be combined with them.

This is done by the procefs named in metallurgy *Eliquation.*   The crude antimony is put into earthen pots, pierced in the bottom.   Thefe are fet upon other pots buried in the ground.   The fuel is thrown in around the pots ; and the fire kept in a uniform ftate, by the judicious ftructure of the furnace.   When the ore is fufed, it runs into the lower pots, and the earthy matters remain in the upper ones.   It is frequently moulded in truncated conical loaves.   As this is the only operation performed on this mineral to fit it for fale, it comes into the hands of the chemifts and druggifts in this ftate, in which the metal is ftill combined with fulphur, and may be confidered as yet in the ftate of an ore.   They were accuftomed to call it *crude antimony,* or fimply, *antimony.*   And the pure metallic part, when feparated from the fulphur, has been commonly named *regulus of antimony ;*

but the later chemists give the name of *antimony* to the pure metallic part; and to this compound, that of *sulphurated antimony*, or *sulphuret of antimony*.

It is therefore by working on this sulphurated antimony, or crude antimony, as it is called, that most of the preparations of antimony are produced, in the art of pharmacy. By what you know already of the nature of its two constituent parts, I mean sulphur, and the metallic substance which I have already described, you will now easily understand the changes it undergoes by the action of heat and air, or of different solvents, and other active substances.

Preparations of Antimony.

Before I begin to describe the different preparations of this mineral, I beg leave to make a remark upon the whole of the subject, which will assist you to understand how these preparations differ from one another, in point of efficacy as medicines; for the general purpose of preparing crude antimony is to give it more or less medicinal efficacy, to make it either a strong and powerful remedy for some purposes, or one that shall be mild and safe for others.

In the first place, there is good reason, from experience, to believe that antimony never acts as a medicine, except when dissolved or combined with an acid, either before it is taken into the stomach, or in consequence of its finding an acid there with which it unites. But in crude antimony, the metal neither is already united with an acid, nor is it much disposed to join with the weak acid which alone it can meet with in the stomach. The sulphur with which it is joined, and the uncalcined state of the metal, are unfavourable to its being easily dissolved by a weak acid. In order to make it easily soluble in the acid of the stomach, we must separate more or less of the sulphur, and *calcine* or oxydate the metal to a *moderate degree*. I say a moderate degree, for if we oxydate it very much, we again diminish its disposition to be dissolved. By taking away, therefore, more or less of the sulphur, and moderately oxydating the metal, we produce preparations which are capable of acting very powerfully; but their action is not always the same, or in proportion to the dose. It depends on their meeting with acidity in the stomach, which is not always present there in sufficient quantity to give them all their efficacy.

The surest way to reduce antimony to a form in which it will act equally and strongly (at least so far equally as the different constitutions of different patients will permit), is to separate the sulphur, and combine the metal with an acid, so as to give it a saline soluble form before it is thrown into the stomach.

And now, having premised these remarks, we shall proceed to describe the different preparations, in the order in which they are arranged in the table of them which I have put into your hands. (See Note 27. at the end of the Volume.)

The first is the *antimonium præparatum*, in both our pharmacopœias. It is crude antimony, reduced to a very fine powder, simply by triture and elutriation with water. I remarked already, that crude antimony has but little efficacy as a medicine, or is an exceedingly mild one. This is a consequence of the state of the metal in it, which is neither already united with an acid nor much disposed to unite with the acid of the stomach, not being prepared by any degree of calcination, and having also the sulphur adhering to it. The antimonium præparatum, is, therefore, never given when we propose to vomit or purge. It is only used to promote perspiration, or to excite the other evacuations so gently that its effect is hardly perceived, and that the use of it may be continued constantly for some time; and to make it act even thus, the greatest levigation is necessary. The antimonium præparatum is, therefore, crude antimony, reduced to this state of a very fine powder ready for use. Kunkel, an eminent German chemist, is said to have cured himself of a rheumatism with it. I once had an opportunity of perceiving that it sometimes excites nausea, and expels worms; a proof that it is not quite inactive. In preparing it, we muft be careful to avoid the bafe or broader end of the conical loaf into which it is moulded for fale. If any impurities, or admixture of other metals are in the mafs, they are in this upper part. They fcorify with part of the fulphur, and being thus lighter than the reft, float above it.

The next preparations are thofe produced by the action of heat and air.

When crude antimony is expofed to heat fuddenly, the moft of it evaporates, efpecially if air be admitted. This is to be expected, as both the metal and fulphur are volatile fubftances. Its evaporation is attended with fome inflammation, vifible in the dark. By careful procedure, the fulphur may be completely diffipated by roafting the powdered antimony. No fuffocating fmell will be perceived. A fudden increafe of heat would mar the operation, by melting the antimony, and thus diminifhing its extenfive furface. As the operation advances, we may increafe the heat without danger of melting. This may be continued till all the fulphur is expelled. The metal now begins to attract oxygen, and grows gray, and at laft white; and in this ftate is a pure oxyd. Thus you fee that by terminating the procefs at different periods, we fhall procure the metal in very different ftates. It is therefore a matter of importance that it be conducted in a very uniform manner.

There is the fame, or perhaps a greater variety, when the heat is fo great as to caufe the fulphur attract oxygen. In this cafe we have fuffocating fumes, &c.

The only preparations which have been produced by the action of heat and air are, the *flores antimonii fine addito*, the *calx antimonii*, the *vitrum antimonii*, and the *vitrum ceratum*, (*Pharm. Edin.*) prepared from the vitrum.

To prepare the flores antimonii fine addito, the fulphuret is put into a crucible, or unglazed earthen veffel. A fet of aludels are fet on it, and a pipe is inferted by a proper opening in the fide or lip of the crucible. The fire being kindled, and grown pretty strong, the fumes begin to arise, and then a pair of bellows are applied to the pipe, and a gentle stream of air is made to play on the surface of the materials, but in the most gentle manner possible. The metallic and sulphurous fumes now rise, and are condensed in the different aludels. The sulphur of the flores, being the most volatile of the fumes, rises highest, and the oxydated flowers are contaminated by it so much the more as they rise higher, or rather they rise so much the higher as they contain more sulphur. The artificial stream of air cannot be managed with such delicacy as to preserve that regular gradation that may be obtained in a small quantity sublimed without the blast. This preparation, therefore, is unequal and uncertain, and is now disused.

Vitrum.——In making the glass of antimony, we begin by separating the sulphur, or the greater part of it, by the action of heat and air, and at the same time oxydate the metal to a moderate degree.

Both these objects are attained by performing the operation practised with ores of metals to free them from sulphur. It is named *ustulation.* The operation of ustulating sulphurated antimony is difficult on account of the fusibility of this mineral, and of the volatility of its metallic part.

To diminish fluidity, some add charcoal dust; others add common salt, which is easily separated afterwards by water.

But the common way is to ustulate it without addition.

It necessarily happens in this operation, that the metallic part is oxydated, while most of the sulphur is evaporated. And, if the operation be properly managed, we have an oxyd of the metal of a gray or ash colour, moderately oxydated, and with very little of the sulphur remaining in it, and therefore in a condition to be easily melted into glass; for we have learned by experience that the presence of a little of the sulphur greatly promotes its melting into glass.

It is therefore by simply melting an oxyd of antimony, prepared in this manner, that the glafs of antimony is obtained.

Vitrum ceratum is prepared (in order to mitigate the action of the vitrum antimonii) by kneading the pulverized glafs of antimony with bees wax, and then burning away the wax.

The next preparations of antimony in the table are thofe prepared with alkaline falts.

The fixed alkalis act readily on crude antimony, on account of the fulphur which it contains, and with which they have a ftrong difpofition to join.

Of this we have examples, when we apply thefe falts to the crude antimony, either in the dry way by fufion, or in the humid way, in the form of watery folution. They act moft readily and powerfully when applied in the way of fufion. The alkali joins itfelf to the fulphur, and forms a fulphuret of potafh,—a compound which has great power to act on the metals in fufion, and to diffolve even thofe with which fulphur alone cannot be combined. The metallic part of the antimony, therefore, does not feparate, or very feldom, and in fmall quantity only. Commonly the whole mixture melts into one uniform vitrified-like mafs. If one part of fixed alkali and five of crude antimony are melted together, and fome common falt added, which promotes their fufion, we obtain a dark coloured vitriform mafs, once known under the improper name of *regulus antimonii medicinalis.* It is eafily ground to powder, and gives a powder of a reddifh brown, or a fort of chocolate colour, which is infipid on the tongue, and not foluble in water. It is a very mild preparation, but has a little more efficacy than the antimonium præparatum.

Larger proportions of alkali form maffes which prove more or lefs foluble in water. Thofe which contain one of alkali to two of crude antimony are foluble in hot water, not in cold. With larger proportions of alkali, they are foluble in cold water as well as in hot. With two of alkali to one of crude antimony they are even deliquefcent, or attract humidity.

The moft proper name for all thefe compounds is alkaline fulphurets of antimony.

When we apply water to them to diffolve thofe that are foluble, none of them are diffolved completely.   There is always a feparation of a reddifh or brownifh fediment, which is more plentiful in proportion as there is lefs alkali in the compofition of the mafs.   This fediment is formed by a part of the antimony, and it is fome of the more metallic part which thus feparates.   The reft, which in confequence of this feparation is more fulphurous, and has more alkali combined with it, remains diffolved.

Befide the way by fufion, I obferved before that we can form compounds of this kind in the way of watery folution, or, according to the language of the chemifts, in the humid way.

Though the alkali, in whatever manner applied, acts by its attraction for the fulphur chiefly, it diffolves in this way alfo the greater part of the metal.   It continues adherent to the fulphur, and is diffolved along with it.   The folution, therefore, which we thus obtain, is a folution of an alkaline fulphuret of antimony, which is rather more eafily decompounded, and requires more water to keep it diffolved, than a plain fulphuret of potafh ; and the caufe of this is, that the metal, by its attraction for the fulphur, diminifhes a little the cohefion of this laft with the alkali.

We quickly decompound this fort of fulphuret, when we add an acid to the folution.   This immediately neutralizes the alkali, and occafions the fulphur and metal to feparate from it.   While they feparate from the alkali, they remain combined together, and form a powder or precipitate of a deep orange or red colour, called formerly *fulphur antimonii auratum*, now *fulphur antimonii præcipitatum.*

In this preparation, the antimony, although it retains much the fame principles as in its crude ftate, is much more active however as a medicine. The diffolution it underwent has divided the particles far more fubtilely than can be done by any mechanical divifion or triture. And the clofe union of the fulphur and metal was diminifhed during the combination, in confequence of the action of the alkali on the fulphur. The metal is therefore more difpofed for being diffolved by the acids of the ftomach, than it is in crude antimony.

But, in order to make the product of this procefs always equal, it is neceffary to add the requifite quantity of acid for faturating the alkali, and precipitating the antimony all at once. If we add only a part, and collect the firft precipitate before we add the reft of the acid, the firft and fecond precipitates will be unequal. The firft will contain more of the metal, and lefs of the fulphur, than the fecond precipitate ; the attraction of the alkali being chiefly for the fulphur of crude antimony. It therefore quits it more flowly than it quits the metallic part.

When we take the potafh in its common ftate of an imperfect carbonat, and diffolve the antimony by it in the fame humid way, it acts much more flowly and with lefs power. And when the boiled folution is allowed to cool, it depofits again the greateft part of the crude antimony which it had diffolved, fhewing thereby that the prefence and affiftance of heat is neceffary to enable fuch an alkali to hold the crude antimony diffolved.

We have an example of this in the procefs for *kermes mineralis*.

This kermes mineralis is reckoned in France one of the capital preparations of antimony. It firft attracted notice while it was a fecret remedy in the hands of a Carthufian monk, who performed feveral furprifing cures with it in the pneumonia, and other violent fluxions or congeftions in the lungs. Upon inquiry it appeared that the monk had got it from one Ligerie, who was poffeffed of the fecret, and who had learned it from Glauber, or from a fcholar of Glauber's. The Duke of Orleans, then regent, was advifed by the king's phyfician to purchafe the difcovery of it at the king's expence, and make it public *. But the younger Lemery, in a memoir, inserted in the *Mem. de l'Acad.* for the year 1720, proves that this medicine was contrived, or at least the process for it published by his father, in his treatise on antimony, and that his father's process is rather better than La Ligérie's. The elder Lemery called it a *sulphur auratum antimonii*, and not improperly, for the kermes mineral, in whatever manner it be prepared, is very much, and almost precisely, of the same nature with the *antimonium*, or *sulphur antimonii precipitatum*, which was formerly named *sulphur auratum* †.

Lemery's process, however, and the product of it, had been neglected, until it attracted notice as a secret remedy, and was purchased for the use of the public, and then it became so much an object of attention, that the most celebrated French chemists have exercised their ingenuity in throwing light on the process, and improving it.

A great inconvenience in La Ligérie's process is, that the quantity of kermes produced by each boiling of the materials is very small.

Geoffroy, who made many experiments to investigate the nature of

---

* *Ligérie's Process.*—Crude antimony broken small, four ounces; deliquium of nitrate of lime, one ounce; and water eight ounces. Boil the mixture two hours. Decant and filtre it while hot. While it cools, the kermes is precipitated. Add to the remaining antimony six drachms of the nitrate of lime, and eight ounces of water; repeat the boiling and filtrate.——Repeat the last part of the process; collect the three precipitates; and edulcorate with water.

† The elder Lemery's original process is this.—Put into an iron pot five or six pounds of a solution of pure fixed alkali, with three or four times as much water. When the liquor boils, throw into it four or five ounces of finely powdered crude antimony. Boil the mixture a short time, stirring it with an iron spatula, and then filtrate it boiling hot through paper. It will deposit the kermes, which must be well washed, dried, and ground to a fine powder,—a gray powder remains, which is regulus which the hepar could not dissolve. The same alkaline liquor may be used several times.

this preparation, propofed a procefs by which it may be obtained in far greater quantity, and with incomparably lefs trouble *.

The precipitated matter, when well walhed and dried, is a foft and tender powdery matter, of a deep brownifh red, or rather a coffee colour.

The liquor from which it has been depofited retains the greateft part of the alkali, together with a fmall quantity of the antimony adhering to it, but retains fo much of the uncombined alkali, that, if boiled with more antimony, it diffolves it again, and by cooling, depofits a frelh quantity of kermes; and this after many repeated operations of the fame kind.

This preparation is much the fame with the fulphur precipitatum. It has a moderate degree of power as an antimonial.

Meuder, a pharmaceutical chemift of reputation, fays that all depofitions by cooling, from the warm wafhings of the hepatic preparations of antimony, may be diftinguifhed from a kermes mineralis by the following marks :

1. The kermes is red, and the others are brown.

. Kermes act much more gently as a medicine. I may obferve, that if the kermes be prepared by employing a cauftic alkali, we have the *fulphur fixum Stabelii.*

And now, having enumerated the preparations of antimony obtained by the ufe of alkalis, as well as thofe produced by the action of heat and air, we fhall next take thofe produced by deflagration with nitre, which are in fome refpects fimilar to thofe already defcribed.

* *Geoffroy's procefs.*—Two parts of antimony, and one of alkali, are melted together and pounded. This makes an alkaline fulphuret, foluble only in hot water. It is boiled with water, and filtered into a large quantity of hot water, and upon cooling, yields of kermes, three-fourths of the weight of the antimony.

In young Lemery's procefs, fine powder of antimony is boiled with the pure deliquium of nitrate of lime; and thus the whole antimony is diffolved.

When crude antimony is mixed with nitre, and this mixture is set on fire, or thrown by degrees into a hot crucible or iron pot, there is more or less deflagration, according to the proportion of the nitre in the mixture.   In this deflagration, the acid of the nitre, by its abundant oxygen, acts most violently on the sulphur, which it changes into sulphuric acid; and the greater part of this acid combines with the alkali of the nitre.   But the acid of the nitre acts also more or less on the metallic part of the crude antimony, and oxydates it either moderately, or to a high degree, according to the quantity of the nitre employed.   A number of preparations made in this way have been contrived, and highly recommended, at different times.   I shall describe a few of them.

First, what Mr. Lewis calls *crocus medicinalis*.   A mixture of one part of nitre and eight of antimony in fine powder, is projected into a red hot crucible.   So small a proportion of nitre can produce no sensible deflagration.   It seems only to assist the fusion and the solubility of the mixture.   It breaks with a glassy fracture, is opaque, and has a surface like polished steel.   It produces a mass of a deep purple colour, like that obtained from antimony acted on by a small quantity of alkaline salts.   This preparation is also very similar in its degree of efficacy to the alkaline sulphuret, which was called *regulus antimonii medicinalis*; and indeed the same name of regulus medicinalis has been given sometimes to this preparation also.   It is insoluble in water, and insipid.

In the other preparations of this kind with nitre, larger proportions of nitre are used, and which produce more active effects; but the chief of them is the *crocus antimonii*, formerly named (often very improperly) *crocus metallorum*. (*Pharm. Lond.*)   In this preparation, since the proportion of nitre (equal parts) is very considerable, the deflagration is very brisk, or rather violent; and the fusion does not require the crucible to be red hot in the beginning; the first projection may be kindled by a bit of lighted paper; and by properly timing the fubfequent projections, the heat rifes to a great pitch, and makes the whole melt very completely. The faline part is feparated and thrown away; the vitrified metallic matter only being the ufeful part. It affumes a yellow colour when ground to powder, whence it has got the name of crocus. The nitre is partly converted into vitriolated tartar, and partly alkalized. If the operation is managed fo that all fhall be fluid at once, the greateft part of the falt collects uppermoft by itfelf. The metallic matter goes to the bottom femi-vitrified; but the glaffy matter at the bottom is not perfectly free from falt. It contains, united with the vitrified matter, a fmall quantity of alkaline falt, produced from that part of the nitre which deflagrated with the metallic part of the antimony: And, befide this alkaline falt, there is even a very fmall portion of fulphur remaining united with the alkali and metallic oxyd. It was, therefore, fome time ago, a practice to feparate as much as poffible this faline matter from the crocus, by reducing the crocus to a fine powder, and boiling it in water, which (after fubfidence) being poured off, the crocus was dried, and was named *crocus lotus*. I find the moft fuccefsful way of proceeding is, to deflagrate in a hot iron morter fet on the fire, projecting very fmall quantities, but in quick fucceffion. Thus you will have a good fufion and feparation, and a fine crocus without further trouble. *Crocus*, or *crocus Rolandi*, according to Meuder, is the deflagrated ftuff edulcorated without previous fufion.

You will find great confufion in the names of thefe productions of antimony. It was formerly called, with the greateft impropriety, *crocus metallorum*. And you will find that the French writers, Lemery and Macquer, choofe to call it liver of antimony,—*foie d'antimoine*; and with as little propriety, gave it the name of *crocus metallorum* after it was wafhed.

The nature of this preparation, as a medicine, is fimilar to that of the glafs, but rather inferior in violence.

When the dofe of nitre is much farther increafed, befides the complete deſtruction of the ſulphur, the metallic part is more calcined; and this may be to ſuch a degree, that its operation as a medicine is thereby diminiſhed, as it becomes leſs diſpoſed to unite with the acids found in the ſtomach. Such is the *antimonii emeticum mitius* of Boerhaave, viz. antimony, one part; nitre, two parts: Such alſo is the *antimonium cal-cinatum*, (Lond.) viz. antimony, one part; nitre, three parts. The matter at firſt taken out of the crucible contains ſulphat of potaſh, and a little unſaturated alkali. Theſe ſalts are ſeparated by hot water, and then we have the pure white oxyd. Four ounces of antimony yield five and a half ounces of the *antimonium diaphoreticum lotum Meuderi*, which is accordingly called *antimonium calcinatum*; as alſo *antimo-nium diaphoreticum* *.

The *antimonium uſtum cum nitro* of the Edinburgh Pharmaco-pœia may be conſidered as ſimilar to theſe preparations, only more active than the antimonium calcinatum. The crude antimony is firſt roaſted to a calx; then mixed with an equal quantity of nitre, and melted, or made red hot one hour; and then edulcorated with wa-ter. It was meant to be an imitation of James's powder; the proceſs being copied from the ſpecification of his patent. But either Dr. James changed his proceſs afterwards, or gave a falſe or diſguiſed de-ſcription of it; for lately it has been clearly proved to be very different, by Dr. Pearſon's experiments, read to the Royal Society of London, in 1791, and publiſhed in the Philoſophical Tranſactions. It now ap-pears to be a combination of antimony with the acid of phoſphorus, or rather with a phoſphat of lime. Nitrous acid diſſolves the phoſ-phat; and the phoſphoric acid remaining in the ſolution is eaſily diſ-

* Geoffroy obſerved a ſingular phenomenon in his operations on this medicine. Having ground an ounce of it with two ounces of black ſoap, he roaſted the mixture to a coal, and then ground it coarſely with another ounce, and expoſed it to violent heat, in a covered crucible. Five hours after all was cold, he took off the cover. The mixture inſtantly took fire; and being very ſpongy, ſo that the air had eaſy acceſs to the interior parts, it took fire all over, and diſſipated with a terrible exploſion.

covered, by means of a folution of mercurial nitrate, or by a nitrate of lead. The antimonial oxyd is difcovered by means of muriatic acid, and precipitation from it by water.

Dr. Pearfon's judicious examination led him to attempt the preparation, on the principles arifing from this analyfis ; and his imitation is fo undiftinguifhable from Dr. James's powder, that it is adopted by the London and Edinburgh Difpenfatories, under the name of *pulvis antimonialis*, or *antimonium calcareo-phofphoratum*. Equal weights of antimony and horn fhavings are calcined till of a very fair gray colour, and then kept red hot in a coated crucible, for two hours ; and when cold, reduced to a fine powder, which, when well prepared, is of a pure white, or has a flight caft of yellow, but not inclining to brown.

Preparations with Acids.

In general, the acids do not act fo well on crude antimony as they do on the regulus, or pure metallic part of it. The fulphur, in fome meafure, protects the metallic matter, but not to fuch a degree as to prevent entirely the action of the foffil acids, nor even fome weak action of the vegetable acids. I fhall now defcribe the preparations made with acids applied to antimony in different ftates. And firft, with the

Sulphuric acid.—There are no preparations with this acid in either of our pharmacopœias. A Dr. Klaunig of Breflaw, in a book entitled *Nofocomium Charitatis*, recommends the following : Diftil vitriolic acid from antimony feveral times. By doing this the metallic part is oxydated by the acid, and combined with a part of it ; and the fulphur is, at the fame time, feparated by fublimation. After which, the fulphat of antimony is taken out of the retort, ground to powder, and fome alcohol burned on it. It is a medicine, of which two grains work gently, by ftool, and vomit, and fweat. And he recommends it ftrongly for quartans. Wertholfs (in his *Obf. de Febribus*) teftifies that it is emetic, and purgative, and diaphoretic. Wilfon's *antimonium catharticum* is probably of the fame nature. Wilfon calls it an infallible purge, and fays that he knew three poxes cured by it. They were probably fome cutaneous difeafes which he miftook for pox.

There are no preparations obtained by the action of nitric acid upon crude antimony; unlefs we choofe to confider the *bezoardicum minerale* as fuch, in fome meafure: But it is different, as I fhall mention prefently.

The muriatic acid, affifted with a little nitric acid, or in form of aqua regia, diffolves the metal, and leaves the fulphur. It therefore thus forms a folution of the metal, that is quite the fame as if the pure metal had been ufed. And the corrofive muriat of quickfilver, on account of the abundance of muriatic acid and oxygen it contains, can alfo be made to act on the metal in fulphuret of antimony, as well as upon the metal in its feparate ftate. The muriat of quickfilver and the fulphuret of antimony being mixed together in powder, and diftilled in a retort, a corrofive muriat of antimony arifes; and the quickfilver unites with the fulphur, and remains in the retort, in the form of a black fulphuret of quickfilver, which, if the heat be very much increafed, fublimes into the neck of the retort, in the form of cinnabar. To this cinnabar the chemifts give the name of *cinnabar of antimony*. This was the procefs moft commonly practifed for preparing the *antimonial cauftic*, or *corrofive muriat of antimony*.

But the London college of phyficians, learning probably that the chemifts had another method, much cheaper, and which is as good, have adopted it in the laft edition of their pharmacopœia. This is, to form the muriat of antimony directly, in the fame, or nearly the fame way, as was practifed for corrofive fublimate. Equal quantities of vitriolic acid and antimonial crocus are mixed with a quantity of common falt equal to their fum, and treated in the diftilling apparatus. I refer you to Mr. Ruffel's own account of the procefs, given in the firft volume of the Tranfactions of the Royal Society of this place.

It may alfo be effected by the oxygenated acid, as it is formed by diftilling the ordinary muriatic acid from the black oxyd of manganefe. But Mr. Ruffel's procefs is much better. Scheele's procefs is not very different.

I obferved before, that this antimonial cauftic, or muriat of antimony, is extremely corrofive. On account of its extreme acrimony chemifts ftudied how to render it milder. The moft fimple way is, by adding to it plenty of water. The greater part of the acid is thus immediately feparated. A fmall portion remains adhering to the metallic calx, and precipitates with it, in the form of a white powder, called *pulvis Algerothi*, from Victor Algeroth, or Algarotti, formerly a phyfician at Verona, who called it *pulvis Angelicus*. It is an oxyd very moderately oxydated, and which ftill retains, as I already obferved, a very fmall portion of the acid combined with it ; and on both thefe accounts, is very foluble in more acid, and alfo very fufible. It melts moft eafily into a tranfparent yellow glafs. Mr. Macquer, and others, will have it to be a pure calx, totally free from acid ; but in this I am perfuaded they are miftaken. Its fufibility is a proof. It diffolves readily in vegetable acids, and in folution of fal ammoniac. It is recommended by Dr. Saunders in this form as a good application to ill-conditioned ulcers ; and alfo as a milder internal medicine.

Bezoardicum minerale may alfo be confidered as a mitigated muriat of antimony. The preparation of this drug feems to me a procefs which has been undertaken very much at random, juft to fee what would refult from it.

To four ounces of butter of antimony twelve or fixteen of nitrous acid are added, by two ounces at a time. Violent fumes and ebullition, or effervefcence, are immediately produced, which muft be carefully avoided, they being extremely acrid or corrofive. The effervefcence is lefs at each fubfequent addition. When all is quiet, the whole is diftilled to drynefs, and the refiduum in the retort is the bezoardicum minerale.

The firft effervefcences are chiefly an oxygenated muriatic acid; and the nitrous acid does not rife till fome time after. This is not the ufual opinion; and it is even thought that this preparation is nearly the fame with diaphoretic antimony, or a perfect white oxyd. It may not be impoffible to bring it to this ftate; but in my own experiments, as well as in the examination of fuch fpecimens as I was well affured that they had been prepared from butter of antimony, I found it to be ftill a *muriat of antimony.* I believe indeed that the antimonium diaphoreticum is often fold for it, being a much cheaper preparation; and, when made with impure nitre, it may be intrinfically the fame. But I have not found that any number of abftractions of pure nitrous acid will free butter of antimony completely from muriatic acid. The miftake, if any, is of little confequence in medicine. But it is perhaps a more difficult matter to explain or account for the deoxygenation of the muriatic acid by means of the nitric, which we know to be the fpeedieft means to oxygenate it. We muft confider it, not as a deoxygenation of what part of the acid remains combined with the antimony, but as a *fuper-oxygenation* of what is expelled from it. The effervefcence is violent, and the fumes are uncommonly corrofive. We know that the acid combined in the antimonial muriat is in its oxygenated ftate. The metal, already faturated with oxygen, cannot decompofe the nitric acid, and thereby occafion the fumes of aqua regia. It is decompofed by the nafcent muriatic acid; and this comes off in a fuper-oxygenated ftate. There are two or three other procefles in which this rare acid is formed in the fame manner.

The only remaining faline preparations are thofe produced with the vegetable acids. The *antimonium tartarifatum;* (Edin. and Lond.) The *vinum antimonii tartarifatum;* (Edin. and Lond.) and the *vinum antimonii.* (Lond.)

The vegetable acids do not act fenfibly upon crude antimony, and but weakly on the pure antimony. To facilitate their action, we muſt take the metal, not only ſeparated from the ſulphur, but mode-rately oxydated, or reduced to its moſt ſoluble ſtate. Accordingly, the compounds of it with the vegetable acids were formerly ordered in our diſpenſatories to be prepared either with the waſhed crocus, or the vitrum antimonii : But the crocus was formerly in moſt gene-ral uſe for this purpoſe, (and it is ſtill uſed at London), until com-plaints prevailed every where of the weakneſs, inefficacy, and inequa-lity of the emetic tartar. This was occaſioned by frauds in preparing the crocus, uſed much for horſes, &c. The great demand for it en-couraged the druggiſts to attempt cheaper preparations ; and inſtead of employing the crocus antimonii lotus, as was directed by the phar-macopœia, which is made by deflagrating equal quantities of nitre and crude antimony, they uſed only half, or three parts, and even leſs, of the nitre, the moſt expenſive article ; and to ſucceed the better in bringing the mixture into thin fuſion, ſo as to make a uniform glaſs, they employed a little alkali. The reſult of all this is a preparation nearly the ſame with the regulus medicinalis ; a drug almoſt inactive, and inſoluble. If the weakneſs of this preparation were all its imper-fection, it could eaſily be remedied by increaſing the doſe. But, by thus ſtinting the nitre, it is almoſt impoſſible to make a uniform maſs ; and different portions of the ſame lump will often be in different ſtates. This ſpurious crocus may be eaſily diſtinguiſhed from the true by its colour. In the maſs the ſpurious is opaque, and almoſt black ; and when reduced to fine powder, it is of a dirty purple colour. The genuine is liver-coloured in the maſs, and a deep yellow when finely powdered.

Now the ſtate of the crocus is moſt conſequential for the prepara-tion of emetic tartar ; for a large quantity of the nitre is neceſſary for deſtroying the ſulphur, and then bringing the metal into that ſtate of moderate oxydation that renders it moſt ſoluble in the weak acids. Leſs nitre will leave ſome ſulphur, which ſheathes the metal ; and the pure regulus has little folubility in the vegetable acids, and requires a little previous oxydation.

In confequence of thefe frauds in preparing the crocus, the emetic tartars and wines, as prepared by different apothecaries, were widely different in ftrength, and often totally difappointed the practitioner.

Different chemifts have exerted themfelves to remedy this inconvenience in a medicine of fo much activity and importance,—indeed the moft valuable of the antimonial medicines; and in confequence of their experiments, and of what has been publifhed on this fubject, the *tartrite of antimony,* or *antimonium tartarifatum,* is now always made, I believe, fufficiently ftrong.

One of the beft proceffes they recommended, and which was lately the procefs of the Edinburgh Pharmacopœia, was to ufe the vitrum antimonii, which is preferable to crocus, becaufe we have no experience yet of any miftakes or frauds committed with regard to it, and we can always know if it be good. We are there directed to beat it to very fine powder, and mix it with an equal weight of the cryftals of tartar, alfo powdered. In the next place, for every ounce of the mixture, or half ounce of tartar, take one pound of pure water, (diftilled water is the beft), and fet it in a furnace to boil. As foon as it boils, throw in the mixture of tartar and vitrum by degrees, until all is in; and continue to boil gently for feveral hours. The number of hours neceffary cannot be fpecified with precifion. It depends upon the degree of pulverization of the vitrum. If it be a very fine and almoft impalpable powder, about four hours, or even lefs: If not fo fine, twelve hours. It is more neceffary to attend to this, when the vitrum is employed, than when we employ the crocus. The firft, being powdered mechanically, can never be fo impalpably levigated as the crocus, which falls a foft powder, completely divided by the feparation of the faline matter. Beaumé, in examining the preparations of eminent apothecaries, found differences in the quantity of metal obtainable from an ounce of the tartarized antimony, which were furprifingly great. In fome he found 150, and in others scarcely 40 grains. Vinum emeticum may also differ in consequence of a difference in the acidity of the wine.

I need scarcely insist on the necessity of much water and a boiling heat. We know that water dissolves only one-twenty-fourth of its weight of tartar with a boiling heat, and that it lets go a great portion of this by a moderate diminution of its heat.

Mr. Macquer would prescribe the mercurius vitæ in preference to the crocus or glass, as more to be depended on for perfect uniformity. But I do not see reason to doubt of the goodness of these preparations; and by prescribing so costly a thing as mercurius vitæ would only occasion more substitutions or adulterations.

Beaumé has further given a caution with regard to the vessel in which the mixture is boiled. He says it must neither be iron nor copper. He finds that these vessels decompound the emetic tartar in some measure. The vessel must be glass, or silver, or earthen ware not glazed with lead. But M. de la Caille assures us that iron may be used.

When the operation is conducted in this manner, the tartar is saturated with the antimonial calx, part of which remains undissolved. And the liquor being filtrated affords, by evaporation, fair crystals, pyramids of three sides,—transparent while wet, but becoming white and opaque in the air. And these crystals are the saturated tartrite; and are powerfully emetic. The dose is from one grain, or one-half, to two, or three at the utmost, for an emetic. This may seem an inconvenience, but it is easily remedied by mixing ten grains, for instance, with three or four times as much sugar; and thus we will have a powder which it will be easy to weigh out in moderate doses: Or a more common way is to dilute it in water,—three or four grains, for example, in six ounces; and a spoonful is taken every half hour until it operate.

But beside the vitrum antimonii, any other preparation in which the metal is freed from the sulphur, and very moderately oxydated, may be employed to faturate the tartar, and produce a medicine equally powerful and certain.

To complete this article, Dr. Saunders added a fimple and ufeful method of examining emetic tartar, fo as to judge if it be properly prepared, and as ftrong as poffible, or what degree of ftrength it poffeffes. On examining the folubility of different tartars, he found, by experiments, that they are more foluble in proportion as they are more completely faturated with antimony; and that the difference among tartars in this refpect is fo .confiderable, that it is eafy to diftinguifh them. Thus, one ounce of water, at a middling temperature, diffolved,

Of the faturated tartar	52 grains.
Apothecaries' hall	38
A London chemift's	32
Edinburgh fhops	lefs.

I only wifh he had fpecified the degree of heat more particularly.

The next fet of preparations we have to mention are the *reguli*, and preparations from them.

The procefs by which antimony is obtained in the largeft quantity and pureft, is fimilar to that followed in extracting other metals from fulphurous ores, viz. evaporation of the fulphur by uftulation, and reduction of the calx. But other proceffes have been commonly followed: Thefe have in general been two. The procefs for fimple regulus of antimony, and that for martial regulus of antimony.

For the fimple metal, or what is called fimply *regulus of antimony*, the crude antimony is melted with about half its weight of black flux, or rather the ingredients for producing black flux, viz. nitre and tartar, which occafions part of the regulus to feparate ; the hepar fulphuris formed by the alkali and fulphur, having rather more attraction for the charcoal of the black flux than for the metal. But this fuperiority of attraction is not fo great as to occafion a complete feparation of the antimony. Part only feparates. The reft remains diffolved in the faline fulphurous fcoria, and forms with it a hepar antimonii, and is employed as fuch in the London Pharmaco-pœia.

The other procefs which I faid is alfo often followed, is that for *regulus antimonii martialis.* Crude antimony and iron filings, or fmall nails, in equal parts, are mixed and melted with a violent heat, and commonly with addition of a fmall quantity of nitre or fixed alkali. In this operation, the fulphur fhews a ftronger attraction for the iron than for the antimony, and forms with the iron a hard fulphurous compound. The antimony finks feparate in large quantity, but not fo pure as by the former procefs. It diffolves fome of the iron. It is purified by crude antimony, and two or three meltings with a little nitre. The fulphur of the crude antimony forms a fcoria with the iron, and floats above; and thus the regulus is freed from the iron. Melting again with nitre deftroys the fulphur brought in by the crude antimony, and not carried up by the iron. Thus it may be made as pure as the other. Many other metals can alfo be ufed, all except gold, zinc, and platinum, having a ftronger attraction for fulphur than antimony has *.

In whatever manner the antimony be feparated from fulphur, if it be made very pure, and caft into a conical veffel, it exhibits a ftar. *Regulus antimonii ftellatus,—regulus ftellatus.* Many myfteries have been fuppofed to be indicated by the figure which the furface of this metal affumes in cooling, it having generally a ftellated or radiated figure,—this is owing to its mode of cryftallization, combined with the

---

* Lehman relates a very fingular experiment made with antimony and arfenic, which being diftilled together, with a violent heat, yield a fulphurous fublimate, which does not deflagrate with nitre; and he fays that a regulus of antimony remained in the retort. I found this refiduum not diftinguifhable from a regulus by its chemical properties, but very diftinguifhable from it by a beautiful changeable colour upon the facets of the metal when broken. I attributed this to a film of arfenic, which being tranfparent, fhould exhibit fuch colours. But no trial that I could put it to fupported my conjecture. The fublimate did not deflagrate, but burnt flowly with a dull heat.   EDITOR.

progrefs of cooling, which proceeds from the fides of the veffel to the centre. It is needlefs to take up your time with proving the folly of thefe fancies.

Preparations from the Regulus.———*Ceruſſa antimonii ; ſtomachicum Poterii ; cardiacum Poterii ; antihecticum Poterii ; tinctures of antimony*

Uſes of Antimony.———It affords, in many of its preparations, medicines of great efficacy to remove many difeafes of the human body, as well as of other animals.

In the arts, it is ufed for refining gold,—and the regulus is mixed with tin and lead to harden them, for the compofition of pewter ; but I apprehend that the greateft quantities are employed to mix with thefe metals for letter founding. In this compofition, great hardnefs, and great fufibility, are important properties. It is alfo peculiarly fortunate, that the latent heat neceffary for the fluidity of this compofition is very moderate, fo that a man can work without intermiffion, when the letters are fmall. The mould lofes fo much heat while opening to fhake out the letter and fhutting again, that the next letter makes no accumulation of heat. Were it otherwife, the mould muft be cooled after a few letters, and the expedition of the work greatly diminifhed.

GENUS VI.—ZINC, OR SPELTER,

Another brittle metal, or femimetal, bears fome refemblance to the laft. Like it, it fhews a plated texture when broken. But it is eafily diftinguifhed. *1ſt.* It is not fo brittle, but can be compreffed by the ftrokes of a hammer ; and even bears fome degree of extenfion under the hammer. It bears extenfion by rollers very well, and can thus be drawn into very thin plates, if carefully annealed between the operations. It alfo bears drawing in the wire plate. It has therefore more cohefion, and is more difficultly broken to pieces. The plates are not fo broad, and its colour inclines to bluifh, when compared with the other two. It is the moſt expanſible of all the metals by heat. It melts before it begins to grow red hot, and flows quite thin at 680. When melted and poured into a veſſel, it may be reduced to a pretty fine powder by ſimple agitation. Its ſpecific gravity is nearly 7,2. It has ſome ſingular properties relative to its ſolid form. This imperfect ductility is accompanied by an odd kind of toughneſs, which makes it extremely difficult to work it in the ordinary way, by filing it. It ſticks in the files ; and ſoon renders them uſeleſs. Lead, though a much ſofter metal, has not this quality. Pure copper has a little of it. It ſticks in like manner to the edges of the cutting tools, with which it is ſcraped or turned in the lathe,—and what is ſtill more ſingular, when mixed with copper which has the ſame quality, it forms a metal which the workmen find to work more pleaſantly than any other metal. Moreover, although annealing it, after every paſſage through the flatting mill enables it to extend much more than it would otherwiſe do, yet if it be hammered in its hot ſtate, it crumbles to pieces with the greateſt eaſe.

If the heat be further increaſed to a vivid red or white heat, in cloſe veſſels of proper form, ſuch as an earthen retort and receiver, the whole ariſes in vapour, which, in cloſe veſſels, may be condenſed again without further change. But when the ſame heat is applied in the open air, it ſuddenly takes fire like an inflammable body, and burns rapidly, with a bright and dazzling flame, and the metal is quickly changed into a calx. This is very white, and ſoft, and rarefied, in compariſon with other metallic calces. The greater part of this calx is accumulated in the veſſel, juſt over the burning zinc, and ſoon impedes the further action of the air on the remaining zinc ; but ſome part of it riſes up into the air, and floats or flies about in it like cobwebs. In this ſtate it was called *pampholix,* and *philoſophic* wool, alſo *nihil album.* This, therefore, is the moſt expeditious method of calcining zinc. The proceſs deſcribed in the Edinburgh Pharmacopœa is this : " Place a large or deep crucible, or other deep

" earthen veffel, in a melting furnace, inclined a little towards the
" door. When heated to the proper degree, let the zinc be thrown in-
" to it in fmall pieces, waiting till the firft is entirely burnt before the
" fecond is thrown in. Thus the oxyd of the zinc is accumulated in
" the crucible ; and being light and bulky, it muft be ftirred now and
" then with an iron rod, that the air may be admitted to the burning
" metal : And when a quantity of it is accumulated, it muft be taken
" out with an iron fpoon before any more of the zinc is thrown in."

The late Profeffor Gaubius of Leyden, who recommended this
oxyd to phyficians as an ufeful remedy for convulfive diforders, defcribed a different way of preparing it, but not fo good.

From thefe phenomena, it is evident that this metal is one of the
moft calcinable of any, when it is expofed to a violent heat ; and it
is reafonable to conclude, from the great brilliancy of its flame, and
the great heat which its inflammation produces, that in its calcination
it attaches to itfelf a very great quantity of oxygen. But it has not a
difpofition to attract oxygen eafily from the atmofphere, and to calcine by means of humidity, without the affiftance of heat, in which it
differs from iron and moft other metals *.

The oxyd of zinc is fo unfufible, becaufe it is probably more highly
calcined than the fufible oxyds of fome other metals.· It is, however,
in fome degree fufible in a very violent heat, and forms a fine yellow
glafs.

The reduction of this oxyd was long a difficult problem. It was

* It is not altogether inactive in this refpect. Its polifh is very eafily tarnifhed by
expofing it to very damp air ; and if wetted, and kept in that ftate an hour or two, its
polifh is entirely taken away, and it becomes of a dull leaden colour. When viewed in
this ftate with a microfcope, we fee it evidently corroded ; but I have not obferved this
to increafe by length of time. This is furely an oxydation, or ruft. It feems completely
to cover the furface, rendering it neutral or faturated, and this defends the interior parts.
But if the metal be made nearly red hot, and water be fprinkled on it with a brufh, it
then acts powerfully on the water, decompounds it, and produces much hydrogenous·
gas. Zinc is oxydated in this way, juft as iron is.

EDITOR.

found fo difficult, that many fuppofed it impoffible; but this proceeded from the improper manner of attempting it. They tried it in the manner ufually practifed with other metals, by mixing the oxyd in a crucible with black flux, or with charcoal and falts. In this way the heat which was neceffary for the reduction, was fufficient for totally evaporating the metal as foon as it was formed. It was therefore all loft.

Mr. Margraaf, attending to this, firft contrived to accomplifh the reduction. He mixed the oxyd with one-eighth of its weight of powdered charcoal, or lamp black,—introduced the mixture into a fmall earthen retort luted to a receiver, and urged it with an intenfe white heat. He thus got zinc again in the form of a metallic fublimate, or mafs, attached to the neck of the retort; and he found it rather more malleable than ordinary zinc.

The fame expedient occurred to Neuman, but he tried too fmall a quantity, (two drachms) fo that the air inclofed was fufficient to burn the zinc again; and he got nothing but half burnt flowers in the neck of the retort, which fhewed, unqueftionably, that they had been once reduced, otherwife the calx would never have been fublimed.

Zinc is eafily diffolved by all the acids, and is united with them by a very ftrong attraction. The acids are therefore more neutralized by it than by moft other metals, and the compounds do not eafily fuffer any feparation of the acid from the metal by large dilution with water.

The fulphuric acid even requires to be diluted confiderably with water, to make it diffolve the zinc well, juft as in diffolving iron. And further, the zinc, while diffolved by this acid, decompounds a part of the water, as iron does; and a great quantity of inflammable air is, in confequence, produced during the brifk effervefcence with which this diffolution is performed. One ounce of zinc produces 356 ounce meafures with one ounce of fulphuric acid. One ounce of iron gives 412 of air with 2 ounces of the acid. The metal is therefore oxydated as well as diffolved, but receives all the oxygen from the water. The compound of oxydated zinc and fulphuric acid thus formed is eafily cryftallized, and gives cryftals that are not deliquefcent. They are an article of the materia medica, and were named formerly *white vitriol ;*—in the new language of chemiftry, *fulphat of zinc.*

The nitric acid, in the diluted ftate of aquafortis, diffolves this metal with the greateft violence and rapidity, and becomes exceffively hot. In the violent effervefcence which attends this diffolution, the acid fuffers a great and violent abftraction of its oxygen, and all the changes which are neceffary concomitants of fuch abftraction. Part of it is therefore changed into red vapours; a great part into nitrous gas; another part into the lefs oxydated nitrous gas, which I lately defcribed; and laftly a part into pure azote.

By fome variations in the manner of diffolving the metal we can modify thefe changes of the acid. If, for example, the aquafortis be largely diluted with water before it be applied to the metal, the gas produced is almoft totally the lefs oxydated nitrous gas, or dephlogifticated nitrous gas of Dr. Prieftley.

The muriatic acid, in its common ftate, alfo diffolves zinc eafily, with effervefcence and production of a great quantity of inflammable air. But when applied in its oxygenated ftate, there is no inflammable air produced. The reafon is obvious. The zinc is fupplied with oxygen, which was very loofely combined with the acid. This muriat of zinc does not give cryftals eafily. It is a deliquefcent compound. The folution of it, mixed with glue, or employed to diffolve it, forms a compound which does not become dry in the air, but has the invifcating quality of birdlime. This birdlime would probably be the beft for catching birds and infects for the natural hiftorian, as it would eafily be wafhed off from the feathers of birds, or limbs of infects, by water.

The vegetable acids act alfo on zinc without difficulty and fome inflammable air is produced during their action. The oxyd of zinc alfo diffolves in all the acids, but without effervefcence. A faturated folution in the acetous acid is a liquor very like olive oil.

The alkalis act on zinc as upon fome other metals. Applied to it in its metallic ftate, they corrode it in fome meafure. Applied to it in the ftate of oxyd precipitated from acids, they in fome cafes diffolve it. Volatile alkali is thought to act on it in a peculiar manner. When powdered zinc is put into a folution of the cauftic volatile alkali, it yields, after a long time, inflammable air. This, however, feems rather to come from the water than from the alkali. Did it come from the alkali, we fhould alfo obtain azotic gas, which I have never been able to difcover by this treatment.

Such of the compound falts as act on other metals, act more readily upon zinc, which in confequence of its inflammable nature, produces remarkable effects on them, fimilar to thofe produced by the inflammable bodies. Thus, melted with fulphat of potafh or of foda, it changes them into alkaline fulphuret, and is changed itfelf into an oxyd : And with the falt of urine, or phofphoric acid, it yields phofphorus. It alfo decompofes fal ammoniac, by grinding them together, as appears by Margraaf's experiments, which I mentioned when delivering the theory of lime. This is unqueftionably owing to its very ftrong attraction for acids, which is fuch, that alkalis decompound the nitrate of zinc with difficulty. To the fame caufe we may afcribe the decompofition of alum by boiling it with zinc. If the fal ammoniac and zinc be treated in the way of diftillation, we obtain firft an incondenfible fuffocating alkali,—then a volatile muriatic acid, in thick white fumes,—in an open fire, white flowers fucceeded ; and at length, a reddifh and a black butter.

Nitre deflagrates violently with zinc ; but it muft be heated fo hot that is not eafy to diftinguifh its deflagration with the nitre from the inflammation to which it is fo much difpofed of itfelf. Its flowers do not fenfibly deflagrate, yet they alkalize double their weight of nitre. The fixed alkali and calx form together a mafs externally greenifh, internally purple. Infufed in water, the alkali diffolves, and with it a part of the calx, tinging the water purple *

Zinc does not unite with fulphur, nor hepar fulphuris, nor with crude antimony, and is therefore purified by the action of fulphur from admixtures of lead, which it frequently contains. Alkaline and calcareous fulphurets, however, readily unite with the calx and diffolve it, forming a fubftance refembling its ores.

None of the earths have been obferved to have any remarkable effect on zinc or its calx.

It unites with all the metals except bifmuth and nickel. It is indeed difficult to combine it with iron, becaufe the great heat that is neceffary diffipates the volatile metal, and it carries off fome iron with it. It has fomewhat of this effect on all the more calcinable metals. Moft of thefe mixtures boil and deflagrate more than zinc alone, and globules of the metal are frequently fcattered about. Hence it is called *metallic nitre*. Copper is but little affected this way, and the *cadmia fornacum* of the brafs-founderies rarely contains any copper. When they are melted feparately, and then mixed, there is frequently a violent detonation, and much of the metal is thrown about. When lead is added to melted zinc, the mixture takes fire, and the zinc burns away. They mix quickly, when the zinc is added to melted lead. Arfenic makes it black and friable.

United with mercury in the form of amalgam, it is the moft powerful exciter of an electric globe. The beft proportion is four parts of mercury to one of zinc.

Mixed with tin, in a fmall portion, it greatly increafes it hardnefs,

---

* With refpect to the other neutral falts, the only remarkable effect that I know, is the formation of a butyraceous fublimate of zinc, like the muriat of antimony, by diftilling common falt and the fulphat of zinc, or by fubliming the folution of zinc in the oxygenated muriatic acid. EDITOR.

and improves its colour, making it like filver; but this mixture is more apt to be corroded by acids. It is ufed however in the manufacture of pewter.

Mixed with lead in the proportion of one to twelve, or even lefs, it improves its tenacity in a moft furprifing manner, making it four or five times ftronger. It would therefore be a great improvement on water pipes.

The moft ufeful mixtures of zinc are thofe with copper, in various proportions. It communicates a yellow colour, notwithftanding its own whitenefs, and it removes the difagreeable toughnefs which makes copper fo difficult to work with the file and in the lathe, while it very little impairs its tenacity and ductility. In a very large proportion to the copper, it makes the *hard*, or *fpelter-folder*, ufed for copper, brafs, and iron. In a fmaller proportion, it makes brafs; and in other proportions it makes more or lefs perfect imitations of gold. Such are *pinchbeck*,—*princes metal*,—*fimilor*,—*Bath metal*,—*tutenag*, &c. Homberg, Geoffroy, Hellot, and Lewis, have made many experiments and ufeful obfervations on thefe mixtures; and to their writings I refer you for farther information on this head. Obferving, in general, that all thefe mixtures lofe fome of the zinc by evaporation in great heats, and that long continuance of it will expel the whole. This is owing both to the volatility and to the inflammability of the zinc. You will always obferve an uncommon brightnefs on the furface of the melted metal, greater than that of the furrounding fuel. This is juft a low flame, not one-tenth of an inch high.

It remains only to mention the operations by which it is extracted from its ores.

For a long time, all the zinc ufed in Europe was imported from China, except a fmall quantity which was obtained in the Hartz foreft in Germany, from an ore which yielded lead, and copper, and filver, and a little gold at the fame time. We have not been informed how the Chinefe zinc is obtained. The procefs by which the German was got was a little uncommon.

In the fide of the furnace, (which is a reverberatory) oppofite to the bellows, the wall is double, confifting of two thin fire ftones, with a hollow between. On a level with the ufual furface of the melted metal, there is a chink opening into this cavity. The zinc evaporates as faft as it is formed; and the vapour is driven into this hollow through the chink by the blaft of the bellows. The fide of the cavity, which is in contact with the air of the hut, is cool in comparifon with the reft, the ftone being thin and being often fprinkled with water. Here, therefore, the fublimed metal attaches itfelf, while the flowers, which are unavoidably formed by the calcination of part of this metallic vapour, rife farther up, and get into the long funnel, where they collect, and are got out from time to time by the name of *cadmia fornacum, Tutia, diaphryges.*

This very fingular procefs was the contrivance of a common fmelter, who had obferved this ftrange metal collecting in the retired corners of the furnace, at Rammelfberg. It is certainly ingenious, and fhewed a fagacious and intelligent mind. Such have been moft of the proceffes in metallurgy; and it is by a collection of thefe cafual arts that our fcience has arifen. The quantity thus got was but very fmall; and no other method was known or attempted in Europe for producing this metal, although its ores, which are very plentiful, appear to have been fufficiently known as fuch from their effect in making brafs.

In the greater part of thefe ores, the zinc is in the ftate of a calx; in confequence of which, they appear more like ftones than metallic minerals; and were actually confidered by many as a particular fpecies of ftone or earthy matter, and called *lapis calaminaris.* The original name, however, by which it was known to the ancients, and mentioned by Pliny, was that of *cadmia.*

Sometimes it is in a lefs calcined ftate, which gives it more or lefs metallic opacity and luftre; and in fome cafes there is fulphur in its compofition, which produces the fame effect. It is then called *black Jack*, by the Englifh miners; by the Germans, *blende;* and by natural hiftorians, *pfeudo-galæna,*—*i. e.* mock-lead ore.

This ore is fomewhat curious. It is tranfparent when pure; and in a very ftrong light, although feemingly opaque, it tranfmits a very deep brown light. It breaks alfo with a glaffy fracture. Notwithftanding thefe marks of being homogenous, a bit of it, flowly diffolved in weak aquafortis, leaves a fpongy mafs, which retains the original fhape completely, and this weighs about one-fifth or one-fixth of the blende. It is pure fulphur, not chemically combined.

Thefe feveral varieties of it have been long employed for making of brafs. The procefs confifts firft of roafting; then cementation with charcoal and copper. The copper fixes the zinc, and acquires the yellow colour.

This was the only ufe made of the ores of zinc in Europe; and the manner in which they produced their effect was not thoroughly underftood, until Mr. Margraaf extracted zinc from them, without the affiftance of any other metal. But now, in confequence of his difcovery, manufactories of zinc have been eftablifhed in England, and elfewhere.

The Englifh procefs is what the chemifts call *deftillatio per defcenfum.* The ore, mixed with charcoal, is put into conical pots, having in the bottom an iron pipe, which paffes through a hole in the hearth of the furnace, and reaches to the mouth of a rude receiver containing water. Fuel being kindled around the pots, (whofe mouths are ftopped with clay) and care being taken that it fhall not be fo hot as to melt their contents, the metallic vapours are expelled downwards, and condenfed in the receivers.

But lapis calaminaris is ftill much more ufed for making brafs, the procefs for which is the cheapeft way of extracting the zinc, while by the fame fire it is alfo mixed with the copper. The ore being cleared from fulphur, and other impurities, by roafting, is ground and mixed with charcoal. The mixture is put into pots, and ftratified with plates of copper. A proper heat being given, the ore is metallized, and the metallic vapour is immediately feized on by the copper,—renders it more fufible, and melts down with it, and lodges in the bottom of the pot, where it is defended from the air by the vitrified flag floating above it. When the proportion is properly hit, the copper gains an addition of one-third of its weight, in becoming brafs.

Although copper fixes zinc to a certain degree, it does not entirely prevent its evaporation and calcination in ftrong heats. Hence when brafs is melted, if it be expofed to air, it is liable to lofe part of the zinc, and by repeated fufions, the whole. Hence *cadmia fornacum,— pompholix,—diaphryges,—nihil album,—tutia,* &c. which are different names for the condenfed vapours which efcape from the furnaces in thefe procefes for brafs.

Ufes of Zinc.—It is chiefly ufeful for metallic compounds,—brafs,— pinchbeck,—pewter,—for tinning,—and for folders; and it is an ufeful article of the materia medica.

Until lately, zinc was hardly ever ufed internally; except fometimes the white vitriol was ufed as a vomit of quick operation. It was chiefly confined to external ufe. It poffeffes ufeful powers as an aftringent and repellent; and is ufed particularly in inflammations of the eyes. Its preparations are,—*Lapis calaminaris præparatus ;* Edin. and Lond.—*Tutia præparata ;* Ed. and Lond.—*Calx zinci,* vulgò, *flores zinci ;* Ed.—*Zincum calcinatum ;* Lond.—*Vitriolum album ;* Ed. made of zinc and fulphuric acid.—*Zincum vitriolatum purificatum ;* Lond. —*Aqua vitriolica ;* Ed.—*Ceratum è lapide calaminari ;* Ed.—*Unguentum è tutiâ ;* Ed. and Lond.—*Unguentum è calce zinci ;* Ed.

But certainly, it might be eafy to fubftitute preparations of this metal, inftead of fome of thefe, which could be much more depended on. For lapis calaminaris is extremely different, (*Vide Effays on different kinds of it, by Margraaf,*) and may accidentally contain arfenic and other minerals.

With respect to tutia, see the experiments of Neuman, which shew it plainly to be artificial. And it is seldom sufficiently levigated for ointments.

Both of these might be excellently superseded by a pure calx of zinc, or pompholix, which is extremely fine: And the vitriolum album by an artificial sulphat of zinc.

When the chemical state of the preparations is once such as may be depended on, they are medicines of very uniform operation, and of confiderable powers. The sulphat in the dose, a drachm or a drachm and a half, is a quick emetic, with little distress or sickness. The calx in much smaller doses, not exceeding five grains, produces sickness and vomiting. We are but little informed as yet as to the internal use of the preparations of zinc. But it has been long known as a powerful astringent, emollient, and cooling application, when we desire to remove the remains of a tedious inflammation, and restore strength to the vessels. The sulphat and nitrate, in the quantity of two grains to an ounce of water, answer very well in such cases.

GENUS VII.—BISMUTH, OR TINGLASS.

BISMUTH has a near resemblance to antimony in external appearance. It is nearly as brittle, and exhibits the same texture when broken. But it is much heavier, its specific gravity being 9,823; and the colour of it inclines a little to red. It is also more fusible, and less volatile than antimony. It melts sooner than lead; and flows the thinnest of all the metals. It also expands in congealing, and therefore takes the finest impressions of its mould.

This metallic substance is easily calcined to a moderate degree, forming in its oxydated state a very thin yellowish glass, one-eighth heavier than the metal. But it is difficult to calcine it further; at least the oxyd of it is always fusible in a moderate red heat; by which quality, and fome others, it refembles the oxyds of lead.   It may be got, either by keeping the bifmuth melted, and ftirring it conftantly, to expofe the different parts of it to the air, or more quickly, by way of fcorification in a more violent heat, which may calcine it fafter, and melt the calx as faft as it is formed ; fo that it may conftantly run off from the furface of the melted metal, and leave it expofed.   The metal fmokes conftantly, while any of it remains uncalcine. ; and a faint blue flame may be obferved on its furface.

It is eafily reduced again, by melting it with addition of charcoal, or of the black flux.

When we try it in mixture with falts, we find that the fulphuric acid acts only when applied ftrong, and affifted with heat.   The metal is then corroded and oxydated, by attracting oxygen from the acid ; but we do not get a foluble compound.

A foluble compound of this metal is beft and moft eafily obtained by the action of the nitric acid, which diffolves it with a ftrong effervefcence, and the production of great heat, red vapours, and nitrous gas, in confequence of the abftraction of oxygen from a part of the acid by the diffolving metal.   The folution, when completed, is almoft colourlefs ; and the compound it contains is liable to fuffer an imperfect feparation of the acid from the metal, when we dilute it largely with water.

This precipitate has been fuppofed by fome authors to be the pearl white, faid to be employed as a cofmetic ; and it is faid alfo that the tranfpiration by the fkin blackens the oxyd, as if by an imperfect reduction, and thus deftroys the fkin.   But I believe that this is a miftake, having examined many fpecimens of pearl white, which I found to be precipitates from a nitrate by a folution of common falt, or of tartar.   Water alone makes but a very imperfect feparation, and leaves fo much acid adhering to the metal, as unfits it for any fuch purpofe.

In confequence of this mutability of colour by the fteams of inflammable fubftances, the folutions of bifmuth, like thofe of fome other metals, form what are called *fympathetic inks ;* that is, inks which are invifible, till fomething has been done to the paper. When we write with the folution of the nitrate of bifmuth, the exhalation from hepar fulphuris, or putrefcent animal fubftances, foon make the writing legible. All bad fmells have this effect.

You will find in Neuman *(p.* 107, 108.) a number of experiments upon the precipitation of bifmuth from aquafortis, by a variety of different additions, which you may confult, if you have occafion to attend to this particular fubject.

The fame author alfo defcribes a procefs for combining bifmuth with the muriatic acid, by means of muriat of mercury, and the confequences of diffolving it with vegetable acids and alkalis, and of uniting it with fulphur, with which it forms a mafs very like crude antimony. To him, therefore, I refer you for thefe particulars, which are not fo important as to require our time. Mr. Pott of Berlin alfo may be confulted, who wrote a differtation upon this metal.

The compounds of bifmuth are remarkably fufible, fo that it may be employed in the compofition of folders for lead or tin. The fufible metal, called *Newton's metal,* has bifmuth for its principal ingredient. The beft proportion for the fufible metal are, eight parts of bifmuth, five of lead, and three of tin. When thus mixed, it will become fluid in a temperature fomething below that of boiling water. It is from the fame caufe, probably, that when diffolved in mercury, it difpofes the mercury to form more fluid amalgams with other metals, particularly with lead, than ordinary. And hence it has been propofed to be added to mercury, the more readily to diffolve lead balls lodged deep in a wound. Mr. Lewis fays *(Notes on Neuman, p.* 93.) that mercury united into a fluid amalgam with one-fourth, one-eighth, or one-twelfth of its weight of bifmuth, diffolves maffes of lead in a gentle warmth, without the neceffity of agitation. Whether this would anfwer the purpofe, I cannot pofitively fay, becaufe the bifmuth, or part of it, feparates in the form of powder while the lead diffolves, which might poffibly prove inconvenient. An abuſe of this quality of biſmuth has been practiſed in adulterating mercury.

The principal uſes of biſmuth are, to mix with tin for proper hardneſs in the compoſition of pewter for folders for lead and tin. It is ſometimes uſed alſo in the compoſition of metal for printers types; and when uſed, feems to be intended for greater tenuity of fuſion, and ſharp impreſſion. The proper degree of hardneſs is always procured by the cheaper regulus of antimony. With lead one part, tin one, biſmuth two, and mercury ten parts, it compoſes a kind of fluid amalgam, which being moved backwards and forwards in a clean glaſs veſſel, leaves a train behind it, adheres to the glaſs, and foils or ſilvers it. It is therefore uſed for foiling the inſide of glaſs globes. *(Vide Boyle's Treatiſe on the Uſefulneſs of Exper. Phil.; and for other uſes, vide Neuman, p. 112.)* But it has never been introduced into the materia medica, nor is there any reaſon yet for introducing it.

Its ore is moſt plentiful in Saxony, near Schneeberg. There is ſome too in Bohemia, and in Dauphiné, and ſome in England. Sometimes the biſmuth is found pure; feldom or never a vein of it pure, but intermixed with other ores, eſpecially arſenical, and particularly the ore called *cobalt*. The biſmuth is eaſily feparated from the ore by eliquation. It feparates in its metallic form, and runs along the inclined hearth into a channel, by which it is conveyed to a receptacle, where it is defended from the further action of heat. Mr. Macquer gives a long account of a very curious tincture, or ſolution, obtained from ores of this kind; but as the biſmuth has no part in its production, I ſhall confider it afterwards in its proper place.

GENUS VIII.—COBALT.

THIS name has been long appropriated to certain minerals or ores, which, when duly prepared, and melted with glaſs or enamels, give them a deep and rich blue colour. In their natural ſtate, theſe ores have commonly the metallic opacity and luftre, and a colour refembling that of iron, although it is various, in confequence of their being more or lefs comp unded with other minerals.

Thofe parts of them which are long expofed to the air are liable to decay, and contract a fort of ruft, which has a paie purplifh red or pink colour, like that of peach bloffoms. It is named *cobalt bloom.*

All the varieties of cobalt generally contain a large quantity of arfenic, the greater part of which is eafily feparated from them by uftulation : After which, the remaining metallic matter is an oxyd of a dark colour like foot, or fometimes it has a violet hue. It ftill retains arfenic, which appears to have been changed by the uftulating procefs into its acid ftate, and in that ftate remains ftrongly combined with the oxyd of cobalt.

The dark coloured oxyd, obtained by uftulating the ores of cobalt, is named *zaffre* *; and when added to glafs, or the m terials for making glafs, it gives it the blue, or deep violet colour, in proportion to the quantity ufed. But it was employed for this purpofe a long time before we knew that it could be reduced to a metallic ftate. Dr. Brandt, of the Swedifh academy, firft fhewed that it may be procured by the help of the common inflammable fluxes, and he called it *regulus of cobalt.* From the experiments he made upon it, as well as the properties of its calx which were known before, there is no doubt but that it is a metallic fubftance,—a femi-metal of a peculiar kind. The following are its moft remarkable qualities.

Its colour approaches to that of iron or antimony. It breaks with a granulated furface, like fteel. Its fpecific gravity is 7,7. It requires a pretty ftrong red heat to its fufion, and can be calcined without difficulty ; but its calcination goes on flowly, like that of copper, and without the appearances of inflammation. The calx produced from

* This zaffre, however, which we get in the fhops, contains but a fourth or fifth part of this oxyd, and the reft is powdered flint.

it is always of a blackish colour, and is not eafily melted by itfelf. But if it be added to glafs, it melts with it, and colours it to different fhades of blue,—from the lighteft to a deep blue, which appears black.

Of the acids, aquafortis diffolves it the moft readily, and with effervefcence. The fulphuric acid will not act upon it, except when applied ftrong, and with the affiftance of heat. A very diluted acid, however, will readily diffolve the oxyd obtained by precipitation from any acid, by alkali or lime. The muriatic acid acts beft upon the calx, or when it is affifted with a little of the nitric acid. The folution has a faint rofe colour, which becomes green when heated.

All the faline compounds thus produced are foluble in water, and capable of being largely diluted without feparation of the acid ; and the folutions are all of a rofe colour, or reddifh.

The muriatic folution, as alfo the faline compound which it contains, is remarkable for changing its colour when gently heated, and refuming its former colour when the heat leaves it. Air deprived of its natural humidity by quicklime, or by fulphuric acid, produces the fame effect ; and breathing damp on it effaces the green, though hot.

This folution is the *fympathetic ink* of Mr. Hellot, and is the moft curious of all the preparations which go by that name. The procefs for it is defcribed by Macquer. Hellot defcribed it as produced from ore of bifmuth ; but this was a miftake, as I mentioned when defcribing the properties of that metal. The calx, or zaffre, is diffolved in aqua regia, (which keeps it much better fufpended than aquafortis) and the folution, when cold, has a pale rofe colour, and becomes green when hot. Writings with this folution are invifible (when recently done) in the cold ; but when held before a fire, they acquire a beautiful leek green colour ; this difappears again when cold, and may be renewed by heat ; and this may be repeated as often as we pleafe, taking care that we do not make the paper too hot. This will render the colour permanent. Frequent repetition has fomewhat of this effect ; and in this refpect, a folution of the calx in muriatic acid is preferable to that in aqua regia. If a drawing of a plant of abundant foliage be washed with a full ftraw colour, which is not very unlike the wither- ed leaf of fome plants, the folution of cobalt being laid over it, raifes it to a very beautiful lively green, and is a pretty fancy for a fire- fcreen. There is another preparation of a fimilar kind, made by boil- ing oxyd of cobalt in fixteen times its weight of diftilled vinegar, till the liquor is reduced to one-fourth;—it is filtered, and again reduced by evaporation to one-half;—muriat of foda is now added to it,—and the muriatic acid takes the oxyd from the acetous. This folution is invifible, like the other, when cold, but when heated, fhews a fine blue, which difappears on cooling *.

In trying to mix cobalt with other metals, it has been found incap- able of uniting with mercury and lead, except in fmall quantity, or with filver, according to Bergmann. It mingles with all the reft, and makes the moft malleable metals quite brittle; but it was thought rather to increafe the toughnefs of iron.

It is precipitated from acids by zinc, but not by iron.

When borax, which has been tinged with cobalt, is boiled in water, the water diffolves the borax, and leaves a firm gelatinous matter, which becomes friable by drying, and of a tranfparent rofe colour, like cobalt bloom.

The only ufe that I know to be made of cobalt, is to tinge glafs and enamels. The calx or zaffre, being mixed with a certain propor- tion of flint or glafs, is melted, and forms a glafs of a rich purplifh blue.

---

* The action of the neutral falts has not been much attended to. It is fomewhat remarkable that it fcarcely detonates with nitre,—lefs than even copper. Yet in its oxy- dation, it feems to combine with more oxygen than any other metal. The oxyd pro- duced from 100 grains weighs 140. (Fourcroy.) Nitre calcines or oxydates it. What becomes of the caloric of the oxygen? The difficulty to reconcile this with the general doctrine is great, but not peculiar to this metal; for the deflagration of the metals with nitre is by no means in the proportion that we fhould expect from their increafe of weight by calcination.

EDITOR.

This is ground to a coarfe powder, and fold by the name of *fmalt*,—in which ftate it is ufed by the fign-painters. Smalt is ground in mills to a very fine powder, which is forted, by its precipitation in water, into parcels of different finenefs, and fold by the name of *powder blue*. The fame compound, ground to the utmoft degree of finenefs, is ufed in painting all the blue that we fee on porcelain, Delft ware, and all other imitations of porcelain. Some of it has a richnefs of colour furpaffing all other works of the pencil. This is particularly remarkable in the Saxon porcelain. The painters in enamel complain that it is fo fufible, that the heat neceffary for raifing their other colours makes the blue fpread juft as ink does on bibulous paper.

GENUS IX.—NICCOLUM.

THE name now given to this metal is derived from the name which was firft given to the ore of it by the German miners, at Freyberg in Saxony, where it is moft abundant.

As the colour of this ore refembles in fome degree the colour of copper, they expected at firft to get copper from it ; but not fucceeding in their attempts to get copper from it by any procefs, they gave it the name of *cupfer nickel*,—the literal tranflation of which into Englifh is *Niccol's copper*.

A laborious inveftigation of the component parts of this ore by feveral of the modern chemifts, efpecially Cronftedt and Bergmann, has fhewn that it contains a peculiar metal, to which the name *niccolum* is now appropriated, but which is intimately and ftrongly combined in the ore with feveral others, and with fulphur. Cronftedt was the firft who made the difcovery. The other metals prefent in this ore are iron, arfenic, and generally fome cobalt. And all thefe ingredients cohere fo ftrongly together, that an attempt to feparate them, or to obtain the niccolum pure, is a tafk of the greateft difficulty and labour. Of this you will be fatisfied, when you look into Bergmann's differtation or effay on niccolum, in the fecond volume of his Chemical Effays. He was at aftonifhing pains with it ; and after all, was not fatisfied that he had completely fucceeded. The moft highly refined niccolum that he was able to obtain by his great labour and fkill was ftill attracted by the magnet, and had a very confiderable degree of toughnefs or malleability. And he remained doubtful whether thefe qualities proceeded from iron ftill adhering to it, or are qualities belonging to this metal as well as to iron.

Nickel refembles bifmuth in its colour,—it being white, with a caft of reddifhnefs ; and its fpecific gravity is about 8,66.

When it is expofed to the action of the air and heat for calcination, it is flowly calcined or oxydated in the fame manner as copper *, but the oxyds got from it are of a green or greenifh colour, whereas the oxyd of copper is footy black.

It can be diffolved by moft of the acids ; and its folutions are all of a pleafant green colour. In thefe the metals is ftrongly combined with the acid ; and no other metal has the power to precipitate it.

It can only be precipitated by the fixed alkalis, and the alkaline earths, and by fome of the compound falts or faline compounds which contain fome of the acids, for which this metal has a ftrong attraction, and which form with it infoluble compounds. The metal is combined with fuch acid in this cafe, in confequence of a double elective attraction, and precipitates with it.

When it is precipitated by alkalis, it is neceffarily in an oxydated ftate ; and in this ftate can be rediffolved by adding more of the alkali. The volatile alkali efpecially diffolves it with facility, and can even act on the pure metal in its metallic form, and always forms with it a blue folution.

In trying to mix the oxyd of nickel with vitrified earthy fubftances, we find that it can be employed to give a colour to glafs ; and the colour it gives is a reddifh yellow. No other property of it relative to the earthy fubftances has yet been difcovered,

* Cronftedt fays that when ftrongly heated it emits fparks, or brandifhes, like iron.

Of the inflammables, fulphur readily unites with it, and adheres to it very ftrongly. It alfo difpofes the nickel and arfenic to be more eafily feparated by uftulation. And a part of Bergmann's experiments for refining niccolum was founded on this fact.

Phofphorus alfo combines with nickel; and they adhere ftrongly, fo as to bear a melting heat. Long expofure to this heat and the action of air diffipates the phofphorus, the metal all the while having a fenfible fluttering glow. When it cools, fo as to freeze, the phofphorus flames out very brifkly. This is probably owing to the emerfion of latent heat.

Nickel can be mixed with moft of the other metals, excepting quickfilver, and a few others to be mentioned afterwards. And it is faid that the Chinefe *packfong*, which is a white metal, or metallic mixture very like filver, is compounded of copper combined with nickel and zinc. We fhall take farther notice of this when treating of copper.

GENUS X.—LEAD.

LEAD is the fofteft of the metals; and has fcarcely any elafticity or found. It is the leaft ductile; and cannot be beat into thin leaf. Its cohefion is weak. A wire one-tenth of an inch in thicknefs breaks with 29¼ pounds; but what is very remarkable, a rod of caft lead becomes almoft four times as ftrong by wire-drawing.

Its fpecific gravity is 11,35.

The fufibility of lead is familiarly known to you all. When paffing from a fluid to a folid ftate, it has a fingular intermediate ftate, in which it can be eafily divided into very fmall parts or grains *. When poured then into a hot iron-mortar, fet fo as to cool very flow-

* This obtains, in a greater or lefs degree, in all bodies, while they congeal. It does not furprife us in the brittle metals, nor in the roaching of alum, and many other cafes, where the folid is brittle, and the fluidity thin; but in vifcid fluids, or even in metals with a clammy or oily appearance, it attracts more attention, and it is even thought peculiar to them. EDITOR.

ly, if it be continually ftirred with a ftick, it comes to a ftate in which the parts cohere fomewhat like wet or greafy fand; and by ftirring it now brifkly, the greateft part can be reduced to the fize of a coarfe fand. This is called the granulation of a metal. An apparatus is made on purpofe for this operation;—a fort of oval wooden box, the infide of which has partial partitions, and is rubbed all over with chalk. The melted lead is poured into this box, the lid fixed on, and the box is violently fhaken for fome time. This dafhes the lead againft the ·fides and partitions, and foon breaks a great deal into grains: But the other method will convert more of it into grains, becaufe it does not cool fo faft.

There is another, and more curious manufacture of lead, in which it is alfo d'vided, when fluid, into very fmall parts. This is the manufacture of fmall fhot. A little orpiment or arfenic is added to the lead, which difpofes it to run into fpherical drops, much more rapidly than it would do when pure. The melted lead is poured into a cylinder whofe circumference is pierced with holes. The lead, ftreaming through the holes, foon divides into drops, which fall into w·ter, where they congeal. They are far from being all fpherical, however, many being fhaped like pears; and muft be picked. This is done by a very ingenious contrivance. The whole is fifted on the upper end of a long fmooth inclined plane, and the grains roll down to the lower end. But the pear-like fhape of the bad grains makes them roll down irregularly, and they waddle, as it were, to a fide; while the round ones run ftraight down. They are received into a fort of funnel which extends from the one fide of the inclined plane to the other, and is divided by feveral partitions, fo that it is really the common mouth of feveral funnels, which lead to different boxes. Thofe in the middle receive the round grains. On each fide are grains of a worfe fhape, but good enough for low-priced fhot: The grains which have gone far afide are melted again. The good ones are forted into fizes by fieves. (*See Note* 28. *at the end of the Volume.*)

Lead is caft into fheets, by letting it run out of a box through a long horizontal flit at the bottom, while the box is drawn along the table, leaving the melted lead behind it to congeal. The Chinefe caft it extremely thin in this way on cloth, for lining their chefts of tea.

Lead is very calcinable. It calcines flowly in a heat a little above its melting one. Its furface is quickly covered with a dirty wrinkled pellicle, which is renewed as faft as it can be raked off. This calcination goes on moft quickly at a heat rather below rednefs. If allowed to get too hot, the furface of the pellicle calcines in a moment to a much greater degree, forming a fluid film, which protects the reft from the action of the air.

Lead calcines much more quickly and perfectly by way of fcorification. Great quantities are changed into calx by this method in the procefs for feparating it from filver, which is generally found in lead ores. This is done in a reverberatory furnace, the bottom of which, called the hearth, is made of materials which can withftand the diffolving power of this calx. The lead is pretty ftrongly heated; and the fluid calx immediately forms on its furface. Two pairs of bellows blow flantingly on the middle of the melted lead, and thus blow this fcum over to the other fide of the hearth, where there is a channel by which it may run off. It collects in a mafs of an unequal flaking texture, juft as we fhould expect from the way by which it is forced to accumulate. By thus blowing off the fcum, the air is continually applied to metallic lead, which is in a fit condition for abforbing its oxygen. But it is very imperfectly calcined, becaufe by running off the hearth, the air no longer acts on it. It is of a greenifh yellowifh colour, with here and there fome ruddy ftreaks, where it has been more oxydated. The calx, in this ftate of oxydation, has fome volatility, and very noxious vapours rife from it during this operation, which produce terrible effects on the human body,—bringing on, fometimes in a moment, an univerfal paralyfis and rigidity, fo that the arm that was lifted up cannot be let down again.

The calx thus formed, when allowed to cool, congeals, as I just now observed, into a mass which is not transparent, but of a yellow colour, and plated texture, and is called *litharge*. The litharge in the apothecaries' shops is the same matter, broken into a coarse powder.

The lead gains about one-tenth of its weight by oxydation. This was the phenomenon which induced the French chemist Rey, in the 17th century, to ascribe the calcination of metals to the combination of them with air; a truth which no body attended to for more than a century. This observation was much more distinct than Dr. Mayhow's, to the same effect, on diaphoretic antimony.

This is not however the most highly calcined calx of lead. Litharge can be calcined more perfectly, by beating it to powder, and exposing it two or three days to a heat hardly sufficient for producing the most obscure degree of ignition. This is done also in a low reverberatory furnace; and the flame of wood rushing along it is beat down by the form of the arch, so as to play continually on the surface of the litharge. The colour of it is thus gradually changed from the yellow to an orange, and from that to a bright red, when it is called *red lead*, or *minium*.

We have here a striking example of the superior power of a moderate heat to that of a stronger one for enabling a metal to attract plenty of oxygen, and become thereby highly calcined; for this very minium, if exposed to a full red heat, gives out a quantity of vital air, returning at the same time to a less calcined state, and becoming a yellow powder, which, if the heat be increased, melts into litharge.

A quantity of vital air can be expelled from minium also, by the elective attraction of the sulphuric acid, assisted by a gentle heat. By this easily separable portion of vital air which minium contains, it has the power to oxygenate to a certain degree the muriatic acid, as Scheele has observed.

All the calces of lead are easily fusible, forming a thin fluid without viscidity; and when pure, concrete like litharge. All these oxyds, by melting them with one-eighth of charcoal, are eafily reduced. Common red wafers burn and yield drops of lead. Red lead on a bit of charcoal is converted into lead in a moment by the blow-pipe. Dr. Prieftley reduced it in ammonical gas; and water was produced at the fame time. He alfo reduced it in hydrogen gas, and alfo had water. By giving its oxygen fo readily to other fubftances, this oxyd greatly promotes the calcination of other metals. Red lead, melted with iron filings, will be reduced, and the iron oxydated or corroded.

In the manufactures of lead work, they prevent the calcination by throwing greafe or pitch upon the furface,—the operation of which is eafily underftood. When a private perfon has occafion to employ a plumber, he may obferve that they never throw greafe upon the melting pot, becaufe the drofs, as they call it, is confidered as ufelefs by the employer, and becomes the property of the workman. He rather encourages its formation, and contents himfelf with clearing the furface with a piece of deal before he cafts the lead *.

The older chemical writers fpeak of a golden and a filver litharge: Thefe names are given to litharge, as it is more of an orange or of a grey colour. The firft is moft abundant when the flame of wood is employed, and the other, when other fuels are ufed. Thefe diftinctions are now neglected, there being but one calx, which when it cools greatly is of a glaffy ftructure, with traces of oblong cryftallizations.

Such is the nature of lead with regard to heat. In trying it with the different acids, we find that this metal is affected by them nearly in the fame manner as feveral of thofe we have already defcribed.

The fulphuric acid will hardly act upon it, except the lead be in thin plates, and the acid applied ftrong, and with the affiftance of heat. During its action, part of the acid becomes volatile and fulphurous, its oxygen being taken by the lead. And when the operation is

* This obfervation is by Dr. Hooke, who faid, in a meeting of the Royal Society, that the plumber can make it all drofs, and much more, by the air with which it combines, (Birch's Hift.)   EDITOR.

managed in a particular manner, part is changed into perfect fulphur. The oxydated lead forms with the reft of the acid an infoluble fulphat of lead.   Leaden veffels are ufeful in preparing vitriol and alum, and even fulphuric acid itfelf, in confequence of the infolubility of the fulphat of lead.

The nitric acid, properly diluted, diffolves lead the moft eafily, producing fome nitrous air.   It forms a foluble compound, which bears to be largely diluted with water without being decompounded ; and it gives cryftals which decrepitate with great noife, but do not deflagrate.

The cryftals of this falt are of a fingular fhape, namely *triangular prifms,* like thofe ufed in optical experiments.   I know no other cryftals of this form.   They have not a fine edge, but a narrow facet fcarcely vifible.   Thefe are frequently fo grouped as to form hexagonal prifms, whofe fides are ftriped or fluted, by the want of every fecond triangular prifm, which is the primitive molecula, or nucleus, of this cryftallization.

This nitrate has its ingredients but weakly united : A moderate heat long continued will expel the acid from it,—leaving a fufible oxyd in the retort.

The muriatic acid, in its ordinary ftate, acts but very flowly and imperfectly on this metal.   Woulfe fays that fome inflammable air is produced during the folution : I have never found this.   It acts much more powerfully in the form of the corrofive muriat of quickfilver,—the lead being then fupplied by the quickfilver, with both oxygen and acid at the fame time.   Or if the lead has been previoufly oxydated in any other way, the muriatic acid very readily joins with it : Nay, it will fometimes be oxygenated by a calx of lead.   Such a calx may then be confidered as fuper-oxygenated.   When, for example, we add muriatic acid to the nitrate of lead, we immediately form a muriat of lead, which is always lefs foluble than the nitrate.

The fame thing happens if we add to the nitrate of lead any neutral falt which contains the muriatic acid.   A double exchange takes place.

The qualities of the muriat of lead are, difficult folubility in water, and volatility. It is remarkably fufible, melting before it is red hot. When cold, it congeals into a femi-tranfparent uniform mafs, called *plumbum corneum*, on account of its refemblance to a muriat of filver, which has long been called *luna cornea*, from its colour, flexibility, and a degree of foftnefs, which allows it to be cut with a knife like horn. Plumbum corneum is the moft powerful aftringent that I know.

It is alfo confiderably volatile, like all the metallic muriats; but has little of the butyraceous appearance which moft of them have.

The acetous acid can alfo be combined with this metal, fo as to reduce it to a faline and foluble form; but it unites much more readily with fome of its oxyds than with the metal in its perfect ftate.

The oxyd which diffolves the moft readily, is a carbonat of this metal, well known by the name of *white lead*, or *cerufia*, formed by a procefs contrived on purpofe.

Vinegar is put into a ftone-ware pot, and at the diftance of two inches from its furface, there is a crofs of wooden bars, on which is fet a roll of fheet lead, of fix feet long, and fix inches broad, and about one-tenth thick. It is rolled up loofe, with the diftance of about one-fourth of an inch between each turn. The pot is fet in a bed of tan, or horfe dung, and covered with a plate of lead. In about three weeks, the whole furface of the fheet is covered with a faline cruft, which is detached by unrolling the lead, and fcrubbing with a wire brufh. The fheet is then rolled up again; and undergoes the fame procefs till the whole is diffolved *.

* This procefs merits the attention of the philofophical chemift. Dr. Black calls white lead, not a fimple oxyd of lead, but a carbonat; and in fact, white lead contains a very great quantity of carbonic acid, and this acid feems to be combined with the lead. The infolubility in water does not prove that white lead is not a combination of the metal with an acid, for we know others, fuch as the fulphat of this metal, not more foluble. How does the lead acquire this acid, and what is the nature of the vapour, after it has furnifhed this ingredient? EDITOR.

Here the acid is found to have greater effect when applied in the form of vapour. A certain quantity unites with the lead, in the form of a white powder, which is afterwards much more easily dissolved. This substance, dissolved in distilled vinegar, affords salt of lead, called *saccharum saturni*, from the sweetish taste with which it affects the tongue, resembling that of sugar. It crystallizes, or rather grains, very readily. If the salt be again dissolved in more vinegar, it combines with a greater proportion of it, becomes more soluble, and affords fine crystals.

With the tartarous acid, lead forms a compound very insoluble. It may be produced by boiling a solution of tartar, and adding any pure oxyd of lead. The superfluous acid of the tartar joins with the lead, as with chalk, and forms an insoluble sediment; and the tartar is thus neutralized, being deprived of the acidulating part of its acid.

Or if, instead of an oxyd of lead, we take a compound of this metal with nitric or acetous acid, and add this compound to a solution of tartar, or any liquor which contains tartar dissolved in it, the tartar is then completely decompounded by a double elective attraction; the whole of its acid uniting with the lead, to form a white insoluble precipitate.

Lead also forms an insoluble compound, in the same manner, with the acid of phosphorus, which can therefore be precipitated from urine by a solution of lead.

When we, in the next place, compare the forces of attraction between lead and these different acids, we find the strongest is that of the sulphuric acid. Other metals in general unite most strongly with muriatic acid; but in the case of lead, it is the sulphuric acid that is most strongly attracted, and when combined with the lead, it always forms an insoluble compound. It therefore precipitates lead from other acids, by joining itself to this metal in their place. This happens, not only with pure sulphuric acid, but with any salt or saline compound which contains the sulphuric acid. In however small quantity the fulphuric acid be thus applied to the diffolved lead, we have a double exchange, and a proportional part of the lead precipitates with it.

The folutions of lead, efpecially the acetite, are therefore, when properly ufed, an exceeding nice trial of the prefence of the fulphuric acid, pure or compounded, in mineral waters, or other mixtures, by occafioning turbidnefs.

After the fulphuric acid, the muriatic is next in force of attraction for lead. As I already remarked, it feparates the nitric acid, and joins itfelf to the lead in its place, forming the plumbum corneum, or muriat of lead.

The muriat of foda, and the muriat of lime or magnefia are decompofed by the oxyd of lead: Hence the corrofion of leaden pipes. A procefs has been followed for procuring the foffil alkali by employing litharge. Accordingly, the decompofition of the fea falt is accomplifhed, and foda is produced. But the expence would be far too great, were it not that the muriat of lead is manufactured into a fine yellow pigment of confiderable demand. The manufacture is chiefly carried on for the pigment, and the foda is a very fecondary object.

Lead has alfo a ftrong attraction for the tartarous acid, and the phofphoric acid, and will quit other acids to unite with them; and they always form with it infoluble compounds. They therefore precipitate the lead from its folutions in the nitric or acetous acid, and the lead precipitates them. I made a number of experiments on fuch precipitations of lead fome years ago, in confequence of which I found it eafy to diftinguifh different kinds of vinegar from one another.

Thefe compounds of the different acids with lead are all decompounded by common mild fixed alkalis, or carbonats of alkalis, and by calcareous earth *. Some of the other metals alfo attract acids

* *Quaritur.*—Is the precipitate of lead, produced by a mild alkali, the fame with white lead? It is furely a carbonat, and white lead is not a pure oxyd.

EDITOR.

more ftrongly, as zinc, which feparates it in fine metallic cryftals. There are alfo two of them that can be decompounded by heat alone, viz. the nitrate and the acetite of lead ; the firft, juft in the fame manner as the nitrate of quickfilver, with this difference, however, that the oxyd of lead is not reduced. The fecond may be decompounded with a heat inferior to red, and is fometimes decompounded by fuch a heat, in order to extract the vinegar from it in a very ftrong ftate. During this operation the lead calx is reduced,—a fmall quantity of inflammable fluid, refembling alcohol, is produced, and found mixed with the vinegar.

From Lemery's account of this diftillation, it does not appear that much pure or ftrong acid is obtained. He rectifies the diftilled liquor afterwards by itfelf, to feparate the more volatile half of it, which is inflammable, like fpirit of wine. The other half, he fays, is called *oil of Saturn*, and is ufed as a detergent for the eyes of horfes.

But by means of the fulphuric acid, an exceeding ftrong vinegar can be got from faccharum faturni. The faccharum being firft diftilled by itfelf, with a moderate heat, gives an ardent fpirit ; after which, fulphuric acid being added, extricates a ftrong vinegar.

Sulphat and muriat of lead cannot be decompounded by heat. The firft melts in the fire, without feparation of the acid, unlefs inflammable matter be made to touch it, which volatilizes the acid. The muriat not only melts, and that very eafily, but alfo proves volatile, and will wholly evaporate without being decompounded, and cannot be reduced by inflammable matter alone.

Alkalis, in their liquid form, efpecially if cauftic, corrode lead.

Nitre calcines it, when applied with a proper degree of heat, and converts it into a yellow powdery fubftance, precifely fimilar to litharge in all its properties.

Common falt is decompofed by calx of lead.

Sal ammoniac, diftilled with granulated lead, is partly decompounded by the lead, but much more eafily by the calces of lead ; and this is the readieft way for plumbum corneum. The volatile alkali is cauftic, or nearly fo.

The muriat of lead made with fal ammoniac, when heated with a covered patella under the muffle, becomes pafty, and in fome degree fluid, diminifhing much in bulk. When it is juft red hot, it begins to fume, and upon increafing the heat ftill more, it becomes perfectly fluid, contracting ftill more in its bulk. It now emits elaftic fumes. When allowed to cool, it is of a greenifh gray colour, with a little tranfparency, and is quite brittle. When viewed with a microfcope, its graynefs appears owing to very fmall globules of reduced lead.

With refpect to the earths, when they are expofed to melting heats with the calces of lead, we have an opportunity here to obferve one of the moft remarkable and ufeful qualities of the metal. This is a power which its calces have to diffolve and melt in the fire the earthy fubftances, very few of which can refift this power; and they form with them glaffy compounds. This property is fo eminent in the calces of lead, that no earthen veffels can be contrived to contain them melting for any length of time. Some chemifts have faid that they are poffeffed of fuch a fecret, but do not choofe to difcover it. Pott of Berlin has made a vaft number of experiments, in order to difcover a compofition of earthen veffels which can refift the action of the glafs of lead. The refult of his refearch is, that pure clay, which is free from iron, is the beft. And as it muft be kneaded with fome dividing matter like fand, to prevent its cracking by fudden changes of temperature, the beft addition is pure clay, which has already been burnt in the moft intenfe fire. This, beaten to powder, and forted into a fand fomewhat coarfe, makes a compound which will refift fufion by the calx of lead much longer than any other: But it alfo will imbibe it, and then will melt, as falt diffolves in water.

In confequence of this property, lead is employed in many proceffes, in which it is neceffary to bring earthy fubftances into fufion, as in working the ores of the precious metals. Alfo, on account of its being tranſparent, it is employed in making the fineſt kinds of glaſs, called *flint-glaſs* and *paſtes*. This name is given to the artificial glaſſes for imitating the gems. The calx of lead is eminently fitted for this purpoſe, becauſe it communicates to the compoſition a greater refracting and diſperſing power than any ſubſtance except the diamond, and to this it approaches very nearly. Unfortunately, theſe paſtes are ſo very ſoft, that they ſoon loſe their brilliancy, being ſcratched by almoſt every thing. This great diſperſive power has enabled us to remove the defect of optical inſtruments by refraction, which had been long deſpaired of. Mr. Dollond, and Mr. Hall, a country gentleman, at the ſame time, and unknown to each other, diſcovered a method of conſtructing *achromatic teleſcopes*, by joining lenſes of crown and of flint glaſs. There remains, however, a great obſtacle to their perfection, which the ſuperior ſkill of the chemiſt only can remove. Flint glaſs is never uniform in its compoſition. The great weight of the calx of lead ſeems to make it fall down towards the bottom of the pot, and when the workman takes it up, and collects it into a maſs at the end of his pipe, by turning it round and round, it is collected in ſtrata of different denſity and refracting power, which greatly diſturbs the formation of a fine image in the optical inſtrument. A premium of ſeveral thouſand pounds is offered for the removal of this defect. It is remarkable that the calces of lead not only increaſe the refractive and diſperſive powers of the glaſſes of which they are ingredients, but that they produce the ſame effect in their watery ſolutions. Sir Iſaac Newton firſt obſerved that a ſolution of ſugar of lead had a much greater refracting power than pure water. Mr. Zeiher, an ingenious chemiſt of Berlin, by increaſing the doſe of minium, compoſed a paſte which ſcarcely yielded to the diamond in brilliancy and refracting power, and exceeded it in the diſperſive quality, by which the different colours of light are ſeparated from each other *. Minium,

* As the difference of diſperſive power in different ſubſtances is very conſiderable, and as it is much more remarkable with regard to ſome colours of light than others, it

with a very fmall proportion of diaphoretic antimony, when mixed with flint, produces a pafte fcarcely diftinguifhable from the fineft topaz.

Calx of lead, or glafs containing lead, is neceffary for making all white opaque compofitions called *enamels*. Arfenic and tin alone will not do.

Flint-glafs and paftes often have *veins*, or *eels*, as they are called, which differ in their refracting and difperfing power from the reft, like a mixture of fyrup and water; and on account of this, are not ufed for mirrors. If the calx of lead be very redundant in fuch glafs, and efpecially if it be not much calcined, it is of a yellow colour. It forms the glazing of common earthen ware, but it is very liable to corrofion by acids, which renders it very unfit for many culinary purpofes. The folution alfo is very hurtful to the conftitution. The common foft glazing on earthen ware is quickly corroded by vinegar, and fuch ware cannot be ufed for holding pickles. There are no veffels fit for this purpofe but thofe of the porcelain kind, or what we call ftone-ware. Till within thefe 30 years it was made in great abundance in this country, and it had no glaze, but a flight fuperficial vitrification of the ware itfelf, by the vapour of foda thrown into the fire-place while the whole kiln was in a white heat. The prefent ware for the table, which has fupplanted the ftone-ware, is a bibulous clay, glazed with lead; and pots for pickles and fweet-meats are made of it. But they fhould be rejected entirely for fuch ufes, and we fhould only ufe the ftone-ware. It is ftill to be had, but of a brown, or dirty colour.

Lead is eafily recovered from glafs which contains it. Such glafs is liable to be fmoked or blackened with the flame of the blow-pipe. This is nothing but the reduction of the lead.

has all the appearance of being owing to elective attractions for the rays of light. It is not improbable, therefore, that there may be found a fubftance that attracts them all alike. Such a fubftance would be a moft precious difcovery to the optician. In the mean time, it affords another ftrong argument for the materiality of light.

EDITOR.

Of the inflammable fubftances, feveral act upon lead.

Many of the oils can diffolve lead or its calces; and it is remarkable that they are diffolved moft readily, and in greater quantity, by the mildeft, or the unctuous oils. The folution of the calces of lead in thefe requires the heat of boiling water to affift it; and when they have diffolved a proper quantity, and are then allowed to cool, they form a mafs that is firm and hard when cold, but which becomes foft, and tough, and adhefive, by being warmed a little. On account of thefe qualities, and its oily nature, in confequence of which it cannot be diffolved or changed by humidity, it is very valuable.

The *emplaftrum lythargyri*, or *commune*, of the fhops, is made in this manner, and forms the bafis of feveral others. A fmall quantity of the calx of lead is alfo diffolved, by boiling in the lintfeed oil, which is ufed as paint. It makes the oil dry fooner, and leave a thicker varnifh on the furface of bodies to which it is applied. Oil-paints, in which the coloured calces of lead are employed, are, for this reafon, the moft durable pigments. White lead and lintfeed-oil contracts a fkin on its furface which is almoft impenetrable by any action of the air and weather. Its only defect is, that fulphurous and putrid fteams foon blacken it. Hence it is that this brilliant white cannot be ufed in water colours in miniature painting.

Lead, either metallic, or in the ftate of calx, has alfo a ftrong attraction for fulphur. Sulphur is added to melted lead, and when mixed with it, readily unites; and forms a compound lefs fufible than lead, and which in cooling produces a mafs internally cryftallized into glittering particles, and greatly refembling one of its ores. It is decompounded with difficulty by uftulation; more readily and perfectly by iron.

Sulphur acts alfo upon the oxyds of lead, and upon the compounds of thefe with acids, when it is applied in the form of a liquid alkaline fulphuret. The fulphur unites with the oxydated lead, and blackens it, or gives it a dark-brown, or blackifh colour.

This effect is also produced by the inflammable air, or sulphurous hydrogen, that is contained in liquid alkaline sulphurets, and which separates from them in the form of a fœtid gas, when acids are added to them, or when they are largely diluted with water. The hydrogen gas having, when pure, very little attraction for water, has a tendency to evaporate from such a diluted mixture; and it carries away a small portion of the sulphur, which, if it meet with any oxyd of lead, is sure to be attracted by it, and to blacken it. This is the origin of what is called *sympathetic ink*, or ink that appears by sympathy. In the course of the numberless experiments of the alchemist on lead, in the fond hope of converting it into silver, it appears that none engaged their fancy so much as those that were disgusting. Human urine, human ordure, rotten eggs, and such things, were favourite objects of research, or means of farther inquiry. When the abominable smell of hepar sulphuris was observed, hepar sulphuris was tortured by every alchemist in Europe, and its mysterious properties were innumerable. Basil Valentine discovered that rags moistened with acetite of lead became black in a few minutes if near hepar sulphuris, and he made a magical trick by it. If you write with a solution of saccharum saturni, and lay the paper between the first pages of a thick book, and put between the last pages a leaf of paper that is moistened with a solution of an arsenical preparation containing hepar sulphuris, the writing will become legible in a few minutes, which was ascribed to a certain inexplicable sympathy between lead and arsenic. Lemery discovered that the arsenical preparation was effectual, only because it contained hepar sulphuris. The hydrogen gas itself, independently of the sulphur which is combined with it, has some share in producing this effect, and it is not improbable that the sulphur alone would not produce it. The hydrogen gas acts by reducing the lead to its metallic state, and thus disposing it to unite the more readily with the sulphur. Of the power of this gas to restore dissolved metals to their metallic state we have many examples in some ingenious experi-

ments made by Mrs. Hulſham. A wet mark made with liquid acc-
tite of lead, when expoſed to the hydrogen gas as it ariſes from iron
filings and ſulphuric acid, blackens immediately.

Mrs. Hulſham diſcovered that wa.er muſt be preſent with the metal
and acid, otherwiſe the reduction will not ſucceed. The effect of the
water probably depends on its attraction for the acid, by which it promotes the ſeparation of the metal, while the hydrogen attracts the
oxygen from it. But the preſence of water is not neceſſary to the
action of ſulphurated hydrogen gas. The metal is then ſeparated
from the acid by the power of two attractions acting at once ; the attraction of the hydrogen for the oxygen of the metal, and that of the
ſulphur for the metal itſelf.

The ſame effects are alſo produced on the oxyds and ſolutions of
biſmuth, and rather more ſtrongly than on thoſe of lead. Pearl white
precipitated from the nitrate of biſmuth by the ſolution of cream of
tartar, being rubbed on paper, and expoſed to the hydrogen gas, is
much more blackened than the ſimilar preparations of lead. And
theſe effects enable us to diſcover the preſence of lead in various liquids in which it may happen to be concealed, as in wines eſpecially,
and cyder, and in oils and other fluids. The anxiety of the French
chemiſts about this matter may be ſeen in the works of Macquer and
others.

Cyder made in Devonſhire was formerly liable to this adulteration ;
but I believe that care is now taken to avoid it in future. It was produced by putty and other preparations of white lead, which were uſed
for filling the ſeams, and ſtopping leaks in the preſs-floors and vats. A
better conſtruction of them has made this unneceſſary.

Relation of Lead to other Metals.

It will not mix with iron or cobalt, but unites eaſily with all the
reſt, and is employed ſometimes to promote the calcination of ſome
which are otherwiſe calcinable indeed, but which calcine more quickly

along with the lead. Of this I fhall point out particular examples hereafter.

The ufeful compounds of this metal with others, are 1/t, Type-founders metal, in which it is the principal ingredient, but mixed with antimony. 2d, Organ-pipe metal, in which it is mixed with tin. 3d, Pewter, in which tin is the principal metal, but there is commonly fome lead as an alloy to the tin; commonly too much. Nails made of a compofition of three parts of tin, two of lead, and one of antimony, are hard enough to be driven into oak without being blunted, and are not rufted by falt water.

Another great ufe of lead is in working of the ores and metallic mixtures containing the precious metals filver and gold. In working the ores of thofe metals, it produces a thinner fufion of all earthy and vitrifiable matters than any other fubftance. In fuch a thin liquid, the fmalleft globule of the precious metal makes its way to the bottom, which it could not do through a more clammy glafs.

Iron, tin, and all other metals, except copper and platinum, are feparated from the gold and filver by means of lead, by actually deftroying them as metals, and reducing them to a thin vitreous flag or fcoria, which is blown off the furface of the melted mafs, in the fame way as litharge is, or is abforbed by the porous veffel or teft in which this fcorification is performed. The gold or filver is left in the veffel, becaufe it completely refifts the action of heat and air, and the fcorifying power of the lead. The fcoriæ which are driven off, or imbibed by the teft, are reduced to metal by pounding and melting it along with inflammable matters.

With regard to the ores of this metal, there is no great variety. There is only one fpecies found in abundance, the *galæna faturni*, which varies a little in its appearance. The richeft is commonly that which breaks with broad and bright furfaces, and into cubical maffes. Some of the varieties of it break with lefs broad and more irregular furfaces, and fome with fuch fmall ones, that the fracture looks more or lefs like that of fteel. But though galæna is the principal, and only abundant ore of lead, fmall quantities of this metal are found in other ftates in the mines which abound with galæna: For example, *1ft*, Lead almoft pure and metallic. 2*d*, Calx varioufly cryftallized and combined with fixed air. 3*d*, And fometimes combined with fome acids, as in green ore, &c. Some colourlefs cryftals of this kind contain arfenical acid, and others the fulphuric.

Treatment of the Ores of Lead to extract the Metal.

If pyrites be mixed with it, the galæna muft be roafted. If it be free from pyrites, it may be melted out without roafting, or roafted and melted at the fame time. This is done in this country in very low furnaces, or a fort of open hearths. In England it is done in reverberatories.

As an article of the materia medica, this metal is reckoned powerful. As an internal medicine, it has an eftablifhed character as a moft powerful aftringent. It is therefore ufed fometimes, after all other remedies have failed, in profufe hemorrhages from the uterus, the lungs, or the ftomach. But it is not employed, except in cafes which appear otherwife defperate, as it is certainly in itfelf a dangerous remedy. It is well known that the workmen employed in many of the manufactories of lead lofe their health by it, and become objects of compaffion;—fuch as thofe who melt it from its ores, and thofe who prepare colours from lead, or grind them for the painters. Such perfons are feized with obftinate conftipation, dreadful griping pains, contractions, and paralytic affections of the legs and arms; and as all comes on very flowly, men are not alarmed; and the profits of their employment makes them continue at it till their cafe becomes diftreffing, and generally incurable.

Mr. Clutterbuck, of the college of furgeons in London, gives an account of a new and fuccefsful method of treating thofe affections. His method is to give mercury, which he confiders as a powerful ftimulant and exciting remedy, and therefore oppofite to lead, which is powerfully fedative, and often hurtful, and even mortal, by its fedative, and weakening benumbing power. A number of cafes are related, in every one of which a cure was completed in fix weeks at fartheft. To obviate the coftivenefs produced by lead, he ufed calomel; a grain or two every night operated more certainly and eafily than any other laxative. He alfo removed the palfy of the hands, by continuing the mercury until the mouth was a little affected.

As external applications, however, the preparations of lead are not only fafe, but in the higheft degree ufeful and falutary. The character it bears in this view, is that of a cooling fedative anodyne, and difcutient medicine. A French author, Goulard, wrote two books upon it, in which he fhews plainly that he is an enthufiaft in his opinion of its value; but relates, in a very fimple and diftinct manner, cafes which are very ftriking proofs. He recommends it highly in moft external inflammations, where there is great pain or heat, and efpecially in paranychia, phymofis, and paraphymofis, buboes, eryfipelas, when tending to mortification; the painful inflammation before mortification; gun-fhot wounds that are long in healing, and all ill-conditioned ulcers; the hard painful fwellings, and the ulcerations of womens breafts; all fort of fiery eruptions of the fkin; and itch. In thefe ill-conditioned ulcers, and painful inflammations, its effects are often very remarkable in relieving the patient from pain, and abating the inflammation, and all its fymptoms; and inducing fleep,—and changing the ftate of ulcers for the better in every refpect, and facilitating the feparation or extraction of extraneous bodies from them. In fome, however, it does not do fervice; and fuch are more relieved by relaxing emollient applications. I was informed by a navy furgeon, that it feldom did fervice in the cafes of failors. A weak folution of faccharum faturni is at prefent found the beft dreffing for fcrophulous fores or tumours.

Mr. Goulard's preparation is made by boiling litharge two or three hours in vinegar ; but as this muſt be of very uncertain ſtrength, according to the different vinegars employed, we find it more adviſable to uſe faccharum faturni, diſſolving a ſcruple, or half a drachm, in a bottle of water. This gives a determinate ſtrength, if the water be pure. But it generally becomes milky, in conſequence of ſome muriat or vitriolic falt contained in it. This diminiſhes the ſtrength a little.

GENUS XI.—TIN—STANNUM.

This is the lighteſt of all the metals, its ſpecific gravity not exceeding 7,3. The fineſt Engliſh tin when melted does not exceed 7,29. Hammering will bring it to 7,3. Therefore lightneſs muſt be conſidered as a teſt of its purity. It is remarkable that when mixed in due proportion with all other metals, except regulus of antimony, the ſpecific gravity of the mixture is much greater than what ſhould reſult from the weights of the ingredients. Nay, mixed with ſome much heavier than itſelf, the mixture has a greater ſpecific gravity than the heavieſt metal. The tin muſt receive the other metal into its pores, in ſuch a manner as not to increaſe the bulk.

The colour of this metal is very like that of fine ſilver ; and it retains its colour and brightneſs in the air better than many others. It has a conſiderable degree of malleability, and it can be hammered into very thin leaves, but it has not enough of that kind of tenacity that fits it for being drawn into wire. When thick pieces of it are bended, it gives a grating ſound, which appears to proceed from ſome deficiency of tenacity, or ſome imperfect union and coheſion of its parts.

It is the moſt fuſible of the metals, and is eaſily melted with a heat conſiderably leſs than the boiling heat of quickſilver ; and when large quantities of it are melted, and then allowed to cool and congeal without diſturbance, its parts concrete into oblong angular maſſes or prifms. This is feen moft diftinctly when a large mafs or block of pure tin is broken in pieces with the ftrokes of a large hammer, immediately after it has concreted from a melted ftate, or rather before the concretion of it is quite completed. It then appears in the form of what is called *grain tin*. We can alfo reduce it to a powder, or to fmall grains like fand, by melting it in an iron mortar, and ftirring the melted tin brifkly and inceffantly until it be congealed. Thus we get the pulvis ftanni of the difpenfatories.

Tin has a ftrong attraction for oxygen, calcining with a very moderate heat; and from the whitenefs and infufibility of the oxyds which it affords, it appears to undergo, in proper circumftances, an high degree of oxydation, which can even be carried fo far by the action of the nitric acid as to convert it into a fort of acid.

The oxyds of tin cannot be melted or vitrified by themfelves, and are even difficultly melted in mixture with other vitrifiable bodies. For this reafon an oxyd of tin, mixed with the ingredients of glafs, forms one of the beft of the white enamels *.

When we diffolve tin in fome of the acids, we find that while diffolving it has fome degree of power to decompound water.

It can, for example, be diffolved by digeftion with fulphuric acid, a little diluted, in which it diffolves flowly; and while diffolving it decompounds a little of the water, and produces fome inflammable air, but far lefs than iron or zinc. The tin is thus flightly oxydated as well as diffolved, or it is combined with a fmall quantity of oxygen taken from the water. But when it has attracted this fmall quantity of oxygen, its attraction for more is diminifhed to that degree that it has not power to decompound any more of the water, although it is capable of attracting a far greater quantity of oxygen when this is

* This, however, cannot be made in perfection without a little lead. A glafs containing lead may be whitened with tin or arfenic, or diaphoretic antimony: But if the calx of tin be added to a glafs that was free of lead, it is diffolved, and becomes tranfparent.

applied to it in a more fimple ftate ; in the form, for example, of oxygen gas, either pure, or as contained in atmofpherical air *.

The fulphuric acid, not diluted, and applied hot, can corrode and diffolve half its weight of tin, and is at the fame time changed in part into fulphurous acid, the diffolving tin attracting in this cafe from the acid itfelf a part of its oxygen. If the action of the acid be made lefs rapid, by diluting it with a very fmall proportion of water before it be applied to the tin, a portion of it can be changed into perfect fulphur, as is the cafe alfo with lead. [" But, largely diluted, it will not " act upon tin." FOURCROY.] †

The nitric acid and tin act on one another in the fame manner as the fame acid and the metal of antimony. The tin is not diffolved, but oxydated to an high degree, by attracting to itfelf the oxygen of the acid. But this action, which is flow when we make the experiment with the metal of antimony, is fudden and violent in the cafe of tin. The acid is violently changed and decompounded ; part is changed into nitrous acid ; part into nitrous air ; and a confiderable part even into azotic gas.

The oxydation of tin by the nitric acid affords another diftinct example of the formation of volatile alkali. Tin, as well as iron, decompofes water,—and we fee that it decompofes this acid in a more

* Is this conformable to the general train of chemical phenomena, or even to the particular phenomena in the oxydation of metals ? It feems to be a doctrinal point, that when once a metal has combined with pure oxygen it is then, and not till then, difpofed to combine with more oxygen clogged with another fubftance in the form of an acid. Is the oxygen in oxygenous gas more fimple than in water ? It requires, in moft cafes, a very high temperature to begin the combination of combuftible bodies with the oxygen in the gafeous form. EDITOR.

† Is the theory quite perfect here, and in fome other fimilar inftances ? Tin difengages hydrogen by reafon of its ftrong attraction for oxygen : Yet if water be added to the fulphuric acid, and thus afford a more copious and eafier fupply of oxygen, the fulphuric acid is, in this cafe, more deprived of its oxygen than when the water fupplied it lefs liberally. EDITOR.

remarkable manner, fince it extricates pure azotic gas. That the hydrogen of the decompounded water unites with the azote of the acid, appears (in the fame manner as in the cafe of iron) by adding a cauftic alkali or lime. This immediately detaches the volatile alkali, and we perceive it by the fmell, or by the cloud which it forms with the vapour of the muriatic acid. We may ufe, of aquafortis three parts, water five, tin leaf three, and flaked lime three. Dr. Auftin was the contriver of this experiment, having firft made it in another form, namely, by moiftening tin leaf with the acid, and rolling it up loofely till it was corroded. He then wetted it with a folution of potafh, which inftantly detached the volatile alkali.

This experiment, therefore, coincides admirably with the others from which we have acquired the knowledge of the compounded nature of the volatile alkali. And it is the more fatisfactory, on account of its fucceeding equally well when made with other metals which have a powerful attraction for oxygen, of which iron is an example *.

The muriatic acid, in its ordinary ftate, diffolves tin very well, and in great quantity, efpecially when affifted with a digeftive heat. It will diffolve one-third of its weight. During the diffolution, fome inflammable air is produced by the decompofition of a fmall part of the water, from which the diffolving tin attracts the oxygen. And if there be the fmalleft quantity of arfenic in the tin, which is often the cafe, it remains undiffolved in the form of a black matter, which is metallized arfenic.

The folvent, however, which has been commonly employed for

---

* Dr. Black has omitted taking notice of what has been called the *acid of tin,* or *ftannic acid,* by fome late chemifts. It is a fingular combination of tin with oxygen, firft obferved by Mr. Hermftaedt. A folution of tin in muriatic acid, being boiled with nitric acid, which has been diftilled from manganefe, till the red fumes have ceafed, is then diftilled to drynefs. The remaining mafs is foluble in water, faturating thrice its weight. This is called *acid of tin.* It melts by a red heat, and is deprived of its folubility and acid tafte,—but long expofure to the air reftores thefe properties.   EDITOR.

diſſolving this metal is the nitro-muriatic, or aqua regia. It has
hitherto been employed by the dyers for producing the ſolution of tin
neceſſary for fixing the ſcarlet dye in the cloth dyed with cochineal ;
and it was univerſally believed, both by the dyers and the philoſophical
chemiſts who wrote on the art of dying, that no other ſolution of
tin was fit for this purpoſe, and that it was neceſſary for giving the
bright red or flame colour to the dye of the cochineal, which when
uſed alone, gives only a purpliſh red.

But Dr. Bancroft, who has benefited the world with a moſt valuable treatiſe on the art of dying, has ſhewn that theſe were very
great miſtakes. He has ſhewn that the bright and florid colour of
the ſcarlet dye is not produced by the ſolution of tin alone, or principally, but by it and a quantity of tartar, which is abſolutely neceſſary in the proceſs, and chiefly by the acid of the tartar. The ſolution of tin and the tartar mutually decompound one another, by a
double elective attraction, and a tartrite of tin is produced, which is
neceſſary for the effect deſired. And the doctor has ſhewn that a
cheaper ſolution of tin than that produced by the nitro-muriatic ſolvent is fitter for this purpoſe. It is made by diſſolving about fourteen ounces of tin, in a mixture of two pounds of the ſtrong ſulphuric acid and three pounds of the muriatic acid. This mixture
may be called the *muriatico-ſulphuric* ſolvent. It is a far cheaper ſolvent than the nitro-muriatic employed by dyers ; and it diſſolves a
much greater proportion of tin.

The nitro-muriatic ſolution of tin is alſo employed for preparing
from the ſolution of gold a fine purple or red colour, employed in
enamels. For this purpoſe it is neceſſary that the tin in this ſolution,
ſhould be as little oxydated as poſſible. This condition is obtained by
diſſolving it ſlowly, a little at a time, in an aqua regia compoſed of
four parts of aquafortis, and one of muriatic acid, diluted with an
equal quantity of vinous ſpirits ;—we muſt keep all very cold. When
enough of tin has been diſſolved, the liquor has a light yellow colour.

This folution muft be preferved from communication with the air, both while the tin is diffolving, and afterwards; for the tin, when diffolved in any acid, efpecially in the muriatic, has a powerful attraction for more oxygen, and attracts it very faft from the vital air of the atmofphere, and from other bodies. This appears from many facts, and efpecially from a remarkable one difcovered by Mr. Woulfe. He found that white arfenic in powder, put into a recent folution of tin in muriatic acid, was metallized.

We have another remarkable muriat of tin, viz. the *liquor fumans Libavii*. The French call it the *oxygenated*, or *fuper-oxygenated muriat of tin*. It has feveral remarkable properties, which we have not time to confider,—and I muft refer you to Mr. Fourcroy.

To make this preparation, amalgamate the tin with one fifth of mercury, and triturate with an equal weight of corrofive fublimate. This mixture, diftilled with a gentle heat, gives out, firft an infipid liquor, and then fuddenly produces abundance of white vapours, which condenfe into a tranfparent liquor, which emits fimilar vapours by mere expofure to the air. I take this to be the muriat of tin overcharged with oxygen; for we have in the retort, befides an amalgam of tin and mercury, the common muriat of tin in a folid form, and needing only a greater heat to volatilize it completely.

Some experiments made by Mr. Adet give us more information concerning the nature of this fingular muriat, and juftify my calling it a fuper-oxygenated muriat of tin. A certain proportion of water, nearly one-third, added to it, caufes it to become concrete. In their combination, a confiderable quantity of common air is difengaged from the water, if it has not been carefully cleared of it before. Heat melts this mafs, and it is then able to diffolve more tin,—and in this folution no inflammable air is difengaged. Thus faturated with tin, the compound will bear a red heat without fublimation; but vapours arife, which are a muriat of tin,—what remains after a ftrong heat, is a white calx.

From thefe facts it would feem, that the compound, faturated with the metal, is an ordinary muriat of tin ; and the fmoking liquor of Libavius is an oxygenated muriat, differing from the other as corrofive fublimate differs from calomel. Mr. Adet's notions about the confolidating effect of water are ingenious, but are not explanatory ; and I think the inferences are not juftly drawn.

I do not know any remarkable effect of other compound falts on tin. Nitre deflagrates with it as with other metals. The vitriolic falts are decompofed by it, and its avidity for oxygen decompofes the acid, and we have fulphur. It acts in the fame manner in decompofing fal ammoniac ; and we obtain inflammable air by the decompofition of fome of the water.

None of the earths exhibit any interefting phenomena in conjunction with tin.

Sulphur combines with it in a remarkable manner. An ounce of flowers of fulphur being added to five ounces of tin in fufion, they form a black compound, much lefs fufible than the metal. The mixture therefore fixes as foon as the chemical union takes place, and the latent heat of both ingredients emerges, and there is a bright incandefcence, and the mafs fuddenly catches fire. When the fulphur is in a much larger proportion, the tin feems to deflagrate.

This combination of tin with fulphur is the bafis of another remarkable preparation, which long amufed the fancy of the alchemifts. —I mean *aurum mufivum*.

Aurum mufivum, or *mofaicum*, is made by various proceffes. The one moft certain of fuccefs is the old one of Woulfe. (*Phil. Tranf.* 1771.) Melt twelve ounces of tin, and add three ounces of mercury ; triturate the mixture with feven ounces of fulphur, and three of fal ammoniac ; put the powder into a matrafs and fet this in the fand-pot of a furnace deeper than the furface of the contents ; employ firft a gentle heat for feveral hours, and then raife it confiderably for feveral hours more. In the bottom of the veffel will be found the golden scaly mafs, of a flippery texture, like black lead. If the heat has been too great, or too long continued, it gets a dark colour and ftriated texture ; farther up from the heat, we have cinnabar, and a compound or muriat of tin ;—a volatile hepar efcapes. The aurum mufivum contains feven parts of tin and five of fulphur, and confiderably exceeds half of the materials. The cinnabar is much overcharged with fulphur. The volatile hepar is but trifling. There is no mercury in the aurum mufivum. This muriatic fublimate greatly excels all folutions of tin for giving a fcarlet dye. The mercury and fal ammoniac are not effential, but improve the colour remarkably ; and aurum mufivum, prepared without mercury, does not make a good amalgam for increafing the power of an electrical machine, for which the other preparation is very remarkable. Chaptal, however, fays, that this pigment, as generally prepared, contains almoft one-fixth of mercury*.

Tin produces fome ufeful mixtures with other metals †. Firft, it is ufeful for tinning iron, and copper, and brafs veffels. *Tinned plate iron*, or *white iron*, is a moft ufeful production of art, as it is fo eafily wrought. The tin not only adheres to the iron, but penetrates it to a confiderable depth, and renders it extremely foft and ductile, fo that it can be forced up into any fhape by the hammer, without cracking, or lofing much of its pliancy.

Fine pewter is a mixture of tin with a fmall proportion of copper, and zinc, and bifmuth. The copper gives hardnefs ; the other two improve the colour.

* *Procefs for preparing a beautiful Aurum Mufivum.*

Precipitate nitrate of tin by a liquid fulphat of potafh. Wafh and dry the precipitate, and mix it with one-fourth of fal ammoniac, and one-half of fulphur. Put this mixture into a retort, and give it a fubliming heat.

† A mixture of tin and lead calcines fafter than any metallic mixture that we know, burning like a dull peat ; and the compound calx is more oxydated than either of the metals can be fingly. Hence the advantage in the admixture of a fmall quantity of lead with the tin for making white enamel.

A metallic mixture, which has the beautiful whiteness of fine silver, is made of tin and bismuth. It is probable that copper or iron may be easily tinned with it. I suppose some of the iron work of chariots is whitened with it.

Tin is also added to copper in different proportions, to serve different purposes, which I shall mention when I speak of copper.

I have already observed that the calx of tin is employed for the composition of white enamels. It is a mistake in the writers on this subject to say that the pure calx of tin is the proper basis of this composition. It must contain a minute portion of lead, or the flux used with it must contain lead, otherwise it will make only a semi-transparent white. Montamy, who has written the most accurately, and indeed excellently, on enamel painting, gives minute directions for the preparation of this article, because it is used with almost every colour, in order to give to each its proper intensity. It is a most tedious process, so that the article must bear a very high price. He prescribes pure tin; but then his *fondant*, or *flux*, with which it is diluted, is *crystal d'Angleterre*, which is our flint-glass, containing lead. The common white glazing used for the cover of Delft ware is a much cheaper composition, being merely the calx of pewter carefully made, and often having an admixture of arsenic. Tassie's medallions are made of this, with a little magnesia,—with flint, and minium, or flint-glass, for a flux.

Tin is very rarely produced by nature in its pure and metallic state; and there is no great variety of its ores. Only one kind is found in plenty. And no part of the world abounds with it so much as Cornwall. There is an exact register kept there of the produce of the tin-mines, and it appears that the average of 20 years has been about 3000 tons weight a-year.

The component parts of the ore are, calx of tin, and some iron, and sometimes a little arsenic. Being thus composed entirely of metallic matter, it is remarkably heavy; and being at the same time very hard;

it is the ore which, of all the ores, bears pounding and washing best, and which remains the longest time in the gravel, and sands, and soil, of the mountainous countries in which the veins of it are found. In such countries great quantities of this ore are got by washing the sands and gravel of the brooks, or the soil that is near them.

After being well dressed, that is, washed clean from the matrix or sand, it is roasted, and then melted in air-furnaces, or reverberatories, with addition of some small coal to promote its reduction to the metallic state. It is afterwards assayed, to ascertain its fineness or purity; and a stamp is put upon it according to the result of the assay.

Grain tin may always be accounted pretty pure, because a very small portion of lead, which is the cheapest adulteration, and most frequently practised, destroys the tendency of the melted tin to assume that particular form. Antimony changes the form, destroying the greasy look of the fracture. Bismuth changes the crystallization still more, and is never mixed for gain.

I find nothing so effectual for clearing tin for medicinal use from the minute portion of arsenic, which it sometimes receives from the matrix, as distillation with sal ammoniac. Mercury also separates it very well; but a very long and careful trituration with water is necessary for causing the amalgam to throw out all that it contains.

Tin is used, in the practice of medicine, as a remedy to kill worms in the intestines: For this purpose it is reduced by granulation to a powder like sand, called *powder of tin*. This powder, however, has been often prepared, not from pure tin, but from pewter: And in the directions given for preparing it, by Dr. Alston, who first recommended it, in the *Edinburgh Medical Essays*, we are expressly desired to take *pewter metal*. But common pewter contains a good quantity of lead, and I think I once observed colic pains, that appeared to proceed from the use of tin powders. I believe this seldom happens now, the tin powder being now commonly prepared with pure tin.

It may be proposed as a question, whether the efficacy of the original powders, in killing or expelling worms, depended on the tin or the lead? The lead, however, never was fufpected. But different opinions have been formed on this fubject. The moft general one has been, that the powder operates mechanically. If this be juft, fand might ferve the purpofe better, being lefs heavy, and more eafily moved by the inteftines: And I have been told that fand is fome-times given to horfes. Mr. Lewis hinted a fufpicion that the tin might act by the powers of a fmall quantity of arfenic contained in it. But late experiments do not fupport this notion. They go againft it, and fhew that commonly tin does not contain arfenic, or fo very little that it cannot produce its effect. And the truth is, that fo very little, if any, of the tin is diffolved or corroded in the inteftines, that it is impoffible the very fmall quantity of arfenic which it has been found to contain fhould act in any perceptible manner: The opinion therefore of its acting mechanically appears the moft probable: Or if it be fup-pofed to act in confequence of its being corroded or diffolved, fome other preparation of it would be preferable.

GENUS XII.—COPPER.

THE obvious qualities of this metal are well known.

Of all the metals, copper is the moft fufceptible of condenfation by hammering, or wire-drawing. For this reafon, it is very difficult to fay what is its fpecific gravity. In Mufchenbroeck's voluminous table of fpecific gravities we have it from 7,243 to 9000. Briffon fays that copper melted is 7,788, and when drawn into wire is 8,878. This is much owing to the manner of cafting. Unlefs poured out in a very liquid ftate, that is, of very great heat, it will not caft either folid or tenacious, but is cavernulous and weak. In its beft ftate it feems por-ous. It is confiderably fofter when juft red hot, and bears the ham-mer. In this ftate it was employed in the hafty coinage of the Ro-mans; frequently in camp, with the general's die.

ABSORBS MUCH HEAT IN MELTING. 637

It is far more ductile and malleable than any of the metals yet described, and therefore is considered as a more perfect metal. Its tenacity is also very great when wire-drawn, exceeding that of all other
metals except iron. It resists the injuries of the weather much better
than iron, and acquires by time an obscure glossy skin, which helps to
defend it, like a varnish. There is a way to give it this sort of superficial colour by art. It is done by first polishing it, and then daubing
over its polished surface the deep red oxyd of iron called colcothar, or
Spanish brown, diluted with water, after which, the copper is heated
to a certain degree, and the colcothar again rubbed off.

Copper, though not much inferior to iron in hardness, will not
strike fire when violently driven or rubbed on flint, or other hard
bodies. This is owing to its being so much less calcinable, that is, inflammable; so that, although when a rag of copper is torn off by the
stroke, it is perhaps red hot, yet it does not *kindle* and become a
source of heat to kindle other bodies. Copper is therefore employed
in the powder mills for every purpose that iron is required for in other
works. The hoops of powder casks, the chizels, mauls, hammers,
and even nails, are all made of copper.

It requires a degree of white heat for its fusion, and in its fluid state
has a bluish green colour. Melting it simply, without any addition,
renders it less malleable; but when melted with a little charcoal on its
surface, its malleability is preserved. When copper is urged by very
great heat, it smokes; and the smoke is found to be metallic copper,
(*Mem. Acad. des Sciences*, 1769, *Jan.*)

It has a great capacity for heat, and requires besides, a great quantity of latent heat to render it fluid. A consequence of which is, that
when melted copper comes into contact with a small quantity of water, or watery humidity, the melted metal communicates heat to it so
rapidly, and in such immense quantity, that dreadful explosions are
often thus produced, by the instantaneous change of the humidity into
highly elastic vapour. The most anxious care is therefore employed,

VoL II.   4 L in founderies of copper, to have the moulds perfectly dry: They are even made red hot for some hours before we venture to cast any large piece of work.

When we choose to calcine copper by the action of air and heat alone, this is best done by exposing it to a red heat far inferior to that by which it is melted. The surface of it is then gradually changed into a blackish crust, which is brittle and pulverable, and gives a powder of a deep red colour. This is an oxyd of the metal, which is but little oxydated, and is very easily reduced again to pure metallic copper, by the usual means.

All the saline substances act with great facility on this metal. The more simple and active dissolve it perfectly: The more compounded or milder salts corrode it into rust.

The sulphuric acid requires the assistance of heat to make it act upon copper. Except when made very hot it has not power to dissolve this metal; and it must also be applied in its strong state. During its action, part of it is changed into sulphurous acid, by the loss of a portion of its oxygen, which is attracted by the copper. And the metal, being thus oxydated, and then combined with the rest of the acid, assumes the form of a whitish mass, which can afterwards be easily dissolved in water, and gives a liquor of a fine blue colour, which admits of large dilution in water, without separation of the acid. From this liquor again we can obtain, by due evaporation, deep blue crystals, commonly named *blue vitriol*.

This is, therefore, the third of the compounds which were commonly called *vitriols*, which have been already mentioned in this course. The first was the sulphat of iron, commonly named *green vitriol*; the second was the sulphat of zinc, commonly called *white vitriol*; and this third one, as I said just now, is commonly known by the name of *blue vitriol*. There is also mentioned, in Mr. Boyle's works, and those of other ancient chemists, a vitriol by the name of *Roman vitriol*. This was a mixed one, accidentally formed of the sulphat of iron and fulphat of copper, cryftallized together, but containing more copper than iron.

The nitric acid, in the diluted ftate of aquafortis, diffolves copper the moft readily and quickly. The acid becomes very warm, and there is a ftrong effervefcence, and production of a great quantity of nitrous air, into which a part of the acid is changed, in confequence of the abftraction of oxygen from it by the diffolving copper. This nitrous air is fo pure, or nearly pure, that Dr. Prieftley commonly ufed it in his experiments for trying and comparing the wholefomenefs of different airs. It is not, however, quite fo pure as that produced by diffolving quickfilver; but copper produces it more conveniently *. The compound formed by the oxydated copper and the remainder of the nitric acid is very foluble, and tinges the folution of a fine fky blue. It bears large dilution in diftilled water, without feparation of the acid and metal. By due evaporation, it yields cryftals of a very deep and rich blue colour, very foluble, and even deliquefcent.

If the cryftals of the nitrate of copper be bruifed, and flightly moiftened, and then be haftily wrapped up in tinfoil, touching it in as much furface as poffible, nitrous fumes immediately come out, the mafs becomes hot, and prefently catches fire, brandifhing and throwing out defultory darts of flame and melted tin. It is not at all clear whence does arife the great heat in this experiment. The only che-

---

* Chaptal fays that when the copper is put into this acid in fmall pieces, fo as to prefent an extenfive furface, the action is exceedingly violent, the acid is rapidly decompofed, and after the eruption of much gas, there is a fudden re-abforption, and ammonia is produced. This is a curious, and a fomewhat embarraffing fact. Whence comes the hydrogen? for we know that a tube of pure copper, red hot, does not decompofe watery vapour. If, indeed, water be decompofed, and hydrogen difengaged, the re-abforption might be attributed to its combination with the azote difengaged from the nitrous acid, had we not learned from Dr. Auftin's experiments that although gafeous azote will combine with nafcent hydrogen, gafeous hydrogen will not combine with nafcent azote.

EDITOR.

mical change that we obferve is a nitrate of tin in the place of a nitrate of copper, and the eruption of nitrous fumes. Some more oxygen feems, therefore, to be neceffary for the new compound ; and we fhould rather have looked for a diminution than an increafe of heat ; unlefs we admit the heat to be produced in the fame way as when alcohol or fugar is added to nitric acid.

Paper, moiftened with a folution of this nitrate, when held before the fire, catches fire at a very low temperature,—not fufficient to melt tin.

Thefe cryftals are very fufible, but it is a fort of watery fufion they undergo ; for they are very retentive of water ; and if we increafe the heat a little too high, the acid begins to evaporate, and carries away a fmall portion of the copper.

The muriatic acid diffolves this metal flowly. The colour of the folution is at firft dark brown, but becomes green by keeping. Sympathetic, or rather changeable ink, of Mr. Macquer, is prepared from this folution.

The folution, when duly diluted, has a light blue colour, and when fpread on paper, exhibits the fame hue as long as any moifture adheres to it. But when flowly dried before the fire, and allowed to attract the moifture again, it returns to its former ftate of an almoft imperceptible blue. In order to infure the difappearance when cold, dip a pencil in the folution (very weak) of muriat of lime. Pafs this over the copper folution,—it will not decompofe it, but, by attracting moifture, will caufe the colour to vanifh almoft entirely. Care muft be taken with this, and indeed with all thofe fympathetic inks, not to let the paper become too hot. This fcorches that part on which the folution is fpread, and makes it a permanent brown.

The fcales of copper, formed by calcination, exhibit nearly the fame appearances in folution in this acid. A recent and brown folution of the fcales, if largely diluted with water, depofits part of the copper in the form of a white precipitate, fuper-faturated with copper. This white precipitate, if immediately diffolved in frefh acid, gives a brown folution, which, like the brown folution of copper, or of the fcales, becomes green, if expofed to the air; abforbing oxygen from the atmofphere, becoming more calcined, and more foluble.

Hence we may conclude, 1ft, that the fcales are very moderately calcined; and that metallic copper undergoes nearly the fame degree of calcination, when forming a brown folution in the muriatic acid. 2d, That copper cannot be more calcined by mere heat and air, its furface being already neutralized. But if the copper be divided by folution, and expofed fimply to the air, it abforbs more oxygen; and the folution becomes green.

The vegetable acids do not diffolve copper readily in its metallic ftate. The confectioners have obferved that acids, fyrups, and pickles, made with vegetable acids, do not acquire a bad tafte from copper veffels, if only boiled quickly, and not allowed to cool in them, or to remain after cooling. It is only when we apply fuch acids to copper in the ftate of an oxyd, that they have power to diffolve it. But an oxyd is very foon formed, if thefe acids, or any faline fubftance whatever, is applied to this metal, while air is allowed at the fame time to act on it; a green or blue ruft is foon formed, which is moftly an oxyd of this metal, in an highly oxydated ftate, and very foluble in vegetable acids. And the reafon why thefe may be haftily boiled in clean copper veffels without being tainted is, that there is not time nor fufficient action of the air for the formation of an oxyd *.

* This co-operation of the air, in the flow folutions, or rather oxydations of copper, is unqueftionable, and is the fource of fome very curious appearances, which have long puzzled the chemifts, but are now pretty well underftood. It is not peculiar to copper, but its effects in the cafe of copper are more remarkable. But it is furely a difficulty in the theory. It is unlike the whole train of chemical phenomena, that a partial combination with any particular fubftance fhould increafe the difpofition to combine with more. There ftill lurks fome yet undifcovered agent.

EDITOR.

Verdigrife is an acidulous oxyd or ruft of copper, produced by art, prepared much in the fame manner as the faccharum faturni. It is alfo prepared in a coarfer way, by taking what remains in the wine-prefs, after the muft has been entirely fqueezed from the grapes, adding fome foured wine, and ftratifying this mixture with plates of copper. As the hufks grow four, the copper is corroded; and after fome weeks of this treatment, the faline cruft is beaten off by the hammer. This coarfe preparation is then diffolved in weak vinegar, with the addition of fome more copper; and laftly the cryftals are purifyed in the ufual manner.

This fubftance diffolves readily in diftilled vinegar, and gives a liquor of a deep green, which being evaporated, gives cryftals of the fame colour; but the powder of them is bright and elegant, like that of all femi-tranfparent bodies. Hence it is ufed by painters, though they find it corrofive. The name they give it is very improper,— *diftilled verdigrife*.

All thefe compounds of copper with acids, when they are thrown into the fire, or mixed with flaming fuel, give a green or blue colour to the flame, and noxious fumes.

If we defire to decompound thefe compounds of copper with acids, we can effect the decompofition of fome of them by heat alone, others can fcarcely be decompounded without the aid of an elective attraction.

The one which is the moft eafily decompounded by heat alone, is the acetite of copper. This muft be done by diftillation of the dry materials, that we may obtain the acid as concentrated as poffible. As a dry powder cannot communicate the heat fo quickly and perfectly to the interior parts, as a fluid body can do, this diftillation always deftroys fome of the acetous acid. The firft part of the produce is tainted with a fmall portion of copper, which gives it a green colour; and before the production of acid vapours is much abated, there is a very fenfible empyreuma. We therefore do not obtain nearly the whole of the acid in a ftate which can be purified. The taint of copper is eafily removed by a rectification; but the empyreumatic fmell cannot be taken from a confiderable portion, obtained about the end of the diftillation, when the bottom of the retort is even red hot. If this be left out, we obtain the acid in a very concentrated ftate, having a very pungent, but agreeable odour; inferior however to that which may be obtained, by a careful dephlegmation of vinegar concentrated by freezing.

From fixteen ounces of cryftallized verdigrife we may obtain three ounces of water, and fix and a half of acid, of which five and a half are (after rectification) free from copper and from empyreuma. There is a refiduum of five ounces and three quarters, and a lofs of nearly an ounce. This lofs is fuftained before the ftrong acid comes over.

Mr. Berthollet has fhewn that it is a more powerful acid, and has a ftronger attraction for alkalis than common vinegar; and that this is in confequence of its receiving oxygen from the oxyd of copper, which is reduced nearly to its metallic ftate during the diftillation, and is generally a pyrophorus. *(See Note 29. at the end of the Volume.)*

Of the other acids, only the nitric acid can eafily and perfectly be feparated by heat*; the fulphuric acid difficultly. The muriatic acid is not feparable by heat alone, but rather renders the copper volatile along with itfelf.

But alkaline falts, and alkaline earths, precipitate the copper; and fome of the precipitations deferve notice.

1/t, Lime-water is fometimes employed to precipitate copper from fome of its folutions, and gives a precipitate, which is an oxyd of the

---

* If the acid fo obtained be condenfed in a receiver containing a folution of cauftic foffil alkali, it forms a nitrate of foda, differing exceedingly from that formed by diffolving foda in nitric acid. I could not bring it into a good ftate of cryftallization; nor would it form cubical cryftals at any rate, but coalefced in flakes, which felt flippery and foft.

metal, combined with a small quantity of the lime, of a pleasant light blue colour, and is called *verditer*. It is employed by the painters. The French call it *cendres blues*,—blue ashes. The copper is first precipitated exactly by quicklime, and then from seven to ten *per cent.* of quicklime is ground with the calx and a little water.

2dly, Mr. Scheele has taught us another way of precipitating the copper, by which we obtain another colour useful for painters *.

When alkalis in their ordinary state are used, the copper is precipitated in form of a very pale or whitish-green oxyd, which can be dissolved again by adding more of the alkali than what is necessary to saturate the acid; the superfluous alkali acting on the metal as a solvent. This is most remarkable when we use the volatile alkali.

The volatile alkali has this power of a solvent with respect to all the purer oxyds or rusts of copper, and acts, though more slowly, even on the metal in its metallic state. And the deepness or richness of the colour is such, that it enables us to detect a very small quantity of copper in any mixture. One-hundredth of a grain will sensibly tinge a pound of water.

The action of ammonia on copper, or its oxyds, presents a very remarkable phenomenon. If a solution of caustic volatile alkali be

---

* Mr. Scheele's process is nearly as follows.—Dissolve two pounds of the blue vitriol in a copper kettle, over the fire, in about six quarts of pure water; dissolve in another vessel two pounds of dry potash, and eleven ounces of pulverized arsenic, (N. B. what is sold ready pulverized is generally adulterated with gypsum) in two quarts of water, also over the fire; when all is dissolved, strain the solution through linen.

Pour this arsenical solution into the solution of blue vitriol, stirring the mixture all the while with a wooden spatula; let the mixture stand quiet some hours, that the green colour may settle; then pour off the clear liquid, and edulcorate the sediment with repeated hot waters. At last the green pigment is obtained by draining it through a linen cloth, hanging loosely between four sticks; and then it is dried on blotting paper, or a chalk stone. By this process we obtain one pound six ounces of the powder.

The most careful edulcoration is necessary; for if it retain any of the alkali, it will not work with oil as a paint. It makes a very elegant and lively green. EDITOR.

poured on copper filings, we have in a short time a solution of a beautiful blue. If a bottle be filled quite full of the solution of alkali, and we put in some filings, and immediately put in the stopple, we shall not have any sensible blue colour till after a very long while. When ammonia is in a vessel not quite full, and has acquired this blue colour, it loses it by long keeping when the vessel is all the while close; but if we open it, the blue colour begins to appear at the surface, and in time it penetrates to the bottom. Closing the vessel causes it to lose the colour completely after some time; and it will again return when the vessel is left open. And this may be repeated several times before it becomes permanently blue. Chemists attempt to explain these appearances, by observing, 1st, That when the alkali has become saturated with copper, the solution is permanently blue; 2d, The solubility in alkali increases, as the metal is more oxydated, to a certain degree; 3d, The action of the air contributes greatly to the oxydation, when there is something at hand ready to take up the oxyd. Therefore the first solution is that of a very minute portion of copper at the surface; all below has such an excess of alkali that it is colourless. When the vessel is now closed, the copper, which is more abundantly dissolved at the surface, diffuses, and is again combined with an excess of alkali, and grows colourless. Now open the vessel,—the superficial parts become more oxydated, a greater quantity is dissolved, and the colour reappears, and will reach the bottom, if the vessel be kept open; if not, the copper again diffuses till the solution is again equable, with excess of alkali, and colourless. This may go on till *all* the alkali is saturated, when the colour becomes permanent.

Alkalis, therefore, have the power to dissolve copper, as well as to precipitate it from acids. They shew this power also when applied to it in the way of fusion, but act more easily on the oxyds than on the pure metal. And not only the alkalis alone, but any glass that contains them, acts upon those oxyds, and dissolves a certain quantity of them. The glass is thus coloured by the copper combined with it, and the colours it assumes are either transparent blues or greens, more or less similar to the colours of the solutions of copper ; or, when a great quantity of this metal is combined with the glass, the colour of the mixture is an opaque red.   The beautiful greens that are seen in the figures and foliage on the glazings of pottery are generally done with copper.

Another precipitation of copper from its solutions, which deserves notice, is the precipitation by iron.   The copper is precipitated in its metallic form, as in many other cases of the precipitation of one metal by another ; the iron attracting both the acid and oxygen at the same time.   This manner of precipitating the copper is practised in some. places with the waters of mineral springs, which flow through veins or mines of copper ore, and which contain a small quantity of blue vitriol dissolved in them.   This is in some places so abundant, that they are treated for the copper they contain, by putting old iron into the gutters in which they run,—the cupreous mud which settles on it is collected and reduced to metal in the usual manner, by a very small quantity of inflammable matter and flux.

Sulphur unites readily and strongly with copper, when they are properly mixed, and made to act on one another with the assistance of heat ; and while they are uniting, a quantity of heat is extricated from them, which suddenly produces in the mixture a remarkable and bright degree of ignition.

Some very remarkable experiments were made by the Dutch chemists Van Trooftwyck, Deiman, and others, which occasioned considerable surprise, and led some to call in question the theory of combustion published by Mr. Lavoisier.   It appeared that bodies could burn without the absorption of oxygen.   For the heat and incandescence produced in these experiments were so remarkable as to be mistaken for a real inflammation in close vessels.   When three parts of copper are mixed with one of sulphur, and are rammed into a small narrow-mouthed phial, and exposed to a heat gradually raised to rednefs, the mixture fwells, and then, almoft in a moment, becomes luminous all over, and burfts out into a violent flame. Zinc and fulphur require a ftronger heat, and give a brighter flame: So do iron and fulphur. This experiment fucceeds equally well in *vacuo*, in carbonic acid, in azotic, or in hydrogenous gas. (*See Exper. Phyf. et Chym. par V. Trooftwyck, Deiman, &c. Amfterd. tom.* iii.) But fince thefe facts have been more carefully examined, it appears that it is not an inflammation or combuftion, but merely an incandefcence, arifing from a copious emergence of latent heat,—the compounds are totally different from thofe formed by real combuftion. They have not, however, been fufficiently examined, though very interefting *.

The compound, when the parts of it are completely united, is a dark leaden-coloured mafs, which is quite brittle, not in the leaft malleable; and I fufpect that a great part of that heat which was extricated from it was the latent heat, which gave toughnefs and malleability to the metal in its feparate ftate. The fulphur is not eafily burnt out of this compound. There is lefs difficulty in freeing iron from fulphur, although iron is fuppofed to have, and has in reality, a ftronger attraction for fulphur than copper has. The reafon why the fulphur is not fo eafily evaporated from copper is, that this metal is not fo calcinable by heat as iron, and can be calcined only to a very moderate degree. In this moderately calcined ftate, it retains the fulphur ftrongly, and parts with it flowly; whereas iron, when joined with fulphur, forms a compound, in which the iron is very eafily calcinable, and to a high

* I think that it may be inferred from this, and many fimilar facts which have already come in our way, that the oxygenous gas is not the fole fource of the heat and light which emerge in the combuftion of bodies. In the union of fulphur with dry filings of iron, copper, zinc, lead, tin, filver, and with running mercury, there is no oxygenous gas, nor gas of any kind in action, and yet a very great deal of heat and light efcape from the bodies. Not to call this combuftion is, at beft, philofophical affectation and faftidioufnefs. It is not oxydation,—nor is the union of azotic gas with oxygen combuftion. So to mifname the one is ignorance, and fo to mifname the other is pedantry.

EDITOR.

degree. It has a ftrong attraction for oxygen or vital air; and when the iron is highly calcined, its attraction for fulphur or the fulphuric acid, is greatly diminifhed,—the calces of metals which are highly calcined, having in general but little attraction for fulphur or acids. But although we cannot fo eafily and quickly burn and evaporate fulphur from copper, it may be burnt out and evaporated from it by a flow and fkilful uftulation or calcining procefs, continued for fome time. The firft effect of it is to change the fulphur into acid, which remains adhering to the copper, and makes it foluble in water into a blue liquor, from which blue vitriol may be obtained by cryftallization; or if, inftead of diffolving the calcined matter in water, we roaft it further with a ftronger heat, we can evaporate the fulphuric acid, and have at laft an oxyd of copper, nearly in a pure ftate.

Some of the other inflammable fubftances alfo fhew an attraction for copper or its oxyds. Several of the aromatic oils can diffolve fo much as to receive a deep green tincture from it. Some of the unctuous oils alfo, efpecially when rancid, eafily diffolve fo much copper, or ruft of copper, as is fufficient to tinge them blue or green *. In the tallow of common candles, a fmall quantity of verdigrife is added to improve the colour.

The ufeful compounds of copper with other metals are, 1ft, *paak fong*, or *paak tong*, the white copper of the Chinefe, which is fuppofed to be copper alloyed with nickel. And when zinc is added to the compound, it refembles filver by its colour. An imitation of filver is alfo attempted with copper and arfenic. 2dly, The metallic compounds of which brafs guns are caft, and bells, and the mirrors of reflecting telefcopes, are copper or brafs, mixed with different proportions of tin. A fmall proportion of the tin gives the copper a yellowifh colour, and makes it very hard and inflexible, although it has ftill fufficient toughnefs, and great ftrength. This compofition is therefore employed in cafting what are called brafs guns. A larger

* Hence the terrible effects produced by copper veffels employed in cookery.

proportion makes it very elaftic, though lefs tough and ftrong.   This is ufed for bells.   A ftill larger proportion of the tin forms a denfe white compound, which receives a very perfect polifh, but is quite inflexible and brittle as glafs.   This is ufed for mirrors of telefcopes. The beft compofition for the mirrors of telefcopes was difcovered by the induftry of Mr. Edwards, and publifhed in the *Nautical Almanack* for the year 1787.   You will find an account of it in *Nicholfon's Chemiftry*, where he mentions the combinations of copper with other metals.   The ancients made ufe of tin very much, as an alloy of copper, and in the compofition of their brafs, for their arms and cutting inftruments, which required great hardnefs.   It is deficient in toughnefs: But this is confiderably improved, as I am informed, by paffing it between the rollers of the flatting mill.

There is fomething curious in the procefs for whitening pins.   It is commonly believed that this is done by mercury.   But it is a coating with tin, incomparably more durable.

Copper, diffolved in any acid whatever, is precipitated completely by tin.   Yet in this procefs the tin is precipitated by the copper. The effect depends on the attraction of pure or metallic copper for pure or metallic tin.   The tin is firft diffolved by a mixed folvent, which contains fome tartar; during which diffolution, the tin is flightly oxydated: And this folution alone does not whiten copper in the leaft.   But if frefh tin, or a fmall quantity of iron, be added while it is boiled with the copper, the frefh tin, or iron, by their attraction for oxygen, deoxydate the diffolved tin, or deoxydate a part of the water, the hydrogen of which deoxydates the diffolved tin, which, before it has time to form coherent particles by itfelf, is attracted by the furface of the copper, and adheres to it.   Mr. Gadolin tinned gold in this manner, although the folvent he ufed could not act in the leaft upon gold.

Of folvents for fuch a purpofe, when a bright and fhining furface is required, rather than an high degree of whitenefs, the folution of tartar is to be preferred. The folution of alum gives merely an exquifite whitenefs; but it is a dead white. If tartar and common falt are added to the alum, they take off the deadnefs of the white, and give fome polifh and brightnefs. One part of white tartar, two of alum, and two of common falt, form the beft folvent. The folvent fhould not diffolve any of the copper. If any part of the copper fhould be diffolved, it would fruftrate the procefs. (Nicholfon's Dict. p. 947.)

This metal is alfo alloyed with zinc, to form pinchbeck, brafs, and fimilor. Copper, mixed with zinc, in the proportion of two parts of copper to one of zinc, forms brafs, which has a light greenifh yellow colour, and when made red hot, is made quite brittle. I have already defcribed the procefs for making this compound by cementation. But with equal parts of zinc, the copper forms a Bath metal, or fimilor, which is malleable when heated, like iron. And if a little iron, or caft fteel be added to this compound, the mixture proves as malleable when heated as iron, and as ftrong and difficult to bend and break when cold. They are, therefore, making trial of it now for bolts in building fhips of war. Copper, in the proportion of three or four parts to one of zinc, makes pinchbeck. Similor has a rich yellow colour, and in confequence of its foftnefs and malleability when hot, is formed into all forts of goods at Birmingham, by ftamping or coining.

Copper is alfo the moft fit metal to be ufed as an alloy to filver and gold, to give thefe metals ftiffnefs and hardnefs, without diminifhing their toughnefs too much.

The relation of lead to copper is fomewhat particular. Copper does not mix with lead, unlefs the lead be heated to the boiling or fcorifying degree. The copper then finks in, and is diffolved very quickly; but if allowed to cool, feparates in fome meafure, fo that the mafs becomes like fand and water. When copper and lead are mixed, they are therefore eafily feparable; firft by eliquation, then by fkimming the lead, which will make it fufficiently pure from the copper; and as the copper still retains a little lead, this is worked off by scorification. Copper is separated from other coarser metals in general by scorification, especially with lead.

Natural History of Copper.

Copper is frequently found in its metallic state, and nearly pure, in the form of plates or fibres, or branched masses, or irregular lumps, intangled in the matrix of the vein. In the inland parts of North America, there are rich veins containing copper in this form. It abounds in the same state in some of the islands that are to the west of North America; and is found more or less in many other parts of the world.

Professor Robison told me that an old officer in the Russian service, who was with Captain Bering in the first voyage from Kamtschatka to America, informed him that on the beach in Bering's island there lay an immense quantity of metal, rounded like pebbles by the washing of the sea, which were all malleable copper. He thought that many ships might be loaded with it. This has probably been a volcanic production; for Kamtschatka, and the whole cluster of islands between it and America, seem to have been raised by the same cause. I also suspect that it was rather an ore, or an oxyd, than the native metal which was seen at Bering's island. The western district of the Siberian mines contains many horizontal strata of copper gravel, in which are found plants and trunks of trees (all exotic) completely penetrated by the copper.

But it occurs much more frequently in the form of ores, of which there is a considerable variety and abundance. Anglesea contains the richest bed of copper perhaps in the world; and of late years, yields about twenty-five thousand tons of metal annually. The vein is about seventy feet thick.

In taking a cursory, but comprehensive view of the ores, they may be distinguished into three principal kinds:

1ſt, The natural oxyds of copper.

2d, The gray ores of this metal.

3d, The yellow ores.

The moſt remarkable natural oxyds are of four varieties:

1. The deep red oxyd of copper,—*minera vitrea rubra.*

2. The blue oxyd,—*minera lazurea.*

3. The green oxyd,—*minera viridis.*

4. The fibrous green oxyd, named *malachites.*

The ſecond general ſection of the ores, the gray ores, comprehends thoſe in which the copper is combined with ſulphur, arſenic, and other metals, but is itſelf the moſt abundant ingredient in the compound. Iron is generally contained in theſe ores, and often ſome ſilver. If the ſilver is in ſuch quantity that the extraction of it is a profitable buſineſs, they are often named *ſilver ores.*

The third general ſection comprehends the yellow ores, or copper pyrites. In theſe, the copper is not the moſt abundant metal, iron being preſent in greater quantity, together with much ſulphur, and often arſenic ; but the copper is in ſuch quantity, that it is the moſt valuable ingredient, and is the only one which gives value to theſe ores. They have a yellow colour and metallic luſtre, like that of braſs, or a little reſembling that of the common pyrites, which contains iron and ſulphur only, but are eaſily diſtinguiſhable from the common pyrites by obvious differences. The common pyrites is ſo hard that a knife or file makes no impreſſion on it ; and it ſcratches glaſs, and ſtrikes fire with ſteel. The copper pyrites is ſo ſoft that a knife eaſily makes impreſſion on it, and ſcrapes or ſcratches the ſurface of it. They are alſo remarkably different in colour. The yellow of the common pyrites is pale ; that of the copper pyrites is deeper and richer, often ſimilar to that of gold, and it is liable to a ſort of tarniſh, which produces the priſmatic colours on its ſurface, as they are produced on the ſurface of ſeveral metals by the action of heat.

We can be ſatisfied by an eaſy experiment, whether or not a pyrite contains copper. We need only to beat a fmall quantity of it to powder, and uftulate or roaft this powder with a low degree of red heat, by which we fhall burn out the fulphur, or the greater part of it. If we then apply liquid ammonia to the remaining matter, this will immediately produce a deep blue folution, if there is copper contained in it.

As the gray and yellow ores of copper are moft abundant, and are found to contain, although in various proportions, the greateft quantity of fulphur, arfenic, and iron, I fhall firft defcribe the procefs by which the copper is extracted from thefe. To effect this feparation, both labour and time are required, on account of the ftrong adhefion of fome of thefe fubftances to the copper.

The firft operations which the richer ores of this kind undergo, are repeated roaftings with a moderate heat, and fubfequent fufions with a ftronger one. Thus the greater part of the fulphur and arfenic are gradually diffipated, and the iron is calcined and fcorified, or vitrified with the earthy and ftony matter that is intermixed with the ore, efpecially if this ftony matter is quartz. This filiceous ftone, when it happens to be mixed with the ore in moderate quantity, greatly promotes and facilitates the extraction of the copper, by its having a remarkable difpofition to unite with calcined iron, and to be vitrified with it, or converted into fcoriæ or flag. And when a rich copper ore is brangled with quartz, fo much that it cannot be pounded and wafhed clean without confiderable lofs, it is ufual to melt the whole, with addition of common pyrites; the iron of which being very calcinable, foon vitrifies along with the quartz, or brings it into fufion, while the fulphur, or a part of it, joining itfelf to the copper ore, increafes its fufibility, and makes it more readily fubfide from the fcoriæ. This manner of melting the poor or ftony copper ores, is called *cryde melting*, or *crude fufion;* and the product of the operation is afterwards treated as a rich ore of copper, by repeated roaftings and melting proceffes. In England, the firft roafting of the copper pyrites is performed in fuch a manner, that a great part of the fulphur is faved, by a contrivance which I defcribed not long fince, when I gave you the natural hiftory of fulphur. The fubfequent roaftings and meltings are performed in low reverberatories, or ovens of confiderable length, heated with flame. This is kept playing on the furface of the metal; and there are tranfverfe ribs in the arch of the oven, which obftruct the direct paffage of the flame to the chimney, and beat it down on the melting matter. The flame carries with it abundance of unconfumed vapour and fmoke, which robs the ore of its oxygen, and leaves it in a metallic ftate. It is ftirred from time to time, through a hole at the far end of the oven, immediately below the upright chimney; thus no cold air gets to the metal, to calcine it or cool the furnace. A tap hole is opened at the fide, at the loweft part of the bed; the metal flows out, with the fcoriæ floating above it, and then falls into a mould, which gives it the form of a triangular prifm. When this mould is full, the metal runs over into another, which it feldom fills, but has flag above it; and the reft of the flag runs over into a receptacle large enough to hold it all. I muft content myfelf with this fhort defcription; for it would employ a feffion to defcribe thefe things minutely.

On the Continent, a variety of furnaces are employed, which you may fee defcribed and reprefented in copperplates, in Schlutter's work on metallurgy, which was tranflated into French by M. Hellot.

When the copper is, by thefe operations, freed from its impurities to a certain degree, but ftill retains a fmall quantity of them, it is called *black copper*, and is very deficient in malleability; and often has not at all the colour of copper. If it contain filver, fome lead is added to it in this ftate, and is then feparated from it again by eliquation. In this procefs, all the lead melts and fweats out of the copper, and runs off, except a fmall portion, which remains inherent in the fpongy mafs of the copper. The filver is taken out by the lead, which has a ftrong attraction for it. The copper then undergoes the laft operation, which is the refinement of it; and the fmall portion of lead which it retains, facilitates this procefs, and makes it more perfect.

The hearth or furnace for refining copper contains two tons, or two tons and a half, of copper, which is refined at once. The diameter of the melted metal is about three feet, the depth about two feet. A little water is sprinkled upon it from a broom. This is dissipated in vapour in an instant, and the surface of the copper is frozen. This skin is lifted off immediately with a pair of tongs, and thrown into a tub of cold water. Its under surface is of a bright rose colour, and somewhat rough and glittering by a sort of cryftallization. It is called *rose copper*. This operation is continued till the whole of the copper is thus lifted off in skins.

The refined copper requires water to be poured on about forty times before the last of it is confolidated. This is a clear evidence of the great quantity of latent heat which melted copper contains, for the first water must cool the metal to its congealing point, otherwise no part of it would be congealed. All the fubfequent additions of water, therefore, are neceffary to extract the latent heat only, the perceptible heat continuing the fame.

We have only now to add further, that in fome of the states or conditions in which copper is found, it does not need all thefe operations to fit it for ufe. Such of it as is found in its metallic state, and nearly pure, requires only the operation of refinement. And the ores in which it is in the state of an oxyd, need only to be first reduced to the metallic state, and then refined.

The very poor yellow ores alfo, which fometimes contain no more than three, or four, or five pound of copper in the hundred weight, are often treated in a particular manner. They are first uftulated with a very low red heat, during which operation the metallic matter is oxydated, and the fulphur burnt, and changed into fulphuric acid, and part of it evaporated; but a part remains adherent to the oxyd of copper; and water being afterwards applied to the uftulated matter, a folution of blue vitriol is formed, which is either evaporated afterwards and cryftallized, to be fold in that state, or inftead of evaporating the liquor, pieces of old iron are put into it, and allowed to remain until the copper is precipitated on them, which is then collected and melted into a mass.

The preparations of copper for medicinal use, are,

> *Cuprum ammoniatum ;*—Edin.
>
> *Pilulæ cæruleæ ;*—Edin.

For external use,

> *Ærugo ;*—Edin.
>
> *Unguentum ex ærugine ;*—Edin.
>
> *Vitriolum cæruleum ;*—Edin.
>
> *Aqua styptica ;*—Edin.
>
> *Aqua vitriolica cærulea ;*—Lond.
>
> *Aqua sapphirina ;*—Edin. and Lond.
>
> *Ens veneris*, formerly in the Edin. Pharm.

SILVER AND GOLD.

We have now considered all the metals which have been commonly known and used, except silver and gold.

These two have been long distinguished by the chemists from the rest of the metals, by the title of the more noble or perfect metals, on account of their having the metallic properties, malleability, ductility, and durability, or incorruptibility, in a much more eminent degree than the other metals.

I shall first consider the qualities by which they thus excel other metals, and afterwards those which distinguish these two.

First, therefore, their amazing ductility and malleability are qualities in which they far surpass all other metals. By experiments made upon silver, it appears that a single grain, which hardly exceeds in bulk the head of an ordinary pin, may be drawn out into a wire near nine feet long. And further, this wire, or any part of it, can be beaten out by hammering, into a thin plate or leaf one inch broad. It can therefore be made to cover 108 square inches.

Of the ductility and extensibility of gold, we have examples in the different branches of the art of gilding, or the art of covering the furface of other fubftances with gold.   One way is with what is called *gold leaf*.   And the art of beating out the gold into leaf, is very well and accurately defcribed by Mr. Lewis.   The degree of extenfion and thinnefs to which gold is brought in common gold leaf is aftonifhing. It is plain from the moft evident and fatisfactory calculations, that the thicknefs of gold leaf is not the $\frac{1}{200,000}$th of an inch.   But gold is much more extended in manufacturing gold lace.   What is called gold lace is not woven of gold wire, but of filver; the furface of which only is covered with a film of gold.   To make the wire, a thick cylinder of filver, weighing forty-eight ounces, has its furface covered with one ounce of gold, and is then drawn out into wire fo fine that fix feet weigh only one grain.   It is then flatted and twifted round filk thread, &c.   While the filver is thus extended, the gold is extended along with it, fo as to cover the whole furface of the filver.   The wire, if examined with a microfcope, does not fhew the fmalleft atom of the filver uncovered at its furface, although the gold is only one-forty-ninth part of its weight, and lefs than the one-eigthtieth of its bulk.   (I find it to be the one-eighty-fixth of its bulk).   But this proportion of gold to the filver is much greater than what is barely neceffary for covering or coating the wire.   It is fufficient for giving it fuch a coat of gold as will bear fome wearing before the filver appears.   Mr. Reaumur had the curiofity to get fome wire drawn with fuch a thin coat of gold, that there was in his wire only one part of gold to 360 of filver.   It was drawn to the finenefs of fix feet to each grain, and then was flattened between the polifhed fteel rollers, which gave it a breadth of one-forty-eighth of an inch, and extended it in length one-fourth.

When our calculations end in fuch numbers as thefe, we lofe all diftinct comprehenfion of our fubject.   In order to underftand the thinnefs of the gold on fuch gilt wires a little more clearly, I thought it worth while to compare it with the thicknefs of paper, in this manner:

——It would take 14,000,000 of films of gold, like that on some gilt wire, to make up the thickness of one inch. *Quæritur.*——What thickness would 14,000,000 leaves of common printing paper make up? *Answer.*——1262½ yards, or near three-fourths of a mile. The thickness of the gold on Mr Reaumur's gilt wire, therefore, bore the same proportion to one inch, that the thickness of a single leaf of printing paper does to three-fourths of a mile nearly, or wanting fifty-eight yards.

Another eminent quality of these metals I said, is their durability, or the power they have to retain their metallic form, and remain unchanged, though exposed in circumstances, and to the action of bodies, which produce great changes in other metals. One example of this we have in the difficulty of calcining them. They cannot be calcined by the ordinary action of heat and air, by which other metals are calcined, nor by that of nitre, nor with the assistance of lead. In a violent heat, they remain perfectly metallic and unchanged. Mr. Boyle exposed some silver and gold in separate vessels to the fire of a glass-house furnace one month. The gold was neither changed nor diminished in weight. The silver was unchanged, but diminished by one-twelfth. Kunkel, making the same experiment, found the silver diminished only one-sixtieth; and when he repeated it with the same bit of silver, there was no diminution; from which he concludes, that the diminution in the first experiment was occasioned by impurity. Cramer also attests the truth of these facts from his own experience. It is said, however, by Fourcroy, that Mr. Macquer, by exposing the same silver in a porcelain crucible twenty times to the fires in which porcelain was baked, found it changed at last into an olive-coloured vitrified matter, which is supposed to have been a calcined and vitrified silver. I do not however find this fact in Mr. Macquer's own writings, though he relates some experiments which he and other members of the Academy of Sciences made, by exposing silver and gold to intense heats, produced by means of the best burning glasses in France. In these experiments, these metals were in part volatilized, or emitted visible vapours. These vapours were formed by the metal in an uncalcined state, as appeared plain from their condensation on the surface of silver or gold. After a long continuation of these experiments, a very small quantity of vitrified matter was formed, but Mr. Macquer justly doubts whether it was simply a part of the metal calcined. It is possible, however, that by mixing these metals with others, and exposing such mixtures to mild calcining heats, continued for a long time, some degree of calcination might be effected. But if they should undergo a slight degree of calcination by such a process, a stronger heat alone makes them immediately resume their pure and metallic form *.

When nitre is applied to silver or gold in a strong fire, provided these metals are pure, there is no deflagration or calcination of the metal. The nitre evaporates without producing any effect.

Neither can they be calcined with the assistance of lead. Lead is often employed to promote the calcination or scorification of other metals, or metallic substances ; and its power depends on two particulars. 1st, The disposition which mixtures of metals have to calcine more easily and quickly than the metals in their separate state. 2dly, The great fusibility and dissolving power of the calx of lead. For when another metal, which produces an unfusible calx, is calcined by itself, its surface is soon covered over with a crust of its calx, which defends it more or less from the further action of the air, and occasions the calcination to go on afterwards slowly and with difficulty. But, if to such metal we add a certain proportion of lead, the mass is disposed to calcine faster, in consequence of its being a mixture of metals. And the calx of the lead dissolves and liquefies the other calx, and occasions it to flow off continually from the surface of the

* The allusion, therefore, in the Bible, to this quality of the precious metal, to illustrate the triumph of a good heart over misfortune, is peculiarly beautiful ; and, as this is to be seen in the book of Job, the discovery must be very ancient. EDITOR.

melted metallic mafs towards the fides, fo as to leave this furface expofed to the conftant action of the air; and thus the calcination, or
fcorification, as it is rather called in this cafe, goes on much fafter.

But when we make this experiment with filver or gold, there is no
calcination or fcorification of either of thofe metals. They mix eafily
with lead in the fire; and when this mixture is expofed to a proper
heat and the action of the air, the lead alone is calcined, or fcorified, and is gradually thrown out of the melted metal; for a calx and
a metal cannot mix together. But all this time, the filver and gold
remain unchanged, and will be found pure and undiminifhed in
quantity, after all the lead is feparated from them by the fcorifying
procefs. The eighth part of an ounce, or even a fmaller quantity
of either of thefe metals, contained in one hundred weight of lead,
may be feparated from it by this procefs.

In confequence of their power to remain uncalcined in fuch proceffes, filver and gold are often refined or purified from the admixture of other metals by nitre or by lead.

Thefe metals are purified with nitre, by fimply mixing it with
them, and melting. It confumes the bafe metal, bringing it to the
top in a fcoria, which is eafily worked off with a little lead. If they
are much contaminated, it is ufual to melt them with lead, which
takes to itfelf the greateft part of all the other metals. The mixture
is feparated from the gold and filver by eliquation; and they are then
treated with the nitre.

The purification by lead is ufually called *cupellation*, from the fort
of veffel in which it is performed. The cupel is a very flat cup, but
of great thicknefs, made of bone afhes burnt to a perfect whitenefs.
Thefe veffels are formed by filling a ftrong brafs mould with the
afhes, previoufly damped a little. The depth of the mould is confiderably greater than the intended thicknefs of the cupel, becaufe it is
neceffary that all the materials be put in at once, and afterwards compreffed by a piece of fteel like a peftle, with ftrong blows of a ham-

mer. If there has not been enough, any addition would form a layer which would afterwards separate in the heat. The face of the pestle must have nearly the convexity of a watch glass, and must be well polished.

The gold or silver is mixed with lead that is known to contain none of either of these metals. They are then subjected to a strong heat under a muffle, and to the action of a free current of fresh air. This calcines the lead, and along with it the baser metals, converting them into a thin glass, which is absorbed by the cupel as it forms. The operation is known to be finished, when the last film of lead quits the surface of the button of gold or silver, which in an instant grows resplendent. This is called its *fulguration* or *coruscation*. The practice with large quantities of metal differs from the above only in this, that the glass of lead, instead of being absorbed by the cupel, is blown off by bellows. The lead is got from the cupels again (which are expensive) by beating them to coarse powder, and reducing the lead with flux and charcoal dust.

In considering these properties of silver and gold, we might perhaps be inclined to conclude that they must be quite different in their nature from the other metals, since we do not perceive in them that affinity with the inflammable substances, which is so obvious and remarkable in the other metals*. But when we study their other qua-

* Is not this affinity, and their perfect inflammability, distinctly seen in their dissipation by the electrical flash? They exhibit all the appearances of heat and light, and they are dissipated in smoke, which has a very peculiar and disagreeable smell, different in each metal.

This smoke should be examined. The dissipation should be made in water, which would enable us to collect it, or form some judgment by the changes which it produces in that fluid. When the combustion (for I must give it this name) is effected between two plates of glass, (even the purest and freest from metallic admixture) some of the gold or silver is incorporated with it. We know that metal and glass will not unite, though their oxyds unite perfectly. These facts have not been sufficiently attended to. If it be thus exploded when surrounded by azotic gas, whence comes the light and heat?

EDITOR.

lities,—thofe which they fhew in mixture with acids,—we plainly perceive that they have a clofe affinity and fimilarity to the other metals, and are at bottom of the fame nature : For we can calcine or oxydate them by means of fome of the acids,—and they produce the fame changes on thefe acids that are produced on them by other metals. Both filver and gold, for example, can be oxydated by the nitric acid, and they convert a part of it into nitrous air. And filver can alfo be oxydated by the fulphuric acid, and converts a part of it into fulphurous acid, or fulphur.

It therefore appears that they are of the fame nature with the other metals, and differ from them chiefly by being more difficultly oxydated, having lefs attraction for oxygen, and refifting completely fome of the agents, by which this change is eafily produced on the other metals. And, when we do bring them into a calcined ftate by particular procefles, they are capable of refuming their metallic form by the fimple action of heat on them, which expels the oxygen in the form of vital air, without their requiring the affiftance of any inflammable matter to attract it from them, as it is attracted by charcoal and fome other inflammable fubftances in the reduction of moft other metallic oxyds.

Thefe are the qualities by which filver and gold differ from the other metals already defcribed. We fhall now take a view of the qualities peculiar to each of thefe two metals.

GENUS XIII.—SILVER.

The appearance of this metal is too well known to need any defcription. Its fpecific gravity is nearly 10,5 or 10,47.

It is never liable to have ruft formed on it, but it becomes tarnifhed if it be not frequently cleaned ; that is, the furface of it becomes darkcoloured, as if it had been fmoked, and fometimes almoft black. This happens the moft certainly, and in the fhorteft time, when the filver is expofed to the vapours of fulphur, or fulphurous gas. And putrid vapours from rotten animal or vegetable fubftances (eggs efpecially) produce alfothe fameeffect on it. Late obfervations, indeed, have fhewn that fuch putrid vapours frequently contain fulphur diffolved in them. Inflammable air alone is faid alfo to tarnifh filver, and to give to the furface of it a purple hue. This fometimes accumulates on the furface of filver to fuch a degree as to form a fcale, which may be detached, and is found to be a compound of filver and fulphur.

Plate, when thus tarnifhed, may be cleaned by foap and water, but more perfectly by foot and vinegar. A folution in water of the camæleon minerale, or nitre alkalized by manganefe, does it in the completeft manner.

The metal is melted by a bright red heat, and when in fufion, has exactly the appearance that pure quickfilver has, reflecting with an equally bright furface the objects around it.

While it is congealing from a melted ftate, it is liable to form fuddenly protuberances from its furface, which are fometimes branched, and are named the vegetations of fine filver. Their formation is accompanied with a remarkable incandefcence, or fudden increafe of ignition in the congealing filver, which incandefcence is named *corufcation*, and happens in the congealing of very fmall quantities of melted filver, although no vegetations are formed on its furface. But thefe vegetations are fcarcely ever formed when the quantity of the congealing filver is very fmall. The caufe of their formation appears to be a power which filver has, as well as many other bodies, to retain the latent heat which gives it fluidity, although it be cooled to a degree confiderably lower than its proper congealing point. And this happens chiefly when it is cooling flowly, and without being difturbed. At laft, however, the latent heat fuddenly breaks out, or emerges from a part of the fluid filver, and affumes the form of fenfible heat, which produces the incandefcence; and that part of the metal from which it is extricated, being at the furface, fuddenly congeals. Thus a thin cruft of filver is inftantly formed on the furface of the mafs.   But this fudden formation of the folid cruft being attended with the increafe of fenfible heat, the fluid filver within the cruft is expanded, and burfts through the cruft in one or two places, ftarting up in one or more protuberances, which fuddenly congeal while they are rifing, and in congealing frequently throw out lateral protuberances.   And thus are formed the branched figures, like cauliflowers, which are fometimes produced, and are preferved by the curious.

Moft of the active falts may be combined with filver in one way or another.   Sulphuric acid acts only when ftrong, and boiling hot.   Sulphurous acid and fulphur are produced.   The falt formed is of little folubility, except it contain abundance of the acid.

Diluted nitric acid is the beft folvent of filver.   It diffolves it with moderate effervefcence, during which the metal is oxydated, by attracting oxygen from a part of the acid; and that part of the acid is changed into very pure nitrous air; the reft unites with the oxydated filver, to form nitrate of filver.   The folution of pure filver is greenifh at firft, a common phenomenon in the folutions by this acid, which decompofe the acid in part; but in a little while it is colourlefs.   If it remain green, the filver has been tainted with copper.   This folution bears dilution in pure water, and ftains hair, bones, and many woods a deep and lafting black; or if weak, a brown.   It gives a permanent ftain even to ftones; as marbles, agates, and jafpers.   This ftain does not appear unlefs the bodies be expofed fome time to the light of day. The way to fee this effect of the fun's light in producing the ftain, is to dip a bit of chalk or marble into the folution of filver.   If we then keep it in darknefs, it will remain white; but if any part of it be expofed to the light of the fun, that part becomes black in a fhort time *.   Thofe bodies to which the folution is thus applied, attract the acid from the calcined filver, while at the fame time this metal is re-

* And, which is remarkable, it is much more ftained by the blue and violet light than by the red and yellow, which are much more luminous and heating,   EDITOR.

stored to its metallic state, or made to approach to that state by the action of the light, which expels from the calx a quantity of vital air. This effect of light, in this and some other similar examples, is well known by experience; but we do not clearly understand how it is produced. We can imagine that the light joins itself to the metal, and expels the air from it. Others think that it joins itself to the air, or oxygenous principle, and restores to it the aëreal form, or gives it elasticity to fly off from the metal. In consequence of this quality of the nitrate of silver, it is employed for staining hair brown, and for stamping cambrics, and lawns, and staining marbles, jaspers, &c.

By evaporation, the nitrate of silver is reduced to crystals, which are very corrosive. They melt like nitre, and they may be moulded into small cylindrical pieces, called *lunar caustic*, used by surgeons as an escharotic, and to corrode fungous excrescences.

This nitrate is perhaps the most powerful antiseptic known. Meat impregnated with it so as to acquire a sensible taste, though never so slight, will not putrefy even in a warm place. In 12000 times its weight of water, it will preserve it from putrefaction for ever; and it will separate in a few minutes by throwing in a pinch of common salt.

The muriatic acid is the most strongly attracted by silver, and will separate it from every other acid; but it has very little power to act while the silver is in its pure and metallic state. The silver must be somehow oxydated to prepare it for joining with this acid. The usual way is, to take the solution of silver in aquafortis, in which the silver has received oxygen from the acid. If we add muriatic acid to this solution, the silver instantly quits the nitric acid to join with the muriatic, and with this last forms a compound nearly insoluble in water. It therefore subsides like a curd or coagulum; and this effect is produced, not only by the pure muriatic acid, but by any saline compound whatever which contains it; such as muriat of soda, muriat of ammonia, and all other muriatic compounds, a double exchange taking place. This compound, when wafhed, to free it from the faline liquid, and dried, is properly called *muriat of filver* by the French chemifts. It contains $\frac{5}{27}$ of acid, and has fome remarkable properties; as,

1ft, It is infoluble in water to that degree, that the moft minute quantity of muriatic acid, or of common falt, or any other faline compound whatever that contains muriatic acid, can be difcovered in water by adding the nitrate of filver.

2dly, In the fame manner as the nitrate of filver, it is changed in its colour, and fome other properties, by the light of the fun, or the light of day. We have an opportunity of feeing this, if we take two portions of it new made, and keep one of them in a dark place, while the other is expofed to the light. The one kept in the dark will continue white and unchanged; the other will affume a bluifh, or dove colour, and in it fome of the filver will be reftored to its pure and metallic ftate. This has often occafioned great embarraffment in the examination of mineral waters. After having indicated the prefence of muriatic acid, by growing milky all over when fet by, they have become brown, and the chemift was made to fufpect the exiftence of fome other taint in the water, and led into a long and fruitlefs inveftigation to find out what it was. It was the change produced by light in the lunar falt floating all over the fluid. The teft of this is, that if a portion of the milky folution be fet by in the dark, its colour will not change. Nor will this falt, perfectly dry and white, change its colour in air that is alfo perfectly dry, although acceffible to light. In an experiment made by Mr. Scheele, fome of it being expofed to the light, with a large quantity of water covering it, or mixed with it, the filver was reftored to its pure metallic ftate, and the muriatic acid which was before combined with it, was found in the water which had attracted it. The evidence for this was unqueftionable. Nitric acid, being poured on the edulcorated black muriat, took up all the black part of it, leaving the reft white. When fome of the liquor which had been decanted from the black muriat, was dropped into this solution, as also into another solution of silver in nitric acid, a muriat was immediately formed.

Mr. Berthollet says that this salt, exposed and coloured in the sun, emits oxygenous gas very copiously, and that this also happens when the nitrate of silver blackens by the action of light under water. But Scheele also remarks, that if nitric acid be poured on the muriat, it does not blacken. The cause of all these changes is not very clearly understood. The action of the sun's light seems to expel oxygen from the solution. If this be taken from the acid, it leaves it in the state of common muriatic acid, unable to dissolve metallic silver; but the presence of a little nitrous acid supplies this, (itself becoming fuming) and thus prevents the deoxygenation of the muriatic, and the reappearance of metallic silver.

3*dly*, It is fusible and volatile, and after fusion assumes a horny flexible appearance, which has occasioned it to be called *luna cornea*.

Mr. Beaumé, however, denies that luna cornea has any flexibility. Indeed it has but little, being commonly shivery, and full of transverse cracks, arising from its shrinking faster in cooling than the substance to which it adheres. Kunkel, however, imagined that this is the substance of which a cup was made which an artist presented to one of the Roman emperors. He threw it on the ground, and the fine transparent glass cup did not break, but was dimpled by the fall. The artist took it up, and putting it again to rights, returned it to the emperor. It was therefore, says Kunkel, a malleable glass; and he proposes the manufacture as a chemical problem, declaring that he was in possession of the secret. The whole is in all probability a fiction, too easily credited by Pliny or Solinus, who were but indifferent chemists. At any rate, luna cornea would make but a poor figure as a glass.

This substance contains the silver in a very pure state. The muriatic acid precipitates no metals from its acid solutions, but silver,

mercury, and lead. Copper, which may have refisted the cupellation with lead, is thus retained in the acid. This method, therefore,
is frequently employed for purifying filver. It requires fome art,
however, to feparate it without lofs from the muriatic acid. They
are not feparated by a low red heat; for the compound partly efcapes
in fumes united with the acid, and partly runs through the veffel. It
muft be ground with an equal weight of fixed alkali, and bedded
and covered with the fame in the crucible. It is then obtained in a
metallic form, and of the higheft purity *.

Sulphuric acid alfo, added to the nitrate of filver, feparates the
nitric acid, and is joined to the filver in its place, forming a fulphat
of filver. If the quantity of the fulphuric acid is fmall, or moderate,
a great part of the fulphat of filver is precipitated, or feparates from
the fluid by a hafty cryftallization; the fulphat of filver, when not
fuper-faturated with acid, being much lefs foluble in water than the
nitrate. But if abundance of fulphuric acid be added, the filver is
fuper-faturated with acid, and forms a much more foluble compound,
which does not precipitate. It has been ufual, however, to attempt
to purify aquafortis by adding to it a fmall quantity of nitrate of
filver; for if there be a little of muriatic acid in the aquafortis, it is
fure to form a muriat of filver, which precipitates. And if a little of
the fulphuric acid were prefent, it was fuppofed that this alfo would
form an infoluble compound with a part of the filver, and alfo fall
down. But this is a miftake; and the only way to have aquafortis
perfectly pure is to add a little of the nitrate of filver to it, and diftil
it. Thus the fmall quantity of both the muriat and fulphat of filver
which had been formed, remain in the retort, along with fome of the
nitrate of filver, when it is more than fufficient for feparating the two
acids.

----

* I obferve it faid, though I do not recollect by whom, that this muriat, dried and
niturated with calx of tin, catches fire.

In thefe different ways can filver be combined with the greater number of the fimpler falts; even with the vegetable acids of wine, of forrel, and of tartar. This was firft done by Margraaf, in the fame way that I mentioned as followed by him with mercury, namely, by applying the acid to an oxyd of filver.

I fhall now inform you how it may be recovered in its pure metallic ftate. It is from its folution in aquafortis that we have moft frequent occafion to recover filver; this metal being often diffolved in aquafortis to feparate it from gold.

The recovery of the filver from aquafortis may be effected by different means; but there is but one method that is commonly practifed, on account of its being convenient, effectual, and quick. This method is to precipitate the filver by copper. The copper attracts from the filver both the acid and the oxygen. The filver concretes in its metallic ftate on the furface of the copper, and cryftallizes into a thick downy covering, which, when viewed with a magnifying glafs, or with young eyes, is feen to refemble in form a thick foliage of vegetables, or the down of feathers. The way to fee this cryftallization the moft diftinctly, is to drop fome of this folution on a plate of glafs a little concave, and then to put a bit of copper in the midft of it. We fhall fee a filver fringe form all around it in a minute; and this will continue to increafe by ramifications in all directions till the whole filver is precipitated.

This forms one of the moft agreeable objects for microfcopic obfervation. A fingle drop of diluted folution being placed before the microfcope, and a particle of copper made to touch its margin, we fhall obferve a ftem arife immediately, protruding its branches by ftarts; and in five minutes, the little bufh occupies the whole field of the inftrument. (*See Note* 30. *at the end of the Volume.*)

Clock dials are filver-wafhed in this way. The diffolved nitrate of filver is applied to them, along with a mixture of the fame with a quantity of common falt and tartar. They are then dried, and rubbed with a foft linen cloth, and immediately varnifhed.

While filver is thus precipitated by copper, a fmall portion of the copper is faid to unite with the precipitated filver; but it is a very fmall one indeed, and fuch as produces no inconvenience in bufinefs. If edulcorated and redifiolved in nitric acid, the folution becomes perfectly colourlefs; a fufficient proof of the purity and fitnefs for the moft nice experiments.

Quickfilver has alfo been ufed fometimes for precipitating the filver; but the difference of their attraction for the acid is fo little, that the precipitation is flow and indiftinct,—they rather mix in a thick folution. The feparation of the filver by quickfilver, when managed in a particular way, is attended with fome amufing appearances.

Into a nitrate, formed by one ounce of filver and pure nitric acid, not greatly diluted, but afterwards diluted with twenty ounces of water, put two ounces of mercury. The filver cryftallizes as it feparates from the acid, forming beautiful ramifications refembling broom. It is called the *tree of Diana*. It requires thirty or forty days, and the moft undifturbed repofe. Much more expeditious proceffes, but inferior in beauty, have been publifhed by Homberg, Beaumé, and others*. The fuccefs depends upon the proportion of the mercury to the filver, and the ftrength and ftate of the folution.

I faid that filver happens moft frequently to be combined with the nitric acid. Sometimes too it is combined with the muriatic acid, by adding common falt to the folution of the filver in nitric acid. A muriat of filver is inftantly formed by a double exchange, and when this muriat of filver is properly wafhed, and the filver recovered from

* Beaumé's procefs is to mix two faturated folutions of filver and mercury,—to dilute the mixture with diftilled water, and to add a bit of amalgam. If a fkeleton of a bufh be made of glafs, and fet up in the mixture, and fmeared over with as much amalgam as can be made to ftick to it, the ramifications will foon cover it, and make the toy very pretty.

EDITOR.

it, we have it in the pureft poffible ftate.—It is recovered from this compound, as I have already obferved, by grinding it with fixed foffil alkali, and expofing it to a melting heat, bedded in this alkali, and covered with it.   It is of great confequence to employ the alkali in a cauftic ftate.   I find that much lefs heat fuffices for the fufion, which is a great advantage, becaufe a very fmall excefs of heat volatilizes the lunar muriat; whereas, by this precaution, I have always recovered the filver without any perceptible lofs.

Silver is eafily feparated from vegetable acids, or from alkalis, when it happens to be combined with thofe falts, which is never except in the way of experiment.   Heat alone is fufficient to deftroy the vegetable acid, and to diffipate the volatile alkali.   And the fixed alkalis are feparated merely by being melted.

The fulphuric acid is feparated, and the filver recovered from it, by fufion with pure fixed alkali.   The cauftic alkali is the fitteft.

The combination of filver with the volatile alkali has fome very furprifing properties, firft obferved, I believe, by Mr. Berthollet, who prefcribed the following procefs.——Let filver be precipitated from a folution in pale nitric acid by quicklime.   The calx is then to be feparated by decantation, and carefully dried for two or three days, by expofure to the *light* and air.   When this is ftirred and mixed with volatile alkali, it feparates in form of a black powder.   Decant the liquor, and allow the powder to dry in the air.   This powder, if touched in the fmalleft degree, fo as to prefs it againft the veffel, and often, if merely parted fo as to break a grain, explodes with aftonifhing violence, dafhing the veffel to pieces.   It far exceeds gunpowder, or aurum fulminans, and requires no heat, the moft trifling agitation or friction being fufficient,—even though produced by a drop of water falling into it.   We dare not attempt to inclofe it in a bottle, and it muft remain in the difh in which it was prepared.   We fhould never attempt to explode more than a grain at once.

The volatile alkali employed in preparing this powder, being exposed to ebullition in a small matrass, cryftallized on cooling. When one of the cryftals was touched under the liquid, in order to feparate it from the glafs, the whole exploded and beat every thing to pieces.

Mr. Berthollet confiders this as a loofe compound of oxygen, filver, hydrogen and azote, having their double affinities fo nearly balanced, that the flighteft alteration of temperature, or even of that pofition of ingredients which produces the equilibrium of affinities, is enough to difturb this tottering equilibrium, in fuch a manner, that the fuperiority fhall incline to that fide which produces the union of the oxygen and hydrogen. The azote of the volatile alkali is difengaged, and the filver is reduced, and water is formed. The explofion is produced therefore by the azote, and from the vapour into which the water is changed in the very inftant of its formation *.

This metal can alfo be joined with earthy bodies in vitrifications; and the colour which it gives to glafs is a beautiful yellow. It is accordingly ufed to produce that colour in enamels, and the art of ftaining glafs. And when more of the filver is ufed, the colour is deepened to a blood red, efpecially with luna cornea, which has been expofed to the light. But the colours thus given to glaffes by filver are liable to change by over-heat. Leaf filver applied to common glafs or plates, and made red hot with them, penetrates into them, or is

---

* The explanation is ingenious. But even in this eafy cafe, we are not yet fufficiently acquainted with the forces which tend to continue the old affemblage, and thofe which tend to the production of the new, to be able to give an explanation on which the mind can reft with confidence. Were we certain that thefe are really the only forces in action, we could fay, from the fact, that thefe are the prevalent forces which unite the oxygen and hydrogen. But this is all *poft factum*, and we are ignorant of the fource from which the heat is to be drawn. The new doctrines conftantly hold forth the oxygen gas as the great fource of all the heat which appears in combuftion. Now in the prefent cafe, the oxygen is in the liquid or folid form already, and has none to beftow. Nor do we know any thing of the capacity of the azote for the heat. Befides, this is not a cafe of double affinity. The filver and azote are not united. The phenomenon is by no means explained on eftablifhed principles.   EDITOR.

diffolved, tinging them of a deep yellow colour. The lowest red heat is fufficient, and the moft proper is fuch as will not fpoil in the leaft any impreffion which the glafs may have received before. A ftronger heat produces a dull dirty colour, and a degree of opacity. All degrees of it may be perceived in the diffipation of a ftripe of filver leaf in a ftrong electrical fhock. The beft procefs for this elegant enamel is as follows: Take glafs of litharge three parts, flint in grofs powder one part. Pour on this a folution of one-twentieth of its weight of filver. Stir the mixture till very uniform, and then dry it, carefully avoiding all duft of an inflammable or combuftible nature. Melt it and pour it out. Grind it into a very fine powder. This is applied to the porcelain previoufly heated, and the ignition is continued under a muffle till it melts, which is obferved by its gliftening. The piece is taken out, and while continued red hot, it is expofed to the fmoky flame of burning vegetables, which brings out the beautiful colour. (*Lewis' notes on Neuman, p. 51.*)

Sulphur may eafily be combined with this metal in the ufual manner, and produces a compound of a colour like lead, but duller, and fomewhat foft and malleable. It is more fufible than filver. This compofition fo much refembles the *minera argenti capillaris*, that it is frequently impofed on collectors of ores as natural, by forming it on fragments of fpar, and other ufual matrices of fuch ores, and by conducting the refrigeration in a particular manner, which caufes it in the moment of congelation to dart up flender capillary filaments, which congeal like a brufh on its furface. The fulphur is feparable by uftulation. Many other metals alfo attract the fulphur, as lead, iron, and others.

From experiments made by mixing it with other metals, it appears that all the other metals will readily mix with filver. Mercury amalgamates moft readily, even in the palm of the hand, and their union produces a very fenfible heat. Nickel will not unite with it, and is, I think, the only metal which it rejects. Lead has a ftrong attraction for it, and will abforb the filver from iron or copper. All of them diminifh its malleability greatly, except gold and copper. Tin does it moft. A moderate quantity of copper does not change its colour much, but increafes its hardnefs a little, in which quality it is rather deficient when quite pure. Hence copper is commonly added, and the quantity regulated by law to form the alloy, is about one-twelfth or one-thirteenth. Silver mixed with this proportion of copper is called fterling or ftandard filver. Nicholfon fays (on what authority I know not) that our ftandard admits but one-fixteenth of copper, that is, one part of copper to fifteen of filver.

When filver is mixed or alloyed with other metals, and we wifh to feparate them, this is done by different proceffes. The art of doing it completely has been ftudied with great attention, on account of the value of this metal, which often happens to be mixed with other metals in their ores, and on other occafions. The procefs is different, according to the nature of the metal combined with the filver. Thus it is feparated,

1. From quickfilver; 1ft, by expreffion, and 2dly, by heat.

2. From all, except gold or platina, by fcorification and cupellation with lead, or by nitre. It is feparated at Birmingham from plated copper, by boiling the plates in ftrong fulphuric acid, to which has been added one-tenth of its weight of nitre. The acid acts *more readily only* upon the filver than the copper, and the plates are taken out when this is thought to be effected. The filver is obtained in the form of a white precipitate, and the copper with which it is ftill tainted is eafily feparated by cupellation.

3. From gold by aquafortis, or aqua regia. This procefs is called the *departing*. Mr. Tillet, of the Academy of Sciences, confidered this procefs by order of Government, with great care, and has publifhed very valuable memoirs on the fubject; and to him I muft refer you, becaufe a proper account of the procefs, fhewing the propriety of its different parts, would take up a great deal of our time, without

much addition to chemical fcience. Aquafortis, or aqua regia, is employed according as the filver or gold exceeds in quantity.

Thefe are the methods followed in practice to feparate it from the metals and from gold; and they bring it to as great a degree of purity as is ever needed for common ufe. But thefe methods do not render it perfectly pure. Scorification with lead, or calcination by nitre, leaves a little copper in it. And in feparating it from gold by aquafortis, fome of the copper employed for precipitating it from the acid, joins itfelf to the precipitated filver. Or if we feparate it from gold by aqua regia, fome gold is left in it.

But the procefs by which it is reduced to the moft perfect purity, is by firft purging it of all the coarfer metals by cupellation with lead, or by melting it with nitre, and then making it into muriat of filver, and reducing it. Thus any gold which might have been in it, or any minute quantity of copper which the lead might have left, is feparated from it. *(Vide Cramer.)*

Natural Hiftory.

Silver is found more or lefs in all parts of the world that are remarkable for metallic veins; moft plentifully, howeyer, in South America, particularly Peru and Potofi.

It is found either,

1mo, In its metallic ftate, and pure, or nearly pure.

2do, In the form of ores, in which it is the moft abundant metal, and which are called *proper ores of filver* : Or,

3tio, As an ingredient, contained in fmall quantity, in the ores of other metals, which are often, on account of the value of the filver, called *filver ores ;* but diftinguifhed by the title of *improper ores of filver.*

When it is in its metallic ftate, it is found formed into fibres, or plates, or branched maffes, often cryftallized into regular figures in the cavities of the vein, or intermixed with the folid matrix. Not unfrequently it adheres to an ore of filver. The finenefs of the fibres is often furprifing. They affect a fingular arrangement, nearly rectangular, forming chequers that are interrupted like a fret-work ornament in architecture, each being a fort of quadrangular fpiral. The geologifts afcribe this formation to a gradual fhrinking of the matrix from without, like what is more diftinctly obferved in the round balls of iron pyrites.

Of the ores of filver, the proper are chiefly three or four in number:

1. *Minera argenti vitrea*—glaffy ore.
2. *Minera cornea*—horny ore.
3. *Minera rubra*—the red ore.
4. *Minera alba*—the white ore.

The firft is very improperly called *vitrea*. It has neither tranfparency nor any other quality by which it refembles glafs. It is filver combined with fulphur, and is exactly fimilar to an artificial compound of filver and fulphur, being foft and malleable, almoft like lead, and of the fame colour with metallic lead, and very fufible. The clean pieces of it contain more than three-fourths of their weight of filver.

The horny ore has been analyfed with great accuracy by Mr. Woulfe. (*Phil. Tranf. anno* 1776.) He found it to be filver combined partly with muriatic, partly with fulphuric acid, and there is often pure metallic filver, or fulphurated filver, intermixed with it, which darkens its colour. It is a rare fort of ore, but when fpecimens of it are found pure and unmixed, it bears a great refemblance in appearance and properties to muriat of filver. It then contains about two-thirds of its weight of filver.

The red filver ore is of a deep red colour, and more or lefs tranfparent, often cryftallized, brittle, and very fufible. Profeffor Bergmann found that 100 parts of it contain 60 of filver, 27 of arfenic, and 13 of fulphur. But the proportions are a little different in different specimens; and some kinds of it are said to contain antimony instead of arsenic.

The white silver ore,—*minera alba*, has the metallic opacity and reflecting surface when broken, and a white or whitish colour, often inclining to yellow or gray. It contains the silver combined with arsenic and with other metals, as iron, copper, &c.; and the proportion of the silver in it is very various; some kinds of this ore being very rich, and others but poor. And it is subdivided by mineralogists into a variety of species. These and the improper or poorer ores of silver are accurately enumerated and described in Mr. Kirwan's *Mineralogy*, and others, to whom I refer you,—and I proceed to give a short account of the processes by which silver is extracted from its ores.

When the silver is found in its metallic state, almost pure, and only entangled in the state of filaments, or thin plates, or grains, or branched forms, in a stony matrix, there is only one way of extracting it, and that is, by reducing the whole to powder, or small grains, by stamping mills, or by other machinery, and macerating and working this powder in tubs with water and quicksilver, so as to give it the consistency of mortar for building. It is kept in this state for some time,—and is the better of being kept in a warm place; after which, it is diluted with plenty of water in large wooden vessels; and a circular motion is given to this water by machinery, while more water is supplied by a small pipe. Thus the water runs out slowly, over the brims of the vessels, and carries away with it the earthy and stony part of the powder. The silver and quicksilver united together, and forming a very heavy compound or amalgama, remain at the bottom of the vessels, and are at last completely freed from the earthy and stony matter. The amalgama, which is in some measure fluid, is then taken out and dried, and being tied up in pieces of leather, is strongly compressed, that the fluid parts of the quicksilver may pass through the pores of the leather. These fluid parts of the quicksilver contain a little only of the filver diffolved in them, and are employed again with frefh ore. The reft of the quickfilver remains in the leather bags, combined with fo much filver that it has the confiftency of a ftiff pafte, or is almoft folid. This rich amalgam is afterwards exposed to a proper heat to make the quickfilver evaporate, and leave the filver in a pure ftate; this operation being fo conducted that the vapours of the quickfilver are condenfed by themfelves in another part of the apparatus.

It often happens that the metallic filver is intermixed in the matrix with fome of the proper ores of filver, and alfo with ores of other metals which contain a fmall quantity, more or lefs, of mineralized filver. When this is the cafe, the whole, after it is reduced to powder, muft be fubjected to a procefs of uftulation, by which the fulphur and arfenic are evaporated from thofe parts of the filver which are in the ftate of ore; and the other coarfer metals are oxydated, and therefore feparated from the filver, which cannot be oxydated or calcined by a roafting procefs, but retains its metallic form, and is ready to unite with the quickfilver, in the procefs of amalgamation. To facilitate this, fome common falt is generally added. It helps to clean the furface of the particles of filver, and thereby promotes the union of this metal with the quickfilver.

All this is detailed by different authors, who have given treatifes on this particular fubject, or who have defcribed the rich mines of South America, in which the filver is almoft univerfally found in its metallic ftate. The moft noted of thefe authors are Alonzo Barba, Don George Juan de Ulloa; but above all Baron Born, who never vifited America, but who, in confequence of an office he held, was induced to ftudy the procefs of amalgamation, as practifed in Hungary, for extracting filver and gold from their matrices. He publifhed a book on this fubject, in which he has propofed fome important improvements, and defcribed the whole of his apparatus, and illuftrated his defcriptions with accurate figures. It is tranflated into our language by the late Mr. Rafpe.

When filver is found in the ftate of rich ores, in which it is combined with fulphur or arfenic, and with other metals, the extraction of it is commonly effected by the action of lead, and on a large cupel. This cupel is made of bone afhes, or often of wood afhes, from which all the falts have been carefully extracted. It is two and a half feet or three feet in diameter, and is formed and fixed in an iron hoop of proper depth, which limits and fupports the fides of it, while fome crofs bars fupport the bottom. This large cupel in its iron frame is placed in a reverberatory furnace fitted to it, to be heated with flaming fuel; and the iron hoop is defended from the deftructive action of the heat by covering it well with plenty of bone afhes, made to cohere with a little water. The flame and heat are applied to the cavity of the cupel only; the under furface is defended from the heat. A good quantity of lead is then melted, and made red hot on this cupel; and the filver ore is added to the lead. A part of the lead immediately unites with the fulphur and arfenic, or other matters with which the filver was combined. And thefe matters occafion or promote the fcorification of a part of the lead, which is alfo promoted by the wind of one or two bellows which play on the furface of the metal. The fcorified lead is a powerful flux or diffolvent for the earthy and ftony fubftances, or other metals in which the filver was involved; and the whole becomes fluid fcorified matter, flowing uppermoft, or around the melted and red hot metallic lead. This laft powerfully abforbs the filver that has been extricated from the ore. The fcoriæ thus produced are foon after drawn off by making a little gutter in the edge of the cupel, on the oppofite fide, from the blaft of the bellows. And the lead, if fufficiently enriched, is further fcorified by heat and air, until it be totally changed into litharge, the filver remaining at laft on the cupel. In fcorifying lead, to extract filver from it in the large way of working, the fcorified lead is not allowed to fink into the cupel, or to be abforbed by it, as is done in refining fmall quantities of filver, or in affays. This at leaft is avoided as much as poffible. The melted litharge is made to run out of the cupel by the little gutter I mentioned, and it drops into a cavity below, where it immediately congeals. The reafon of this is, that when the litharge is pure, it is very eafily reduced again into lead. But when it is abforbed into the bone afhes, it is difficult to reduce it; and it cannot be reduced without confiderable lofs.

Thefe proceffes are fometimes varied in other ways, the reafons of which you will eafily underftand from what has been already explained to you. If an ore, for example, containing fome filver, is very much brangled or difperfed through a large quantity of ftony matter, the whole is melted among the charcoal; fome other kind of fparry or ftony matter being added to promote the fufion of that which involves the ore; and they add alfo fome litharge or fcorified lead which has been produced in other proceffes, fuch as the cupellation laft defcribed.

GENUS XIV.—GOLD.

This metal has been diftinguifhed by the chemifts with the titles of *fol* and *rex metallorum*, on account of its excelling all the reft by the perfection of its metallic qualities; but chiefly, I imagine, becaufe it is the moft highly prized in fociety.

When pure, it is foft and flexible, not elaftic or fonorous. I already noticed its great ductility and malleability. Mr. Lewis confiders its fpecific gravity with great accuracy. It is about nineteen times as heavy as water; and a cubic inch of it weighs 4902 grains, or 10 ounces and 102 grains. It requires for its fufion a very ftrong red heat, or almoft a white heat, a heat ftronger than that of melting filver; and when in fufion, it emits or reflects a bluifh green light from its furface.

Its furface always remains bright in the fire; and if foul or tarnifhed before, it is always cleaned and brightened by a red heat.

When we try it in mixture with falts, we find that none of the acids, applied in their ordinary ftate, produce any effect on it. The fulphuric acid, or the nitric, or the common muriatic, if applied feparately, either in their watery form, or in the much more active ftate of dry and burning hot vapour in cementation, make no impreffion on it, or fhew any power to diffolve it in its pure and metallic ftate. Upon this is founded a trick for defrauding the purchafers of gold. The bars or ingots of impure gold are cemented for a while with proper materials, which refine it to a depth proportioned to the continuance of the cementation. Trinkets treated in this manner bear to be burnifhed and polifhed; and when tried by the touchftone appear perfectly fine. When trinket gold appears of this extreme finenefs by the touchftone, it is always to be fufpected as being only a cruft covering bafe gold. An ingot, fo treated, appears fine, even when cut through with a chizel, becaufe this carries a film along with it from the furface, which covers the reft. It is proper, therefore, to cut it only half through, and then to break it.

The faline folvents which act on metallic gold, are only the oxygenated muriatic acid, and the nitro-muriatic folvent, which is therefore called *aqua regia*. The oxygenated muriatic acid is effectual in confequence of its power to oxydate the gold, which the common muriatic acid cannot do: And the nitro-muriatic acid has the fame power.

This mixed acid may be prepared and applied by different ways, fo as to be effectual in diffolving gold; as

1*ft*, By mixing the pure nitric and muriatic acids together.

2*dly*, By adding common falt, or fal ammoniac, in moderate quantity, to the nitric acid, and diftilling this mixture.

3*dly*, The mixture I juft now mentioned can be employed very well without diftilling it.

4*thly*, Alum, nitre, and common falt boiled with water, form a folvent of gold, though not a ftrong one.

Scheele, the firft difcoverer of the oxygenated muriatic acid, was of opinion, that the nitric acid dephlogifticated, or, in other words, oxygenated the muriatic, and thus enabled it to diffolve the gold.

But Dr. Prieftley found that *nitrofe* acid, and efpecially the moft red and volatile vapour of it, which he calls nitrous vapour, when added to the muriatic acid, occafions a more rapid diffolution of the gold than the more perfect nitric acid does, which cannot eafily be explained upon Scheele's principle.

Another fact, analogous to this obfervation of Dr. Prieftley's, is, that when the folution in aqua regia is going on in a flow and languid manner, the addition of a fmall quantity of æther, or of alcohol, makes the folvent very active. The folution goes on brifkly, and is accompanied by much effervefcence. This alfo feems very unlikely to produce an oxygenation of the muriatic acid. It feems more fitted for depriving it of oxygen ; and it is not clear how the vapours of nitrofe acid fhould fupply oxygen more readily than nitric acid, which contains more of it. Yet, from what experience I have in the diffolving of gold, and by attending to the odour of the acid when it is acting properly, I am perfuaded that Scheele was right. While the acid is acting on the gold, it always gives the diftinguifhing pungent difagreeable odour of the oxygenated muriatic acid. When it is faturated, the nitrofe fumes are plainly diftinguifhable, and then the addition of a little muriatic acid immediately renews the action, and yields the fmell of the oxygenated muriatic acid. The addition of a little nitric acid, when the folution flags, does not produce the fame effect. From this I think the inference is plain. The muriatic acid is the bafis of the folvent, but requires more oxygen, and the faturated folution is a folution in oxygenated acid. Therefore more nitric acid can do nothing. More muriatic acid can, becaufe there is enough of nitric in the mixture to oxygenate it.

But there is alfo reafon to be of opinion that the acid of nitre itfelf, when it has loft a part of its oxygen, becomes in fome meafure a folvent of gold, or affifts in diffolving it, and that in the nitro-muriatic folvent, both acids act.

It has been found by experiment, particularly by Dr. Brandt, that a ftrong nitrofe acid, although employed alone, will diffolve a fmall proportion of gold, when boiled with it fome time. But, in truth, it is only a very fmall quantity that is thus diffolved ; and even this fmall quantity is liable to feparate again from the acid fpontaneoufly. Moreover, when gold is diffolved in the ufual manner in aqua regia, the muriatic acid adheres the moft ftrongly to the gold, and is the principal folvent. Nitrous vapours arife from the falt while evaporating to drynefs. Mr. Lewis remarks, in his Notes on Neuman, that when the folution of gold is diftilled, the nitrous acid eafily paffes over into the receiver, leaving the muriatic alone in poffeffion of the metal. And when gold is obtained in the ftate of an oxyd, by precipitation from its folution, or any other procefs, the muriatic acid alone, and even a weak one, can eafily diffolve it, and forms a folution which has all the properties of the common folution of gold.

It is a fufficient anfwer to any doubts about the real folvent of gold, that the compound refulting from the folution in the oxygenated muriatic acid is perfectly the fame in all refpects with the falt formed by any of the compounded folvents. The nitric, or even nitrous acid may perhaps oxydate the gold, but it does not combine with it in a faline form.

The folution of gold is always of a rich yellow colour : But when applied to animal or vegetable fubftances, it produces an indelible ftain, of a reddifh purple. By evaporation, it can be made to form yellow cryftals. In thefe the muriatic acid adheres fo ftrongly to the gold, that it cannot be completely feparated by heat. When they are expofed to heat in a retort, part of the gold rifes with the acid, and is condenfed into a red or deep yellow fluid, or into cryftals or flowers of the fame rich hue, in the neck of the retort, called by the alchemifts the *red lion*.

We can eafily, however, precipitate the metal from its folvent by various additions, fuch as alkaline falts and other metals.

Fixed alkalis and lime precipitate gold of a ruddy yellow, and the oxyd is foluble in all the acids.

Nut galls, or the gallic acid, alfo precipitate gold of a purple colour; and it was difcovered by Mr. Monnet, that this precipitate diffolves moft readily in nitric acid, giving it a fine blue colour.

The volatile alkali is one of the beft precipitants of the gold, and produces the remarkable precipitate named *aurum fulminans*. It is an oxyd of the gold, which has a dull yellow colour, and is thus named on account of its difpofition to explode with aftonifhing violence, when expofed to a very moderate degree of heat.

To fee this explofion properly in its full energy, we muft wafh away the faline matter from the precipitate carefully with pure water, and then dry it well in a cool place. After this preparation, if we expofe it to a gradual heat, it firft becomes dark coloured, and foon after, if the heat be increafed, it explodes with an obfcure and momentary flafh, vifible in darknefs. And when the experiment is made with a view to know what becomes of the gold, by covering it, for example, with a glafs bell, it is found difperfed in fmall particles, and all of it reftored to its metallic form.

The explofion of this preparation of gold is fo exceedingly violent, that great caution is required to prepare it with fafety, or to make experiments with it; terrible confequences having in fome cafes proceeded from the want of care or fkill in handling it. Its force is fuch, that when fo fmall a quantity as ten grains are exploded on a thin metallic plate, it makes an impreffion like that of the ftroke of a hammer, or fometimes breaks a hole through it. And after it is rightly prepared, it is dangerous to treat it in any manner by which it may be heated, or expofed to friction or compreffion. Grinding it in a mortar, nay even the friction of the glafs ftopper on a grain left in the neck of the phial in which it is kept, has been known to occafion its explofion.

It is remarkable that the heat produced in this explosion is very small. It will not singe paper. If a little be rubbed into a bit of soft spongy paper, and held over a candle, or before a hot fire, it will turn dark coloured, and then go off in a succession of cracks, with scarcely any light, and without changing the colour of the paper. I imagine that even the light arises from an electrical concussion of the air. (*See Note* 31. *at the end of the Volume.*)

The nature of this surprising preparation of gold, and the cause of its explosion, has been but lately even guessed at with any probability, for I cannot say that it has been discovered: And the sagacious conjecture was made by Mr. Scheele. Formerly it was supposed to contain nitrous salts, combined with the gold, and that these gave it this fulminating quality. And it is in fact a little heavier (almost one-fourth) than the gold from which it is prepared. But neither can any nitrous salts be found in it, nor is the heat at which it explodes sufficient for the explosion of nitrous compositions.

When I was engaged in making my experiments on magnesia and other alkaline substances, I had a notion that the explosion of aurum fulminans depended on carbonic acid communicated to the gold by the precipitating alkali;—and this conjecture was founded on my observing that gunpowder and pulvis fulminans contained, or produced in the explosion, a great quantity of this gas. But the important discoveries which have been made since that time shew clearly that it proceeds from a very different cause.

Professor Bergmann, in a dissertation on aurum fulminans, proves that it cannot be made without volatile alkali. If the gold be dissolved in a solvent composed of the pure nitric and muriatic acids, and we then precipitate it with a fixed alkali, we get an oxyd totally destitute of the exploding power; but if we digest this oxyd with liquid volatile alkali, we then give it that property. And further, the pure or caustic volatile alkali serves rather better for making good aurum fulminans than the carbonat of ammonia. All this, and many other facts to the fame purpofe, were afcertained by Profeffor Bergmann.

And afterwards Scheele in Sweden, and Berthollet in France, gave us more complete knowledge on this fubject, by demonftrating the prefence of volatile alkali adhering to the gold in this preparation, and fhewing by experiments, that this volatile alkali is decompounded when the explofion takes place. Mr. Berthollet, equally judicious in his choice of decifive experiments, and ingenious in his conclufions from them, made two experiments on thefe compounds, which leave little doubt as to their chemical conftitution. He expofed aurum fulminans in a copper tube to a very gentle heat, gradually increafed, and obtained from it great abundance of alkaline gas, by which its weight was diminifhed confiderably, and its exploding power entirely taken away. He exploded dry aurum fulminans in a proper apparatus, and obtained water and azotic gas,—and the gold was reduced to the metallic ftate.

The volatile alkali is now known to be a compound of hydrogen and azote. The precipitate of gold alfo contains the oxygen which the metal received from the nitric acid. The attraction, however, of the oxygen for the metal, and of the hydrogen for the azote, prevents them from acting on each other until heat be applied ; but when heat is applied, the oxygen and hydrogen unite with a rapid or momentary combuftion, and form vapour of water with an explofion ; and the azote alfo affuming the elaftic aëreal form, the explofion is thereby increafed *.

Another difcovery which has been lately made, and which is connected with the fubject, is, that filver can alfo be prepared fo as to have the properties of fulminating gold, and that in a far higher degree than the gold itfelf. This difcovery was alfo made by Mr. Berthollet, and has been confidered already.

* I obferve that fome French and Italian chemifts confider this phenomenon as the combuftion of azote ; but this is inconfiftent with the experiments of Mr. Berthollet.

EDITOR.

As to the formation of this curious compound I muſt obſerve, that when the gold is precipitated in this manner from the aqua regia by alkaline ſalts, it may be rediſſolved, like the precipitates of ſome other metals, by adding to the mixture a ſuperfluous quantity of the alkali. The fixed alkalis diſſolve it better than the volatile alkali; and, according to Margraaf, the phlogiſticated alkali is better for this purpoſe than the common or pure fixed alkali. Theſe ſolutions are not-permanent, the gold being depoſited from them, eſpecially from the volatile alkali.

If we chooſe to deprive the precipitate of its exploding power, and bring back the gold to its metallic ſtate, we can effect this by ſeveral different methods; as,

1ſt, By calcining long, with a very gentle heat, inferior to that which makes it explode. Thus it ſlowly aſſumes a dark purple colour; and the volatile alkali is thus evaporated from it. This method requires great caution to avoid friction or concuſſion.

2dly, By heating it in a ſtrong and cloſe veſſel *.

3dly, If the particles of it be ſeparated from one another by the interpoſition of any powdery ſubſtance mixed with it on purpoſe, as ſulphur, or any of the neutral ſalts, or earthy powders, it may be heated ſecurely, until it is deprived of the power of exploding, and can afterwards be melted into a maſs.

4thly, If ſome ſtrong ſulphuric acid be boiled with it, and evaporated from it, the gold remains bereft of its exploding power.

5thly, The vitriolic æther, by digeſtion, ſimply reduces a part, and diſſolves a part, which is alſo afterwards reduced.

* A ſmall quantity of aurum fulminans was ſhut up in an iron ball, of which it completely filled the cavity. This was expoſed to a great heat, but did not explode. (See Fourcroy, vol. iii. p. 369. Edin. 1788). Can this have happened becauſe external preſſure prevented the chemical combination, as we ſee it prevent the abſorption of caloric by water, ſo as to make it boil?

EDITOR.

6thly, Muriatic acid, digefted with it, diffolves it; and then, by precipitating with the fixed alkali, we obtain it no longer fulminating.

You will find it alleged in books, that wafhing aurum fulminans with much water, or boiling it in water, took away its exploding power; but this is a miftake. The more perfectly it is wafhed, it explodes the better: And if we neglect to wafh it, and thus leave in it any quantity of the falts formed in precipitating it, it will not explode well.

Such are the effects of alkaline falts on the folution of gold.

This metal can alfo be precipitated in fome meafure from its folution, by fome inflammable fluids, as fome aromatic oils, and fpirit of wine. Thefe liquors act by attraction for the acid.

The effects of the vitriolic æther have been thought more remarkable. It feizes on the gold immediately, and often reduces it to the metallic ftate, in fine films, which are rendered buoyant, even in the æther, by the adhefion of imperceptible bubbles of vital air. By adding æther to a folution of gold in aqua regia, the whole of the metal may be brought to the furface, forming a ftratum a-top of the æther; and neither of the fluids below (the æther nor the aqua regia) contain a particle of it. Generally, however, the gold remains for fome time diffolved in the æther (but completely abftracted from the acid), tinging it of a rich yellow.

This effect of the æther has been the more admired, on account of its being the lighteft fluid of any we know, and therefore was thought the lefs qualified to fufpend fuch a ponderous metal. But the power of chemical attraction performs greater wonders than this; for gold can be even rendered volatile by fome of the falts which have the ftrongeft attraction for it. And in this cafe of the diffolution and fufpenfion by æther, the æther acts more by its attraction for the acid with which the gold is united, than for the calcined gold: But it appears alfo to have fome attraction for the oxydated gold itfelf. Mr. Lewis and others fay that aurum fulminans can be diffolved by it at leaft in part. Still, however, I am inclined to fuppofe that the æther acts, even in this cafe, by its attraction for the oxygenous principle, and the fmall quantity of faline matter combined with the gold; for it is certain that when we apply it to pure gold in its metallic ftate, it has not the leaft power to act on it, or to diffolve it. And when it diffolves the aurum fulminans, it does not hold it long fufpended, but depofits it in its pure and metallic form *.

Other metals alfo can be employed to precipitate the gold, and there are feveral that precipitate it, although they are themfelves combined with an acid. They produce this effect in confequence of the ftrong elective attraction which they have for the muriatic acid, and for the oxygen, both of which they feparate from the gold.

The metallic folutions which have been applied to this purpofe, are,

The folution of fulphat of iron.

The folution of nitrate of quickfilver;—and

The folution of tin in the nitro-muriatic folvent.

It is eafy to underftand how the two firft produce their effect.

---

* Since we fee the æther feparate the gold from its folvent in the metallic ftate, we can fcarcely expect any action of æther on metallic gold. Experiments on this metal being fo expenfive, they have not been made on large enough quantities for obferving with accuracy what changes are induced on the æther, in this experiment. The phenomena accompanying the feparation of gold by aromatic oils, give confiderable information, and very conformable to Dr. Black's opinion. Thefe oils are infpiffated in the fame manner as by the contact of vital air. We fhould expect a fimilar change on the æther to what pure vital air produces on it, forafmuch as it deoxydates the gold; and we fhould expect fome combination of the æther with the acid. If we obferve the æther changed as it would be by fo much aqua regia, we fhould afcribe the feparation of the gold entirely to the attraction of the æther for the acid. If the change be different, we fhould afcribe it, at leaft in part, to its attraction for pure oxygen. The phenomena give indications of both. The æther mixes with the acid only to a certain degree, and the acid, thus combined with æther, has no further action on metallic gold.

EDITOR.

And I fhall only obferve, that the precipitate by the mercurial nitrate is very readily diffolved by mercuy. It is ufeful in this way for gilding upon glafs or porcelain.   But the precipitate by martial vitriol is far preferable, affording a much richer colour, feemingly owing to a fmall quantity of copper contained in the vitriol.

The effect of the folution of tin depends on this particular circumftance, that though the tin be already combined with a quantity of the muriatic acid in the nitro-muriatic folvent, and alfo with a quantity of oxygen, it is not faturated with either of them.   It has ftill an attraction for more, and therefore takes them from the gold.

The precipitate of gold, thus obtained, is the moft remarkable of any by its fine and rich colour.   The colour is fo deep, that this experiment is a way to difcover the fmalleft quantity of gold in a folution.   And a red or purple appears to be the colour natural to gold, when very fubtilely divided,—by electrical fire, for example, or other means.   The purple ftain given by the folution of gold to animal and vegetable fubftances, is another example.

This precipitate is valued, and very much ufed, as a fine purple for enamel colours.   It has the advantage of enduring the fire without undergoing any change of its colour, to which many other enamel colours are liable.

I have already defcribed the procefs for preparing the folution of tin that is employed in the preparation of this purple precipitate of gold, known by the name of the *purple of Caffius*.   It was defcribed as a folution of tin in aqua regia, and with every precaution to have the metal as flightly calcined as poffible, and the acid completely faturated with tin.   I muft now take notice of the circumftances which muft be attended to, in order to have the folution of gold in the moft proper ftate.   It is found often to fail, when the gold is diffolved in a mixture of the two acids; feldom, when the aqua regia is made by adding common falt, very pure, to nitric acid; and fcarcely ever, if fal ammoniac be employed. The greateft nicety lies in the degree of dilution of the folution of tin. As this depends on the oxydation of the tin, it is beft to determine it by trial. Having diluted the tin folution with eighty times its bulk of water, put three or four portions of it into glaffes, and dilute each of them differently; then, dipping a glafs rod to a certain depth in the folution of gold, rinfe it in one of the glaffes; do the fame to another; to a third, &c.; then notice in which glafs the precipitate has the richeft and moft beautiful tint of purple, and dilute the whole according to that ftandard. The precipitate both forms and falls down very flowly, being in fome degree gelatinous; by long reft, however, it all falls down, and then the clear liquor may be decanted off, and the precipitate cleared by edulcoration and a filtre.

Orfchall one of the celebrated older chemifts, fays that he obtained a very fine precipitate by means of the fuming liquor of Libavius; alfo by means of a folution of tin with corrofive fublimate made by deliquefcence; and a ftill more beautiful precipitate than what tin can produce, by means of mercury diffolved in aqua regia.

But Orfchall knew only the preparation of the enamel colour, which has great body, as the painters call it, but little tranfparency. The tranfparent red and purple is of great value for *ftaining* glafs. It is made by diluting the purple of Caffius with pure cryftal or glafs. This is a preparation ftill more capricious and uncertain. Frequently the glafs has no colour whatever; but if a rod of fuch glafs be made red hot, and held in a fmoky flame of wood, it becomes purple in an inftant. But this is merely fuperficial; and if the glafs is to be formed into any other fhape, the colour vanifhes in the working, and it again requires the affiftance of the oily flame. This is called *Jew's glafs.*

Neuman fays, and Dr. Lewis confirms it in his Notes, that the preparation of the purple precipitate never fails, if made by fimply putting pure tin into the folution of gold greatly diluted.

Margraaf has publiſhed, in the Mem. Acad. Berlin, 1779, a ſeries of moſt judicious experiments for determining the beſt proceſs for this valuable precipitate ; and I refer you to him for farther information. I muſt obſerve, however, that our artiſts call the purple of Caſſius a tender colour, becauſe a conſiderable heat makes it tranſparent, and therefore of a different tint, according to its thickneſs on the ware. It does not ſuit enamel, therefore, ſo well as ſtaining of glaſs, becauſe the other colours by copper, cobalt, &c. require much higher heats in order to bring them to their full colour, and are therefore melted ſeveral times in crucibles before they are ground to powder for the painting in enamel. Even then, they require more heat than the purple of Caſſius bears without riſk of loſing its body.

Inflammable Subſtances with Gold.

Sulphur, which unites ſo readily with moſt of the other metals, cannot be combined with gold. But if we firſt join the ſulphur with a fixed alkali equal in quantity to the ſulphur, the ſulphuret thus formed, if applied to gold in thin plates or leaves, and in the way of fuſion, very readily combines with it, or diſſolves it ; and if the compound be immediately poured out of the crucible, and ſoon after diſſolved in water, a part of the gold will be diſſolved along with the ſulphuret, while the reſt remains in the ſtate of a very ſubtile powder. Dr. Stahl had a notion that this proceſs, which is certainly an ancient one, was known to Moſes, and was practiſed by him when he made the children of Iſrael drink the golden calf. It is indeed true that this potion is extremely nauſeous, having a pungent bitterneſs not to be felt in the ſimilar preparations of other metals.

The effects of alcohol, æther, and aromatic oils, on the ſolution of gold have been mentioned already.

The only other inflammable ſubſtance remarkable for its effects on gold, is phoſphorus, the powers of which, with reſpect to metals in general, have been ably inveſtigated by M. Pelletier. When thrown into melted gold, a certain quantity unites with the metal, forming a phofphoret of gold, which is more fufible than the gold by itfelf. But if the compound remain in the fire. and air be admitted, the phofphorus is gradually burnt, and changed into acid, which feparates from the gold.

Relation of Gold to other Metals.

Gold may be mixed with any of the other metals; and it is by the admixture of fome of thefe that it is made to appear of thofe different colours which are feen in the inlaid gold of trinkets and toys. Copper inclines the colour of gold to red; filver makes it pale; and if the filver be one to four of the gold, a greenifh hue is produced. Pure gold is a full yellow. The Venetian chequin has the richeft colour of any gold; and the art of giving it this high colour is kept a fecret in the mint of Venice. I made a piece of fine gold acquire the fame colour, (but it was only fuperficial) by keeping it long red hot under charcoal duft.

In its pure ftate, it is thought too foft and flexible for making toys and utenfils, or coin. And the general practice is to add fome of the other metals, which give it more ftiffnefs and hardnefs. This addition is named the *alloy* of the gold. It is commonly filver and copper, of which one part is added to eleven of the pure gold, for the ftandard of thefe kingdoms. One pound avoirdupoife of ftandard gold, is worth $56\frac{1}{2}\frac{1}{3}$l. Sterling.—1000l. Sterling weighs (avoirdup.) 7 pound $9\frac{1}{10}\frac{4}{10}\frac{1}{10}$ ounces.

In fpeaking of the quantity of alloy in gold, the term carat has been ufed. A carat is the twenty-fourth part of a mafs of gold, great or fmall: And each carat is fubdivided into twenty-four parts, denominated grains. Our coin then is of twenty-two carats.

The relation of gold to quickfilver, and the art of gilding metals founded on it, has been noticed already in treating of quickfilver.

Other methods are practifed for gilding, in which mercury has no fhare. As almoft all the metals precipitate gold from its folution, and precipitate it in the metallic ftate, any of thefe metals, put into a folution of gold, will be immediately covered with the depofited gold. This covering wi'l rarely be made to adhere, b caufe the furface of metal which we would gild is oxydated. But, by methods refembling what I defcribed for whitening or tinning pins, the pure metals may be applied to each other, and then the burnifher makes them adhere, fpreading, at the fame time, the gold over the parts where none was depofited. The beft of thefe proceffes are kept fecret by the poffeffors. There is a family at Nuremberg, which has preferved one a fecret for upwards of two centuries; and their gilding, even on the moft common work, has a folidity and richnefs of colour that is not equalled by any other artifts in Europe.

There is ftill a more fimple way of flight gilding. Linen rags, foaked in the folution of gold, are burned to afhes;—a fmooth cork, fuperficially charred, is wetted and dipped among thefe afhes, and then rubbed carefully over the piece to be gilded. The reduced gold among the afhes is thus preffed on the work fo as to adhere; and by going over it feveral times, it is completely covered; and bears to be rubbed very hard with a fine linen rag, ftrained on a bit of cork. This gives the work a fine polifh, and great brilliancy; and is pretty durable on the infide of cups, and other fituations which do not require often fcouring.

You muft have often heard the terms of *tried gold*, and of the *trials of gold*. Thefe are trials, or proceffes, by which we can affure ourfelves whether a metal which refembles gold be gold or not; and if it be gold, whether it is pure or alloyed; and if alloyed, what proportions of pure gold and of alloy it contains. The operations which have been long in common ufe for thefe purpofes are five in number:

1. The ufe of the touchftone, *(lapis lydius)*.
2. Cementation.
3. Refinement with antimony.

4. Cupellation with lead.

5. Parting, or the depart.

1*ft*, The touchstone shews whether a metal be gold or not; and if gold, of what fineness nearly, by needles. The piece of gold is rubbed on a black stone of the jasper kind, having a fine siliceous grain and argillaceous cement. Some of the hardest of the antique, or of Wedgewood's black pottery, answers the same purpose. The metal leaves a trace on the stone. A set of touch needles are made, consisting of gold, with various proportions of alloy. The tryer makes a stroke with one of the needles close by the one made by the metal under trial, and changes the needle, till he has got a stroke exactly like it: Thus he judges of the proportion. To be more exact, he draws over both strokes a pencil dipped in aqua regia. This dissolves all the gold, and leaves the silver. The proportion of alloy is thus more clearly seen.

This method will not detect adulteration, when the piece has been cemented in the manufacturing. This makes it fine superficially, though internally base. If the stroke shew very fine gold, we may be almost certain that the piece has been cemented.

2*d*, Cementation is seldom used, as requiring repetitions. Although a small quantity of base metal, concealed in a large quantity of gold, is thereby shielded from the acids acting in the humid way, they cannot resist their action in dry burning vapours. Therefore adulterated gold, being first laminated, is stratified with a mixture of calcined martial vitriol and nitre, or common salt, (not both), and a quantity of powdered brick, and exposed to a cementing heat for some hours; the piece is taken out, melted, and again laminated, and again cemented,—" seven times tried in the furnace." By each operation part of the base metal is destroyed.

3*d*, Crude antimony is more frequently used. It is better than pure sulphur, because pure sulphur is too volatile, and metallic antimony washes down the gold; but the crude antimony absorbs and scorifies all the baser metals.

4th, Cupellation is ftill more frequent as a means of purging the gold of copper, and all bafer metals; and an addition of filver facilitates this operation, and makes it more perfect. If there be no filver in the mixed metal, the gold retains, in the end of the cupellation, a fmall portion of copper, which it defends from the action of the lead and heat.

5th, But after thus cupelling the gold with this admixture of filver, we muft have recourfe to the operation of parting, to have the gold pure. This parting is the feparation of the gold from the filver, performed by aquafortis, or aqua regia. Aquafortis is commonly employed, and the ufe of it is quite fimple and eafy. I fhewed you an example when I diffolved filver in aquafortis. The fmall portion of gold which the filver contained remains undiffolved, and retains its fhape, and it needs only to be boiled with a little frefh aquafortis to make it quite pure; for if this be not done, it retains fome of the filver. In order to enable the aquafortis to act properly on this metallic mixture, there muft be at leaft twice as much of the filver as of the gold, and the aquafortis muft be fufficiently ftrong. It is alfo ufual to reduce the metallic mafs into fmall grains and fragments, by melting it, and pouring it into cold water. This is done to increafe the furface of contact between it and the aquafortis. Or if we have but a fmall quantity of it, as in affaying, it is rolled out into a thin plate, between two fmall fteel rollers: And this plate is twifted into a fpiral.

When the mixture contains more gold than filver, in the proportion of two to one at leaft, we may ufe aqua regia to feparate thefe two metals. It will diffolve the gold, and leave the filver at the bottom, in the form of muriatic filver. The folution of gold being then carefully feparated from the filver, the gold is commonly precipitated with folution of quickfilver in aquafortis. This occafions the gold to fall in the form of a dark-coloured powder, which, after it is well wafhed, is eafily melted into a mafs with a little borax.

There is another method of refining gold, which is now commonly practifed, when the gold contains a fmall quantity of alloy only, and that of fuch a nature as to deprive it of toughnefs and malleability, fuch as iron, tin, brafs, or any of the femi-metals. The method I mean is, to melt it, and add to it repeatedly, while in fufion, fmall dofes of the corrofive muriat of quickfilver, until it is fine or tough. This corrodes the alloy, and foon brings it to the furface as a fcoria, where it works off.

When a fmall proportion only of gold is contained in a metallic mixture, the procefses for extracting it are often different from thofe already defcribed. In general it is feparated from all other metals, except copper, in the fame manner as filver is extracted from them. But when the gold is contained in copper, it cannot be extracted by the fame operation which ferves for extracting filver from copper, which is the addition of lead and the eliquation of the lead. The gold is not brought out by this procefs: It has a ftronger attraction for copper than for lead. A fuccefsful method was, however, difcovered in Germany, which was very profitable to thofe who practifed it firft; and it ftill continues to be the beft method for extracting a little gold from metallic copper. It is done by melting the copper with fulphur and lead at the fame time. The copper unites with the fulphur, and is thereby difpofed to quit the gold to the lead. The procefs is defcribed in Lewis' *Commercium*, and Cramer's book on *Affaying*.

A fomewhat fimilar procefs, by which gold was extricated from filver that contained but a fmall quantity of it, had long been practifed before the chemifts thought of employing it in the cafe of copper. Cramer defcribes this procefs with filver very particularly.

Natural Hiftory of Gold.

It muft be acknowledged that gold, though it be not the moft ufeful of the metals, has fome admirable qualities. Its rich colour and luftre, which are never obfcured with tarnifh or ruft, and its aftonifhing extenſibility, which enables us to employ it in the embelliſhment of the works of art at a very moderate expence, are one foundation for the value that is ſet on it. The principal cauſe, however, of the high price of gold, is the difficulty of procuring it. And yet there is more of it produced by nature than is commonly imagined; but it is generally diſperſed through ſuch immenſe quantities of other matter, that it cannot be collected without great labour and expence.

The Spaniſh and Portugueſe parts of America, and ſome parts of India and Africa, afford the largeſt quantities of gold. Amazing reports have been publiſhed of the abundance of this metal in ſome of the Spaniſh poſſeſſions in America; but theſe reports are publiſhed in late accounts of ſome newly diſcovered places where gold has been found, and the accounts of ſuch new diſcoveries are commonly amplified *.

Wherever gold occurs, it is found much more commonly in its metallic ſtate, and nearly pure, than in the ſtate of an ore. This is probably a conſequence of its having no attraction for ſulphur, and very little for arſenic, and of its reſiſting the action of the mineral

---

* Some parts of the new kingdom of Grenada (which is a high inland country eaſt of the Andes, and in the north end of South America) are rich in gold, which is all waſh-gold. On a riſing ground near Pamplona, ſingle labourers have collected in a day what was equal in value to 1000 peſos, or to 225l.=57 ounces 4 drachms 42 grains. A late governor of Santa Fe brought with him to Spain a lump of virgin gold eſtimated to be worth 740l. Sterling. (The weight of it muſt have been about 189 ounces, or 23 merks and 5 ounces, even ſuppoſing it gold equal in purity to Engliſh ſtandard.)

At Cineguilla, in the province of Sonora, the Spaniards found a plain fourteen leagues in extent, in which they found waſh-gold at the depth of only ſixteen inches, the grains of ſuch a ſize that ſome of them weighed nine merks, and in ſuch quantities, that in a ſhort time, with a few labourers, they collected 1000 merks of gold in grains, (equal in value to 31,219l. 10s. Sterling) even without taking time to waſh the earth that had been dug, which appeared to be ſo rich, that perſons of ſkill computed that it might yield gold to the value of a million of peſos; which is equal to 225,000l. Sterling. In one place, called the *Mine Yecorata*, in Cinalod, they found a grain of gold 22 carats fine, which weighed 16 merks 4 ounces 4 ochavas. It is now depoſited in the royal cabinet at Madrid. This grain is worth 498$\frac{1061}{111}$l. Sterling.

acids. I believe all the gold collected in America and Africa is found metallic, and uncombined with the common mineralizing substances. But in Europe there are some mines which yield *ores*, containing a small quantity of it. These ores are not, properly speaking, ores of gold, they are ores of other metals; but the small quantity of gold which they contain along with the other metals, occasions their being considered as ores of gold. In Hungary, a considerable quantity is extracted from some ores of quicksilver, and from a pyrites of iron and sulphur. In the Hartz forest in Germany, there is an ore which affords zinc, and lead, and silver, and a small quantity of gold. And both there and in other parts of Europe, there is found in some places a black sand like small grained gunpowder, which is an ore of iron, and which contains some gold.

In some of those ores, it is either intimately combined and mineralized with the materials of which the ore is principally composed, or, if it be in its metallic and pure state, it is in such minute atoms, that it is not discoverable by vision until it be collected together by itself. There is reason, however, to believe that it is always in this state. When the Hungarian pyrites is dissolved with aquafortis, it is said that the gold is left by that acid in the form of minute atoms, and fine films, which are in a metallic state. And as this pyrites varies much in the quantity of gold it contains, and some of it does not contain any, there is reason to believe that all the gold in it is metallic and pure, and only dispersed through it in very minute films not mineralized. However that be, by far the greater part of the gold which is collected in different quarters of the world is found free from mineralizing substances, and nearly pure, or contains only some silver, and sometimes a little iron or copper. And this virgin gold, as it is called, is found in two states or conditions, that is, either in the solid veins of the mountains in which it seems to have been originally formed, or deposited by nature like other metals, and from which it is in this case cut out by mining in the usual manner,—or

mixed with the loofe foil and rubbifh, on the furface of fuch mountains, or in the plains that are below them, or near them.

In the firft of thefe ftates (I mean in the original veins) it is found
but rarely. By far the greateft part of it is collected from among
loofe foil, and gravel, and fands,—efpecially from the gravel and fands
of brooks in fome mountainous countries, and from the foil which is
near to fuch brooks. Alfo in the foil of plains formed by the rivers
into which fuch brooks empty themfelves, and in the fands of fuch
rivers.

And when we examine the gold found in thefe different places,
we find reafon to be fatisfied that it is brought down from the mountains by the gradual and long continued action of water and the atmofphere on the materials through which it is difperfed. The proofs
of this are, that the nearer to the mountains it is found, the grains of
it are in general fo much the larger, and the rougher or lefs worn ;—
while fuch of it as is found at a diftance from the mountains, or from
the higheft parts of the country, always confift of grains fmaller in
their fize, and evidently fmoothed on their furface, and worn by the
attrition they have fuftained from the fand and gravel, during the
great length of time required for their being brought down fo far.
You have examples of this general fact in the gold which is found in
the fands of fome of the rivers of France, of which an elegant and
entertaining account is given by Mr. Reaumur in the memoirs of the
academy for the year 1718.

We have alfo an example of the fame fact in this country. Gold
is found at Lead Hills, and in a diftrict there called Ettrick Foreft.
The gold is all in the form of fmall grains, difperfed through the gravel of the brooks, and in the foil that is near them. It was thought
an object of value a long time ago, when gold was dearer than now ;
and great numbers of hands were employed in collecting it. From
fome regifters, it appears that in one year 48000l. Sterling value of
this gold was coined in the Scotch mint. The grains are in gene-

ral the larger and rougher the higher up the country they are found. The farther we defcend, they are more difperfed and of a fmaller fize, and more worn and fmoothed. The fame thing alfo was obfcrved by Baron Born in Hungary and Bohemia. This points out very evidently from whence this gold comes, which is thus found in the foil, gravel, or fand, of thefe particular diftricts or rivers. And there cannot be a doubt that the gold which has been lately found in Ireland has had the fame origin. In thofe hills that confine the valley and the brook around which it is found, there are certainly mineral veins containing gold: And they cannot be far diftant, if I may judge from a fpecimen I faw of the gold grains. They are fo rough, and have fo little appearance of being worn, that they cannot be far removed from their original veins.

From the manner in which gold is depofited in fands, &c. it will often appear to be accumulated in particular foils and fands into much richer collections than what can be found in the veins themfelves from which it defcended. The depofits of this kind in fome of the lately explored diftricts of Spanifh America, as they have been defcribed by the vifitors, exceed all belief. According to fome of them, a man can gather to the value of 200l. in a day, by wafhing the common foil in fome of the valleys. The foil in which it is found muft be confidered as the relics of the rubbifh of mountains, depofited upon mountains which have been demolifhed and wafhed away. One of the ftrong proofs of the great antiquity of this globe is the confiderable quantity of gold found in the foils and fands of fome diftricts. It is a clear proof of an operation or procefs which muft have required a length of time that is far too great for our comprehenfion.

The fize of the gold grains is extremely various. The greateft part of them are very fmall; fome as large as the feeds of apples, and fome much larger. The grains of the Irifh gold are uncommonly large. Reaumur reports that a piece was fhewn to the French Academy, faid to weigh 56 marks, or 448 ounces. Feuillée fays he faw one in the cabinet of Antonio Portocarrero which weighed upwards of 66 marks. Both pieces were affayed and found of different fineneſs in different parts of the maſs. One was of $23\frac{1}{2}$ carats, 23, 22. The other of 22, 21, $17\frac{1}{2}$. It is, however, rare to find pieces one ounce in weight. The largeſt in the Britiſh Muſeum is only 15 pennyweights. In Chili 5000 pound weight of the richeſt ore yields only 12 ounces, which is not quite one grain per pound. Ore is wrought there without loſs, if it contains one-ſixth of this. On the coaſt of Guinea, a man can gain ſeven ſhillings a day by waſhing the common ſands on the river banks. From half a ton of the richeſt part of the ſoil at Lead Hills, waſhed under my inſpection, the produce of gold was $11\frac{1}{2}$ grains.

In this manner, therefore, is gold ſeparated from its original matrix, and collected in certain places by nature. It remains to deſcribe how man provides this metal for himſelf, by imitating more or leſs, or by completing this operation. This deſcription will be extremely ſhort, becauſe the account already given of the peculiar properties of this metal, and of the ſtate of its ores, and of the operations in metallurgy, requires nothing but general indications of what is to be done.

The firſt operation is an imitation of nature. The native gold being in ſmall grains, either among the ſands and gravel, or perhaps bedded in the ſtony matters, the whole is pounded by mills, and then the ſand ſo formed is agitated with much water in buckets or baſkets or ciſterns, by ſtirring it with rakes; and by a ſlight of hand, acquired by practice, the lighter ſand, occupying the upper part of the water, is daſhed over the brim. The gold particles remain with the heavier ſand. This is alſo daſhed over, after ſtirring; but it falls on a long ſloping table covered with rug. The ſtream carries the lighter particles over all the rug, and off the table. The heavier matter ſticks in the rug; and a few grains of gold are found towards the upper end of it. Theſe are ſhaken or waſhed off from time to time, and added to the heavieft particles which remained in the bafkets or cifterns. Thus all the grains are at laft collected; but along with them much bafer metals, or their ores, which are alfo very heavy.

The next operation is roafting this duft, to drive off fulphurous and arfenical matter which is combined with the bafe metals.

The whole is now triturated in tubs with a quantity of mercury, which diffolves the gold and filver, and forms an amalgam. This operation is expedited by adding a quantity of water and of common falt. The water facilitates the labour, and alfo affifts the mercury in throwing out the bafe metals. The common falt decompofes the vitriolic falts which were produced by the roafting and burning of the fulphur.

In the next place, the greateft part of the mercury is feparated, nearly pure, by fqueezing the whole thin pafte through porous leather. The firm mafs which remains contains all the gold and filver, and a good deal of mercury. The laft is got off by diftillation.

The laft operation is the feparation of the filver from the gold, and the refinement of both, by fuch of the proceffes already defcribed as is fuited to the proportion of the two metals, and the impurities with which they may be tainted.

And when the gold is found combined with the ores of other metals, that is, when ores of other metals are treated for the gold which they contain, proceffes muft be employed for fcorifying thefe metals, in one way or another, whether by fulphurating them, or by lead, till the gold and filver are left alone, to be treated as now mentioned. For a more particular account I muft refer you to Agricola, Schlutter, and Born.

GENUS XV.—PLATINA, OR PLATINUM.

THIS metal was firft brought to England in the year 1749 or 1750, by Dr. Brownrigg, who prefented it to the Royal Society.

He had got it from Mr. Charles Wood, affay mafter in Jamaica, who told him it came from fome of the Spanifh provinces in America, and that it had feveral of the qualities of gold, in confequence of which it was very difficult to feparate it from that metal; and that the name given to it by the Spaniards was *platina*, or *platina del Pinto*, and *Juan blanco*. Platina is the diminutive of *Plata*, filver; and Pinto is fuppofed to be the name of a mountain or river or perfon.

Soon after it was known in England, fpecimens of it were brought to other parts of Europe; and the attention of many chemifts was engaged by it. They made many experiments, and wrote differtations on this metal; of which we have examples in thofe publifhed by Mr. Lewis in the Philofophical Tranfactions for the years 1754 and 1755, and in his *Commercium Philofophico-technicum*; Mr. Scheffer's paper in the Tranfactions of the Swedifh Academy for the year 1752; Mr. Margraaf's differtation in the Memoirs of the Royal Academy of Berlin for the year 1757, printed in the 1759; and a memoir by Macquer and Beaumé, in the Memoirs of the Royal Academy of Sciences at Paris for the year 1758, printed in the year 1763; and many more fince that time.

Platinum, as brought from America, in loofe grains, of the fize of coarfe fand, moft of them fmooth on their edges and fides, having the metallic opacity and luftre, and a dingy white colour, not brighter than that of iron; and they are all attracted, more or lefs, by the magnet. But, intermixed with thefe grains of the platinum, there are generally others of a fmaller fize, which are plainly iron ore, or iron fand. In fome fpecimens alfo, a few fmall grains of gold are found; and Mr. Lewis alfo obferved a little quickfilver adhering to the gold. Mr. Lewis therefore conjectured that the platinum is got in wafhing fome foils or fands for gold.

By the numerous experiments that have been made with this metal, it is now afcertained that in thefe grains it is intimately and ftrongly united with about one-third of its weight of iron, and it is not eafily refined. When rendered free from all admixture, it has a whiter colour, and is the heavieft metallic fubftance known at prefent. Refined gold is little more than nineteen times as heavy as water: But refined platinum is twenty-two times the weight of that fluid. Even in its natural ftate, the fpecific gravity of it is very uncommon. It is fixteen or feventeen, or when the picked grains are tried, eighteen times as heavy as water.

When refined, it is alfo very malleable and ductile; although in its natural impure ftate it has but a very fmall degree of malleability.

One of the moft extraordinary qualities of this metal is its refiftance to the action of heat, when applied with the purpofe of melting it, without the addition of other metals. Macquer and Beaumé kept it in the greateft heat of a porcelain kiln, and in a glafs-houfe furnace, without making any change on it. The fufion of it was at laft accomplifhed, however, by Mr. Macquer, by employing the intenfe heat of Vilette's fpeculum, and the great lenfe belonging to the Academy, both of which exceed the beft furnaces, in the intenfity of their heat. Mr. Lavoifier alfo fucceeded with the blow-pipe, and with other fuel, by employing oxygen gas.

It poffeffes another fingular property, that of compacting or welding, by the blows of a hammer, in the fame manner as iron; and thus it may be brought from the fpongy ftate in which it is left by the proceffes for refining, to the utmoft denfity and compactnefs.

In the further profecution of thefe experiments with heat, and in thofe that were made with folvents, platina was found to have the fame degree of power to retain its metallic form and purity, and the fame difpofition to recover them when loft, that filver and gold have.

It fuffers no change from the action of air and heat; nor is it affected by nitre in the fire. Crude platinum, or the grains in their natural ftate, are corroded and calcined by melted nitre; but it is evidently the iron which they contain that is attacked by the nitre, and, when the platinum is once completely refined, the nitre has no power to calcine it.   This metal alfo refifts the calcining powers of heat and air, in cupellation with lead.   It not only remains unchanged, like filver and gold, while the lead is changed into litharge, but defends a portion of the lead from being calcined equal to about one-fifth of its own weight.

Thus, by its refiftance to the action of oxygen, it bears a refemblance to filver and gold.   But, upon the whole, it agrees more with gold than with filver.

It is fimilar to gold by refifting the action of pure fulphur, and confequently of the fulphuret of antimony, which can therefore be employed to affift in refining it from iron; but the metallic part of the antimony adheres to it afterwards very ftrongly.

But, though the platinum cannot be united with pure fulphur, it can, like gold, be diffolved by the melted fulphuret of potafh, though not fo eafily.

None of the acids, in their ordinary ftate, act on platina, in whatever manner they are applied, whether in their common form, or that of hot and dry vapours, by cementation.   Like gold, it can only be diffolved by oxygenated muriatic acid, or by aqua regia; commonly by aqua regia compofed of equal parts of aquafortis and muriatic acid. Such an aqua regia will diffolve about one-twelfth, or one-fixteenth of its weight by digeftion, or one-eighth by cohobation, according to Lewis.   It has little caufticity, and it ftains the fkin an indelible brown.

When this folution is evaporated for cryftallization, it affords fmall cryftals like grains of fand, of a deep yellow or red colour, and fometimes opaque.   When they are wafhed and dried, they require more boiling water to diffolve them than five-hundred times their weight. The folution is yellow, and depofits a pale-coloured and very light fediment, fuppofed by Bergmann to be iron.

The folution of platina fhews alfo very particular properties, when we add to it alkaline falts, in order to precipitate the platina. Profeffor Bergmann made many accurate experiments with it in this way, and has been much more fuccefsful than Mr. Lewis in difcovering the manner in which alkalis act on the folution of platinum. You will fee his account of thefe experiments in his differtation on this metal.

I fhall at prefent mention only the moft remarkable properties by which this folution differs from that of gold.

1/t, One remarkable difference is obferved in the effect of fal ammoniac on it. The folution of fal ammoniac, or muriat of ammonia, in water, produces no perceptible effect on the folution of gold. But when a faturated folution of this falt is added to the folution of platinum, an orange-coloured precipitate is inftantly formed, which experiments have fhewn to be the platinum, or a part of it, combined with a fmall portion of the ammoniacal muriat. This precipitate is ftill diffolvable, though with difficulty, in a large quantity of water.

2d, We find a difference between the folution of platinum, and that of gold, when we mix them with alcohol, æther, or the aromatic oils. Thefe fluids make the gold feparate fooner or later from the acid in its bright metallic form. But when they are added to the folution of platinum, no fuch effect is ever produced.

3d, This folution alfo differs from the folution of gold, by the manner in which it is affected by other metals, or their folutions, particularly by the folution of the fulphat of iron, and the folution of tin. Both of thefe precipitate gold, and the folution of tin, in particular, precipitates in the form of the fine purple for enamels. But neither has any effect upon the folution of platinum. Thefe differences are the foundation of the methods by which thefe two metals can be completely and exactly feparated from one another.

One of the points to which Mr. Lewis was moft attentive in ftudying the natu e of this metal was, to learn the confequences of mixing it with other metallic fubftances. As it was not fufible alone, the only manner in which it could be applied to ufe, appeared to be by mixing it with other metals, which might have the power to diffolve it in the fire, and to unite with it, fo as to form ufeful mixtures. But he got no products that promife to be ufeful, except with copper, a mixture of moderate toughnefs, extremely hard, and which does not tarnifh; and with iron, a mixture of extraordinary ftrength and hardnefs.

Such mixtures, however, afford us the only means of manufacturing it. They render it fufible, and thus it may be moulded. The other metals may then be abftracted, by various proceffes, leaving it pure, but porous or fpongy. In this ftate it may be compacted and rendered perfectly folid, by forging, and thus manufactured into any fhape. By fuch proceffes platinum becomes one of the moft valuable metals that we know.

It unites readily with bifmuth,—moft eafily of all with tin, flowing very thin; alfo with lead; and forms a metal which will not fcorify, nor be abforbed by the cupel.

Dr. Lewis's principal purpofe, by thefe mixtures, was to learn the confequences of mixing platinum with gold, in different proportions.

The reafon of his having this object in view was, that the platinum was confidered as capable of being mixed with gold, fo as to commit great frauds in traffic, without the poffibility of detection by the common trials of the purity of gold, or of the quantity and nature of its alloy.

He therefore mixed it by fufion with gold, in many different proportions. But, in general, it debafed the colour and the malleability of the gold fo much, even when it was only one twenty-fourth part of it, that he concluded that no great frauds could ever be committed by employing the grains of platinum for the adulteration of gold. And methods are now well known, by which we can eafily refine gold that has been adulterated with platinum. We can eafily therefore detect fuch frauds, and learn the amount of them. The beſt of theſe methods is the one invented by Mr. Scheffer of Sweden. The adulterated gold muſt be diſſolved with its uſual ſolvent, the nitro-muriatic: And we muſt then add a ſolution of the ſulphat of iron. Every atom of the gold is quickly precipitated, while every atom of the platinum remains diſſolved. We muſt then wait until the gold has completely ſubſided; and afterwards, pouring off the liquor, we muſt waſh the precipitated gold with repeated waters, and great care to avoid loſs. And laſtly, being dried, it is eaſily melted into a maſs with a little borax and nitre.

It may be remarked, however, with reſpect to Mr. Lewis's experiments on the mixtures of platinum with gold, that they are imperfect on this account. He made uſe of the platinum in its native ſtate, in which ſtate it is always mixed with iron; and the gold which he alloyed with platinum was therefore alloyed with iron at the ſame time. Later experiments have ſhewn that the quantity of iron in native platinum amounts to one-third of its weight: And theſe metals are ſo ſtrongly and cloſely combined together, that it is only of late that methods have been diſcovered for ſeparating the iron.

The moſt ſucceſsful methods that I have heard of are theſe:

One way is, to diſſolve the grains in the nitro-muriatic ſolvent, and precipitate the iron from the ſolution, with Pruſſian alkali. The pruſſiat of ſoda is the beſt for this purpoſe. This is a moſt effectual precipitant of iron, and produces no effect on the diſſolved platina, provided no more be uſed than enough to precipitate the iron. The ſolution of platina, thus freed from the iron, may next be evaporated to dryneſs; and the dry matter, expoſed to a violent heat, gives the platina pure.

A ſecond way is, after diſſolving the grains of platina, to precipitate the ſolution with a ſolution of ſal ammoniac, and then expoſe the orange-coloured precipitate to a violent heat, by which the ſalts that adhere to it are evaporated, and along with them any ſmall portion of the iron which may adhere to the precipitate.

A third method which has been practised with success, and is the cheapest, is to refine it with arsenic. Platina, white arsenic, (or rather the arseniat of potash) and tartar, in equal quantities, are melted by a violent heat, and form a very brittle white mixture, of metallic lustre. This is exposed to heat and air under a muffle, which dissipate the arsenic and the iron in vapour, leaving the platina pure, to be compacted by the hammer.

Fourthly, Mr. Macquer found that it could be refined with lead by cupellation. But a most intense and long continued heat is required to evaporate the last portion of the lead from it. This method can only be applied in the refinement of a very small quantity of it. From a larger quantity it would be impossible to evaporate the last of the lead. (See Lewis's Commercium Philosophico-technicum.)

And lastly, some of Mr. Pelletier's experiments with phosphorus point out another method of refining it, which he says is easily practicable. He heated the grains of platinum in a crucible, and then threw in some bits of phosphorus. The phosphorus instantly penetrated the grains, and formed a fusible metallic compound,—a phosphoret of platinum. If this compound be exposed to a strong heat for some time, the phosphorus is gradually burnt, and with it the iron, which forms a scoria or slag with the melted acid of the phosphorus. The platinum is thus purified, but is no longer fusible; and it remains very spongy, with the scoria adhering to it, partly on its surface, and partly in its pores. He therefore gives it a strong heat, and suddenly compresses it in that state with the blows of a very heavy hammer, such as is employed in the process of refining iron. This effectually compacts the platinum, forces the scoria out of its pores, and makes the parts of it unite together *.

This method of compacting platinum, and uniting the parts of it by

---

* Would not cementation with charcoal and bone ashes, or the stone mentioned in page 268, accomplish this purpose, without the expence of preparing the phosphorus, —and thus afford a very easy method of manufacturing this useful metal?

EDITOR.

percuffion, when ftrongly heated, was firft fuggefted by Mr. Beaumé ; and it muft be employed in every cafe in which this metal is refined. We cannot unite the parts of it by fufion with any heat that furnaces can give.

When by thefe proceffes platina has been purified from all admixture, it has the colour of pure tin, is very malleable, growing rigid by the hammering, but may be foftened by annealing. In this ftate, its denfity or fpecific gravity is from 20 to 22.

It is now to be hoped that this metal will be imported more freely into Europe, and that the mining grounds in America in which it is found will be opened again. The Spanifh government at firft prohibited the collection and importation of it, from apprehenfions that it would embarrafs their commerce in gold, by giving opportunity for frauds not eafily detected. But there is no reafon now for fuch apprehenfions and precautions, fince methods have been difcovered for detecting the fmalleft quantity of platinum mixed with gold, and for eafily feparating thefe metals from one another. And were the properties of platinum more fully inveftigated, methods for working it eafily into utenfils might be difcovered, and then it might juftly be confidered as one of the moft ufeful of the metallic fubftances. Its refiftance to the action of fire, together with its incorruptibility and cleanlinefs, would render it valuable for domeftic purpofes. The chemifts already employ it on account of thefe qualities in fome of their nice experiments and proceffes. *(See Note* 32. *at the end of the Volume.)*

CLASS V.

─────────

OF WATERS.

IN former parts of this courfe, we have had already opportunities to notice the important and numerous ufes of water in the fyftem of the univerfe.

It is faid of fome of the ancient philofophers, that they fuppofed all things derived their origin from water. And it is true that the exiftence of all the numerous tribes of vegetables and animals depends upon it. And when we examine the materials of which they are compofed, water is always the principal part. But muft we alfo fuppofe that water has given origin to the more folid and durable parts of this globe, the various earths and ftones, the rocks and minerals of which the mountains and land are compofed? Have all thefe been produced from water?

Whatever opinion we may form of the origin of thefe, it is at leaft certain, that water has been the agent employed to give the regular arrangement in which many of them are found.

The appearance of numerous relics of fea productions, and other traces of this fluid, are fo frequent in the ftrata of which this globe, fo far as we can obferve, is principally compofed, that there is no room to doubt that the materials of thefe ftrata have been arranged in that manner by the fea. But fome have concluded from thefe phenomena, and fome other facts, that water has even furnifhed the matter of the earthy and ftony bodies,—that water is actually convertible into earth.

The firft author who has mentioned the convertibility of water into earth is Mr. Boyle. He writes, that a friend of his was in poffeffion of three-fourths of an ounce of earth, produced from one ounce of water by the action of a long continued heat. And he relates fome experiments made by himfelf, which convinced him that earth could be produced from the pureft water by heat.

I repeated Mr. Boyle's experiment, by boiling and circulating boiling water a long time in a glafs veffel luted up quite clofe, and I faw the appearance by which he was deceived.

Godfrey, the defcendant of Mr. Boyle's affiftant, thought he converted water into earth by triture. Wallerius alfo relates, in the Swedifh tranfactions, experiments in which, by triture, or agitation, or both, a quantity of filiceous matter was obtained from water. And Mr. Margraaf obtained it by diftillation.

Lavoifier's opinion is, that the earthy matter got by boiling of water in glafs veffels, is a part of the glafs corroded by muriatic acid, prefent in all waters.

My opinion is, that this matter is a part of the glafs corroded or penetrated by the water itfelf, which alone has the power to penetrate it, and even to diffolve it, by length of time and affiftance of heat. This explains the form and appearance of this matter, which is always in fine minute and thin fcales, and alfo the change obferved in the water, when the experiment has been continued a long time. It is faid to boil more like oil than like water. This proceeds from its holding a fmall portion of the glafs, or of the alkali of the glafs, perfectly diffolved. By my obfervations on the Geyfer fprings of Iceland I found that water held filiceous earth in folution.

Another opinion however has been lately formed with regard to it, which is of greater confequence, as it enables us to underftand and explain a number of chemical facts. The opinion I mean is that which we have already adopted and applied in former parts of this courſe ; that water is not a ſimple elementary ſubſtance, but a compound.

This idea of the nature of water was ſuggeſted by Mr. Watt. *(Phil. Tranſ.* 1784.*)* Mr. Cavendiſh, however, was the firſt who gave it ſolid foundation and credibility, by his accurate examination of the conſequences of ſetting fire to a mixture of hydrogen gas and oxygen gas, in a veſſel in which they were confined. Mr. Lavoiſier and other French chemiſts, ſeeing the vaſt importance of this experiment in their ſyſtem, immediately repeated it, with much larger quantities of the materials, and with an excellent apparatus and the moſt ſcrupulous accuracy. They added ſeveral other experiments, which concur to eſtabliſh the opinion of the compounded nature of water. They alſo ſoon perceived and pointed out the important improvements which might be deduced from this diſcovery in the explication of many chemical facts, which were thus ſet in the cleareſt light. And ſince that time, the laſt finiſhing, we may ſay, has been given to this happy chain of experiments ; and the compounded nature of water has been ſtill further illuſtrated, by the capital experiment of the Dutch ſociety of chemical philoſophers, who reſolved water into its two conſtituent parts, without making any addition to it whatever, but the matter of heat, collected and condenſed by electrical operations *.

All this, however, has been already noticed in the preceding parts of this courſe. At preſent we muſt confine ourſelves to the examination of the variety of waters found in nature, and to an account of the means which chemiſtry furniſhes for inveſtigating their qualities and contents.

* I obſerve ſome authors complimenting the ſagacity of Sir Iſaac Newton, by ſaying that he conjectured from its great refracting powers, that it contained an inflammable ſubſtance. I have never met with this aſſertion by Sir Iſaac Newton ; and it is ſurely a miſtake,—for the fact is, that water has the ſmalleſt refracting power of any liquid that we are acquainted with. Sir Iſaac indeed ſays, that the great refracting power of the *diamond,* makes him imagine that it contains inflammable matter. EDITOR.

The natural varieties of waters are produced by the union of various other fubſtances, in different proportions, commonly very minute, with this fluid.   And it is never found perfectly pure from thefe, though in fome cafes it is very nearly fo.

When we have occafion for water quite pure, to be employed in nice chemical operations, we muſt procure it by art.   We muſt choofe good fpring water, or rain, or fnow-water, collected at a diſtance from the fmoke of many houfes, and diſtil it until we have got two-thirds, or three-fourths, and reject the remainder.

The varieties of water found in nature may be all comprehended under the fix denominations of

   1. Rain water.

   2. Fountain water, and well water.

   3. River water.

   4. Water of lakes.

   5. Water of marfhes and fhallow pools.

   6. Sea water.

Of thefe natural waters, rain water, when properly collected, comes neareſt in purity to diſtilled water.   It is water diſtilled by nature; and would be quite pure, but for the vapours of the atmofphere.   To have it in the greateſt degree of purity of which it is capable, it muſt be collected at a diſtance from the fmoke of houfes, or other caufes which occafion vapours or effluvia to mix with the air.   The rain water which runs from roofs, however, is very impure and footy. Sometimes too it has happened, that fhowers of rain have fallen mixed with the ſtaminal duſt of plants, which has been miſtaken for fulphur.

The waters of fprings and wells confiſt of rain water, which has foaked through crevices and the more porous parts of the earth, until it has been ſtopped in its progrefs downwards, by impenetrable ſtrata of clay or other matter, and breaks out again in fome lower part of the earth's furface.

The opinion which was once formed by some philofophers concerning the origin of fprings, is very abfurd,—that they are fed by the fea; that the fea water filtrates through the pores of the earth, and afcends through the interior parts of the high lands and mountains; that the falt is feparated from it during this filtration, and that the water breaks out at laft to form the fprings. It is abfurd to fuppofe that the water can afcend in this manner, contrary to its gravity, or that the falt can be feparated by filtration. It is plain that rain water falls moft plentifully upon the mountains and high lands; and as a great part of it is abforbed, and penetrates into the porous parts of them, it is abundantly fufficient to account for the origin of fprings. And we have a proof from the experience of miners: The deeper they go they meet with lefs water.

Spring water muft therefore neceffarily be lefs pure than rain water, as the water may find various fubftances in its way through the earth, which it will diffolve and partake of. Some fprings, however, give a water remarkably pure. Such are thofe which are filtrated through fandy and gravelly foils, or through mountainous countries, compofed of harder and more undiffolvable ftones. Others contain a variety of matters; and thefe fometimes in confiderable quantity, and of fuch a nature as to give them medicinal efficacy. For this reafon, after we fhall have fhortly confidered the other more general varieties of natural waters, we fhall return to the varieties of fpring waters, and explain them more particularly, with a view to point out the proper method for examining mineral fprings, and for difcovering the principles they contain.

The water of rivers is compofed entirely of the waters of fprings, and of fuch rain water as runs along the furface of the ground, without finking into it. It is therefore various in different places. Thus the water of rivers which pafs through very large cities, as the Thames does, becomes fo loaded with animal and vegetable matter, as to be fufceptible of a high degree of corruption. In general, however, the water of thofe large rivers which are not expofed to fuch caufes of impurity, though often troubled with atoms of mud and clay, when refined from thefe by reft or filtration, are rather more pure and wholefome than the water of the greater number of fprings or wells. The reafon feems to be, that they are compofed partly of rain water, which has not penetrated the foil, partly of the water of fprings, which, while it runs a confiderable way along the furface to join great rivers, becomes generally purer. Some of the principles it contains are volatile; a greater number are diffolved in the water by means of thofe volatile ones, and are depofited when they evaporate. Whether the purity of river water depends on thefe circumftances or not, it is certainly very remarkable in fome cafes.

The water of large lakes, having the fame origin with river water, is equally pure, except in a few examples of lakes fituated in countries which abound with falt; and is in general much more tranfparent, in confequence of its ftagnation, which allows all the muddy particles to feparate to the bottom. But fmall pools and marfhes, and other fhallow collections of ftagnating water, are in general very impure and unwholefome. The heat of the fun has a ftrong effect upon thefe, and occafions quick and multiplied fucceffions of animal and vegetable productions in them. The vulgar have been aftonifhed and terrified at the fight of a red colour which fuch waters fometimes affume, as if fuddenly changed into blood. This arifes from infects. The putrid damps which arife from fuch waters in hot climates, or rather from the marfhes which they leave when they dry up, are caufes of numerous difeafes, agues, dyfenteries, &c.

The laft variety of water which we enumerated, fea water, is ftill farther removed from the ftate of purity than any of the former. Near the fhores efpecially, it is not only rendered impure by the muddy particles brought into it by the rivers, and thofe ftirred up and kept afloat by the waves; but as it contains innumerable animals, thefe muft contribute, by their different excretions, to its impurity.

The admixture, however, by which it is the moſt remarkably diſtinguiſhed from pure water, is that of different ſaline ſubſtances, which
it contains in very conſiderable quantity, in conſequence of which it
has a ſtrong taſte, and is totally unfit for the uſe of man, and other
land animals.

The nature of theſe ſalts, and ſaline compounds, which ſea water
contains, has been inveſtigated by different chemiſts.

Firſt, Dr. Gaubius of Leyden, who made his experiments upon 50
pound of the ſea water, and has publiſhed them in his *Adverſaria.*—
From 50 pound of 16 ounces Troy weight of ſea water, Gaubius got,

					oz.	dr.	gr.
Common ſalt, ſome of it impure	-		-		20	4	1
Selenite, and what he calls *alumen muriaticum*		-			1	1	0
Sal dictum Glauberi	-		-	-	2	4	0

Bergmann examined ſea water, taken up for him by Mr. Sparmann
from a depth of 60 fathoms, in the latitude of the Canary Iſlands.
He obtained from it,

				oz.	dr.	gr.
Common ſalt	-	-	-	37	5	26¼
Magneſia muriata	-	-	-	10	2	14
Gypſum	-	-	-	1	1	44⅘

He examined it very ſcrupulouſly for magneſia vitriolata, but found
none.

Dr. Higgins got from ſea water of our climate,

				oz.	dr.	gr.
Common ſalt	-	-	-	23	3	36
Magneſia muriata	-	-	-	10	1	12
Gypſum	-	-	-	1	1	44

To imitate ſea water, ſuppoſe 50 pound of it to contain 25 ounces
of common ſalt, and 10 ounces of muriat of magneſia. Each gallon
Engliſh will contain four ounces of common ſalt, and 13 drachms of
muriat of magneſia, which 13 drachms of muriated magneſia will
probably be contained in 26 drachms of what is called oil of ſalt.

Dr. Higgins remarks that the magnefia falita appears to be the moft efficacious ingredient in fea water. The proportion of it to the fea falt is as eight to 18½. And as perfons are cured of difeafes by fea water, who eat fea falt every day of their lives with their food, there is no reafon to impute their recovery to the fea falt.

It is ufual, therefore, in the practice of medicine, to imitate fea water with a mixture of common falt and purgative falts diffolved together; and the folution may be made fo ftrong that two fpoonfuls, or a little more, may prove a dofe. But to refemble fea water, it fhould always be greatly diluted when taken.

Sea water proves a ufeful medicine in fome cafes, though taken in fmaller quantities than what prove purgative. It is thought, when taken in this manner, to have fome efficacy in removing obftructions and tumours of the lymphatic glands. And fome have imputed thefe effects to its feptic power, whereby it diffolves concretions of the fluids. It contains lefs than Sir John Pringle's feptic quantity of common falt. But its medical efficacy does not appear to me in this light. Putridity weakens the power of the vital or animal principle, and increafes all difeafes. I fhould rather impute it to a general ftimulus, by which it promotes all the fecretory motions in the body. It is certainly in fome meafure by a ftimulus exerted upon the fkin, and by which it promotes perfpiration, that bathing in fea water has been fo beneficial in thefe and other difeafes.

Such is, therefore, the general outline of the variety of natural waters. To return now to the waters of fprings.—

In order to give you a general view of the variety which occurs among fpring waters, I muft, in the firft place, enumerate the different fubftances which have been found in them by chemical experiments, and mention fome remarkable qualities which fome of them are known to poffefs.

Thus, 1ft, Some are warmer than the medium heat of the latitude in which they are found.

The water of common fprings always has at its fource the medium temperature of the climate in which it is found. In this country, this medium temperature is 48°; in London 52°; about Paris 55°; farther fouth, higher. But fprings are found in many places, the heat of which is confiderably above thefe medium temperatures; and there is great variety in refpect of this fuperior heat. Some are boiling hot, others fcalding hot, others only tepid, and fo on. The moft furprifing particular in the hiftory of thefe fprings is the amazing quantities of hot water which fome of them throw out, without the leaft apparent variation of this quantity, or of the heat; a certain proof that they receive their fupply of water and heat from a great depth. Such are the fprings at Bath.

The caufe of this heat is a fubject of many opinions. There are, doubtlefs, chemical changes going on in many watery folutions in the bowels of the earth, in which heat is extricated. But I cannot conceive any fuch, where the refult is pure water, fuch as flows from the Briftol hot well. In fuch cafes, the caufe of the heat is certainly fubterranean fire. No other caufe is adequate to the effect; and we know that thefe fires exift, and have exifted at all times. In agreement with this opinion, hot fprings are moft frequent in volcanic countries. There are many in Italy and in Iceland. Geyfer, in the neighbourhood of Hecla, in Iceland, emits an immenfe body of water by feveral openings. Two of them are very remarkable. The one, called the old Geyfer, throws out water by a perpendicular pipe, nine feet in diameter; and it fometimes throws the water up in the air 200 feet. Sir Jofeph Banks, &c. faw it rife about 90. Alfo Mr. Stanley. (*Vide vol.* ii. *of Edin. Tranf.*) Some hot fprings are at a great diftance from volcanoes. But how is fubterranean fire fupplied with air? I may anfwer, that water carries air with it to thefe fires, and is converted into air by chemical operations; fome of which are known to us, while others are equally certain, although our chemical knowledge cannot yet explain them.

To proceed. Befide this heat, found in fome fprings, we find alfo in many, as was formerly obferved,

2*do*, Carbonic acid, the quantity of which is very various. Some are fo much faturated with it, that it comes out of them as from fome fermented liquors. It is generally attended by other admixtures.

3*tio*, The foffil alkali is not uncommon, fometimes combined in part with carbonic acid, but much more frequently with the muriatic. Some authors alfo mention the volatile alkali as having occurred. I believe, however, that if this ever happen, it is exceedingly rare.

4*to*, The fulphuric, or the muriatic acid, or both together, are very often prefent in thefe waters; but they are always combined with the foffil alkali, or with fome alkaline earth, or with a metallic fub-ftance.

5*to*, The only alkaline earths hitherto difcovered in fpring waters are the calcareous and magnefia. They are either in the form of car-bonats, that is, diffolved in the water by means of the carbonic acid, or they are combined with the fulphuric or muriatic acids.

6*to*, The argillaceous earth is rarely found diffolved in thefe waters, but when it occurs, it is always combined with the fulphuric acid.

7*mo*, Geyfer is the only example known to me in which the flinty earth is found in fpring waters. It is diffolved by means of fixed alkali.

8*vo*, Of the inflammable fubftances, we fometimes find fulphur, or fulphurous gafes, and more rarely, fome of the foffil oils. Thefe laft appear moftly in drops upon the furface of the water. The fulphur, when it occurs, is more intimately combined or diffolved, and this by means of fixed alkali, or of hydrogen, fo as to form inflammable gas.

9*no*, Some of the metallic fubftances are the laft which we need to enumerate among the contents of fpring waters. And no more than two have, in my opinion, been certainly demonftrated in their com-pofition,—iron and copper.

Iron occurs very often, and is commonly diffolved in the water by the fulphuric acid, and fometimes, as is fuppofed, by fixed air. Copper but rarely appears, and is always diffolved by the fulphuric acid.

Some authors mention alfo zinc and arfenic as occurring fometimes, but I have not met with proofs of this. The ores of zinc are not liable to that vitriolization by the action of air, to which thofe of iron and copper are often difpofed, and which converts them into compounds of metals and fulphuric acid. And as for arfenic, though of itfelf foluble in water, it does not diffolve without boiling heat. Were it ever to make its appearance in waters, the prefence of it in their compofition would be but too manifeft by their pernicious effects.

Thefe are the different fubftances which have been plainly demonftrated in different fpring waters. And they are found in the different fprings in great varieties of number and proportion, only that in general the proportion of them to the water is very fmall. Thofe that predominate in any particular fpring, give to the water particular qualities and a particular character, in confequence of which, the fprings reckoned medicinal have been diftinguifhed by phyficians into a number of different kinds, and by different denominations, which I fhall next explain.

1ft, Thofe that are warmer than the middle temperature of the place where they are found, called hot or tepid fprings.—*Thermæ*.

2d, Thofe in which carbonic acid predominates, fo as to give them a fourifh tafte, and brifknefs.—*Acidulæ*.

3d, Thofe that contain an alkaline falt not completely neutralized with acid.—*Foffil alkaline waters*.

4th, Thofe which contain fuch a quantity or variety of faline matters, as render them purgative.—*Purging fprings*.

5th, *Chalybeate waters*.

6th, *Sulphurous waters*.

Thefe are the general diftinctions of medicinal fprings among phyficians, who thus difcriminate them by fome of their fenfible qualities or medicinal powers.

But fome other qualities have been obferved in fpring or well waters, which have occafioned other diftinctions and denominations of them among common obfervers. Thus,

7th, Some are called *hard*, or fprings or wells of *hard water*. Thefe do not diffolve foap well, but produce a greafinefs on the furface; and they are ill qualified for boiling vegetables, and for penetrating and diffolving their foluble parts. The infufions of tea, or other vegetables, in thefe are weak and bad. They contain a compound of fome foffil acid with other matter, by which it is not fo completely neutralized as to prevent its acting on the alkali of foap.

8th, Some are called *petrifying waters*. They contain calcareous earth diffolved by carbonic acid.

9th, Some are called *falt fprings*. They yield common falt with profit.

10th, *Coppery fprings* contain copper feparable by iron.

11th, *Bituminous fprings*, are fo called,—drops of bitumen being commonly found floating on their furface.

And now I have faid enough to enable you to form fome general notions of the variety of fpring waters which occur in nature. We fhall next attend to the manner of examining them, in order to learn which of thefe fubftances now enumerated, or how many of them, they contain.

In examining mineral fprings, 1mo, Obferve the fituation of the fpring, and nature of the foil and country around it, and whether any minerals are near it, or within the diftance of a few miles. They often receive fome of their contents from a mineral vein.

2do, Obferve its heat with a good thermometer, and alfo the colour, tafte, fmell, and other fenfible and obvious qualities of the water. Deep glaffes are the beft for fhewing colour or clearnefs.

3tio, In order to the examination of the more fixed ingredients, evaporate one or more gallons of it with a gentle heat to drynefs.

This evaporation may be performed, at last, in a small China bowl, that the extract of the water may be the better collected together, and the more easily taken out ; and the bowl should be heated only with the steam of water. The quantity of the dry extract can be exactly ascertained by weighing the bowl before and after the evaporation ; and the different ingredients of it are to be investigated by an analysis.

We must next investigate the materials which compose this dry extract, by the application of different solvents and other agents, such as 1/t, Alcohol. 2d, Distilled water applied cold. 3d, Distilled water applied hot. 4th, Acids. Alcohol dissolves the compounds of magnesia or calcareous earth with muriatic acid, but leaves untouched all other compounds. A moderate quantity of cold water dissolves all the other saline compounds, except gypsum ; and gypsum is dissolved and separated from the mere earthy matter, by 500 times its weight of boiling water, or that quantity of distilled water boiled with it some time.

These different solutions may afterwards be evaporated for crystallization, or we may easily learn, by other trials to be mentioned presently, what acid and what quantity of it they contain, and with what matter this acid is joined.

4to, A set of experiments are to be made on the water itself in its entire state : And they are made by adding different chemical agents, which produce remarkable effects on the various substances which the water contains, or are affected themselves remarkably by them, to such a degree that we are enabled in this manner to detect the most minute quantity of these substances. Of these experiments I shall mention some examples, while I describe the manner of investigating the contents of mineral waters.

I.—Experiments to discover the Carbonic Acid.

1. Attend to the obvious qualities which it produces, viz. briskness like that of a fermented liquor, and a pungent sourish taste, both of which are lost by exposure to the air, especially with agitation.

2. It ftrikes a red colour with litmus, which difappears by expofure to the air for fome time.

3. Lime-water, being mixed with fuch aërated water, precipitates the lime, which will be rediffolved if there be much of the air, and if we have not employed enough of the lime-water.

4. To meafure the quantity, a bladder is tied to a bottle, and the air is expelled by heat.

This is the method which has been often practifed, and which may be practifed when we cannot do better, but it is not exact. A part of the air is very liable to efcape at thofe places where the bladder is tied; and fome of it penetrates through the bladder, or is abforbed by the water with which it is foaked. A more exact way is to fit a cork very clofe to the mouth of the bottle, and then, making a hole through the cork, fix in it a bended glafs tube. The bottle being then quite filled with the water, put in the cork, and make the extremity of the bended glafs tube dip into a ciftern of quickfilver, in which ftands inverted a cylindrical veffel, alfo filled with quickfilver. The air expelled from the water by heat, will rife through the tube into this cylinder, and difplace the quickfilver; and thus it is both collected and meafured at the fame time. And in order to learn the quality of it, we can apply cauftic alkali, or lime and water to it, to know how much of it is carbonic acid gas, which will be quickly abforbed; and we can afterwards examine the remainder *.

II.—*Experiments to difcover Alkaline Salts.*

To learn, in the next place, whether or not there is any alkali not completely neutralized with foffil acids, or combined only with the carbonic, we may ufe fome of the vegetable tinctures, or infufions of

---

* We may alfo learn its quantity with confiderable accuracy, by the weight of the precipitate which it occafions from lime-water,—having taken care that an excefs of lime-water has been employed. The air may be eftimated at eleven-twentieths of the dried precipitate. EDITOR.

flowers, that are the moft eafily affected by alkalis. The fyrup of violets is that which is commonly ufed; but I find an infufion of the flowers of mallow more convenient. The colour is produced exactly the fame, whether the alkali be fixed or volatile; but we can eafily diftinguifh which of the two it is, by a fubfequent experiment with the muriat of quickfilver. And if we find it to be a fixed alkali, we are fure it muft be the foffil alkali. The quickfilver, when oxydated, has a remarkably ftrong attraction for volatile alkali,—a fmall portion of which unites with it in this cafe; and forms with it, and with a fmall part of the muriatic acid, a mercurius dulcis, infoluble in water.

Thefe experiments, therefore, give indications of alkaline falts not completely neutralized with foffil acids. But if alkalis be prefent, combined with one or more of thefe into neutral falts, they are beft difcovered and diftinguifhed by examining the extract of the water obtained by evaporation, and the different faline compounds into which it can be analyfed.

III.—*Experiments to difcover Sulphuric Acid.*

This is often prefent, combined with fome other matter. With whatever fubftance it be combined, it can always be detected by the muriat or nitrate of barytes, the barytes having the ftrongeft attraction for the fulphuric acid, with which it forms a compound infoluble in water. To make this experiment quite fatisfactory, however, we muft add a fmall quantity of nitric or muriatic acid, perfectly pure, or free from fulphuric acid; for the barytes may be precipitated by a carbonat of alkali; but if this happen, the fmall addition of pure aquafortis will rediffolve it.

IV.—*Experiments to difcover the Muriatic Acid.*

This acid can as certainly be difcovered by the folution of filver with excefs of acid.

We do not learn, however, by thefe trials, what matter thefe acids are combined with; but this appears, either from some of the other trials we have to mention, or from the examination of the fixed matter obtained by evaporation.

I observed before, that these acids are hardly ever found in the waters in a non-saturated or pure state; though this is said by some authors, who have probably been imposed upon by the taste and qualities of the acidulæ or aëreal waters. But turnsol would detect them.

V.—*Detection of Earths in Mineral Waters.*

Earth, when present, is seldom any other than calcareous earth, or magnesia, or both, combined with some solvent, either carbonic acid, or some of the fossil acids. I have always observed, that in the waters containing those carbonats, there is also a quantity of the earth combined with some other of the fossil acids. The part dissolved by carbonic acid, is discovered by boiling the water in a clean vessel. The acid is expelled, and the earth precipitates.

Bergmann says that the calcareous earth, suspended by aëreal acid, soon separates from the water by moderate boiling: But that magnesia, in the same state, separates gradually, during the whole time that the water is boiled or evaporated to dryness. Water fully impregnated with carbonic acid will dissolve more than one-fortieth of its weight of magnesia, every ounce holding above twelve grains. The tufa formed by the evaporation of such waters, has been generally considered as calcareous; but it always contains magnesia, and sometimes a great deal. Strata of limestone are met with, containing so much that they are unfit for manure, and even pernicious. This is the case in the neighbourhood of Doncaster.

The nature of the earth which separates in this manner is determined by combining it with the sulphuric acid.

There is another method, better fitted for detecting the smallest quantity of earth suspended by means of carbonic acid. This is done by dropping into the water some solution of the acetite of lead. The lead is precipitated, and the earth fufpended by the acetous acid. But as the lead would alfo be precipitated by a fulphat, we muft add fome more acetous acid. This will rediffolve the lead, if it has been feparated by means of an earth. But obferve that faccharum faturni does not fhew a fmall quantity of earth, if there be much carbonic acid in the water.

When the earth prefent in water is diffolved by fome of the foffil acids, it may be feparated by carbonat of fixed alkali, which however is beft added after the water has been evaporated to one-fortieth, or one-fiftieth, or rather after it has been evaporated to drynefs; the foluble parts of the refiduum being rediffolved with diftilled water, and the alkali added to this folution.

The nature of the earth is determined, 1ft, By diftilled vinegar applied cold, which will diffolve calcareous earth, or magnefia, but not argillaceous earth.

2dly, By fulphuric acid, which precipitates the calcareous earth from the vinegar in form of gypfum, but with magnefia forms a foluble compound.

Profeffor Bergmann alfo recommends the oxalic acid, or the acid of fugar, as a means for detecting the fmalleft quantity of the calcareous earth in particular, and in whatever manner it be diffolved. It is fure to unite with this acid, for which it has a much ftronger attraction than for any other, and always forms an infoluble compound. And this is perfectly true. The precipitation, however, is not fo quick, or fo remarkable, as in fome of the experiments already mentioned, efpecially with the acetite of lead.

If there be any filiceous earth diffolved in the water, we fhall be fure to find it in the matter which the water affords when evaporated to drynefs. After we have extracted the faline compounds and falts from that matter by alcohol, and by diftilled water, applied firft cold, in moderate quantity, and afterwards hot, in large quantity, mere earth will remain, which muft be treated firft with diftilled vinegar, which will diffolve calcareous earth or magnefia ; afterwards with fome of the foffil acids, which will diffolve the argillaceous earth, if there be any. But if a quantity of tender, light, and fpongy earth remain, on which the acids produce no more effect, there is reafon to think that it is filiceous, and we can affure ourfelves that it is fuch, by mixing it with an equal weight, or one-half of its weight, of dry carbonat of foda, and heating this mixture ftrongly with the blow-pipe in a platina fpoon. If the earth is filiceous, it will melt into glafs.

VI.—*Detection of Sulphur.*

Having now fhewn the means for difcovering the prefence of falts and of earths, in the compofition of mineral waters, the means for detecting the prefence of fulphur are next to be defcribed.

Sulphur has been found in two ftates in fulphurous waters : 1ſt, Either combined with an alkali, and forming an alkaline fulphuret, (this I believe is exceedingly rare): or, 2dly, in the ftate of a fulphuret of hydrogen, or fulphurous hydrogen gas, which, you know, is foluble in water.

The nature of this compound was firft invwhen combined with water as to form a fulphurous water of this volatile kind.

The indications of a fulphurous mineral water are,

1ſt, Odour of hepar fulphuris, perceived on approaching the fpring. This is a fure and nice indication.

2dly, A common trial is to put a piece of filver into the water, which is quickly tarnifhed and blackened. I find that other metals are fitter, e. g. lead is quickly blackened.

3dly, A mark made with acetite of lead, or tartrite of bifmuth, on paper, is exceedingly effectual and fure. A true hydro-fulphuret, diffolved in water, (that is, a volatile fulphurous water) blackens thefe fubftances very fpeedily, even when brought very near them ; and they do this without suffering any decomposition, indicated by milkiness or turbidness. But a sulphuret of potash or soda dissolved in water, though it blackens these substances as quickly, is always decompounded by it. To learn the quantity of the sulphur,

1mo, If any part of the sulphur is dissolved in the water, and kept suspended by an alkaline substance, a little of the sulphuric or muriatic acid will precipitate the sulphur, which, being afterwards collected on a filtre, can be weighed by itself. But these acids never precipitate the whole of the sulphur from such waters; a part of it being combined with inflammable air, and dissolved in consequence. And many waters have all the sulphur which they contain dissolved in this form. In this case the sulphur cannot be precipitated by diluted sulphuric or muriatic acid.

2do, But we have other acids that can precipitate the sulphur in this case, and these are the strong red nitrous acid, and the sulphurous acid. The first was recommended by Bergmann, and the second by the French chemists. This precipitation is supposed to be produced by the action of the oxygen of the acid on the hydrogen of the gas. These two principles unite together to form water, and the sulphur, when thus left alone, is not soluble in water. But there is a difficulty attending this account of the matter: for, if it be well founded, we cannot find a reason for this inefficacy of the nitric and sulphuric acids to precipitate the sulphur *equally well, or better*, as they contain more oxygen than the nitrous or the sulphurous acid. Moreover, when the oxygen is taken from the sulphurous acid, why is not the sulphur which that acid contains precipitated also, along with that of the water? As this does not happen, Mr. Berthollet and Fourcroy suppose that only a very small portion of oxygen is taken from the sulphurous acid, or supplied by it; and accordingly, when the water is examined after the precipitation of the sulphur, the sulphurous acid is found still existing in it.

3tio, Several metallic oxyds also can be employed to attract the sul-

phur from thefe waters.—cerussa, or white lead, for example Litharge alfo has the fame power, and the oxyds of quickfilver, and
fome others, particularly white arfenic. Thefe oxyds act partly by
their oxygen, which unites with the hydrogen of the gas, and partly
by their own attraction for fulphur.

4to, Some folutions of metals in acids are ftill more convenient in-
dicators, as they enable us to afcertain the quantity of the fulphur.
One of the beft for this purpofe is the folution of muriat of quickfilver. An æthiops, or fulphuret of quickfilver, is produced, from
which we can eafily feparate the quickfilver by acids, to have the fulphur pure; or we can firft weigh accurately the precipitate, then
revive the mercury, and, by the quantity of it, know how much fulphur was joined with it. You know that the muriat of quickfilver
contains this metal highly oxydated. The oxygen unites with the
hydrogen of the gas, and the quickfilver thus metallized unites with
the fulphur.

VII.—The only other inflammable fubftance which has been found
in waters, viz. bitumen, is obvious to the fenfes.

VIII.—Metals are the laft of the fubftances we enumerated, as
found in waters,—and iron is the moft common. It is often combined with fulphuric acid, fometimes, however, with carbonic acid.

The poffibility of diffolving iron in water by carbonic acid, or fixed air, was firft difcovered by Mr. Lane, and communicated in the
Philofophical Tranfactions of the Royal Society.

In authors you will find the Pruffian alkali recommended for detecting metals in general, and even for diftinguifhing them. It precipitates iron blue, copper of a red or coppery colour, and zinc white.
There is, however, fome uncertainty in the indication of this teft It
is a very difficult chemical problem to prepare this alkali altogether
free from iron We have a better teft of this kind by combining the
tinging matter. not with alkali, but with lime. Lime-water boiled a
little on Pruffian blue, will completely deprive about one-thirtieth of

its weight of all the colouring matter, without taking up any martial oxyd, and is a very fure and expeditious teſt of the prefence of iron. For this purpofe, it muſt be carefully kept in well ſtopped phials, and even ſcreened from the light.

The appearance of ochre in the channel of the little ſtream formed by the water flowing from the ſpring, is alſo a very nice indication of iron.

Aſtringent vegetable matter, and eſpecially the gall-nuts of the oak, will difcover an aftoniſhingly fmall quantity of iron in thefe waters. It fometimes happens, however, that the dark colour does not appear when the powdered galls are added to a water which contains iron. This proceeds from an excefs of acid, and is eafily obviated, by adding a very fmall quantity of alkali. We can make the colour appear and difappear repeatedly, in a mixture of this kind.

As for copper or zinc, their appearance is very rare, and we need not have recourfe to this trial to difcover them. There are others that are better.

For the art of imitating the different mineral waters which are efteemed in Europe, and made objects of commerce, you may confult Profeffor Bergmann, whofe work on this fubject, and on the manner of analyfing mineral waters, is one of the moſt valuable he has left us. More lately, Mr. Fourcroy has publiſhed a laboured and accurate analyfis of a fulphurous water in France, at Enghien, in which he has given many judicious and ufeful remarks on the ufe of precipitants in examining thefe waters.

NOTES AND OBSERVATIONS

BY THE EDITOR.

[*Note* 1. *p.* 41.]

I MUST mention here that it was difcovered by Mr. Milner in 1787, that if this alkali be made to pafs through a red hot tube, along with vital air, (oxygenous gas) we obtain nitrous acid. The experiment was this : He put a quantity of a fubftance called manganefe (much employed in glazing the coarfer black earthen wares) into a gun barrel, and making this red hot, he fent this alkaline gas through it. It is well known that the manganefe yields only oxygen gas by this treatment. I am informed that the French, availing themfelves of this difcovery, procured faltpetre for their military operations, by pafling common air (which contains oxygenous gas) through vapours which contain this alkaline gas. It appears, however, in this, and in many other inftances of fuch mixtures of gafes, that the combination cannot be effected, unlefs one or other of the gafes be mixed in the very act of its formation. The gafes themfelves, in many inftances, will not mix, even when red hot, when completely formed. This alkali alfo combines with fulphur, forming a gas of a moft abominable fmell.

[*Note* 2. *p.* 63.]

The word was firft employed by Van Helmont, and particularly to exprefs the vapour which efcaped from liquors in the vinous fermentation. He afcribes to this gas the effects of the Crotto del Cane, in Italy, fo named from the number of dogs killed in that cavern by breathing it. With a fkill and juftnefs that is furprifing, he explains many changes which happen in animal fubftances by the extrication of gafes. He fays that thofe gafes, into which many bodies are completely refolved, do not exift in them in an elaftic, but in a liquid, or even a concrete and folid form. He gives them the general name gas, but diftinguifhes feveral kinds, fuch as gas filveftre, (an epithet borrowed from Paracelfus), flammeum, ventofum, pingue, &c. By gas, then, we are to underftand a perfectly invisible elastic fluid, that can be contained in a vessel, which expands by heat and contracts by cold, but is not condensed by cold into a liquid or solid, as the vapours of water or of camphor. It is proper to add the first character, to distinguish gases from fire, light, and the supposed elastic atmosphere of magnetic and electric bodies. Some chemists would except respirable air from the gases, and consider them as all distinct from air. But it is not improper to employ gas as the object of chemical examination, and to call them airs, when we examine them mechanically.

[Note 3. p. 108.]

It is indeed a very improper denomination, on the principle by which it is pretended that the French nomenclature is regulated. It neither follows the rule adopted for the simple substances, nor that for those of a second order. It is not distinctive. For almost every aëreal fluid, except the oxygenous gas, is azotic, i. e. extinguishes the life of the warm blooded animals. Moreover, knowing the wonderful augmentation which the respiration of the nitrous oxyd makes in the vivacity and energy of animals, it seems extremely incongruous to call its chief ingredient azote. Lavoisier intended another name for it. Some experiments of Berthollet had made him hope to establish it as the alkaline principle, as vital air was assumed by him as the acid principle, and thus to embrace, in one proposition, the whole round of chemistry; but Mr. Cavendish's experiments put an end to this, seeing that it *must* be adopted as the *distinctive* ingredient of the nitrous acid. Yet he would not call it nitrogen, because it was discovered to be also one of the two ingredients of volatile alkali. He gave this reason, and some hint of this little history, to Mr. De Luc, in conversation. But Mr. De Luc could not get more out of him, though few inquirers are more pertinacious. Perhaps Lavoisier's hopes of making it the alkaline principle were not yet extinguished.

[Note 4. p. 109.]

It is not improbable that electricity has something to do here. The light of the electrical spark, visible in a great extent of air between the discharging balls, is, I think, the indication of some chemical action going on in the whole of that extent; indeed, on the principles of the new chemistry, it cannot be any thing else. There is no such thing as the transference of something luminous from the one body to the other. There is a decomposition of oxygenous gas taking place wherever we see the light. It seems to be simultaneous, but it is successive and amazingly rapid. There is a smell accompanying all electrical experiments. This smell is also a strong indication of some chemical action. But this is very distinct, or peculiar, and has no resemblance to the smell in any process with the nitrous acid. I suspect therefore that electricity acts otherwise than merely by exciting a great heat. Great heat will produce a combination of oxygenous and azotic gafes, if either of them be taken in its nafcent ftate, *i. e.* in the very act of its extrication; but not when they are completely formed. This may arife from the conftituent parts of the completely formed gas. When nitrous acid is decompofed by detonation, much light and heat are difengaged. It may happen that, in the formation of thefe gafes, the mutual attraction between them may be very great, while both or either of them are unprovided with the heat or light required for their gafeous form; but may be much weaker when they are furnifhed with it. This will even be analogous to the greateft part of chemical combinations. The effect of electricity may be to difcharge part of this heat and light from the air in which it is feen. But, on the whole, there are feveral phenomena in the relations of azotic gas which are not eafily reconciled with our prefent notions, and even affect confiderably the whole fyftem of Lavoifier[*].

I may here mention another peculiarity of this gas. It communicates the green fæcula to plants growing in the dark, which would otherwife be white. This appears from a feries of experiments by Sennebier.

[*Note* 5. *p.* 122.]

The order of compofition expreffed by this termination *et*, (*etum*, in Latin), has not been yet explained by Dr. Black. It means, in the French nomenclature, the combination of a radical, fuch as fulphur, carbon, phofphorus, &c. with any fubftance except oxygen, without an intermedium. Thus, we have the *fulphuret of potafh*, the *carburet of iron*, the *hydrogenous phofphoret*, &c. But when the fame radical fubftance is combined with potafh by the intervention of oxygen, the compound is then a fulphat of potafh, or a fulphite of the fame, according as the proportion of oxygen is great or fmall.

The fulphat of potafh is not, however, confidered as a compound of thefe ingredients, but as a compound of potafh united with fulphuric acid, which is occafionally confidered as a fubftance *fui generis*, having affinities and properties truly diftinctive, in which the properties of its ingredients, fulphur and oxygen, are in fome meafure dormant. This was undoubtedly the way in which Mr. Lavoifier confidered thefe things. I apprehend that this is the only accurate notion that can be had of a true chemical combination, and that there is no fuch thing as a compound of three ingredients, in which the primary properties of each are immediately efficient. Such compounds, however, are frequently fpoken of by the new chemifts, efpecially in their attempts to point out the procedure of nature in the fermentations,—vinous, acetous, and putrefactive. But I know no fubftance in which the ingredients exert the fame combining energies as if they were

[*] Many chemifts of eminence imagine that its radical part is the fame with that of oxygenous gas, and that the difference in its chemical qualities proceeds from a difference in the proportion of caloric combined with it. Others imagine that it has not caloric combined with it, but the matter of light; but thefe are all conjectures.

all feparate. They take this method, becaufe it gives them a vaftly greater latitude in their explanations. But once this liberty is taken, you will fcarcely fee two chemifts explain the phenomenon in the fame way. Had the political ambition of thofe whom Lavoifier affociated with himfelf in his labours and honour, fuffered him to remain at their head, he would have faved his followers from many embarraffments in which they have involved themfelves by this manner of proceeding.

[Note 6. p. 213.]

I cannot omit mentioning in this place, that my colleague Dr. Daniel Rutherford read, in the year 1775, to the Philofophical Society of Edinburgh, a differtation on nitre and nitrous acid, in which this doctrine is more than hinted at or furmifed. By a feries of judiciously contrived experiments, he obtained a great quantity of vital air from ~ ric acid; about one-third of that quantity from the fulphuric acid, as contained in alum; and a fmall quantity (and this very variable and uncertain) from the muriatic acid. The manner in which it came off from the compounds, in various circumftances, led him to think that the different quantities obtained did not arife from the different proportions in which it was contained in thofe acids, but merely in the different forces with which it was retained. He therefore concluded that vital air was contained in all acids, and thought it likely that it was a *neceffary* ingredient of an acid; and, feeing that it was the *only* fubftance found, as yet, in them all, he thought it not unlikely that it was *by this that they were acid*, and he points out a courfe of experiments which feems adapted to the decifion of this queftion. I was appointed to make a report on this differtation; and I recollect ftating as an objection to Dr. Rutherford's opinion, " that it would lay him under the ne-" ceffity of fuppofing that vitriolic acid was a compound of fulphur and vital air," which I could not but think an abfurdity. So near were we at that time to the knowledge of the nature of the acids!

[Note 7. p. 243.]

Dr. Black's notes, from which he was accuftomed to lecture on this fubject, for the laft four years of his teaching, are extremely imperfect, and confift of nothing more than references to experiments by Prieftley, Auftin, Milner, Berthollet, and an intimation that the French chemifts practifed a certain procefs to procure faltpetre for the army, and here and there a flight thread of reafoning from the experiments. I think that I have found out all the experiments to which he refers, and I have put them down in the order in which they ftand in the notes; but I confefs that unlefs I greatly exceed any authority derived from the manufcript, I cannot give that clearnefs that will fatisfy a cautious mind. Nor can I adhere to the rule laid down in the beginning of this chemical hiftory, namely, to employ in argument no fubftance of which the properties have not been previouſly diſcuſſed. I have availed myſelf of the notes taken by a young gentleman who attended the two laſt courſes of Dr. Black's lectures, ſo that I preſume that what I have inſerted in this place does not differ much from what he delivered.

[*Note 8. p. 246.*]

It muſt be confeſſed, that the evidence for this compoſition of ammonia is in a great meaſure hypothetical, even when the compoſition of water and of nitrous acid is fully acquieſced in ; and the followers of Lavoiſier differ much among themſelves in the ways in which they explain the phenomena which are adduced as arguments. Having a number of ſubſtances before us, which exert mutual actions, it is plain that we can match or pair them in a variety of ways, and may ſelect that double exchange which ſuits our purpoſe. We do not, in many caſes, ſee clearly why either azotic or hydrogenous gas is not *always* produced. Mr. Berthollet obtained azotic gas, and Dr. Prieſtley obtained hydrogenous gas from ammonia by the electric ſpark. Nitrous ammoniac does not always yield azotic gas and water, but another gas conſiderably different. We aſk what becomes of the hydrogenous gas in Mr. Berthollet's experiment ? It is ſaid that the mercury is always covered by a film of mercury combined with oxygen, and that this is detached by the hydrogenous gas and water formed. It is alſo ſaid that nitrous acid was formed in Dr. Prieſtley's experiment ; but there is no proof of this offered ; it is an aſſumption *on the authority of a previous theory;* and this, without having aſcertained the elective attraction which that theory neceſſarily ſuppoſes. The ſame gratuitous procedure is obſervable in many explanations given by the followers of Mr. Lavoiſier, in all caſes where they employ the decompoſition of water. They have not aſcertained, by experiments inſtituted on purpoſe, the double exchanges that are poſſible among the ſubſtances employed. I can point out inſtances which cannot both be poſſible,—and yet I ſee both employed.

[*Note 9. p. 261.*]

Two ounces of ſlaked lime, and one drachm of phoſphorus cut into very ſmall bits, were made into a ſoft paſte, which muſt be haſtily put into a ſmall earthen retort, having a ſwan neck tube luted into its neck. This is introduced into the ordinary pneumatic apparatus, as it is called, and heat is cautiouſly applied, on account of the exploſions which frequently happen. The gas ſoon comes over, and continues to diſtil for a great while. This quantity of materials will yield three Engliſh quarts of gas.

If a quantity of it be let up through water into the air, it gives a bright flaſh, and a fine ring of white ſmoke riſes from it. This combination was made by Dr. Raymond.—*Ann. de Chym.* x.

Dr. Pearſon at London lately made an experiment, which ſhews ſtill more diſtinctly the nature and hydrogenous origin of this gas. He combined phofphorus with quick-lime, and put the compound into water. While it flowly penetrates, and partly diffolves it, a fmall part of the water is gradually decompounded, the diffolved phofphorus attract-ing its oxygen, and allowing the hydrogen to efcape ; and this is collected in the upper part of the apparatus in its ordinary form, containing (but not always, nor in equal quan-tity) a fmall portion of phofphorus. The gas, accordingly, will not always kindle by fimple contact with atmofpheric air. It rarely fails to kindle in pure vital air ; and in all cafes it kindles by any fpark. The inference is plain.

This combination of phofphorus and alkaline fubftances, is in many refpects fimilar to the combination of fulphur with thefe falts, and like it, produces a hepatic gas, having fimilar properties. This experiment of Dr. Pearfon illuftrates all thefe hepatic pheno-mena.

[Note 10. p. 274.]

The proofs of this not being adduced in Dr. Black's manufcript, and its properties being remarkable, and chiefly becaufe the mode of its formation is one of the complicated applications of the new theories, it feems neceffary to mention fome of the more direct and fimple arguments.

Dr. Higgins, by paffing the fteam of water over melting fulphur, produced fulphurous acid and hydrogenous gas.

Mr. Gingembre, bringing the focus of the fun's rays through a large lenfe on a piece of fulphur inclofed by mercury in inflammable air, produced this gas in perfection.

Dr. Prieftley, by paffing a ftream of inflammable air through ftrong vitriolic acid boil-ing hot, produced this hepatic gas.

Dry hepar fulphuris, made with cauftic alkali or lime, when treated in a retort with great heat, does not afford this gas, but fulphur ; but if it be moiftened, or has been prepared in the humid way, it yields hepatic gas in great abundance. Dr. Auftin preci-pitated fulphur from it by the electric fpark. The remainder was inflammable air, of the lighteft kind.

[Note 11. p. 275.]

This obfervation of Dr. Black's is very juft, and the language of the French chemifts in explaining thefe phenomena is a proof that they proceed without diftinct notions of the fubject. Fourcroy, for example, fays, " L'acide fulphureux decompofe le gaz hydro-" gene fulphuré, en feparant le foufre, parceque l'oxygene, en partie libre dans cet acide, " fe porte facilement fur l'hydrogene de ce gaz. L'acide fulphurique ne produit pas le " meme effet." *Supp. p.* 76. Now, furely in their theory, the fulphurous acid has *no* " *oxygene en partie libre*"—on the contrary it has *fouffre en partie libre.* I may fay the fame thing of the nitrous or fuming acid. He fays in the fame page, that the dephlogifticated marine acid (acide muriatique oxygené:) decomposes this hepatic gas. Here I grant the " oxygene en partie libre." Surely in as far as the decomposition depends on oxygen, the operation of the sulphurous and nitrous acids should be the reverse of that of the oxygenated muriatic acid. In truth, this decomposition embarrasses those chemists ; and they differ exceedingly in the way in which they conceive it to be effected. Fourcroy, and the Dutch chemists of whom Dr. Black speaks in the next paragraph, do it in one way, and in the *Annales de Chymie*, vol. 14. page 313, we have quite a different account of the procedure. Gren, a German chemist of great reputation, is of opinion that the theory is imperfect without another substance which he calls *firematter ;* and with the help of this he solves all the phenomena very easily and elegantly. But there is no end to substitutions founded on such fancies as this. It is like the æthers and other invisible fluids, which the mechanicians have introduced, because their explanations by impulsion cannot go on without them. Before we can proceed with safety in those explanations which employ the decomposition of water, a series of experiments should be made for ascertaining the elective attractions of the radicals of the gases with a precision equal to what we have attained in the more familiar substances. Till this be done, we may form ingenious conjectures concerning the hidden operations of nature in bodies of a complicated nature, but we gain no confident knowledge. The followers of Lavoisier, grasping at every thing, find no difficulty of giving what may be called a *narration* of this internal procedure, taking the combinations and decompositions in such order as suits their purpose ; and this is contrived so as to terminate in the ultimate combination which we all observe. This is not a very difficult task, even when we limit our combinations to two substances, and produce them by simple affinities. If we take compounds of three ingredients, as in the present case, and employ double affinities, our means of·accomplishing the *desired* end are increased prodigiously, and there is nothing that can escape us. The new phraseology is so significant, that every epithet indicates an operation ; so that our explanation has great appearance of a real knowledge of the facts. But that all this is little better than conjecture, and very precipitated and unwarranted conclusions, appears from this,— that two eminent chemists, explaining the same ostensible phenomenon, and employing the same agents, give quite a different *story*, merely by taking a different order of succession in the steps of the internal procedure. They therefore employ quite different combinations and decompositions. This cannot be, if the affinities of the substances, when in the same situations, be constant. Lavoisier has rarely taken much liberty this way ; but his more zealous followers set no bounds to their theories.

[*Note* 12. *p.* 312.]

It will be of some service to us in our future consideration of this subject, to recollect the well known fact, that any plant, the sugar cane for example, will grow from a minute seed to maturity, yielding sugar and every other production competent to its nature, if planted in pure fand, fed with diftilled water, and having a free communication with the air and light, and a proper degree of warmth. The fand in which it grew is neither diminifhed nor changed in the fmalleft degree. It would feem, therefore, that we muft expect nothing from alcohol but what is fupplied from thefe fources, viz. the water, the atmofphere, light, and heat; and that it is to thefe that we are to look for the oxygens, hydrogen, azote, carbon, alkali, and every thing that we obtain from alcohol.

[*Note* 13. *p.* 317.]

Dr. Black has attended to thofe circumftances only of the procefs which relate to the preparation of æther. It is therefore proper to mention, that, at the time when the white cloud prefcribes the damping the fire and changing the receiver, another fingular product begins to form. This drops from the fpout of the retort into the water in the receiver, and is the *fweet oil of wine*, an oily, fragrant, and fweet tafted fluid. With whatever care the procefs is now conducted, the matter in the retort being now much hotter, the vapours of fulphurous acid now rife, and muft be allowed to efcape,—more oil comes over, but horribly tainted with thefe vapours, and growing more and more thick and dark coloured,—at laft quite dirty. This is owing to a perfect charcoal, now formed in the retort, fo light as to rife with the vapours. At laft, the matter in the retort becomes a dry, pitchy-like, fpongy mafs.

[*Note* 14. *p.* 318.]

It would never appear in the fluid form in thefe climates, but in that of elaftic vapour, did not the preffure of the air keep its particles from flying afunder. This muft bring to our recollection Dr. Cullen's experiments, which were of fo much ufe to Dr. Black for eftablifhing the doctrine of latent heat.

If a wide glafs cylinder, fitted with a pifton, which is put half way down, be filled with the pure vapour of æther, kept of fuch a temperature that it juft balances the preffure of the air, and the pifton be then fuddenly drawn to the top, the cylinder will be filled with a white cloud, which lafts for a fecond or two, and then difappears. This fhews that it is not merely *capacity* for heat which fits the vapour for appearing in its more expanded form. This form is the confequence, and not the caufe, of the abforption of heat.

[*Note* 15. *p.* 388.]

I cannot but think that peat, or the black mofs of the moors, is an approximation to coal. Peat is not found in many places; and no where abounds fo much as in Scotland and Ireland. It is by no means enough for the formation of peat that the place be a wet marfh, abounding in vegetable matter. In the immenfe diftriéts of Europe and America, fuch fituations are common; and we have impaffable m r ffes and fwamps of vaft extent, but thefe are not filled with peat, nor is the mud which fills them very inflammable. Accuftomed to the bogs of Scotland, and little informed in natural hiftory, I was much furprifed at not finding fimilar fituations in the Canadian woods without peat; and this made me examine with attention the matter contained in thofe bogs. Even where the vegetable remains were very abundant, and conftituted almoft the whole mafs, I found it very little inflammable, and altogether unfit for a fuel. And what I took particular notice of, the fmell in burning was altogether unlike the fmell of burning peat. This is quite peculiar to peat. I ever faw peat in any part of North America, except in the neighbourhood of Louifburg,—and there it was but a very fcanty mixture of peat-earth with the moorifh foil.

While the fmell of all burning peat has a charaéter by which it may always be known, there are confiderable varieties; and thefe varieties feem to me to be fuper-additions to the diftinétive fmell of peat. This is confiderably like that of the moft inflammable lean coal, and ftill more like to that of jet, but not near fo offenfive. The blackeft, hardeft, heavieft peat, when the matter is almoft an impalpable pulp, is the moft inflammable, and leaves the fmalleft quantity of afhes. This kind of peat has the heavieft fickening fmell. Such is the peat at Canifbay, in the north extremity of Scotland, juft by John-a-Groat's Houfe. This, when dried, is fo fine in its texture as to break with a fort of polifh, like a jafper. Its fmell in burning is not very diftinguifhable from that of cannel coal. The fmell of the beft Dutch turf, which is taken up from the bottom of falt water, refembles that of the peat now mentioned very much.

I am inclined to think that a certain juice is neceffary for the formation of a bog into peat. Perhaps this juice is the primitive bitumen. I fufpeét alfo that it is always accompanied by vitriolic matter. Peat afhes always contain a very great proportion of iron. I have feen three places in Ruffia where there is fuperficial peat mofs, and in all of them the vitriol is fo abundant as to efflorefce. One in particular, hard by St. Peterfburgh, fhews it every morning on the clods, when the dew has dried off.

Peat moffes form very regular ftrata, lying indeed on the furface; but if any operation of nature fhould cover this with a deep load of other matter, it would be compreffed, and rendered very folid; and remaining for ages in that fituation, might *ripen* into a fubftance very like pit-coal.

[*Note* 16. *p.* 407.]

Perhaps the faét would be more properly expreffed by faying that when the effervefcence produces hydrogenous gas, the French chemifts, combining their theory of combuftion with Mr. Cavendifh's difcovery of the compofition of water, infer that the metal aéted only on the water, and attraéting its oxygen, fets the hydrogen at liberty. When

azotic gas, or nitrous air, is yielded by the effervefcence of a folution of metal in nitric
acid, they fay that the metal acted on the acid, attracting its oxygen, and thus liberating
the azote. Before Mr. Cavendifh's difcovery, the folution of metals in the fulphuric and
muriatic acids prefented difficulties which greatly embarraffed the partifans of the new
doctrine. This accounts for their anxious and laborious repetition of the experiments of
Mr. Cavendifh. Till this doctrine was firmly and accurately eftablifhed, the theory of
acidification was quite unfatisfactory. Nor did even the compofition of water make it
of extenfive influence, till the proportion of the gafes was exactly afcertained.

It muft be acknowledged that the order in which metals appear to attract the acids in
folution correfponds pretty well with the order in which they are oxydated by the action
of heat and air. Zinc and iron precipitate all other metals from the nitrous acid; and
thefe metals are more calcinable than moft others. Tin, however, does not fall into its
place in the order of folution: It is very calcinable. It is alfo pretty conformable to
the fame principle that zinc, iron, and tin, produce inflammable air, that is, decompofe
water by their ftrong action on the oxygen. But were this the fole efficient caufe of
this peculiarity of thofe metals, we fhould expect it in a much more remarkable degree in
cobalt, magnefium, and tungften, which calcine fo rapidly by mere expofure to the
atmofphere. Nay, the white oxyd of magnefium fhould do it. But none of thefe
fubftances produce, as far as I can learn, inflammable air from diluted fulphuric
acid, &c.

I would fay farther that the application of this theory to the complicated cafes of animal and vegetable fubftances is, in a great meafure, gratuitous, till we fhall have afcertained, not only the proportion of the ingredients, but alfo their elective attractions in
all different temperatures. This feems peculiarly neceffary with refpect to the bafis of
the three gafes, which act fuch important parts in thefe changes. Till this knowledge
be attained with confiderable precifion, it will always be eafy, by ringing the changes
(fo to fpeak) on oxygen, hydrogen, and azote, and by taking their feveral actions in any
order of fucceffion that we pleafe, it will, I fay, be eafy to explain any phenomenon
whatever. I muft add that fome of thofe elective attractions appear to me to be
fuch as to render fome of the favourite explanations of thefe phenomena inadmif-
fible. I think that I fhall be able to give inftances, before we get to the end of thefe
lectures.

[*Note* 17. *p.* 414.]

Thefe Dutch gentlemen do not choofe to confider thefe facts as examples of combuftion, becaufe they fay that there is no decompofition or change effected in either of the
ingredients. It is a mixture, like that of fulphuric acid and water, in which alfo there is
a vaft extrication of calorique; becaufe the capacity of the mixed is lefs than the fum
of the capacities of the ingredients while feparate. Or it is the extrication of latent

heat. They explain thefe facts in the fame manner. There is indeed another cafe, in the mixture of metal with fulphur, in which this explanation is very admiffible, namely, the incandefcence (and actual inflammation, if in the air) of mercury and fulphur, in the preparation of cinnabar. But in the facts mentioned above there are very effential differences of circumftances.

I have not feen the accov . which the Dutch chemifts have publifhed of thefe experiments, and have read only Van Mons's repetition of them, of which a very diftinct abftract may be feen in the *Analytical Review* for October 1795. I do not know, therefore, the precife ftate of the compounds. This is certainly a chief circumftance, with refpect to the theory of the phenomenon. It was furely a very unexpected thing that zinc fhould exhibit fuch an appearance of inflammation, feeing that this metal contracts no union with fulphur.

We muft at any rate conclude from thefe experiments, that oxygen gas is not the fole fource of the light and heat which appear in the combuftion of bodies, which feems to be a point of doctrine in the antiphlogiftic fyftem of chemiftry. The great heats produced during the folution of metals in acids, notwithftanding the eruption of much gas containing oxygen, prefent, I think, a difficulty in this fyftem. It is fomewhat leffened by thefe Dutch experiments. For we fee fo much heat extricated as to produce ignition ; and therefore may more eafily conceive the change of capacity in folution to be adequate to the production of the fmaller heats which appear in that procefs.

The heat and light alfo, in all deflagrations with nitre, is another indication of a great quantity of caloric combined with oxygen and with azote in the nitre, in their folid form.

From fuch facts, it may perhaps be concluded, that the emiffion of heat and light is more copious in ordinary combuftion than in the other cafes before us, only by the emiffion of what was *further* combined as indifpenfable articles of the gafeous form in which oxygen and azote exift in the atmofphere, that is, by the emiffion of the latent heat of Dr. Black ; a ftate of the caloric different from that which requires a proper third fubftance, along with a high temperature, to difengage it by fuperior affinity.

But, although thefe confiderations give a greater confiftency to the new doctrines, and leffen fome of their difficulties, the intelligent reader muft, I think, perceive that the whole becomes hypothetical, and all the properties, affinities, and other relations of thefe fuppofed bafes of different kinds of gas, are mere interpretations of the phenomena ; or rather are accommodations and corrections of fuppofed properties, till the hypothefis is made to tally at laft with all the phenomena. The hypothefis feems to have nearly the fame rank in fcience with the magnetical hypothefis of Æpinus. Both are ingenious and elegant, in the higheft degree, and have fuch a comprehenfive refemblance to the phenomena, that the hypothetical principle becomes an excellent principle of arrangement or claffification of the phenomena, almoft equivalent to a juft theory, and, in all probability, extremely near to it. Other cafes occur in the fubfequent lecture which will bring this fubject again before us.

[*Note* 18. *p.* 478.]

Dr. Black having only given the general conclusions which have been drawn by these eminent chemists concerning this very singular substance, it seems necessary to mention the chief and most simple facts by which these opinions are supported.

If we boil the prepared alkali, called the lixivium sanguinis, or Prussian alkali, with a diluted sulphuric acid, vapours come off which are extremely volatile. When the distillation is skilfully conducted, the first vapours are found much more volatile than water, and cannot be condensed, unless water be mixed with them, or presented to them in the receiver. This vapour has a strong smell, not disagreeable, and is extremely inflammable.

This substance contains the colouring principle, and the lixivium now possesses none. Its taste is sweetish and astringent. But as this watery solution of it unites with all alkaline substances, and with metallic oxyds, and may be detached from one of them by means of another, in the same manner as acids, it is considered, chemically, as an acid, and has been called the PRUSSIAN, or PRUSSIC ACID, and its compounds have been called PRUSSIATES. That compound called the lixivium sanguinis, prepared, or Prussian alkali, &c. is the PRUSSIATE OF POTASH.

The prussic acid, prepared in the way just now mentioned, is but impure. Mr. Scheele procured it in the greatest purity, by first forming of this impure acid a prussiate of mercury, and then putting some filings of iron into it and a small quantity of sulphuric acid. This instantly decomposed the mercurial prussiate, and even greatly weakened the attraction of the Prussic acid for the iron, by the excess of sulphuric acid. The mixture being now distilled with an extremely gentle heat, and the first vapours only preserved, the result was a prussic acid perfectly pure. With this, or such as this, all the experiments for ascertaining its properties are tried.

It combines, as I have said, with all the substances which combine with acids, with various degrees of elective attraction. Lime water digested on Prussian blue combines with this matter alone, without taking up any iron, as the solution of alkali does. It acquires a yellow or straw colour, and no longer manifests its alkaline qualities. It does not affect the test paper. It has lost its alkaline taste, and its attraction for carbonic acid. This compound, called the *prussiate of lime*, is a much fitter test of the presence of iron, than the lixivium sanguinis, or *prussiate of potash*, because it contains no iron.

The union of this colouring matter with the various bases, is exceedingly weak; for every acid, even the carbonic, detaches it from them all. Nay, Scheele found that carbonic acid rendered blood incapable of imparting the colouring quality to the lixivium sanguinis. But, as observed by Dr. Black, the addition of a small quantity of iron, either pure or very slightly calcined, so fixes this acid in the lixivium sanguinis, that it now resists the action of any diluted acid. It is found, however, that strong sulphuric acid disfolves the blue precipitate, and forms with it a brown liquor,—but it becomes blue by diluting with water.

Mr. Berthollet has greatly augmented our knowledge of the pruffic acid difcovered by Scheele. He found that in Scheele's decompofition of the mercurial pruffiate by means of iron, the iron took from it not only its pruffic acid, but alfo its oxygen ; for the iron was now an oxyd, and the mercury was revived.

His moft curious obfervations related to the fuper-oxygenation of the pruffic acid. He found that when mixed with oxygenated muriatic acid, it reduced the muriatic to its ordinary ftate, while the pruffic had now acquired the abundant oxygen of the other, and its properties were greatly changed. It is now much more volatile, and if expofed to the fun's light, it acquires the fmell of aromatic oil, and even collects into a fluggifh oil like liquor, heavier than water, and it is no longer inflammable, nor precipitates iron. It is faturated with oxygen, which is equivalent to its being burnt already. In this oxygenated ftate, it does not form a blue, but a green precipitate, with the folution of vitriol. This is rot owing to the mixture of a really blue precipitate with yellow oxyd of iron ; for muriatic acid does not diffolve this oxyd, nor change the colour. It may be brought back from this ftate by fulphurous or volatile vitriolic acid. This is evidently by abftraction of part of the oxygen by the fulphurous acid. And, conformably with this account of it, we obferve that the green precipitate mentioned juft now, is made blue by the volatile vitriolic acid. The fame effect is produced by expofure to the light. Now we know that in all cafes of fuper-oxygenation, light detaches the redundant oxygen.

Scheele's preparation of Pruffian alkali by means of fal ammoniac, chalk, alkali, and charcoal, fhews that volatile alkali, or its component parts, hydrogen and azote, enter into the compofition of the pruffic acid. And if cauftic fixed alkali or lime be mixed with oxygenated pruffic acid, we have the vapours of cauftic volatile alkali. Moreover, if the alkali or lime be now detached by fulphuric acid, we do not obtain the pruffic acid again : It has been deftroyed ; and the volatile alkali juft now mentioned has been part of it. But further, this addition of fulphuric acid produces effervefcence of fixed air, or carbonic acid. This was not feen before. It would feem, therefore, that this is the other ingredient of the pruffic acid, and that it confifts of volatile alkali and fixed air, or of their component parts, hydrogen, azote, carbon, and oxygen ; perhaps the laft does not exift in it, but is furnifhed by the water in the diftillation of pruffic acid, and in that procefs combines with the carbon, forming the bafis of carbonic acid.

After all the inveftigations of this eminent chemift, there is ftill a great uncertainty as to the real ftate of things in Pruffian blue. I think that there may be ftrong objections made to the exiftence of any triple compounds in chemiftry, and that all the examples offered are really mixtures of the third, with a true compound of the other two. Mr. Berthollet fays, that there are two kinds of the pruffiate of iron. One is the common Pruffian blue, containing the iron or its oxyd, faturated with the tinging matter. The other confifts of oxyd not faturated, or precipitate of iron with excefs of oxyd. We obtain this by digefting an alkali on Pruffian blue. It evidently takes to itfelf part of the colouring matter. But the remainder is not, as commonly thought, an ordinary oxyd; for when an acid is poured on it, we obtain Pruffian blue.

Alkali prepared in this manner, by digefting on Pruffian blue by the affiftance of heat, contains a confiderable quantity of the fecond kind of precipitate. It will, by cooling for a confiderable time, depofit this in form of a yellow calx. When thus freed from it, evaporated to drynefs, and then redifolved in water, it is called *purified Pruffian alkali.* Mixture with vitriol produces no blue, if kept in the dark; but the light caufes the blue precipitate to form gradually, till the whole alkali is decompounded.

[*Note 19. p.* 483.]

I forgot, when confidering the folution of iron in the muriatic acid, to mention a curious obfervation of the Duke D'Ayen. When the martial muriat is expofed to a mild heat, water only flightly acidulated is difengaged,—then part of the acid rifes, in almoft incoercible fumes, but pure,—then a compound of the iron and acid rifes, and condenfes in elegant lancet-fhaped cryftals. When the more fixed matter which remains is urged by the moft violent heat, there rifes a vapour which feems truly metallic, condenfing in regular hexagonal cryftals, having the full opacity and luftre of polifhed fteel, and ftrongly attracted by the loadftone.

This is unqueftionably a volatilization of the metal; and I think points out fome cementing proceffes; fuch as with common falt, which fhould make us expect valuable refults. If this fublimate be a pure iron, it is furely in a ftate that we are not acquainted with. I fhould wifh that in this procefs of the Duke D'Ayen, a magnet had been placed in the way of the vapours, or even applied externally by one of its poles, to the part of the receiver on which the fublimate appears to accumulate moft fpeedily.

[*Note 20. p.* 492.]

Thefe two remarkable maffes of malleable iron mentioned in the text have greatly puzzled the naturalifts who have attempted to account for them. Both were found fo far from all traces of habitation, that there is no fufpicion that they have been brought into that ftate by the operation of man. And indeed the appearance of the Siberian mafs is fuch as could not have been induced by our greateft heats in our beft furnaces. The whole has been in a ftate of moft violent ebullition, there being many bladders in it larger than an egg, and the furface is all bliftered. The whole is covered with a cruft like iron-ftone, but internally, it is good red-fhort malleable iron. The mafs found in South America is about fifteen tons weight, and of fimilar ftructure, but not fo folid. This, and another of ftill larger fize, and branched like a ftumped tree, found in the the fame woods, three hundred miles from all appearances of iron, or even of ftone, were lying on the furface of the earth. *(Phil. Tranf.* 1788.) A mafs of the fame kind and appearance, near feven tons weight, was found under the pavement at Aken near Magdeburg, and is excellent iron.

In all thofe inftances, it is remarkable that the fufion has been more complete than we can produce in our beft furnaces. Alfo the glaffy matter is tranfparent, and altogether unlike the flags which are formed in our metallurgic operations. No appearances of fulphurous admixture are obferved in thofe maffes.

Profeffor Chladni, an author of great refpectability, is much difpofed to think that all thefe maffes have been portions of thofe fire-balls which are often obferved to move in our atmofphere with immenfe velocity,—with a dazzling flafhing light, frequently with fmoke, making a ruftling noife, and fometimes burfting in pieces with a loud explofion. The Siberian Tartars worfhip the lump above mentioned, as a thing come down from heaven. Many accounts have been given of red hot maffes which have fallen through the air; and thofe which have been fhewn as fuch are all of the fame kind, confifting chiefly of malleable iron. Chladni gives a circumftantial account of three. One example at Hrafchina, near Agram, in Hungary, is attefted by many witneffes, before the Bifhop's confiftory at Agram. It was feen coming from a great diftance, with a brandifhing unequal light, like a long flaming chain; and it fplit into two pieces, which fell about a mile afunder, making holes *more than an ell wide*, and three fathoms deep. One weighed feventy-one, and the other fixteen pounds; and they are now in the Emperor's mufeum at Vienna. They are of the fame kind exactly with thofe already defcribed. That fuch trifling maffes fhould make fuch excavations, can be accounted for, only by fuppofing them to be in a ftate of continual explofion or deflagration, blowing in all directions, like the inftrument of deftruction called a carcaffe, difcharged by a mortar. This will alfo give them a great apparent diameter while moving through the air. *(See Chladni's Differtation on the Mafs of Iron found in Siberia, Leipzig,* 1794; *and the Art of Mining and Metallurgy,* by Mr. Stultz, keeper of the mineral mufeum at Vienna, 1789. See alfo fome valuable papers by M. Howard, in the Philofophical Tranfactions, 1802, &c. or extracts from them in Nicholfon's Philofophical Journal. In thefe papers every thing has been collected which has a relation to the fubject; and Mr. Howard's obfervations are eminently ingenious and inftructive.

[Note 21. p. 506.]

Dr. Black's notes contain no account of this procefs. I believe that it is not generally known; and although Jars has publifhed what he calls the procefs carried on in Sheffield, we get little information from it. From the circumftance that the ores which afford a metal nearly in the ftate of fteel, are generally fpathofe, or analogous to them, it feems not unlikely that the carbon which it wanted to form fteel may be had in the fixed air.

Accordingly, many of the old proceſſes for caſe hardening iron, and for making little bits of ſteel, employ crude calcareous earth. The carburet (ſo plumbago is conſidered) burns very ſlowly. Perhaps even carbonic acid may be decompounded by iron, by the help of ſome double elective attraction ; and the ſecret flux, of which Mr. Jars ſpeaks, may be a mixture containing limeſtone, and a proper reagent.

[*Note* 22. *p.* 522.]

I confeſs that this appears to me to be an inaccurate, or at leaſt an imperfect account of the matter. If indeed it relates only to the heat produced in the union of the nitrous and the vital air, to form nitrous acid, it is ſatisfactory : But if it be meant to explain the heat which is obſerved in the ſolution, when the gaſes are forming, it is ſurely inaccurate. Indeed this heat has always ſtruck me as a very great difficulty in the whole theory. Mr. Lavoiſier unqueſtionably derives his explanation of the heat produced in combuſtion from Dr. Black's theory of fluidity and vapour, and ſuppoſes that a gas conſiſts of its ra- dical, or diſtinguiſhing ingredient, combined with calorique, according to the ordinary laws of chemical affinity. This being ſuppoſed, the ſolution of metal in an acid, ſo far from producing or extricating calorique, ſhould abſorb it from the materials, and pro- duce a cold incomparably more intenſe than any of our freezing mixtures. For ſuch ſo- lutions, whether the metal be oxydated by the acid or the water, are always accompanied by the eruption of gaſeous fumes. Iron in the diluted ſulphuric acid produces an immenſe volume of gas. Thoſe metals which are oxydated by decompounding the acid, produce gaſes which ſtill contain much oxygen. And it may be remarked that, in general, thoſe ſolutions which produce fumes moſt deficient in oxygen produce the greateſt heats. This fact is favourable to Mr. Lavoiſier's explanation. The oxygen remaining in its concrete form, does not expend any of the calorique extricated by other circumſtances of the pro- ceſs. Suppoſing that no more latent heat is neceſſary than for the production of as much watery vapour as ſhould have the ſame denſity, the quantity is very great, when com- pared with that which occaſions the cold in our freezing mixtures. It muſt not be ſaid that the quantity neceſſary for this gaſeous combination may be ſmall ; for in this caſe, we ſhould often have gaſes inſtead of ordinary vapours, whereas, we know that an incom- parably greater ſupply of calorique is required for the formation of a gas. Beſides, the Lavoiſierian theory of combuſtion ſuppoſes a vaſt accumulation of heat in oxygenous gas ; this being, according to that celebrated philoſopher, the ſource of all the heat extricated in that operation of nature. It muſt alſo be obſerved here, that the oxygenous gas gives out this heat, and the oxygen is combined with the inflammable body, in a ſtate very ſi- milar to its condition in the preſent experiment. Therefore, we unqueſtionably have a prodigious quantity to account for, the oxygen in the diſſolving or combining ſubſtances being unprovided in the quantity neceſſary for its becoming a ſource of heat in ſome fu- ture combuſtion.

I would now afk in what ftate is the calorique contained in the materials of an acid and a metal, when they act on each other? Some of the materials muft contain it in a ftate that is unneceffary for their appearance in the ftate of a folution, of an oxyd, or of a metalline falt. When all this calorique has emerged, the oxygen in nitre ftill contains a great ftore of it, feeing that it is extricated from it in deflagration with inflammable fubftances. This only increafes the difficulty. For this great ftore of calorique muft remain in the folution, and in the metallic falt which it produces. Heat is extricated in the folution, and gas containing oxygen is produced. This gas, by uniting with vital air, again detaches calorique, and produces nitric acid. This acid will diffolve metal, and again detach calorique. This may be continued without end. This circumftance alone fhould convince us that there is fome error in our theory, becaufe this endlefs generation of heat is impoffible in the nature of things. We cannot fay, with any well grounded confidence, whether more calorique is extricated from oxygen, when, in the gafeous form, it caufes the combuftion of fulphur, or when, as an ingredient of nitre, it contributes to the deflagration with the fame fulphur. I grant that I think that more is extricated in the firft cafe. But it fhould be an immenfe deal more. For methods may be found for transferring the oxygen of the fulphuric acid, formed in the firft cafe, to azote, and of thus forming nitric acid, and nitre, which will again deflagrate with fulphur.

All this is myfterious and intricate. I do not fay incompatible; but I am not able to reconcile them by means of any known facts. The fame, or greater difficulties, occur in almoft all the fpontaneous inflammations; in the deflagrations of nitrous acid with effential oils, and in many detonations; and in particular the heat and light which we call glow, or incandefcence;—efpecially fuch as appears in the Dutch experiments, mentioned in page 642, on the mixture of fulphur with feveral metals. I acknowledge that I never was fatisfied with the explanations given of this fubject. Indeed it is rather kept out of fight by the French chemifts. I am informed that Mr. Meunier, who was one of Mr. Lavoifier's chief affiftants, tried many experiments, in company with Dr. Sommering of Mentz, and that they communicated their obfervations with Lavoifier and the chemifts of Paris, and that thefe gentlemen were fo little pleafed with the refults, that they were never mentioned in the Academy. I am difpofed to affign a very different fource of the heat in all thefe operations; and fhould this work have a fecond edition, I may probably have fo far matured my notions on the fubject as to think them not unworthy of the public attention. At prefent they are by no means in fuch a ftate.

[*Note* 23. *p.* 528.]

The very ingenious Mr. Davy, chemical lecturer in the Royal Inftitution in London, has publifhed, in *Nicholfon's Philofophical Journal, vol. III. p.* 515. *and vol. V. p.* 281, the outlines of a chemical examination of this gas, which he very properly denominates NITROUS OXYD. He there mentions, in very general terms, its relations to heat, water, acids, alkalis, inflammable substances, &c. the methods of producing it, deduced from a due confideration of thofe relations, and the appearances of its compounds. This fketch is the performance of a fagacious obferver and good reafoner.

With regard to its production, his experiments confirm Dr. Black's conjecture very clearly. He fays that the moft uniform and expeditious method of changing nitrous air into nitrous oxyd is to expofe it to the action of a fulphite of potafh or foda. In an hour's time, 100 grains of dry fulphite of potafh converts 16 cubic inches of nitrous air into eight inches of nitrous oxyd, and is itself partly changed into an ordinary fulphat.

But the method of preparing this oxyd, which he found the moft convenient of all, was by means of nitrous ammoniac, mixed with thrice its weight of fand. This falt, the *nitrum flammans* of the older chemifts, notwithstanding the fingularity of its moft obvious properties, has fomehow efcaped the pertinacious fcrutiny of the chemifts for a long while. Mr. Davy has made fome very judicious experiments on it, and obferved properties which are undoubtedly of great value, by the information they feem to give concerning the changes made on the affinities of chemical fubftances by a change of temperature. If nitrous ammoniac be kept fteadily in a temperature very little exceeding 320 of Fahrenheit, it melts, and decompofes completely, refolving itself into water and the gas now under confideration. If raifed much above 400, and approaching to 500°, the conftituent fubftances take other combinations, and produce the ordinary nitrous gas, and azotic gas, not combined. Probably, a very nice management of the heat might determine it either to the one or the other of thofe products, along with the water, which feems to form in every temperature, once the falt has melted. Should the temperature rife to 700°, another combination takes place, and the oxygenous gas is decompofed, and we have a detonation. It muft be remarked that the formation of nitrous gas and azotic gas is accompanied by a luminous appearance in the veffels. This is probably analogous to the low inflammation of phofphorus, fulphur, and other inflammable fubftances, and fhould induce us to examine thofe phenomena, to fee whether they alfo are accompanied by fuch a change in their gafeous productions.

This is faid to be the procefs employed by the fociety of chemifts who have lately attracted fo much notice by the employment of this gas for medicinal and dietetical purpofes. Its effects, when breathed for fome time, are very wonderful, and were firft difcovered, I believe, by Mr. Davy. To thofe who are not hurt by the fight of folly, they are alfo very amufing.

Mr. Davy has alfo very ingenioufly combined this gas with alkalis. He triturates a dry cauftic alkali with a fulphite, and expofes the mixture to the action of nitrous gas. In this way the alkali acts on the nitrous oxyd, fo as to combine with it in the very inftant of its formation. He had found that the oxyd, already in a gafeous ftate, contracts no union with alkalis, if dry, and scarcely if in a watery solution. The compound tasted almost like a caustic alkali, but with a very peculiar pungency. The oxyd is expelled, in its genuine form, by any acid; also by a temperature exceeding 400°.

Inflammable substances decompound nitrous oxyd, much in the same way that they decompound nitrous acid, but the decomposition requires a much higher temperature. Sulphur burning slowly, with its weak blue flame, is extinguished in it. So are charcoal and phosphorus, if in their lowest temperature of combustion. But if any of those substances be made to burn briskly, producing a white light, they decompound nitrous oxyd with great rapidity, with an enlarged flame, and even deflagration. Therefore a candle burns with an enlarged flame, having a beautiful rose-coloured light. The sulphurous pyrophori burn vividly, but require a considerable heat to begin the combustion.

There is a peculiarity attending the burning in this gas. Nitrous acid is produced. Mr. Davy imagines that this arises from a new arrangement of principles, occasioned by the ignition of a *part* of the gas that is not immediately contiguous to the flame. He explains in the same way, some other phenomena which distinguish this gas from nitrous acid.

By comparing a variety of appearances Mr. Davy concludes, that 100 parts of this oxyd consist of thirty seven-parts of oxygen, and sixty-three of azote, in a much denser state than that of their separate existence.

[*Note* 24. *p.* 545.]

Mr. Howard has communicated to the Royal Society of London a very curious account of a preparation of mercury, which explodes by a spark or an electrical shock, in the same way as gunpowder, but with incomparably greater force, and with some peculiarities, which are remarkable and instructive. (*See Phil. Transf.* 1800. *p.* 204.)

The red oxyd of mercury was mixed with alcohol, and nitric acid was poured on it. It dissolved the oxyd before it acted remarkably on the alcohol; and after some time there was an ebullition, which sent forth a dense white smoke, smelling strongly of æther. The mixture deposited a precipitate, at first of a dark brown, and it gradually grew whiter. When separated by filtration, it appeared crystallized. Sulphuric acid was poured on this saline mass. A violent effervescence ensued, and the mixture exploded. When the sulphuric acid was much concentrated, the explosion often happened at the first contact.

Three or four grains of this salt in dry powder were laid on an anvil, and were struck with a smart blow with a hammer, in the manner practised by Fourcroy and Vauquelin with several salts. It detonated with great violence, and the faces of the anvil and hammer were very sensibly indented. The same effects were produced by sending the shock of an electrical battery through five or six grains of it. The report of two grains is always as loud as that of a musket.

Two or three grains, lying in a fort of cup of tin-foil, were thus made to float on fome oil. This was heated gradually; and the powder exploded when the temperature rofe to 368°.

Mr. Howard then examined the explofive or expanfive power of this fubftance when fired in the chamber of a piece, in the manner of gunpowder. It burft the barrel in which it was fired in a very extraordinary manner. Seventeen grains of it, however, were fired in the chamber of a fowling-piece, without deftroying it. There was no recoil that was remarkable. The report was feeble; and the ball had about half the force which an ordinary charge, or 68 grains of the beft gunpowder, would have given it. When the experiment was repeated with 34 grains, the barrel was deftroyed, being torn into many pieces at the breech.

The expanfive power of this fubftance was compared with that of gunpowder, by firing half an ounce of each in two chambers, bored with the fame tool, in two blocks of hard wood. The gunpowder fplit the block into three pieces, driving them from each other till the flame had room to efcape. The mercurial powder burft its block in many directions, but without fenfibly feparating the fragments;—the parts adjoining to the powder were pounded to duft.

A box of caft iron was made to fit exactly the chamber of a twelve-pound carronade. It had a cavity which held three and a half ounces of the mercurial powder. The box confifted of two parts, firmly fcrewed together. It was put into the carronade, and three twelve-pound balls were put above it. The powder was fired through a fmall touch-hole. The fore part of the box was blown out. The other part was fhivered into many pieces; fo were the two undermoft balls; and the uppermoft was cracked through the centre. The report was extremely feeble. The entire ball ftruck the mark with very little force.

Ten grains of the mercurial powder were fired by an electrical fhock, in the centre of a glafs ball feven inches in diameter, and half an inch thick. The glafs withftood the fhock; and its infide was coated with metallic mercury. Only four cubical inches of gas were found added to the air in the glafs. They were a mixture of azote and carbonic acid. Ten grains of gunpowder were fired in the fame manner, but no meafure was taken of the gas. Indeed little could be inferred from the gafes remaining when all is cold, becaufe we are ignorant of what is abforbed. What is moft remarkable in this experiment is, that the inflammation which takes place in this explofion, will not inflame gunpowder mixed with it in a loofe ftate.

Mr. Howard afcribes the great force of the explofion to a greater rapidity of inflammation. He might have added, in all probability, a more complete inflammation; for fome grains of the gunpowder were not inflamed. Indeed there is ufually a very great portion that is driven out of the way without inflaming, when the powder lies loofe, as in this cafe it did.

The moft important circumftance in the experiment is the rapid diminution of the expanfive force by an enlargement of bulk ; it feems irrefiftible, while the particles of the gas are very near each other, but moderate, as foon as they have receded to a fmall dif- tance. The effects of gunpowder agree very well with the fuppofition of a repulfive force between the particles, and acting only on the immediately adjoining particles, in- verfely proportional to their diftance. The repulfion of the particles of this mercuriated gas muft decreafe much fafter, and may perhaps be as the fquare of the diftance inverfe- ly. But be it what it will, it has a different law, and therefore is a different gas from that of gunpowder, perhaps containing the vapour of mercury itfelf. There are experi- ments, by which it appears that this vapour is indeed very expanfive. It may alfo have very little latent heat.

This differtation is full of the moft acute obfervations ; and contains a moft ingenious analyfis of the phenomena, and much chemical fcience.

[*Note* 25. *p.* 544.]

As this is the firft example which we have, in Dr. Black's arrangement of the metals, of the feparation of one metal by another, it may not be improper to pay a little more at- tention to the general phenomena.

The firft obfervation I have to make on it is, that there is not that violent effervefcence accompanying the folution of the added metal, as when that metal is put into the pure acid. In fome cafes there is no effervefcence whatever, and in every cafe the effervefcence is extremely flight.

I may alfo obferve, that the prefent example, the precipitation of mercury by copper, is the firft fact appealed to by Dr. Stahl, as a ftrong argument for his doctrine of phlo- gifton. He appeals to it as a thing long and well known. There are many others equal- ly diftinct. Thus, lead diffolved by acetous acid is precipitated by zinc, in fine metalline cryftals, like fugar candy. Silver is precipitated from the nitrous acid by copper, alfo in metalline cryftals. This precipitation forms a very beautiful object, when viewed through a microfcope, as it goes on in a drop of the folution, when a particle of copper is placed in contact with it. The cryftals will be feen ftarting, as it were, into exiftence, or fhoot- ing out in fine ramifications.

According to the doctrine of Stahl, when a metal is diffolved in an acid, the acid is fuppofed to unite with the calx and expel the phlogifton. Therefore, when an alkali is thrown into the folution, the precipitate muft be conceived to be a calx, perhaps combined with a portion of the acid, or the alkali, or both. But when one metal is employed to pre- cipitate the calx of another, the feparation muft be afcribed to the fuperior attraction of the acid for the calx of the added metal. For when this metal is diffolved in the pure acid, its phlogifton is fuppofed to efcape. To abide by the example in the text, the acid muft be fuppofed to have a ftronger attraction for the calx of copper than for that of mercury. Therefore, when mercury is thus separated by copper, the phlogiston of the copper must be supposed to be expelled from the calx. It does not quit the mixture, but unites with the precipitated calx of mercury, and changes it into a metal.

Dr. Stahl accordingly asserted this double exchange, and says that it is confirmed by the want of effervescence in most cases, and by the very faint effervescence in all. This small effervescence he accounts for in a manner sufficiently satisfactory, by saying that the precipitated metal did not require for its metallic form all the phlogiston which the acid expelled from the other.

Dr. Black, while he acquiesced in the doctrine of Stahl, did not content himself with a mere expression of this acquiescence, but was accustomed to exhibit this doctrine as not only plausible, but as extremely ingenious and elegant; and was at some pains to obviate the objections which were made to it. He saw clearly that it clashed with no principles of philosophy, to say that a substance might be positively light; and when I reminded him of the experiments on pendulums, by which Sir Isaac Newton proved that all sublunary matter is *equally* heavy, he considered these experiments with care, and thought that they were not susceptible of the accuracy necessary for deciding the question; because he observed that the exact determination of the centre of oscillation of such pendulums as Newton must have employed, was a thing which could scarcely be accomplished. It was this objection by Dr. Black which made me repeat Newton's experiments, in a form where the position of the centre of oscillation is of no consequence. (*See Encycl. Britan. Supp.* § *Astronomy.*)

Impressed with this favourable notion of the doctrine of Stahl, Dr. Black was much offended by the contemptuous terms in which more than one or two French chemists spoke of it,—calling it disgraceful to science, and what no man of common sense could adopt for a minute. He thought Stahl, Margraaf, Cramer, and other German chemists, by no means deficient in common sense; and used to say, that the French chemists forgot, in their dashing lessons to the rest of Europe, that Macquer, Geoffroi, Beaumé, were their countrymen. No man had a higher opinion of the genius, penetration, and sagacity of Mr. Lavoisier, than Dr. Black; and I have often heard him lament the loss which science sustained by his death. He admired particularly Lavoisier's quick sight of the importance of Mr. Cavendish's discovery of the composition of water, and his employing it immediately, not only to get over the difficulties which all the acids, except the nitrous, brought into view, but even to extend his doctrine far beyond the first conceptions of it. Dr. Black adopted all Lavoisier's doctrines, but he did not like the officious (as he called it) interference of some of his coadjutors; and he said that Berthollet was the only one of them in whose judgment and caution he had full confidence. With all this candour, he never would allow Stahl's doctrine to be spoken of in a slighting manner.

But to return from this eulogy on Stahl :—We must examine the manner in which the separation of one metal by means of another is explained in the doctrine of the French chemists. It is extremely easy, and pretty satisfactory, but it requires a little more thought and reflection to underſtand the whole internal procedure ; and there are ſome caſes in which there are difficulties which I do not think eaſily ſurmounted.

What is called the ſolution of a metal in an acid is, in general, rather the ſolution of a metallic oxyd, and we muſt regard this ſolution and the previous oxydation as two ſeparate operations. When a metal is firſt oxydated, and then diſſolved in an acid, we do not ſee where the one operation terminates, and the other begins ; the one operation ſeems juſt the continuation of the other. If we attempt to reſolve theſe queſtions by having recourſe to attractions and repulſions, and to conceive this ſucceſſion of changes in a mechanical way, as is done by many who would be thought philoſophical chemiſts, we ſhall find ourſelves immediately put to a ſtand. For in mechaniſm, the law of continuity ſuffers no exceptions. But chemiſtry itſelf helps our conceptions of this matter, by ſhewing us other caſes, where a ſimilar ſucceſſion of phenomena appears. Thus, the continued addition of heat to a ſubſtance ſhews us firſt the ſolid,—then a fluid,—then a vapour. And in each of theſe changes of conſtitution and properties, there is a great abſorption of heat. We have another example ſtill more analogous. By gradually evaporating any brine, we at laſt produce cryſtals of ſalt, ſtill conſiſting of the ſaline matter combined with water. Here the formation of a ſaline cryſtal, by the pure ſalt *firſt* attaching to itſelf the water of cryſtallization, which it certainly will do, if no more water be ſupplied to it, and then its ſubſequent ſolution in more water, are facts which very much reſemble the formation of an oxyd, and its ſubſequent ſolution in more acid. Careful obſervation may diſcover many more examples of ſuch a ſeries, and it may be known as a very common phenomenon,—as a general though not univerſal law. But this is all the explanation that pure chemiſtry affords ; and it is by a judicious obſervation and employment of ſuch general laws, that it has become ſo comprehenſive and important a department of natural knowledge.

Being more familiarly acquainted with the cryſtalline and briny forms of a ſalt, we ſeem a little entitled to reaſon from theſe to ſimilar ſubjects which are leſs intimately known ; and we are led to believe, that the oxygen which forms part of the oxyd, is more firmly united with the mercury, than what is afterwards combined with it as an ingredient of the acid. And this opinion is confirmed by obſerving, that in moſt caſes, we can eaſily detach the acid, but that it is very difficult to detach the pure oxygen. In many caſes, the moſt intenſe heat will not detach this laſt. Alſo, although we do not in the leaſt underſtand why or how the previous union of the oxygen with the azote of the nitrous acid ſhould make its diſpoſition to combine with the mercury any leſs, we know that this is a general fact in chemiſtry. And laſtly, though ignorant of the efficient reaſon, we know that in proportion as more of an ingredient is already united with another, all ſubſequent additions are made with leſs force, and the union is leſs firm, or more eaſily deſtroyed.

Having theſe notions of the affinities of ſubſtances, we muſt ſuppoſe that the copper firſt attaches to itſelf as much oxygen as will render it an oxyd ſuſceptible of union with the nitrous acid. This muft certainly be taken from the mercurial nitrate, leaving the mercury ftill in the ftate of oxyd. It muft therefore be taken from the nitrous acid in the mixture. This muft be the order of procedure, becaufe it is this portion of the oxygen that is eafieft detached. The acid is already in the form of *nitrous* acid, having been decompofed in part when diffolving the mercury. Or, if we allow the acid of folution to be nitric acid, the firft effect of the copper muft ftill be to decompofe fome acid. The copper oxyd is at length formed, and now diffolves, by abftracting, not oxygen, but nitrous or nitric acid from the nitrate. This it continues to do more and more, reducing the mercury to the ftate of an oxyd, and laft of all leaving it in its primitive form of a metal. This order of procedure is confirmed by obferving, that in fome cafes the metal is left in the ftate of an oxyd, or in a ftate very little removed from it.

Now this procedure has, I think, confiderable difficulties. We do not fee, in any cafe, how the affinity of the copper for acid or oxygen, which continually diminifhes as it attaches more to itfelf, is at laft able to deprive the mercury of that portion of oxygen that is moft ftrongly attached to it. We cannot fuppofe the oxydation of the mercury to be the firft thing effected, being already perfuaded that the copper, which can decompofe pure acid, will more readily decompofe it, when the combination of the parts of the acid is weakened by its union with the mercurial oxyd. It muft therefore take the oxygen from the acid, and not from the oxyd; and the mercurial oxyd muft be the laft thing decompofed. The firmeft combination muft be overcome by the fmalleft force.

Nor is it without its difficulties, to conceive how a metal becomes foluble in an acid, only by being already combined with one of its ingredients. I do not know any fimilar fact,—and I have little confidence in any other mode of chemical inveftigation, and leaft of all in reafonings depending on the confideration of attractions and repulfions.

Farther, if the copper is oxydated by decompofing the nitric acid, we fhould have an effervefcence of nitrous air as ftrong as when copper is put into pure acid. We cannot fuppofe that the redundant azote unites with the oxygen of the oxyd, and changes it into nitric or nitrous acid, becaufe the mercury was oxydated by means of an affinity for oxygen, ftronger than that of azote for the fame oxygen. I think that this effervefcence fhould be obferved in every inftance of feparating one metal by means of another; even when the added metal is oxydated by decompofing the water. For I do not think that the redundant hydrogen will unite with the oxygen of the oxyd. Now thefe phenomena are not obferved,—what effervefcence is obferved is always weak. We have not inflammable air in all the cafes where the added metal decompofes the water. In fuch cafes we are led to think that the hydrogen is expended during the decompofition of the oxyd of the diffolved metal. But this alfo obliges us to fuppofe that the decompofition of one oxyd, and the formation of the other, are the firft parts of the whole procefs, a thing which does not agree with our notions of the affinities.

I do not think it will be difficult to decide thefe queftions by proper experiments, but none fuch have yet come in my way.

[*Note 26. p. 553.*]

As the notes found among Dr. Black's papers are not in such a condition that I can lay them before the reader, so as to answer the purposes mentioned in the lecture, I must content myself with giving the chemical arrangement of the medicinal preparations of mercury that was followed by him in his discourses, with such of his observations on particular medicines as I thought might be of use.

Table of the Preparations of Mercury, as drawn out by Dr. Black.

Hydrargyrus praeparatur ad Usus Medicos.

I. Destillatione, ut purus fiat.

Hydrargyrus purificatus. Lond.

II. Triturâ, ut in atomos invisibiles attenuetur.

Merc. gummos. Plenckii.
Pilulae Hydrargyri. Ed. et Lond.
Hydrargyrus cum cretâ. Lond.
Emplastrum Hydrargyri, sive coerul. Ed.
Emplastrum Lithargyri cum Hydrargyro. Lond.
Emplastrum Ammoniaci cum Hydrargyro. Lond.
Unguentum Hydrargyri, sive coerul. Ed.
Unguentum Hydrargyri fortius et mitius. Lond.

III. Calcinatione Ignis Ope et Aëris.

Hydrargyrus calcinatus. Lond.
Olim Mercurius praecipitatus per se.

IV. Viribus Salium.

1. Cum Acido Vitriolico.

Hydrargyrus vitriolatus flavus, vulgo Turpethum minerale. Ed.
Hydrargyrus vitriolatus. Lond.

2. Cum Acido Nitroso.

Guttae albae Wardi. *(a)*

(a) Dr. Black has given the following account of the process for preparing this medicine, known by the name of *Ward's white drop* :

Unguentum Hydrargyri nitrati.   Ed. et Lond.   *(b)*
Hydrargyrus nitratus ruber.   Ed. et Lond.

3. Cum Acido Muriatico.

Hydrargyrus muriatus corrosivus.   Ed.
Hydrargyrus Muriatus.   Lond.
Hydrargyrus Muriatus mitis.   Ed.
Calomelas.   Lond.
Hydrargyrus Muriatus praecipitatus.   Ed. ⎱ *(c)*
Hydrargyrus Muriatus mitis.   Lond.   ⎰

4. Cum Acido Acetoso.

Hydrargyrus Acetatus.   Ed. et Lond.
Pilulae Keyseri.   *(d)*

A mixture being made of sixteen parts of pure aquafortis, and seven of volatile sal ammoniac, four parts of mercury are added for every sixteen of the mixture; and the mercury being dissolved, as much more is added as the liquor will dissolve.

This is evaporated till a pellicle just appears on the surface. It is now allowed to cool.

When cold, the heavy, uncrystallized, or fluid part, is drained off. The crystallized salt is dissolved in thrice its weight of rose water, which forms the white drop. The dose is in two drops, each dose containing half a grain of mercury.

(b) A variety of this ointment is used with success in the Manchester infirmary. It is prepared with the oil of butter, which has had no salt whatever. This preparation is found to preserve its colour unchanged.

(c) Dr. Black has written, on a copy of this table, an account of Scheele's process for this medicine, *viâ humidâ*, as follows:

Half a pound of mercury, and as much aquafortis, are put into a small matrass or cucurbit, having a pretty long neck. The mouth of it being covered with a bit of paper, a digesting sand-heat is given to the mixture for some hours. When the action of the acid becomes languid, the heat must be increased to nearly a boiling heat, and continued three or four hours,—agitating the mixture from time to time. At last, make the solution boil gently, for a quarter of an hour; at the end of which there should remain a small quantity of mercury undissolved.

While the above is going on, dissolve four and a half ounces of pure sea salt, in six or eight pounds of water. Mix this solution, boiling hot, with the solution of mercury, also boiling,—stirring violently while they are mixed. Then allow the precipitate to settle; and afterwards edulcorate it with repeated washings of hot water,—and dry it. We obtain a *mercurius dulcis* in impalpable powder, amounting to eight and a half ounces.

The efficacy of this process is explained by Dr. Black, on the principles advanced in the lecture, namely, that the solution of mercury, made as here directed, is a sort of compound solution, containing a quantity of the metal, not oxydated, sufficient completely to saturate the muriatic acid. In proof of which, he says that this solution is precipitated black (a mark of crude mercury) by an alkali; whereas if the solution has been continued only till the cessation of effervescence, an alkali occasions a yellow precipitate, &c.

(d) Dr. Black refers for the preparation of this medicine to *Observ. de Medicine, de Mr. Richard, au Louvre,* 1772; and to the *Encyclopedie.*

5. Deturbatione ex Acidis vi Alkalinorum.

Hydrargyrus praecipitatus cinereus. Ed. } *(e)*
Mercurius praecipitatus fufcus.

Calx Hydrargyri alba. Lond. *(f)*
Unguent. Calcis Hydrargyri albae. Lond.

V. Conjunctione cum Sulphure.

Hydrargyrus Sulphuratus niger. Ed.
Hydrargyrus cum Sulphure. Lond.
Hydrargyrus Sulphuratus ruber. Lond.
Pilulae Hydrargyri muriati mitis, five Calomelanos, compofitae. Ed.

[*Note* 27. *p.* 563.]

Medicamenta parantur ex Antimonio, vel Sulphurato, vel Sulphure privato.

Ex Antimonio Sulphurato.

I. Triturâ.

Antimonium praeparatum. Ed. et Lond.

II. Ope Ignis et Aëris.

Flores Antimonii fine addito.
Vitrum Antimonii. Ed.
Antimonium vitrificatum. Lond.
Vitrum Antimonii ceratum. Ed.

III. Vi Salis Alkalini.

Hepar Antimonii mitiffimum.
Regulus Antimonii medicinalis.
Hepar ad Kermes minerale *Geoffioii.*

(e) This, as prepared in the fhops, is faid to be frequently very acrimonious, which Dr. Black afcribes to impure aquafortis in the preparation.

(f) This white precipitate may be made exactly fimilar to calomel, by ufing plenty of volatile alkali for the precipitation, fo much that the mixture may fmell of it. This alkali will take from the metal a confiderable quantity of the muriatic acid. Alfo by its hydrogen it deprives the quickfilver of the oxygen with which it was combined; The precipitate is therefore rendered infoluble in water, and fcarcely foluble in vegetable acids. It will be very different if we are fparing of the alkali.

Dr. Black remarks in this note, that the foffil alkali, though aërated, cannot be made to precipitate mercury from a folution of corrofive fublimate, of a white colour, becaufe the quantity of it which is fufficient for faturating the acid does not contain enough of air to faturate the mercurial precipitate, and render it white. He confirms this explanation by numbers, flating according to Mr. Kirwan's experiments, the quantity neceffary for this purpofe.

Hepar ad Tincturam Antimonii.
Kermes Minerale.
Sulphur Antimonii praecipitatum. Ed. et Lond.

IV. Vi Nitri Salis.

Crocus Antimonii mitiſſimus,
Vulgo Regulus Antimonii medicinalis.
Crocus Antimonii. Ed. et Lond.
Antimonii emeticum mitius. *Boerh.*
Antimonium uſtum cum Nitro, *vulgo* Calx Antimonii Nitrata. Ed.
Antimonium calcinatum. Lond. *Vul. o diaphoret.*
Antimonium Calcareo-phoſphoratum, ſive Pulvis antimonialis. Ed.
Pulvis Antimonialis. Lond.

V. Viribus Acidorum.

Antimonium vitriolatum. *Klaunig.*
Antimonium catharticum. *Wilſon.*
Antimonium muriatum, vulgo Butyrum Antim. Ed.
Antimonium Muriatum. Lond.
Pulvis Algerothi, ſive *Mercurius Vitae.*
Bezoardicum minerale.
Antimonium tartariſatum, vulgo Tartarus emeticus. Ed.
Antimonium Tartariſatum. Lond.
Vinum Antimonii tartariſati. Ed. et Lond.
Vinum Antimonii. Lond.

Ex Antimonio Sulphure privato.

Hoc Metallum, prout diverſis modis liberatur a Sulphure, dicitur Regulus Antimonii
ſimplex, Regulus Antimonii Martialis, Regulus Jovialis, &c. Ex eo parata ſunt;

I. Ignis Vi et Aëris.

Flores Argentei, ſive Nix Antimonii.

II. Vi Nitri Salis.

Ceruſſa Antimonii.
Stomachicum. *Poterii.*
Antihecticum. *Poterii.*
Cardiacum. *Poterii.*

Medicamenta ab Antimonio nomen mutuantia, quae tamen ejus Metalli vix quicquam
retineat.

Cinabaris Antimonii.
Tinctura Antimonii.

[*Note* 28. *p.* 604.]

The grains of small shot made in this way are very often hollow, or have a deep pit in one side, which is frequently rugged within. It is owing to the sudden congelation of the outside by the water. This forms a hard case, while the interior is still fluid, and contracting as it cools, a part is left empty. As this greatly hurts the value of the small shot, many attempts have been made to prevent it, which have been more or less successful. Some manufacturers have mixed other metals with the lead,—others have kept oil on the water,—others receive the lead into boiling hot water covered with melted tallow. I believe that the most successful method has been that of the patent shot manufacture in Southwark, London, where the furnace is at the top of a very high tower, not less than one hundred feet, and the shot is gradually cooled as it falls through the air. The chief effect of this, however, must be the incomparably greater number of spherical grains. I do not see how this will much prevent the hollowness of the shot, even although the tower were four times as high. The shot would fall two hundred and fifty feet in four seconds, and four hundred feet in five seconds, neither of which would sufficiently cool the shot. Its latent heat requires a much longer time than this for its absorption by air. The pear-like shape is occasioned by the fluid lead within breaking through the crust, and freezing as it comes in contact with the water. Metals afford very curious forms, by dropping them when fluid into water. When this is done from a very small height, the drops descend slowly through the water. The fluid within breaking the crust by its great weight, runs out, freezing as it goes down, and often leaves a pretty round thin cup with a taper thing below it like an extinguisher. Some metals thus form themselves into nails with large heads,—others take other shapes, very uniform, and very unexpected. Great differences are produced by different liquors instead of water, and by different heats of the metals, and by the heights from which they are poured, &c.

[*Note* 29. *p.* 639.]

It is by no means clear that the acid has increased its proportion of oxygen by deoxygenating the copper. It is true that the copper remaining in the retort scarcely needs any addition of inflammable matter to reduce it; but there is adhering to it a large proportion of extractive coaly matter, which answers the same purpose; and the fact is, that when this is urged by a strong heat, we obtain a great deal of fixed air. Of this I have had repeated experience. It is indifferent whether this be conceived as composed of the oxygen of the cupreous oxyd or from the acid. In neither case can the acid be said to have increased its proportion of oxygen in any other way than by leaving behind a portion of carbon. I see nothing that induces me to think even this to be the case, or that there is not as much oxygen taken from the acid in this operation as belongs to the extractive or carbonic matter. Mr. Chaptal *(Ann. de Chymie,* vol. 28.) thinks that he fees plain differences between what he choofes to call acetous and acetic acid. But they appear to me very unfatisfactory. The fuperior acidity of the acid obtained by the prefent procefs is fufficiently accounted for by its dephlegmation and freedom from mucilaginous and other extractive matter. At any rate, this is by no means the moft eligible procefs for obtaining from this falt the greateft quantity of acetous or acetic acid, free from empyreuma; and, as Dr. Black has not mentioned any of the other methods of decompounding it, I truft that the reader will not be difpleafed with fome account of the moft approved of them.

The fimpleft and moft obvious of all is to employ the fuperior affinity of another acid, and of thefe the moft fixed is to be preferred. Concentrated fulphuric acid being poured on the acetite of copper, it detaches the acid with great facility, and as agreeably fragrant as the *beft* of the dry procefs, or what comes over after the portion tinged with the copper, and before any empyreuma can be obferved. It is alfo extremely ftrong, and of a pungent odour, and the laft portion is inflammable. This was firft obferved, I think, by Count de Lauragais; and he employed it to form an acetic æther, which exceeds others in fragrance. There is generally fome carbonated matter left adhering to the fulphat; and this has been adduced as a proof that the acetic acid, by depofiting carbon, has become redundant in oxygen.

It will be faid that the acid obtained by this procefs is not in a ftate of complete dephlegmation, as in the other. But I may obferve here, that the other procefs cannot be eafily conducted without fome water. Without this addition, in order to tranfmit the heat more readily to the centre of the mafs, the acid cannot be expelled from thence, without over-heating the exterior parts.

A ftill better procefs is that of Mr. Lowitz. He mixes three parts of the acetite of copper with eight parts of a fulphat of potafh furcharged with acid, (prepared by diftilling fulphuric acid from potafh, to drynefs). This mixture, in dry powder, contains as much redundant fulphuric acid as is fufficient for faturating the copper or its oxyd, and for extricating the whole of the acetous acid, with a very moderate heat. He affirms that in this way we obtain it, with all the fragrance poffible, from this preparation.

Nitric, or the oxygenated muriatic acid, or aqua regia, might be employed, with proper precautions, to decompound the cupreous acetite, and feem the fitteft for enabling us to judge whether the acetous acid can be fuperoxygenated.

[*Note* 30. *p.* 665.]

I muft not omit this opportunity of making fome remarks on this mode of concretion of the filver. It muft happen in confequence of an attraction which brings together, in a certain determinate manner, the minute atoms into which the filver is divided. The fame thing is obfervable in numberlefs inftances of precipitation, and is probably univerfal. Often, indeed, the precipitate is a fine impalpable powder; but each particle of this is a mountain, in comparifon of the atoms in their ftate of feparation, while chemically united with thofe of the folvent. Often are we able to fee with a microfcope, that each of thefe particles of powder is a group of minute regular forms or cryftals. Nay, even when the precipitate is gelatinous, as in luna cornea, I am perfuaded that its ftructure is not uniform, but plated or fymmetrical. This I conclude, from obferving that the formation of fuch jellies proceeds in a way which refembles the more fenfible cryftallizations. When a folution is juft ready to fhoot into cryftals, if we touch it with a particle of the cryftal, the change begins there, and fpreads from it as a centre. I obferve the fame thing in the formation of fome gelatinous concretions by infpiffation. But, to return to this precipitation of filver;—it is particularly to be remarked, that, although the concretion proceeds from a bafis fending out ftems, from which proceed branches, which again produce fubordinate ramifications, and the concretion has therefore been called a vegetation, it does not at all proceed in the manner of a growth, or a gradual protrufion from a parent trunk. This trunk does not grow or change at the fame time. Nor does it appear to contribute any thing whatever to the increafe or ramification. It remains of the fame length, the fame diftance remains between every branch, and all the increment is plainly an accretion of external matter applied to its extremities, or to different parts of its furface in fucceffion. The additions are of matter previoufly floating about. The little fhrub, therefore, is not produced by the gradual developement of a feminal body, nor by the action of its organical ftructure on the fluid matter which it pumps along its canals, affimilating it as it goes along, and fecreting or rejecting it according to fixed laws. In the prefent cafe, every atom feems to have the fame powers, and to be equally capable of becoming either the ftem or any one of the branches. When we obferve the gradual formation with attention, affifted by a microfcope, we obferve each little addition to ftart into exiftence in a moment; and we can fcarcely tell when or where to expect the next. They are all accretions,—appofitions, of filver previoufly floating about in the folution. Sometimes, indeed, a little branch employs a fenfible moment in attaining its full length; but even then, I prefume, that this is a fucceffion of many inftantaneoufly formed plates; for we know that all cryftals are of a tranfverfely plated ftructure. I cannot but think that the atoms of filver continue to be fufpended, or to float about for fome little time, after their feparation from the acid, in their individual and invifible form; and that they are thus carried to fome diftance from the copper (in contact of which only they are detached from the acid) before they arrange themfelves in fome little fymmetrical group, fitted to form a new cryftal, obferving fuch a law of attraction, that it muft attach to thofe already formed, in one way, but not in another. Or, fhall we fay that there is a power refembling magnetifm, electricity, or galvanifm, propagated from the central bit of copper? This laft (galvanifm) feems to offer the beft means for *illuftrating* the phenomenon; for it cannot be called an explanation, but it is of the same kind with all other explanations purely chemical. None of them aim at afcertaining the proximate caufe, and its manner of action. It is thought a fufficient explanation, if we fhew that the phenomenon is one of a great clafs already known, although we are equally in the dark with refpect to each of them. The philofophical chemift, indeed, frequently attempts fomething more, and to fhew *how* the chemical attractions operate in producing the phenomenon. This is precifely the difficulty here. The queftion is not chemical, but mechanical. We are required to fhew how certain centripetal, or centrifugal forces feparate certain particles, and bring them together again in another determinate manner. The chemift afcertains, from obfervation, the affinity, that is, the cafes in which fuch feparations and reunions are invariably effected. When he goes further, and fpeaks of his greater and fmaller attractions, of one attraction oppofing another, &c. he fpeaks as a mechanician, and generally without any diftinct conceptions of the fubject. In queftions of this kind, the object before the mind is an atom of *moveable matter*, under the influence of a *moving force*; and all difcuffion muft now proceed on the acknowledged laws of motion. I take it that the prefent phenomenon is one of thofe which fets the infufficiency of thofe ufual modes of reafoning in the cleareft light. I cannot conceive any action between the copper and the *adjoining* acid which can produce the depofition of a *remote* atom of filver from a *remote* atom of acid, which will not be produced by the action of the already formed folid filver, adjoining to that particle of acid which depofits its filver. All that I can do, is to recollect that fomething very like it happens when I bring a magnet near a parcel of iron-filings, which were lying confufed, and without any mutual influence,—every individual rag of iron becomes in an inftant active, turns about, and prefents one extremity to one adjoining rag of iron, and the other to another; they adhere together; and when I draw one away, it drags another after it, &c.; and all this activity is at an end whenever I remove the magnet. Here then is a confiderable refemblance of effect, but it is ftill very imperfect. The mere prefence of the copper is not enough; it muft be in the very act of diffolving in the *contiguous* acid. The galvanic phenomena have a much clofer refemblance to what we are now confidering. Indeed they are almoft the fame. We have at one wire an oxydation going on, and a deoxydation going on at the other, and the two have an invariable dependence on each other. The one cannot happen unlefs the other alfo happen; and we cannot tell which is the prior event. But this is only an illuftration; and the immediate actors and powers remain as much unknown as ever. Can the phenomenon be conceived in this way? It is probable that an unfaturated folution of copper, and a faturated folution of filver in the fame acid, may remain mixed together without changing. We know this to be true with faturated folutions of filver and mercury. When copper is put into fuch a mixture, it *may be* that a greater dofe of it being acquired by the acid in contact with it, liberates fome filver, (though I cannot tell how) which liberated filver floats off, and is carried by the fluid till it meets with a cryſtal of ſilver already formed, acting with polar forces, like the particles of iron-filings, and there it ſettles ſymmetrically.

It is to be wiſhed that chemiſts would endeavour to acquire ſome more preciſe notions than they ſeem to content themſelves with at preſent, of thoſe internal procedures. Such unbounded uſe is made of the change of partners (ſo to expreſs myſelf) between oxygen, hydrogen, azote, and carbon, in the chemical phenomena of vegetable and animal ſubſtances, that every tyro from the chemical claſſes finds himſelf in a condition to write a ſyſtem of chemiſtry, which leaves nothing unexplained, and admits of no contradiction. The author of a *Pyrologia*, by bringing caloric into the chemical dance, finds himſelf qualified to make a world. It is time to check ſuch unprofitable ſpeculations. Dr. Black's maxim of ſetting one foot firm on ground which we know familiarly, before we attempt to make another ſtep, is the only ſafe way of proceeding in this noble but difficult ſcience. An example, ſtill more remarkable and puzzling, occurred to me very lately. I had ſet up a galvanic pile, of 240 pairs of ſilver and zinc plates, having pieces of woollen cloth ſoaked in a ſaturated ſolution of ſea ſalt. It ſtood on a table in a warm room; and the evening ſun ſhone on it. Chancing to look at it from the other ſide of the room, I obſerved a moſt beautiful nimbus or glory ſurrounding a great part of the column, exhibiting fine priſmatic colours. I went immediately to take a nearer view of it, and it vaniſhed. Returning to my chair, I ſaw it again, but not if I moved my head much to either ſide. I went to it again, and viewing it with a reading glaſs, I obſerved that each piece of cloth had, all around, fine cryſtals ſtanding out almoſt horizontally. They were more ſlender than the fineſt human hair; and many of them (no thicker than the reſt) were two inches long. They were perfectly ſtraight, except a ſmall bending by their own weight. I could perceive, by their manner of reflecting the light, that they were angular priſms, and that they were jointed tranſverſely, like the *equiſetum lacuſtre*.

Now I would aſk how theſe cryſtals of ſoda (for ſuch they are) were formed? I let ſome fibres of cotton fall on them; and could obſerve that the part *beyond* the fibre increaſed, without the fibre itſelf being carried farther from the ſoaked cloth by an increaſe of what was between them.—It is very myſterious.

[*Note* 31. *p.* 681.]

When we reflect on the manner in which it explodes, when lying looſe on a plate of metal, where it is perfectly open to expand upwards and laterally, and on the very ſmall portion of this ſmall quantity of matter that is hindered from expanding downwards, and on the impreſſion that it makes, it gives us ſome notion of the immenſe velocity with which it expands. I found that ſeven grains, lying on a plate of braſs, and exploding by heat, made as great an impreſſion as a piece of iron, two and a half pounds weight, falling on it with the velocity of 25 feet in a ſecond. We cannot compute the downward action at more than one-fourth of the whole expansion. We must therefore calculate the velocity of 1¼ grains which will give as great a blow as the piece of iron, which weighed 17,500 grains. We shall find this velocity to be 250,000 feet in a second, which is nearly 45 miles in a second. Yet this is probably very small in comparison with the velocity of expansion of the fulminating silver, perhaps not the 100th part of it. Thus we approach to some of those astonishing forces that operate in the emission and refraction of light, and many other phenomena of nature.

[*Note* 32. *p.* 707.]

No observations, made with any immediate purpose of employing them in the lecture, are to be found among Dr. Black's papers, on several substances which seem to be pretty generally admitted as metallic; namely, those called *molybdenum, tellurium, chromium, tungsten, uranium, titanium,* &c. It appears that he did not consider the very imperfect knowledge acquired of them as of importance enough to occupy any of the time allotted for a course of elementary instruction in the *science* of chemistry. The few distinguishing properties established with respect to each of them are not such as have any extensive influence, either on doctrinal points, or the conduct of chemical processes, or the improvement of the chemical arts. Some of them, particularly the one called titanium, seems to have engaged his attention a good deal: But the state of his health precluded all labour on his part to acquire a more perfect knowledge of its properties

END OF VOLUME II.

Printed by Mundell & Son, Edinburgh.

INDEX.

A.

C.

F.

G.

H.

PAGE.   NOTES

O.

P.

www.ingramcontent.com/pod-product-compliance
Lightning Source LLC
Chambersburg PA
CBHW030013220326
41599CB00014B/1794